LABORATORY STUDIES IN INTEGRATED ZOOLOGY

LABORATORY STUDIES IN INTEGRATED

ZOOLOGY

FRANCES M. HICKMAN

Emeritus Instructor of Zoology
Depauw University
Greencastle, Indiana

CLEVELAND P. HICKMAN, Jr.

Professor of Biology
Washington and Lee University
Lexington, Virginia

SIXTH EDITION

with **377** illustrations

TIMES MIRROR/MOSBY COLLEGE PUBLISHING

ST. LOUIS · TORONTO · SANTA CLARA 1984

Editor: Diane L. Bowen
Assistant editor: Susan Dust Schapper
Editorial assistant: Jacqueline Yaiser
Manuscript editor: Stephen Dierkes
Design: Staff
Production: Linda R. Stalnaker, Barbara Merritt

Front cover art *Monkey and Grasshopper* by Mori Sosen,
late eighteenth century—early nineteenth century,
ink and color on silk, Shinenkan Collection courtesy Joe D. Price.

SIXTH EDITION

Previous editions copyrighted 1957, 1963, 1968, 1974, 1979

International Standard Book Number 0-8016-2178-X

Printed in the United States of America

GW/VH/VH 9 8 7 6 5 4   02/C/226

PREFACE

This laboratory manual is the outgrowth of the senior author's many years of working directly with students in the zoology laboratory. Its objectives have always been clarity of expression, thoroughness of coverage, and convenience for both student and instructor. Perhaps most important of all is the expectation that through direct laboratory experience students will learn to test basic biological principles, become familiarized with standard techniques of animal study, and learn through dissection of selected specimens the distinctive anatomical features distinguishing each group of animals. Only in the laboratory can students come to understand the common architectural themes and adaptations that unite groups of animals.

Although this manual was written to accompany a particular textbook, *Integrated Principles of Zoology* (ed. 7), in a full-year course in general zoology, it can easily be adapted for use with the shorter sister text, *Biology of Animals* (ed. 3), or any other introductory text and with a variety of course plans. By judicious selection of exercises it has long been used not only in full-year courses but in many one-semester courses in vertebrate and invertebrate zoology as well as in one-semester introductory principles courses.

As in the other Hickman texts, there is an emphasis on biological principles and evolutionary relationships and advances that point the way to more highly organized phyla. Every effort has been made to give clear, lucid descriptions and instructions, and enough background material has been included to create interest in and understanding of the subject matter. Many illustrations are used to complement the written word.

This manual is more comprehensive than most manuals because we believe that students deserve more than a mere dissection guide. Students should have important concepts presented to them while the specimens that illustrate those concepts are before their eyes. This can be more effective than any number of descriptive lectures. We also believe that being able to name the anatomical parts of an organism is of limited value unless the functions and adaptations of those parts are also understood.

We emphasize without apology that this manual embraces more material than can be covered in a single zoology course. This is an asset to the instructor who is thus free to make selections to fit his or her own or the department's needs and preferences, influenced of course by the time and materials available. For example, few zoology courses have time to cover all of the vertebrates (shark, perch, frog, and fetal pig) included here; but whatever the choice, the instructor can be assured of complete and careful coverage of the subject. The fetal pig coverage is, we feel, the clearest and most practical to be found in any introductory guide on the market today. We prefer the fetal pig to the frog for its close similarity to the human and for the opportunity it provides to study mammalian reproduction and fetal circulation.

Convenience for both student and instructor has always been a major consideration. Instructions have enough detail that students can work with a minimum of help from the instructor, thus freeing the instructor from lengthy introductory explanations. For the convenience of the instructor lists of needed materials are given with each exercise, and suggestions for demonstrations and student projects follow each exercise. Appendix B includes many aids for the teacher, including instructions on the handling of laboratory animals; preparation of solutions and reagents; methods of culturing, narcotizing, and preserving; starting and maintaining of aquariums and terrariums; sources of living material, and more.

There are many aids for the student. Throughout the book working instructions are clearly set off from descriptive material. Classifications, where appropriate, are included with the exercises. Function and physiology are explained along with anatomy. Topic headings help the student organize the material mentally. Metric tables and definitions are placed on the inside front cover for convenient use. The ever-popular taxonomic chart, which has again been updated, can either be glued into the back of the manual so that it can be unfolded for use, or it may be hung on the student's wall. In either case the student will find it a concisely and conveniently arranged review of the animal phyla and an invaluable addendum to the text. Pages are perforated for student drawings, observations, and reports. In some cases drawings would be especially tedious, so illustrations are included for the student to label. Much of the new artwork has been designed to assist the student with difficult dissections.

Traditionally there has been too little use of living materials in zoology laboratories. It is easy for the student, confronted with weeks of dissection, to forget that colorless, preserved specimens were once alive. This manual encourages the study, whenever possible, of living specimens in the laboratory. Each exercise, if appropriate, begins with a paragraph on the natural history and habitat of the animal to be studied, followed by a discussion of its behavior. Marine aquarium systems, so widely available now, can bring new life and excitement into the laboratory. For this reason we have stressed the use of living marine organisms. To observe the feeding action of a delicate sea anemone, the graceful movements of a nudibranch, or the movements of the sea star's podia on the side of the tank can reveal more to a student than volumes of reading material or hours of dissection. The addition of one or two "living rocks" to a marine tank will introduce a surprising assortment of strange and interesting marine forms. By seeing a wider variety of life than that presented by the type of specimen being studied, students come to appreciate the diversity of animal life and adaptive radiation within animal groups. Zoology comes alive for them.

THE REVISION—CHANGES AND ADDITIONS

This sixth edition has been extensively revised, and many new drawings and photographs have been added that should increase the manual's usefulness immeasureably. For the most part this edition follows the same general plan of previous editions, but we have tried to enhance the manual's effectiveness by regrouping chapters and rearranging some of the material so that the manual's organization more closely parallels that of the *Integrated Principles of Zoology* (ed. 7). Four new chapters with exercises in animal physiology have been added.

In the interest of space fewer pages have been included for student drawings or observations, but new, labeled drawings of internal structure have been added, for example, of planaria, liver fluke, *Ascaris*, earthworm, crayfish, grasshopper, clam, and frog. As an added convenience the student tear sheets have, where possible, been placed at the end of each exercise rather than at the end of the entire chapter.

Part One includes the six introductory chapters, most of which have been revised and reillustrated. Chapter 1, on the use of the microscope, contains new sections on the use of the oil-immersion lens and the electron microscope and a reworked section on mag-

nification and the use of the stage and ocular micrometers. The exercise on cell structure and function in Chapter 2 was expanded to include descriptions of electron micrographs. Although beginning students seldom have an opportunity to use an electron microscope, they should at least be able to interpret electron micrographs and relate them to slide specimens. Chapter 3, on cell function, incorporates a new exercise on the action of enzymes using amylase. The sections on liquids in mixtures, diffusion, and osmosis were revised. Chapter 4, which now includes gametogenesis and the early embryology of both the sea star and the frog, has been revised and reillustrated. Animal classification, now Chapter 6, was moved to immediately precede the survey of phyla. It has been revised, and the key to amphibians was expanded to embrace a broader spectrum of species. The chapter on field trips has been moved to Part Three.

Part Two, "The Diversity of Animal Life," comprises a survey of animal phyla in Chapters 7 through 20. Classifications have been updated, illustrations have been added, and a number of additions and corrections have been made. New sections on *Plasmodium*, the stony coral *Astrangia*, soil nematodes, hookworms, filter feeding and digestion in the bivalve, and earthworm copulation have been added. In Chapter 14 the exercise on the grasshopper and the honeybee has been rearranged for greater clarity and convenience, new sections have been added on the killing, relaxing, and displaying of insects, and the "Key to Orders of Common Insects" has been entirely rewritten to include a larger number of orders. The section on *Amphioxus* has been revised and reillustrated. The exercise on human blood, formerly located in the mammalian chapter, has been moved to the chapter on circulation and respiration and is incorporated as Exercise 22A.

Part Three, "Activity and Continuity of Life," includes four new chapters on animal physiology in addition to the three closing chapters on animal behavior, genetics, and field trips. The four new chapters, 21 to 24, contain a series of exercises on muscle physiology, circulation, respiration, digestion, excretion, and reflex physiology. These were selected to illustrate basic principles of physiology, using readily obtainable supplies and simple apparatus. None of them is technically demanding. The exercise on field trips has been thoroughly revised, and to it has been added a reference list of guidebooks and keys that should be useful in identifying animals collected on field trips.

The "Key to the Major Taxa" (Appendix A) has been revised to bring the classifications into agreement with the text, as has also the popular chart, "Chief Taxonomic Subdivisions and Organ Systems of Animals" (Appendix C).

ACKNOWLEDGMENTS

Over the years a great many of our colleagues and fellow teachers have enriched this work with their comments and helpful suggestions. To all of these, those named in past editions, those named here, and others too numerous to mention, we extend our special thanks. It is through their experience and their kindness in sharing it with us that we are kept aware of users' needs and preferences and, sometimes, our errors. We are grateful to them all.

Particularly helpful in this edition were the following: John R. Baker, Iowa State University; Stephen W. Wilson, California State University; Brent B. Nickol, University of Nebraska; Martyn L. Apley, Western State College of Colorado; and Michael D. Johnson, DePauw University. We extend our thanks also to Yevonn Wilson-Ramsey, who prepared most of the new illustrations for this edition.

Frances M. Hickman
Cleveland P. Hickman, Jr.

CONTENTS

General Instructions, 1

PART ONE   INTRODUCTION TO THE LIVING ANIMAL

1   The Microscope, 7

Exercise 1A. Compound microscope, 8
Exercise 1B. Stereoscopic dissecting microscope, 13
Exercise 1C. Electron microscope, 14

2   Cell Structure and Division, 21

Exercise 2A. The cell—unit of protoplasmic organization, 21
Exercise 2B. Mitosis—the division of cells, 26

3   Cell Function, 33

Exercise 3A. Diffusion phenomena, 34
Exercise 3B. Action of enzymes, 43

4   Gametogenesis and Embryology, 45

Exercise 4A. Meiosis—maturation division of germ cells, 45
Exercise 4B. Early embryology of the sea star, 53
Exercise 4C. Frog development, 55

5   Tissue Structure and Function, 61

6   Introduction to Animal Classification, 77

PART TWO   THE DIVERSITY OF ANIMAL LIFE

7   The Protozoa, 85

Exercise 7A. The Sarcodina—*Amoeba* and others, 86
Exercise 7B. Subphylum Mastigophora—*Euglena* and *Volvox*, 93
Exercise 7C. Class Sporozoea—*Gregarina* and *Plasmodium*, 101
Exercise 7D. Phylum Ciliophora—*Paramecium* and other ciliates, 104

8   The Sponges, 113

Exercise 8A. Class Calcispongiae—*Scypha*, 114

9   The Radiate Animals, 123

Exercise 9A. Class Hydrozoa—*Hydra, Obelia, Gonionemus*, 124
Exercise 9B. Class Scyphozoa—*Aurelia*, a "true" jellyfish, 133
Exercise 9C. Class Anthozoa—*Metridium*, a sea anemone, and *Astrangia*, a stony coral, 135
Exercise 9D. Phylum Ctenophora—*Pleurobrachia*, a comb jelly, 139

10   The Acoelomate Animals, 143

Exercise 10A. Class Turbellaria—the planarians, 144
Exercise 10B. Class Trematoda—the digenetic flukes, 157
Exercise 10C. Class Cestoda—the tapeworms, 160

11   The Pseudocoelomate Animals, 163

Exercise 11A. Phylum Nematoda—*Ascaris* and others, 164
Exercise 11B. A brief look at some other pseudocoelomates, 170

12   The Molluscs, 175

Exercise 12A. Class Bivalvia (Pelecypoda)—the freshwater mussel, 176
Exercise 12B. Class Gastropoda—the land snail, 185
Exercise 12C. Class Cephalopoda—*Loligo*, the squid, 188

13   The Annelids, 191

Exercise 13A. Class Polychaeta—the clamworm, 192
Exercise 13B. Class Oligochaeta—the earthworm, 196
Exercise 13C. Class Hirudinea—the leech, 204

14   The Arthropods, 207

Exercise 14A. The chelicerate arthropods—the horseshoe crab and garden spider, 208
Exercise 14B. The aquatic mandibulates, 213
Exercise 14C. The terrestrial mandibulates, 224

Exercise 14D. Collection and classification of insects, 236

Key to orders of the more common insects, 239

15   The Echinoderms, 249

Exercise 15A. Subclass Asteroidea—the sea stars, 251

Exercise 15B. Subclass Ophiuroidea—the brittle stars, 259

Exercise 15C. Class Echinoidea—the sea urchin, 261

Exercise 15D. Class Holothuroidea—the sea cucumber, 267

Exercise 15E. Class Crinoidea—the feather stars and sea lilies, 269

16   Phylum Chordata, 271

Exercise 16A. Subphylum Urochordata—*Molgula*, an ascidian, 272

Exercise 16B. Subphylum Cephalochordata—amphioxus, 276

17   The Cartilagenous Fishes—Lampreys and Sharks, 283

Exercise 17A. Class Cephalaspidomorphi (= Petromyzontes) The lamprey (ammocoete larva and adult), 283

Exercise 17B. Class Chondrichthyes—the dogfish shark, 291

18   Class Osteichthyes—the Bony Fishes, 299

Exercise 18. The yellow perch, 299

19   Class Amphibia, 307

Exercise 19A. Behavior and adaptations, 308

Exercise 19B. The skeleton, 311

Exercise 19C. The skeletal muscles, 313

Exercise 19D. The digestive, respiratory, and urogenital systems, 319

Exercise 19E. The circulatory system, 325

Exercise 19F. The nervous and endocrine systems, 331

20   Class Mammalia—the Fetal Pig, 335

Exercise 20A. The skeleton, 336

Exercise 20B. The muscular system, 340

Exercise 20C. Preliminary dissection and the digestive system, 348

Exercise 20D. The urogenital system, 355

Exercise 20E. The circulatory system, 360

Exercise 20F. The nervous system, 370

Exercise 20G. The respiratory system, 375

PART THREE   ACTIVITY AND CONTINUITY OF LIFE

21   Muscle Physiology, 379

Exercise 21A. Contraction of glycerinated skeletal muscle, 380

Exercise 21B. Physiology of the myoneural junction, 381

22   Circulation and Respiration, 385

Exercise 22A. Study of human blood, 385

Exercise 22B. Capillary circulation in the frog, 395

Exercise 22C. Small mammal respiration, 399

23   Digestion and Excretion, 403

Exercise 23A. Distribution of digestive enzymes, 403

Exercise 23B. Glomerular filtration in the mudpuppy, 409

24   Nervous System, 413

Exercise 24. Spinal reflexes, 413

25   Animal Behavior, 419

Exercise 25A. Habituation in earthworms, 420

Exercise 25B. Aggressive behavior and the dominance hierarchy in crickets, 421

26   Genetics, 427

Exercise 26A. Inheritance in the fruit fly *Drosophila*, 428

Exercise 26B. Human inheritance, 430

Exercise 26C. Problems in genetics, 434

27   Field Trips to Study Habitats and Living Animals, 445

Appendix

A   A key to the major animal taxa, 450

B   Laboratory aids, 455

C   Chief taxonomic subdivisions and organ systems of animals, foldout

LABORATORY STUDIES IN INTEGRATED ZOOLOGY

G E N E R A L

I N S T R U C T I O N S

EQUIPMENT

Each student will need to supply the following equipment:

Laboratory manual
Dissecting kit containing scissors, forceps, scalpel, dissecting needles, pipette (medicine dropper), probe, and ruler, graduated in millimeters
Drawing pencils, 3H or 4H
Eraser, preferably kneaded rubber
Colored pencils—red, yellow, blue, and green
Box of cleansing tissues
Loose-leaf notebook for notes and corrected drawings

The department will furnish each student with the following:

Compound microscope
Microscope slides and cover glasses
Package of lens paper
Tripod hand lens
Syracuse watch glass
Box of brass dressmaker pins
All other supplies and equipment needed during the course

AIM AND PURPOSE OF LABORATORY WORK

In the zoology laboratory the student will handle living organisms, make dissections of preserved specimens to visualize structural relationships, ask questions about functional processes, and gain an appreciation of biological principles and the scientific method.

The scientific method requires patient, accurate, and critical observations. It demands orderly and systematic classification of facts, repetition of technical procedures, and frequent checking of observations and results. The conclusions drawn from scientific studies must be made with caution, and results should be reported with as little prejudice and personal bias as possible. Conscientious effort on your part in the performance of the laboratory work should provide training in making observations, collecting facts, and drawing conservative conclusions about them.

GENERAL INSTRUCTIONS FOR LABORATORY WORK

Prepare for the laboratory. Before coming to the laboratory, read the entire exercise to familiarize yourself with the subject matter and procedures. Read also the appropriate sections in your textbook. Good preparation can make the difference between a frustrating afternoon of confusion and mistakes and an experience that is pleasant, meaningful, and interesting.

Follow the manual instructions carefully. It is your guide to the exploration and understanding of the organisms or functions you are investigating. Its instructions have been written with care and with you, the student, in mind, to help you do the work (1) in logical sequence, (2) with economy of time, and (3) to arouse a questioning attitude that will stimulate interest and curiosity.

Use particular care in making animal dissections. A glossary of directional terms used in dissections will be found inside the front cover. The object in dissections is to separate or expose parts or organs so as to see their relationships. Working blindly without the manual instructions may result in the destruction of parts before you have had an opportunity to identify them. **Learn the functions** of all the organs you dissect.

Record your observations. Keep a personal record in a notebook of everything that is pertinent, including the laboratory instructor's preliminary instruction and all experimental observations. Do not record data on scraps of paper with the intention of recopying later; record directly into a notebook. The notes are for your own use in preparing the laboratory report later.

Take care of equipment. Glassware and other apparatus should be washed and dried after using. Metal instruments in particular should be thoroughly dried to prevent rust or corrosion. All materials and equipment must be put away in their proper places at the end of the period.

Your laboratory grade will reflect your attitude, initiative, and commitment, the quality and completeness of your dissections, drawings, and reports, and the results of practical quizzes.

TIPS ON MAKING DRAWINGS

You need not be an artist to make laboratory drawings. You do, however, need to be **observant.** Study your specimen carefully. Your simple line drawing is a record of your observations.

Before you draw, locate on the specimen all the structures or parts indicated in the manual instructions. Study their relationships to each other. Measure the specimen. Decide where the drawing should be placed and how much it must be enlarged or reduced to fit the page (read further for estimation of magnification). Leave ample space for labels.

When ready to draw, you may want first to rule in faint lines to represent the main axes, and then sketch the general outlines lightly. When you have the outlines you want, draw them in with firm dark lines, erasing unnecessary sketch lines. Then fill in details. Do not make overlapping, fuzzy, indistinct, or un-necessary lines. Indicate difference in texture and color by stippling. Stipple deliberately, holding the pencil vertically and making a neat round dot each time you touch the paper. Placing the dots close together or farther apart will give a variety of shading. Avoid line shading unless you are very skilled. Use color only when asked for it in the directions.

Label the drawing completely. Print labels neatly in lower case letters and align them vertically and horizontally. Plan the labels so that there will be no crossed label lines. If there are to be many labels, center the drawing and label on both sides.

Indicate the magnification in size beneath the drawing, for instance, "×3" if the drawing is three times the length and width of the specimen. In the case of microscopic slides, indicate the magnification at which you viewed the subject, for example, × 450.

A reduced sample of a student drawing is given below.

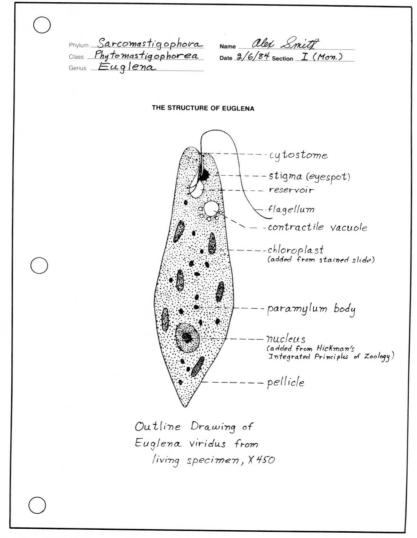

Phylum *Sarcomastigophora*
Class *Phytomastigophorea*
Genus *Euglena*

Name *Alex Smith*
Date 2/6/84 Section I (Mon.)

THE STRUCTURE OF EUGLENA

- cytostome
- stigma (eyespot)
- reservoir
- flagellum
- contractile vacuole
- chloroplast (added from stained slide)
- paramylum body
- nucleus (added from Hickman's Integrated Principles of Zoology)
- pellicle

Outline Drawing of Euglena viridus from living specimen, X 450

Example of a labeled student drawing. When drawing any structure not actually seen on your specimen, indicate in the label the source of your information.

Estimating the Magnification of a Drawing

A simple method for determining the magnification of a drawing is to find the ratio between the size of the drawing and the actual size of the object you have drawn. The magnification of the drawing can be expressed in the following formula:

$$\times = \frac{\text{Size of drawing}}{\text{Size of object}}$$

If your drawing of the specimen is 12 cm (120 mm) long, and you have estimated the specimen to be 0.8 mm long, then $\times = 120 \div 0.8$ or 150. The drawing, then, is $\times 150$, or 150 times the length of the object drawn.

This same formula will hold good whether the drawing is an enlargement or a reduction. If, for example, the specimen is 480 mm long, and the drawing is 120 mm, then $\times = 120/480$ or ¼.

PART ONE

INTRODUCTION TO
THE LIVING ANIMAL

CHAPTER 1

THE MICROSCOPE

MATERIALS

Microscopes, compound and binocular dissecting
Microscope lamps
Blank slides and coverslips
Prepared slides
   Typewritten letters (*e*, *a*, *h*, or *k*)
   Colored threads
   Fluke, tapeworm, or other whole mount
Salt or sugar crystals
Distilled water
Materials suitable for wet mounts
Pipette
Gum arabic solution
Pond water
Crayfish gills, pieces of sea star test, and so on
Finger bowls or watch glasses or both

For a biologist the compound microscope is one of the most important tools ever invented. It is indispensable not only in biology but also in the fields of medicine, biochemistry, and geology, in industry, and even in crime detection and many hobbies. Yet even though the microscope is one of the commonest tools in the biologist's laboratory, too frequently it is used without any effective understanding of its construction and operation. The results may be poor illumination, badly focused optics, and misleading interpretations of what is (barely) seen.

Both the compound microscope and the binocular dissecting microscope (stereoscopic microscope) will open up a whole new world for you if you will make the effort to become proficient in their use. Learn their possibilities and use them to the greatest advantage. Take good care of the microscope. Microscopes are expensive instruments and, although they are sturdily built and will stand many years of usage, they are precision instruments and are delicate enough to require careful treatment.

You will be assigned a microscope. Other students will use the same microscope on other days. It is your responsibility to see that it is in good condition at the end of the laboratory period. If you notice anything wrong with your microscope at any time, please report it to the instructor at once.

EXERCISE 1A
Compound microscope

The compound microscope may be either monocular or binocular, with either vertical or inclined oculars.

Handling the Microscope

☞ Use both hands to carry a microscope (Fig. 1-1). Grasp it firmly by the **arm** with one hand and support the **base** with the other. Carry it in a fully upright position. Set it down **gently** with the arm toward you. Before using it become familiar with its mechanical parts and how they operate.

Parts of the Microscope

If you are not familiar with the parts of the microscope, please study Fig. 1-2.

The **image-forming optics** consist of (1) a set of **objectives** screwed into a **revolving nosepiece** and (2) a **body tube**, or head, with one or two **oculars (eyepieces)**.

Each **objective** is a complex set of tiny lenses that do the major part of the magnification. Your microscope may be provided with two, three, or four objectives, each with its magnification, or power, engraved on the side. These usually include a 3.5× or 4.5× **scanning objective**, a 10× **low-power objective**, and a 43× or 45× **high-power objective**. Some microscopes also carry a 97× **oil immersion lens**, which must always be used with a drop of oil to form a liquid bridge between itself and the surface of the slide being used.

☞ Revolve the nosepiece and note the clicking sound when an objective swings into place under the tube.

The lenses in the **ocular** further magnify the image formed by the objective. The image then appears to your eye as though it were 10 inches away, but the eye should be kept perfectly relaxed, as though you were viewing the image at infinity. The ocular most often used is the 10×. A 6× or a 15× ocular may also be provided.

The **stage** is the platform with clips to hold the slide in place. Some microscopes are equipped with

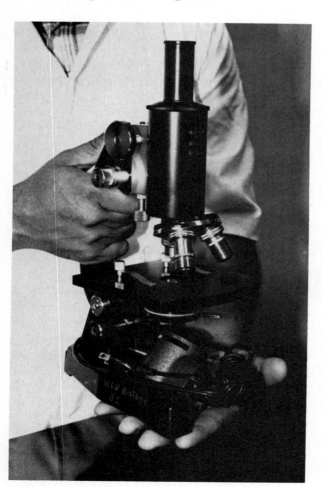

Fig. 1-1. Correct way to hold microscope for transporting. (Photograph by C.P. Hickman, Jr.)

a mechanical stage possessing knobs for moving the slide back and forth or up and down.

An adjustable **iris diaphragm** under the stage regulates the amount of light passing through the **aperture** of the stage. Learn to use the iris diaphragm. Your microscope may have a **substage condenser** with a set of lenses to concentrate the light on the object above. The iris diaphragm would be built into the condenser in this case. On some microscopes the light is regulated by a revolving disc with perforated holes of various sizes.

Many microscopes are provided with a built-in, low-voltage **substage illuminator**. Raise or lower the substage condenser to adjust the light to the desired intensity and then close down the iris diaphragm gradually until the field of view that you see through the ocular is evenly illuminated and filled with light.

Some microscopes lack a substage light; these employ an adjustable **reflecting mirror** that reflects light from a microscope lamp up into the tube. The **concave surface** of the mirror is used with artificial light when there is no condenser on the microscope. The **plane surface** is used with a substage condenser. Set a microscope lamp a few inches in front of the mirror. With the low-power objective in place adjust the mirror to bring a bright, evenly distributed circle of light through the lens. Now rotate the iris diaphragm or adjust the substage condenser to see how it focuses the light.

The **body tube** is the optical housing for the lenses. In some microscopes the body tube and oculars are inclined for greater comfort of viewing. The tube is raised or lowered by means of two sets of adjustment knobs. The **coarse-adjustment knob** is for low-power

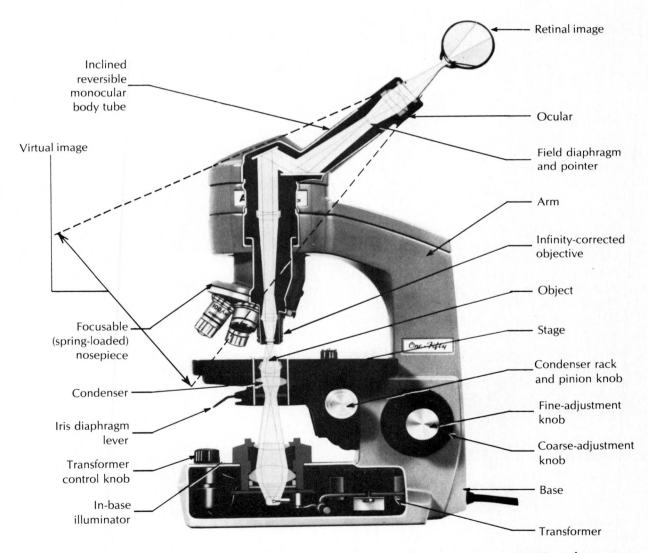

Fig. 1-2. Optical and mechanical features of compound microscope. (Courtesy American Optical, Buffalo, N.Y.)

work and for initial focusing. The **fine-adjustment knob** is for final adjustment and varying the plane of focus for viewing an object at different depths.

Turn the coarse-adjustment knob and note how it moves the body tube. Find out which way to turn the knob to raise the tube and which way to lower the tube. *Never* use the course-adjustment knob when a high-power objective is in place. Turn the fine-adjustment knob. This moves the tube so slightly that you cannot detect it unless you are examining an object through the ocular. The fine-adjustment knob works the same as the coarse-adjustment knob. To focus downward, turn the knob in the same direction as you would to focus down with the coarse adjustment. Practice this. **Always use the fine adjustment when the high-power objective is in place.**

Preliminary Use of the Microscope

You need not wear glasses when using the microscope unless they are for correction of severe astigmatism. Nearsightedness and farsightedness can be corrected by adjusting the microscope.

First clean the lenses of both the ocular and the objectives by wiping them gently with a clean sheet of lens paper. Do this each time you use the microscope. Never touch the lenses with anything except clean lens paper. Do not remove an ocular or objective unless told to do so by the instructor. Dust the mirror with a cleansing tissue.

Keep both eyes open while using the microscope. If this seems difficult at first, hold a piece of paper over one eye or tear a small hole in a piece of paper and place it over the ocular so as to shield the eye you are not using. When one eye becomes tired, shift to the other one.

How to focus with low power. Turn the low-power objective until it clicks in place over the aperture. Open the iris diaphragm. Turn the appropriate side of the mirror toward the light source. While looking into the ocular, adjust the mirror until the field of vision is completely and evenly lighted. It is important to learn how to obtain the best light and how to regulate it.

Obtain a slide containing the letter **e** (or **a**, **h**, or **k**). Place it, coverslip up, on the stage with the letter centered under the objective lens. While watching the objective and the slide, lower the objective with the coarse adjustment until it is close to the slide **but not touching it** (about 0.5 cm). Now look through the ocular and slowly raise the objective by turning the coarse adjustment toward you until the object on the slide is in sharp focus. Is the image upside down? Is it reversed?

If you do not see an image of the object at all, it may be because the object on the slide is not properly centered under the objective. Check the centering of the slide and repeat the process as before.

Shift the slide very slightly to the right while viewing it through the ocular. In what direction does the image move? Move the slide away from you. What happens to the image? Turn the fine adjustment knob toward and then away from you and observe the effect on the image.

How to focus with high power. To be studied with high power, the object on a slide must be covered with a cover glass and must be thin enough to transmit light. **Focus the object first with low power;** then, while watching the objective lens, slowly rotate the high-power objective into position. If the microscope is **parfocal,** the object in focus with low power will be nearly in focus under high power. Turn the **fine-adjustment** knob to bring it more sharply into focus.

Never use the coarse adjustment while looking at an object under high power; you may ram the objective into the slide. This may damage the slide or, more importantly, ruin the lens of the objective.

If the microscope is not parfocal (your instructor will tell you), focus first with low power and then raise the tube by turning the coarse adjustment knob one-half turn. **With your eye at the level of the stage** carefully swing the high-power objective into place, raising the tube a little further if the objective touches the slide. Now, **still watching the high-power objective,** lower it slowly to about 1 mm from the cover glass. Then, **looking through the ocular,** raise the tube with the fine adjustment until the object is in focus. Do this several times to acquire skill in focusing.

If you cannot find anything at all with high power, it may be that, because of the small size of the high-power field, the object is not in the field at all. Turn back to lower power, adjust the slide so that the object you want to find is in the very center of your field of view, and then return to high power. If you still cannot find the object, it may be that it is too far out of focus to be seen. Use the fine-adjustment knob. You will have to learn by experience whether to rack up or down to bring the slide into sharp focus.

The light decreases when you switch to high power because the diameter of the high-power lens is so much smaller than that of the low-power lens. Adjust the light with the mirror and iris diaphragm. Do not tilt the microscope. Keep it vertical. This is particularly important with wet mounts.

Any object you see on the microscope, no matter how thin, has some depth, and you must view it at different focal planes to bring out all the details. To do this, **keep your hand on the fine adjustment** and constantly focus up and down. This is essential to really understand the nature of the material on a slide.

How to use the oil-immersion objective. Occasionally a project or demonstration exercise requires the use of an oil-immersion lens; this is usually $93\times$ but sometimes is $100\times$. Used with a $10\times$ ocular the total magnification ($930\times$ or $1000\times$) approaches the resolving power of the microscope. To deliver as much light as possible to the lens system, oil must be used as a bridge between the slide and the lens.

To use oil immersion, first bring the specimen into focus with the low- and high-power lenses. Carefully center the point of interest in the field of view; then rotate the nosepiece to move the high-power objective off to one side. Place a single drop of immersion oil on the coverslip at the point where the objective will come into position. Now move the oil-immersion objective carefully into position, watching from the side to be certain that the lens clears the coverslip. The oil should now form a bridge between lens and coverslip. **Carefully** adjust the fine focus to bring the specimen into focus. Adjust the iris diaphragm or substage condenser to increase light as required.

When finished, clean the lens with a lens tissue wetted with xylol. **Never** use alcohol, which will dissolve the cement around the lens system. If the lens is to be used again soon (within a day or two), it is best not to clean the lens face. Residual oil will not harm the lens unless it is allowed to harden over a long period without use.

Some Helpful Suggestions

Always check to see that a slide is placed with the coverslip **up.**

Always use **low power first,** to locate the material on the slide, and then switch to high power.

If images of windows, lights, or the like appear in the field of vision, use the concave surface of the mirror or lower the condenser if there is one.

Always focus downward with your eyes looking at the objective; focus upward with your eye(s) looking through the ocular(s).

Keep your hand on the fine adjustment. Frequent change of focus gives depth to your field of vision and helps you understand the material on the slide.

If black spots appear in the field of vision, there may be dirt on the slide, on the objective lens, or on the lens of the ocular. If they move in a circle when you rotate the ocular, they are on the ocular. If they move when you move the slide, they are on the slide. Use only special lens paper to clean the lenses. The slide may be cleaned very gently with a soft damp cloth or damp cleansing tissue.

Avoid direct sunlight for illumination. A north light is the best natural illumination. Too much light is as bad as too little. Transparent objects are often clearer in reduced light. Regulate the iris diaphragm and condenser.

If the fine adjustment refuses to turn, the knob has reached the limits of its range. To correct this, give the fine-adjustment knob several turns in the opposite direction, and then refocus with the coarse adjustment.

Microscopes must be kept dust free. Return the scope **carefully** to its box or cupboard when finished with it. Before putting it away, put the low-power objective in place and raise the tube a little. Be sure not to leave a slide on the stage.

The prepared slides you will be using in the laboratory were made at the cost of great skill and patience and are quite expensive to buy. They are fragile and should be handled with great care. If a demonstration of slide making is not given in your laboratory you can get an idea of how such slides can be made by reading the account on pp. 13 and 14.

Magnification of the Microscope

How much your microscope will magnify depends on the power of the combination of lenses you are using. Your microscope is probably equipped with a $10 \times$ ocular, which magnifies the object 10 times in diameter. Other oculars may magnify $2 \times$, $5 \times$, or $20 \times$. The objectives may be designated, respectively, $3.5 \times$ (scanning objective), $10 \times$ (low-power objective), and $45 \times$ (high-power objective). The total magnifying power is determined by multiplying the power of the objective by the power of the ocular. Examples of the magnification of certain combinations follow:

Ocular	Objective	Magnification
$5 \times$	$3.5 \times$	17.5 diameters
$5 \times$	$10 \times$	50 diameters
$10 \times$	$3.5 \times$	35 diameters
$10 \times$	$10 \times$	100 diameters
$10 \times$	$45 \times$	450 diameters

How to Measure Microscopic Sizes

It is often important to know the length of an organism or object that you are viewing through the microscope, because size is often a diagnostic characteristic. Or, if you are looking for a particular species of protozoan in a mixed culture and know that that species is usually about 400 μm long, it saves time to know just how large 400 μm will appear at either low or high power. For such estimates of size you need to know the diameter of the field of view that you see with both the low and the high power. Alternative methods are described as follows:

1. Measuring Objects with Ocular and Stage Micrometer Calibration. An ocular micrometer has been fitted into the microscope's eyepiece. It is a disc on which is engraved a scale of (usually) either 50 or 100 units. These units are arbitrary values that always appear the same distance apart no matter which objective is used in combination with the eyepiece. Therefore the ocular micrometer cannot be used to measure objects until it has been calibrated with a stage micrometer.

The stage micrometer resembles an ordinary microscope slide but bears an engraved scale on its upper surface, usually about 2 mm long, divided into 200 units of 0.01 mm (10 μm) each.

Place the stage micrometer on the microscope stage and focus on the engraved scale with the low-power objective. Both scales should appear sharply defined. Rotate the eyepiece until the two scales are parallel. Now move the stage micrometer to bring the 0 line of the stage scale in exact alignment with the 0 marking of the ocular scale. The scales should be slightly superimposed.

To calibrate the ocular scale for this objective, use the longest portion of the ocular scale that can be seen to coincide precisely with a line on the stage scale. For example, suppose that 31 units of the 50-unit ocular scale equal 18 units (180 μm) of the stage scale. The value of each unit on the ocular micrometer is therefore $\frac{180 \ \mu m}{31} = 5.8 \ \mu m$. To measure any object available to you in the laboratory, it is only necessary to multiply the number of divisions covered by the specimen or part thereof by the micrometer value you have determined (5.8 μm in this example). Note that the micrometer value applies only to the objective with which the calibration was made. Repeat the calibration procedure with the high-power objective.

2. Measuring Objects with Transparent Ruler Calibration. This alternative is not nearly as accurate as the preceding, but it is a serviceable substitute when ocular and stage micrometers are not available.

With the scanning objective in position, place a transparent ruler on the microscope stage and focus on its edge so that you can see the scale. Move the ruler right or left so that one of the vertical millimeter lines is just visible at the edge of the circular field of view. Count the number of millimeter lines spanning the field. You will probably have to estimate the last fraction of a millimeter. This gives the diameter in millimeters of the field of view for the scanning objective. Since most microscopic measurements are expressed in micrometers rather than millimeters (1 μm = 0.001 mm), you may convert the field diameter to micrometers by multiplying by 1000.

To obtain the diameter of the fields of view for your low- and high-power objectives, it is more accurate to calculate these fields from the measurement you obtained with the scanning lens than to measure them directly with a ruler. To do this, multiply the field diameter you obtained with the scanning lens by the stated scanning lens magnification and then divide the product by the low-power objective magnification. For example, if you obtained a field diameter of 4.6 mm (4600 μm) with a 3.5× scanning objective, the field of view for the 10× objective is:

$$\frac{4.6 \text{ mm} \times 3.5\times}{10\times} = 1.6 \text{ mm} \ (1600 \ \mu m)$$

For the 45× objective the field is:

$$\frac{4.6 \text{ mm} \times 3.5\times}{45\times} = 0.36 \text{ mm} \ (360 \ \mu m)$$

Exercises with the Compound Microscope

> **DRAWINGS**
>
> Any sketches required in the following exercises may be made on p. 19.

1. Obtain a slide of three different colored threads. Focus with low power to determine which colored thread is on top, in the middle, and on the bottom. This gives you the consciousness of depth in the objects you are viewing.

2. Examine crystals of salt or sugar. Sketch what you see. Estimate the size of the crystals.

3. Place a hair from your eyebrow on a slide. Examine with low and high power. Estimate the diameter of the hair.

4. Examine the letter **e** made on onionskin paper by two different typewriters and see if you can find individual differences between the two typewriters. Sketch.

5. Prepare a temporary wet mount. Select a clean slide and cover slip. Mount a piece of insect wing or appendage, a bit of feather, or some similar object on the slide. Add a drop of distilled water with a pipette. Hold the coverslip at an angle with one edge on the slide and the other held up with a teasing needle. Slowly lower the raised edge onto the drop of water. This helps prevent air bubbles from forming. Examine with low power and sketch what you see in the field of vision. Switch to high power without moving the slide, and draw the portion now visible.

6. Make a wet mount using a drop of water from a solution in which gum arabic has been dissolved and shaken up. With low power, look for large and small air bubbles, focusing up and down on them until you are sure you will always recognize an air bubble. This may later prevent you from interpreting a bubble as a small organism or other small object. What is the characteristic outline of an air bubble?

7. Mount a drop of pond water, cover, and look for living organisms in it. As you watch the movement of the swimming animals, remember that movement as well as size is magnified by the microscope. Sketch some of the organisms. On your drawing state the estimated size of each animal and the magnification of your drawing (see pp. 3 and 11).

REFERENCES

The following selected references deal with the theory and practice of microscopy and microtechnique.

Galigher, A.E., and E.N. Kozloff. 1971. Essentials of practical microtechnique. ed. 2 Philadelphia, Lea & Febiger.

Gray, P., ed. 1973. The encyclopedia of microscopy and microtechnique. New York, Van Nostrand Reinhold Co.

Möllring, F.K. 1978. Microscopy from the very beginning, Oberkochen, West Germany, Carl Zeiss.

Wilson, M.B. 1976. The science and art of basic microscopy, Bellaire, Tex., American Society for Medical Technology.

EXERCISE 1B
Stereoscopic dissecting microscope

The stereoscopic dissecting microscope (Fig. 1-3) is as indispensable to the laboratory as is the compound microscope. It enables you to study objects too large or too thick for the compound microscope. It furnishes a three-dimensional view of objects at a very low power (5× to 50×, depending on the microscope). The image is not inverted, and there is ample space for manipulation and dissection under the lens. The microscope stage can be illuminated either by reflected or transmitted light. Some microscopes have a substage mirror or a substage lamp. Focusing is done with very little adjustment.

How to focus. (1) Place an object (coin or other) on the center of the stage. (2) If the microscope has a revolving nosepiece, swing the low-power objective into place and lower it as far as it will go. (3) Note how the two oculars can be moved toward or away from each other. Looking through the oculars, adjust them to fit the distance between your own eyes so that you are looking through **both** oculars at once and seeing a single field of vision. (4) Looking through the oculars, raise the objective lens until the object comes into view. If the focus is not sharp, you may have to focus for each eye separately. The microscope probably has one fixed and one adjustable eyepiece. Look through the fixed lens first and adjust the focus to suit that eye. Then adjust the other ocular to suit the other eye until the focus is sharp for both eyes.

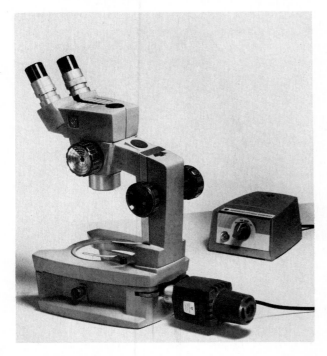

Fig. 1-3. The stereoscopic binocular microscope with illuminator. (Courtesy American Optical, Buffalo, N.Y.)

Move the object on the stage away from you. Which way does the image move?

Move the object to one side. Which way does the image move? How does this compare with the compound microscope?

Exercises with the Dissecting Microscope

Try the following or any similar exercises with the dissecting microscope and **keep a record** on p. 20 of the results of different lighting and background effects with different types of material. This will save time for you in later studies.

1. Examine some pond water. How many kinds of organisms can you see? Are they more distinct with a white or a black stage beneath them?

2. Examine a prepared slide of the whole mount of a fluke, tapeworm, or similar slide. Try it with reflected lighting, moving the light source about for the best effects. Then try transmitted lighting. If the microscope lacks substage lighting, try placing the slide over a small microscope lamp with a piece of writing paper between the slide and the lamp. Which method gives better illumination?

3. Examine the surface of a preserved sea star or the gills of a crayfish. Study the material first without water; then study it submerged in a finger bowl or watch glass of water. In which preparation do you see the most detail? Why? Try the gill both with top lighting and with transmitted light. Can both methods be used with the sea star? Why?

Be sure to make good use of the dissecting microscopes that are available for your use in the laboratory. You will find them invaluable.

DEMONSTRATIONS

1. Demonstration of steps in the making of permanent slides. If an organism is too small to see with the unaided eye, it may be mounted whole on a slide or sectioned into small pieces and mounted. To differentiate the various parts, a stain is applied that usually has more affinity for certain of the structures than for others. Details of cells and tissues can be studied only when thin sections of them are prepared and mounted on slides. The steps in the preparation of such a permanent slide may be outlined as follows:

a. As soon as an animal is killed, small pieces of the desired tissue (such as liver, brain, or kidney) are quickly cut out and immersed for several hours in a fixing solution. The fixation process is perhaps the most important step in the process, for, if it is done properly, "artefacts" or abnormal structures do not occur.

b. The excess fixation is washed out of the tissue by water or other means.

c. The tissues are dehydrated by immersion in a series of gradually increasing alcoholic solutions, such as 35%, 50%, 70%, and up to 100% alcohol.

d. Then the tissues are put into a clearing agent such as xylene or toluene. This is done so that the tissues can next be put into paraffin, which is soluble in the clearing agent.

e. The tissue is placed in melted paraffin in a warm oven until infiltrated with paraffin.

f. The tissue is removed from the melted paraffin and embedded in fresh melted paraffin, which is then hardened. This causes the tissue to keep its shape when cut into thin sections.

g. The block of paraffin containing the tissue is placed on a microtome, on which very thin sections are cut. It is possible to cut cross sections, longitudinal sections, or other planes.

h. The sections are mounted on glass slides and affixed to the glass by albumen.

i. After the slides have dried, the paraffin is dissolved off by a clearing agent such as xylene. Now only the tissues are left on the slide.

j. The clearing agent is removed by alcohol, and the alcohol is replaced with water.

k. The staining process is next. Many kinds of stains may be used.

l. After the tissue is stained, the water is removed again by alcohol, which in turn is removed by a clearing agent. A drop of balsam or other mounting medium is placed over the stained sections, and a cover glass is applied.

m. After the mounting medium has hardened, the slide is ready to be studied in class. Such preparations may last for many years if they are properly taken care of.

2. *Demonstration of the use of an electron microscope.*

3. *Demonstration of phase-contrast microscope.*

4. *Demonstration of photomicrographic equipment and methods.*

EXERCISE 1C
Electron microscope

Although it is unlikely that you will have an opportunity to use an electron microscope, electron microscopy has contributed so greatly to our understanding of cell structure and function that you should have a summary understanding of its principles. In this system the visible light source of an optical microscope is replaced by a tungsten filament that, driven by high voltage, emits a beam of electrons. The beam is shaped by magnets or electric fields, and the image is projected onto a fluorescent screen for direct viewing.

The sample to be examined is specially treated with electron "stains" containing heavy metal ions that block the electron beam to varying degrees. The sample must also be cut extremely thin because the electron beam has a very low penetrating power, and it must be free of scratches, dirt, or other imperfections. Once the specimen is prepared (and this is the greatest challenge to successful electron microscopy), it is placed in the microscope's specimen chamber, which is evacuated. Because specimens must be dehydrated and evacuated before being placed in the specimen chamber, one obvious limitation of electron microscopy, at least at present, is that living material cannot be viewed.

Image formation depends on electron scattering by objects in the electron beam. If structures in the specimen contain chemical groups that react with the electron stains to precipitate heavy metal ions, these structures appear dark (electron dense) in the pho-

tograph. If these ions are not precipitated, the structures or regions appear light (electron lucent) in the photograph.

The kind of electron microscope described is called a **transmission electron microscope** (abbreviated TEM) because the electron beam passes through the specimen. Fig. 1-4 compares a light micrograph of liver cells, photographed with an optical microscope, with the greatly magnified view of a single liver cell photographed with the TEM. Note the much greater **resolving power** of the TEM. The resolving power of a light microscope is determined by the mean wavelength of visible light and the optical properties of the objective lens and is about 0.2 μm. In other words, points closer together than about 0.2 μm cannot be distinguished as separate points. Resolution can be improved somewhat by using shorter wavelengths. Therefore the ultraviolet (UV) microscope, with special quartz optics, has about twice the resolving power of the ordinary light microscope because of the shorter wavelength of ultraviolet light used to illuminate the specimen.

The electron microscope has a practical resolving power with biological preparations of about 1.5 nm, or about 130 times that of the best light microscope (1 μm = 1000 nm). However, unlike ordinary light microscopy electron microscopy is not even close to the theoretical limits of its resolving power. As more advanced electron microscopes are designed, we can

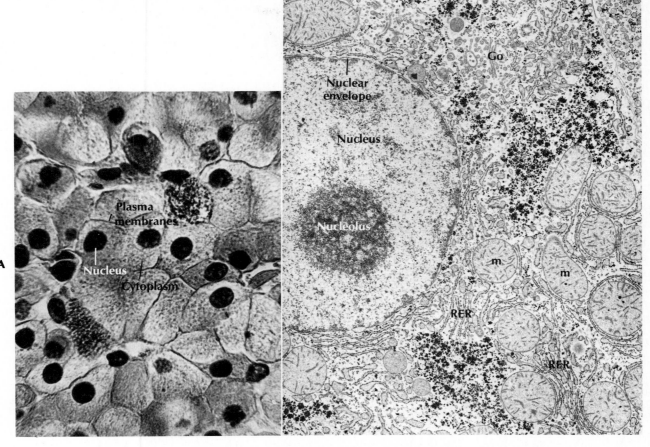

Fig. 1-4. Comparison of light photomicrograph (**A**) and electron photomicrograph (**B**) of liver cells. The abbreviations are *Go*, Golgi apparatus; *m*, mitochondrion; *RER*, rough endoplasmic reticulum; *g*, glycogen granules (Light micrograph × 160; electron micrograph × 7000.) (**B** courtesy G.E. Palade, The Rockefeller University, New York.)

expect to resolve even smaller objects than is possible now.

There are several specialized applications of the TEM that permit shadows to be cast on the specimen at precise angles. These techniques enable the microscopist to see the general shape and profile of the object. One of these techniques, called **freeze fracturing,** is illustrated in Fig. 1-5. The tissue to be freeze-fractured is first impregnated with glycerol and then quick-frozen in liquid Freon at 130° C. The frozen tissue is then fractured with a steel knife, splitting membranes of organelles within the cell along their centers. Water is then removed, and the specimen is "shadowed" with a combination of platinum and carbon, which piles up in front of pits and behind projections from the surface (Fig. 1-5, *B*). The tissue is then floated away, leaving an exact replica of the fractured surface of the specimen. Then it is photographed with the TEM. Note the three-dimensional image of the liver cell shown in Fig. 1-5, *A*. Notice how the pores in the nuclear envelope appear on the surface of the nucleus in the freeze-fractured specimen. Nuclear envelope pores also can be seen in the sectioned specimen (Fig. 1-4, *B*), but they appear differently. Can you find them? Notice also how the sectioned specimen reveals the internal structure of the mitochondria, whereas these organelles appear as rounded projections or pits in the freeze-fractured specimen. No internal structure of the mitochondria is visible in the latter. Why? Compare the appearance of rough endoplasmic reticulum (*RER*) and glycogen granules in the two specimens (glycogen granules are labeled *g* in Fig. 1-4, *B*, and shown with an asterisk in Fig. 1-5).

Scanning electron microscopy (SEM) has become an increasingly popular biological tool. In this technique, which requires an electron microscope of different design from the TEM, the surface topography of the specimen is revealed in great detail. The resolution of the SEM is not as great as that of the TEM (usually whole cells or tissues are viewed with the SEM), but the television-like image that is produced

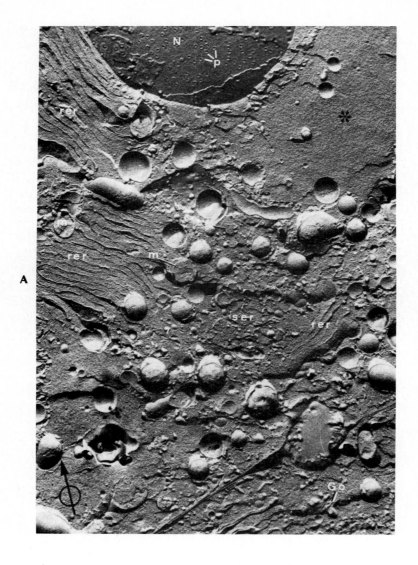

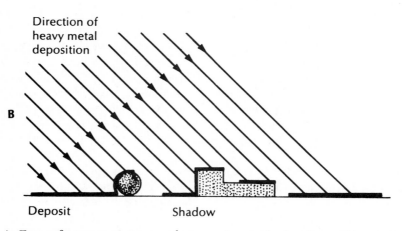

Direction of
heavy metal
deposition

B

Deposit   Shadow

Fig. 1-5. A, Freeze-fracture transmission electron micrograph of rat liver. Abbreviations are *ser,*
smooth endoplasmic reticulum; *N,* nucleus; *p,* pores in the nuclear envelope; asterisks, glycogen
deposits. Other abbreviations are same as in Fig. 1-4. **B,** Method of shadow-casting, showing how
metal and carbon build up before projections and behind depressions. (×17,000.) (**A** from Orci,
L.: J. Ultrastruct. Res. **35:**1-19, 1971.)

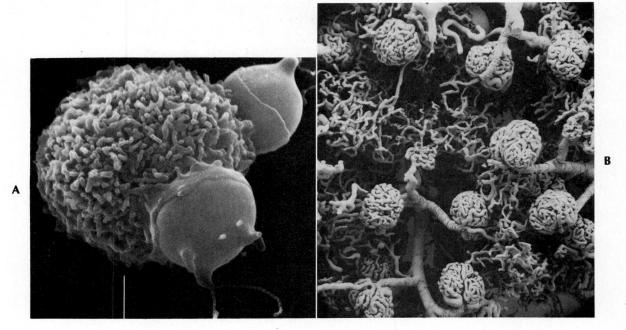

Fig. 1-6. A, Scanning electron micrograph of macrophage ingesting damaged red blood cells. (×5250.) **B,** SEM of microcirculation of mammalian kidney. Note spherical glomeruli, the kidney's pressure filters where urine formation begins. (×80.) (**A** courtesy J.P. Revel, California Institute of Technology; **B** from R.G. Kessel and R.H. Kardon: 1979. Tissues and organs. San Francisco, W.H. Freeman & Co.)

is of great value in understanding the shape, size, and organization of biological material. Two SEMs are shown in Fig. 1-6: one of single cells and the other of a tissue composed of numerous cells. To make these pictures, the specimens were first coated with metal ions and then mounted on a support, which is placed in the beam path of the SEM. The beam scans rapidly back and forth across the specimen, producing bursts of secondary electrons. These impinge on a charged scintillator at the side of the specimen where the electrons are converted to electric pulses. These are passed to a television viewing screen. A second cathode-ray tube is used for photographing the image.

Examine Fig. 1-6, *A*, showing a macrophage (a phagocytic cell) ingesting two damaged red blood cells. Notice the ruffled surface of the macrophage; the cytoplasmic extensions behave like pseudopodia when the macrophage moves about through tissues. You can also see thin cytoplasmic extensions spreading out to envelope the red blood cells. Fig. 1-6, *B*, is an SEM of the microvasculature of a mammalian kidney. It shows several ball-like glomeruli (filtration units) and the arterioles that serve them. Note also the extensive capillary plexus that surrounds the kidney tubules in the living organ (the tubules have been digested away, so they cannot be seen). This specimen

is called a vascular replica and requires special preparation. First, all blood is flushed out of the circulation of the anesthetized animal and replaced with a synthetic compound that polymerizes, becoming solid. Then the tissue is excised and placed in a corrosive alkali that digests away all the tissue, leaving behind a hardened cast of the microcirculation. Finally, the casts are washed, dried, and coated with metal for study with the SEM.

Scanning electron micrographs are of great value in helping us to understand complex relationships within tissues and organs. Fig. 1-6, *B*, for example, allows us to picture the circulation of a kidney instantly, something that is very difficult to do from examination of sectioned material.

DEMONSTRATIONS

Examine the photographs or books placed on demonstration. Two books in particular are valuable sources of electron photomicrographs: Fawcett, D.W. 1966. The cell: its organelles and inclusions. Philadelphia, W.B. Saunders Co. (an excellent source of TEMs); and Kessel, R.G., and R.H. Kardon. 1979. Tissues and organs: a text-atlas of scanning electron microscopy. San Francisco, W.H. Freeman & Co., Publishers.

PRELIMINARY STUDIES WITH COMPOUND MICROSCOPE

CHAPTER 2

CELL STRUCTURE AND DIVISION

EXERCISE 2A
The cell—unit of protoplasmic organization

MATERIALS
Prepared stained slides
   Immature sea star eggs
   Amphiuma liver
Slides
Cover glasses
Toothpicks
Stain (methylene blue, methyl green, gentian violet, or other)*
Microscopes

SOME EXAMPLES OF CELLS
The cell is the basic structural and functional unit of all living organisms. In most protozoans (unicellular animals) single cells exist as individuals. In metazoans (many-celled animals) cells may be associated together as tissues specialized for certain functions. Some cells are themselves highly specialized, such as nerve cells for conducting impulses, muscle cells for contraction, or eggs or spermatozoa for reproduction. But regardless of its shape, size, or special function, each cell is a living dynamic entity, capable of maintaining and propagating itself.

☞ Study the following preparations with both the low power and higher power of the compound microscope, unless otherwise directed.

Unstained Squamous Epithelial Cell from the Inside of Your Cheek

☞ With a clean toothpick gently scrape the inside of your cheek. Disperse the scraping (epidermal cells that are being shed) in a drop of water on a clean slide. Add a cover glass and examine with lower power. The cells are so thin and transparent that you will need to reduce the light intensity to see them. Locate a small group of cells and switch to high power, adjusting the light intensity as necessary.

*For preparation of stains, see Appendix B, p. 455.

Note the flat shape of the **epithelial cells.** Each cell contains a **nucleus,** which controls the development and function of the cell. The nucleus is surrounded by translucent **cytoplasm.** Although it appears structureless with the optical microscope, it consists of numerous **organelles** and **inclusions** suspended in a **ground substance.** The metabolic and synthetic activities of a cell reside mostly in the cytoplasm. You will look more closely at the cell's organelles in your study of liver cells.

The cell is enclosed by a **plasma membrane** 7.5 to 10.0 nm thick—too thin to be resolved with the light microscope. You can, however, determine its location

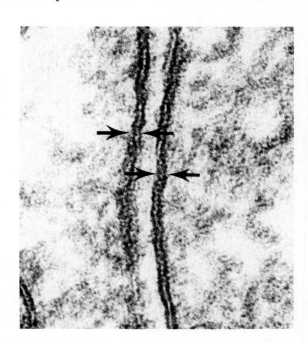

Fig. 2-1. Plasma membranes of two adjacent cells. Note that each membrane (between arrows) shows a typical dense-light-dense staining pattern. ($\times 325,000$.) (Photograph by A. Wayne Vogl, University of British Columbia, Vancouver, Canada.)

by differences in refraction between the cell and the surrounding water.

Through the electron microscope the plasma membrane appears as two dense lines separated by a less dense layer (Fig. 2-1). The membrane consists of a bimolecular layer of mixed lipids between two layers of proteins; the two dark lines seen in an electron micrograph are thought to represent the protein layers, and the less dense center line represents the lipids. The proteins on the surface of the membrane are of several kinds, some floating on the surface (extrinsic proteins) and others penetrating all the way through the membrane (intrinsic proteins). The most widely accepted model of membrane structure, the fluid mosaic hypothesis of Singer and Nicolson, views the membrane as a rather restless and fluid structure. The plasma membrane functions as the cell's gatekeeper, controlling its permeability and regulating the passage of materials in and out of the cell.

Stained Squamous Epithelial Cell

☞ Repeat the preparation you made for the unstained squamous epithelial cell but this time add 1 or 2 drops of a dilute stain (methylene blue, methyl green, gentian violet, or other) before adding the cover glass.

Note which part of the cell stains most intensely. What advantages does the stained preparation have over the unstained? Look for bacteria, which are tiny rod-shaped bodies found in every mouth.

DRAWINGS

On p. 31 sketch some squamous cells, top view, each about 2 cm wide.

Egg Cell of Sea Star

☞ Obtain a slide of immature sea star eggs (Fig. 2-2). If a slide of sea star cleavage stages is used, look under low power among the various stages of cleavage development for an isolated spherical cell with a distinct nucleus. This is an immature or unfertilized egg. Center the egg in the field of vision and focus up and down to study its general shape and structure. Then carefully turn to high power and study the parts of the cell.

The most easily differentiated parts of the cell are the central spherical **nucleus** and the surrounding **cytoplasm.** Which of these makes up most of the cell? Does the cytoplasm appear to be granular or fibrillar in structure? Do you notice any small clear spaces in it? As just mentioned, the cytoplasm contains many organelles and inclusions too small to be visible in this examination. Surrounding the cytoplasm is the **plasma membrane.** Identify the **nuclear envelope** enclosing the nucleus. Within the nucleus are the darkly staining **chromatin granules,** which contain DNA and which at certain stages of the cell cycle form the chromosomes. The chromosomes carry the genes, which are responsible for hereditary qualities. Find within the nucleus a deeply stained spherical **nucleolus.** The nucleolus is rich in RNA and is the active site of ribosome synthesis. Ribosomes pass from the nucleus into the cytoplasm, where they serve as important sites for protein synthesis.

Liver Cells

For the study of liver cells the liver tissue of *Amphiuma* is preferred because the cells are large. However, mammalian liver may also be satisfactorily used. In this tissue you are not studying isolated cells but cells associated together in a glandular organ.

☞ Examine a stained prepared slide of liver tissue first with low power and then with high power.

Refer to Fig. 1-4, *A*, and notice the mosaic appearance of the polyhedral cells, each with six or more surfaces. Note that the thin-cut section on your slide has passed through the nuclei of some cells but has passed above or below the nuclei of others; the latter appear to contain only cytoplasm. Do you think the massing of these cells together has any effect on their shape? Examine one of the cells closely and identify the **plasma membrane**, the **nucleus**, the **cytoplasm**, and the **nuclear envelope.**

DRAWINGS

1. On p. 31 draw and label the egg cell of a sea star, about 4 cm in diameter.
2. Draw a group of liver cells, each about 2 cm in diameter, showing typical shape and nuclei.

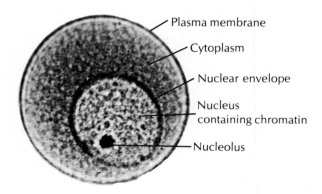

Fig. 2-2. Unfertilized egg of sea star. (Courtesy Carolina Biological Supply Co., Burlington, N.C.)

Fine Structure of Liver Cell. Examine the electron micrograph of a liver cell (Fig. 2-3). This photograph clearly shows some of the organelles and inclusions found in a typical animal cell; these include numerous **mitochondria, endoplasmic reticulum,** and **glycogen granules.**

Nucleus. In liver cells the nucleus is a spherical structure consisting of a double-layered nuclear envelope enclosing the **nucleoplasm.** The nuclear envelope is broken at intervals by gaps, which permit continuity between the nucleoplasm and the cytoplasm surrounding the nucleus. These gaps, or "**pores,**" are clearly evident in Fig. 2-4 and in the freeze-fracture preparation in Fig. 2-5. Messenger

RNA and transfer RNA pass from the nucleus to the cytoplasm through these pores.

Refer again to your slide and locate the **nucleolus** inside the nucleus. Do some of the nuclei have more than one nucleolus? In cells that are actively growing and dividing and in cells that are actively synthesizing protein, the nucleoli are large and often multiple.

The electron microscope shows that the nucleolus does not have a membrane (Fig. 2-6) and is composed of a granular material and a fibrous network. The nucleolus forms at a particular site on a particular chromosome in the nucleus (or chromosomes if more than one nucleolus is present). The DNA in the nucleolus directs the synthesis of ribosomal RNA, which com-

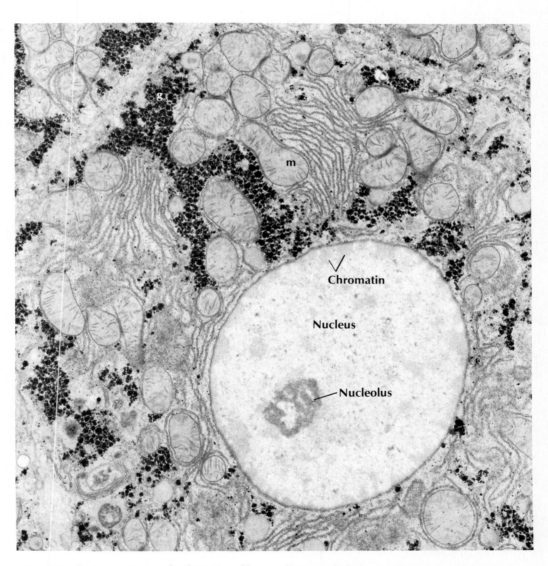

Fig. 2-3. Electron micrograph of portion of liver cell, magnified about 10,000 times. Note single large nucleus containing nucleolus, mitochondria *(M)*, rough endoplasmic reticulum *(RER)*, and numerous glycogen granules *(g)*. (From Morgan, C.R., and Jersild, R.A.: Anat. Record **166:**575-586, 1970.)

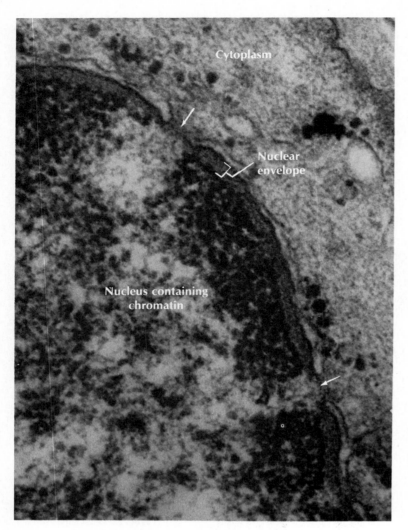

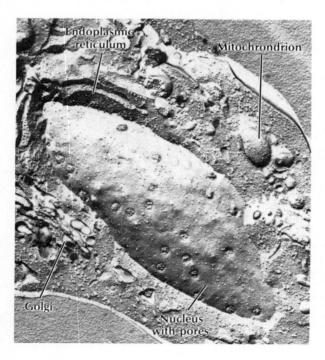

Fig. 2-4. Nuclear envelope showing two gaps, or "pores." Note double-layered nature of nuclear envelope. (×120,000.) (Photograph by G.E. Palade.)

Fig. 2-5. This freeze-fracture preparation dramatically shows pores in the nuclear envelope. The outer half of the double membrane has been fractured away, exposing inner membrane and the pores. The pores penetrate both membranes. (×25,500.) (Photograph by Mary E. Todd.)

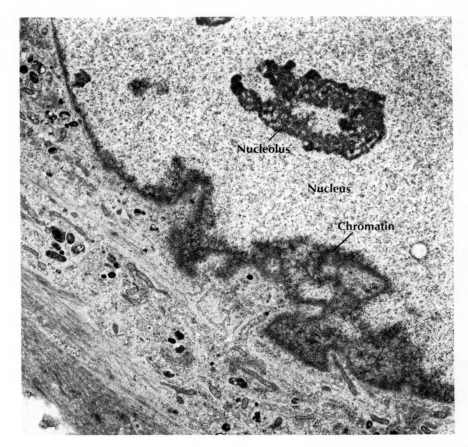

Fig. 2-6. TEM of smooth muscle cell showing nucleolus (×8,300.) (Photograph by Mary E. Todd.)

bines into preribosomal particles. These move out through the nuclear pores into the cytoplasm, where they become ribosomes.

Endoplasmic reticulum and ribosomes. The cytoplasm contains an extensive network of channels called the **endoplasmic reticulum** (ER). You cannot see these on your slide, but they are easily seen with the electron microscope. Refer again to Fig. 2-3. The endoplasmic reticulum in the photograph is labeled *RER*, meaning "rough" endoplasmic reticulum, because at high power the membranes of the ER appear studded with numerous particles, the **ribosomes.** The ribosomes carry out the synthesis of the cell's proteins.

Mitochondria. Mitochondria are also abundant in Fig. 2-3. These vesicle-like organelles, often called the powerhouses of the cell, are involved in many energy-producing and other metabolic functions within the cell. A mitochondrion is actually composed of two membranes. They are difficult to distinguish in Fig. 2-3, but in Fig. 2-7, at higher magnification, they show up clearly. The outer membrane is smooth, and the inner membrane is much infolded to form **cristae.**

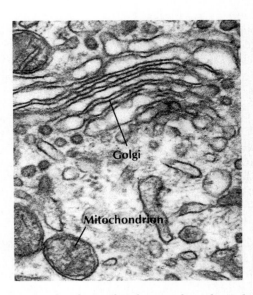

Fig. 2-7. Mitochondria and Golgi complex. The Golgi complex is a network of flattened, hollow discs in the cytoplasm, usually stacked in parallel as shown here. (×60,000.) (Photograph by Mary E. Todd.)

These look like partitions inside the mitochondria in Fig. 2-3, and they greatly increase the surface area of the inner membrane.

Golgi complex. The Golgi complex (also called Golgi apparatus) is a unique network of membrane-lined, flattened channels (called **cisternae**), usually stacked together in parallel rows (Fig. 2-7). The Golgi complex is important in the storage and packaging of secretory products, especially enzymes.

The cytoplasm contains many other organelles and inclusions, such as lysosomes, peroxisomes, and various fibrils and filaments. You have been introduced to some of the most obvious and important organelles of the typical animal cell in this brief introduction.

PROJECT AND DEMONSTRATION

1. Study of unstained cells from a freshly killed frog. The aim of this project is to demonstrate that all tissues are made up of cells that do not require elaborate techniques to be observed. However, slices of tissue must be cut *very thin* and spread out thinly, or the cells teased apart, to permit light transmission. Examine each of these preparations under the microscope and look for cells.

a. Drop of blood on a slide.

b. Thin slice of fresh liver teased on a slide in a drop of salt solution.

c. Thin slice of cartilage from the end of a bone at the joint.

d. Piece of freshly shed frog skin mounted in water.

e. Piece of thin mesentery mounted in water.

f. Strip of mucosa (inner lining) of the stomach mounted in water.

2. Demonstration of cyclosis. An *Elodea* leaf mounted in 10% sugar solution will show **cyclosis**, the streaming movement of the protoplasm.

EXERCISE 2B
Mitosis—the division of cells*

MATERIALS
Stained slides showing mitotic stages:
    Onion root tip, or
    Whitefish blastula, or
    Fertilized *Ascaris* eggs, or
    Temporary slides of root-tip mitosis, using squash technique (see Demonstration, p.30)
Compound microscopes

If student-made slides are used (see Demonstrations, p. 30) the following will be necessary:

Root tips from sprouting onions or broad beans
Microslides
Coverslips
1N HCl (see footnote p. 30)
0.5% aqueous solution of toluidine blue
Large corks
Blotting tissues
Nailpolish or melted paraffin

All cells come from preexisting cells. Most organisms start life from a fertilized egg. Beginning as a single cell, the complex metazoan is the result of repeated cell division. Cell division is involved in growth, regeneration of lost parts, development, wound healing, and many other processes in the body. Every day the living organism sheds about 1% to 2% of its tissue cells. To offset this loss, the tissues make as many cells as they lose. This is accomplished by binary fission of cells or mitosis.

By mitosis animal and plant cells multiply in number. Each mitotic division is an orderly sequence of events in which a mother cell gives rise to two daughter cells, **each having the same number of chromosomes as the mother cell.** The chromosomes carry the hereditary factors, or genes.

Meiosis, unlike mitosis, is the special maturation and division of germ cells that results in gametes with only half the normal number of chromosomes. When fertilization occurs and the egg and sperm unite, the normal (diploid) number of chromosomes is restored. Meiosis will be studied later (exercise 4A). We are concerned now only with mitosis.

The number of chromosomes is generally constant for the cells of any particular species but differs greatly among different species, ranging anywhere from 2 to 200—usually between 12 and 40 chromosomes. The onion has 16, *Ascaris* has 4, and humans have 46.

How long does it take a cell to divide? This de-

---

*See Hickman, C.P., Jr., L.S. Roberts, and F.M. Hickman. 1984. Integrated principles of zoology, ed. 7. St. Louis, The C.V. Mosby Co., pp. 88-92.

pends on the kind of organism, the type of tissue, the physiological condition of the organism, the temperature, and other factors; but in general the duration of the process is somewhere between ½ hour and 2 or 3 hours.

The various stages of mitosis and the end result of mitosis are quite similar in animal and plant cells. There are a few differences, largely correlated with the difference in the structure of animal and plant cells. These differences will be mentioned later.

To see a large number of dividing animal cells, one needs to study a tissue that is growing rapidly, such as the rapidly growing embryo of an animal. The fertilized eggs in the uterus of the roundworm *Ascaris* and the very young embryo of the whitefish are commonly used for such studies. The epidermis of young tadpoles is another good source. These tissues are killed, fixed, cut into very thin slices, mounted on slides, and stained. Each cell will naturally show only the phase at which it was killed and fixed. But by studying many cells killed at various stages of division

you can get a good picture of the whole process. Try to think of the stages as parts of a continuous process, as though they were stills taken from a motion picture.

☞ Before you start to study your slide, read over the following account of the various phases of mitosis in order to understand the process.

Interphase. A cell that is not dividing or preparing to divide is said to be in interphase (Fig. 2-8). This sometimes has been incorrectly called a "resting stage," but the cell, even when not actively dividing, is still carrying on all of the essential life processes. During the interphase period between divisions the nuclear envelope appears well defined, and the chromatin material appears as an irregular network rather heavily stained. One or two nucleoli may be present. Although you cannot observe it on your prepared slide material, we know that the chromatin material (the DNA and associated histone proteins) becomes duplicated, so that, when mitosis begins, the cell already has a double set of chromosomes. Even with the great resolving power of the electron microscope

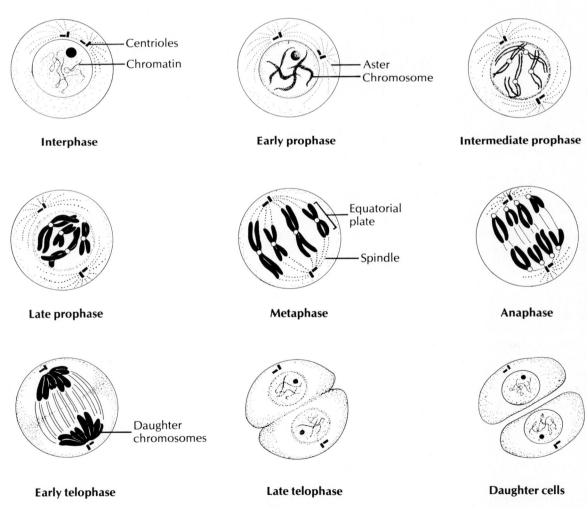

Fig. 2-8. Mitosis in animal cell.

27

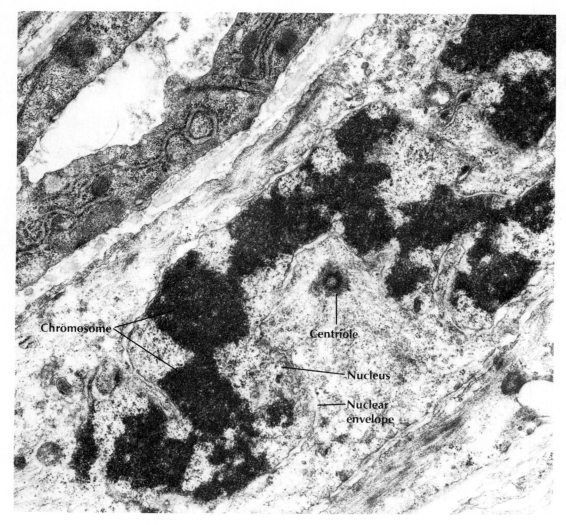

Labels on figure: Chromosome, Centriole, Nucleus, Nuclear envelope

Fig. 2-9. Electron micrograph of smooth muscle cell in mitosis, showing metaphase chromosomes and centriole. (×22,000.) (Photograph by Mary E. Todd.)

we can only see chromatin masses during interphase that represent segments of chromosomes (Figs. 2-3 and 2-4).

In animal cells there is located near the nucleus a small, dark, dotlike body called the **centriole,** which also duplicates itself (Fig. 2-8). Centrioles are difficult to see except in specially prepared slides, but they often show up on electron micrographs (Fig. 2-9).

Prophase. At the start of prophase the centrioles begin to separate, each forming around itself a system of fibrils called astral rays (Fig. 2-8). These radiate out to form a spindle between the centrioles and an aster around each centriole (Figs. 2-8 and 2-10). This is typical of animal cells. Plant cells such as those of the onion root tip, however, have no asters, but the spindle will form in the characteristic manner. The chromatin condenses into long, thin, coiled threads of genes (chromosomes). The nucleolus and nuclear envelope begin to disappear.

In the middle prophase the chromosomes begin to shorten and thicken.

In the late prophase the chromosomes are distinct, and the characteristic number of chromosomes can be seen (Fig. 2-8). (In animal cells with large chromosome counts, such as human cells, special preparation is required to spread out the chromosomes so that they can be counted.) Each chromosome is actually composed of two identical parts called chromatids, which lie close together and are joined at one point by a centromere. Recall that chromosome doubling occurred during interphase; but even in prophase this double nature is not apt to be visible on your slide material. By now the nuclear envelope and nucleolus have disappeared, and the spindle is faintly visible.

Metaphase. The chromosomes then arrange themselves near the center of the cell in an equatorial plate. The chromosomes are arranged in a circle with the apexes of the chromosomes pointed inward. Fibers of

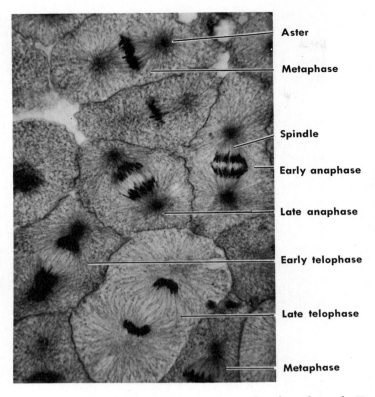

Aster

Metaphase

Spindle

Early anaphase

Late anaphase

Early telophase

Late telophase

Metaphase

Fig. 2-10. Mitosis in whitefish embryo. (Courtesy General Biological Supply House, Inc., Chicago.)

the spindle become attached to each chromosome so that the chromosomes appear V-shaped or J-shaped (Fig. 2-8).

During metaphase chromosomes appear as long, dense objects with the electron microscope (Fig. 2-9). However, electron micrographs never show chromosomes to advantage because the thin sections never pass through more than a few chromosomes, and only occasionally does a section coincide with the long chromosome axis. Chromosomes can be studied much more profitably with the light microscope.

Anaphase. Early in the anaphase stage the centromere divides. Each daughter chromatid has its own centromere and is now called a chromosome. The duplicate chromosomes begin to separate and move toward opposite poles of the spindle, their centromeres pointed toward the poles.

Telophase. The telophase stage begins as the two groups of chromosomes reach the opposite poles of the spindle. The chromosomes crowd together at the poles, and the reorganization of a nucleus begins within each group. In animal cells furrowing occurs; it is a constriction forming in the plasma membrane near the center of the cell and deepening until the cell is divided into two daughter cells (Figs. 2-8 and 2-10). In each daughter cell the chromosomes become again a network of chromatin material, the spindle disap-

pears, and the nuclear envelope and nucleolus reappear.

In plant cells a cell plate starts to form across the equatorial plate of the spindle. The plate extends in length; finally in late telophase it splits, and a new cell wall is formed between them.

Each daughter cell has a complete set of chromosomes identical to those of the parent cell. After the completion of mitosis the new cells go through a period of growth in the interphase stage before preparing to divide again.

Examining the Slides

☞ Read the brief description of the type of slide you are to study. Then examine the slide, first with the naked eye to locate the tissue slices and then with the lower power of the microscope. Scan the tissue carefully until you are familiar with the appearance of the cells and can recognize the difference in the appearance of the nuclei of various cells. When you find a phase you can recognize and identify, center it, and then change to high power for study. Sketch the stage you have found in the proper place on p. 32, and then find another stage.

It is not necessary to try to locate these stages in any particular order. Sketch them as you find them.

When you have finished, you will have all the phases in the proper order on the page.

Not all the cells will demonstrate typical stages of mitosis; in the cutting of the tissue many of the cells will have been cut in such a way that all or part of the nuclear material will be missing.

Mitosis in the Whitefish Blastula

The blastula is a very early stage in embryonic development. Each slide contains several very thin sections cut through one of these blastulas. Hold the slide over a piece of white paper and note the small spots, each of which represents one section cut through a blastula. Study a section with low power first. You will probably find all stages of mitosis represented among the cells there (Fig. 2-10). Note the shape of the cells and the appearance of the asters and spindles. Look for a typical stage; then switch to high power to study and sketch it.

Mitosis in the Fertilized *Ascaris* Egg

The cells of the intestinal roundworm of horses, *Ascaris megalocephala bivalens,* have only 4 chromosomes, and they are very large and easy to see.

In the female ascaris worm the eggs are fertilized in the paired uteri after copulation. Soon after fertilization each fertilized egg (zygote) acquires a shell, secreted by the walls of the uterus. The zygote within its shell now begins to divide—the first of many cell divisions that will finally produce a new young worm. To prepare these slides, longitudinal sections were made through the uterus and its eggs soon after mating had occurred, and the sections were mounted on slides and stained. Each slide section will contain many small spherical eggs in various stages of mitosis.

First, examine the slide with low power. Find the walls of the uterus, which are made up of darkly stained cells. The cavity of the uterus is filled with sections of eggs. Study an egg under high power and identify its structures. Note the thick pale **shell** on the outside. The wide clear **perivitelline space** between the shell and the cytoplasm of the egg is in life filled with fluid. The **cytoplasm** is stained, appears granular, and contains a number of small vacuoles. The cytoplasm is enclosed in a very thin **plasma membrane.** The **nucleus** of the fertilized egg is made up of two **pronuclei** one of which was contributed by the sperm at the time of fertilization. The chromatin of the nucleus is darkly stained.

Scan the slide carefully with low power until you are familiar with the appearance of the eggs; then select a stage that appears typical and switch to high power for study.

DRAWINGS

On p. 32 draw as many of the stages of mitosis as you can find, labeling the drawings adequately.

DEMONSTRATION

Demonstration of mitosis by the "squash technique" of preparation. In a fairly simple and brief operation the student can prepare excellent temporary slides of dividing cells that show mitotic figures containing entire chromosomes. These can be made from the root tips of either the onion or the broad bean *(Vicia fava).*

Position onions over glasses or beakers of water of such size that only the base of the onion is in the water. Smaller onions may be held in place by toothpicks stuck into the sides of the onion and rested on the edges of the glass. In a few days there should be root tips 2 to 3 cm long (unless the onions have been previously treated to prevent sprouting). Seeds of the broad bean will sprout if placed in moist peat moss or vermiculite.

Cut off 1 to 2 mm of the root tip and place on a clean slide. Cover with a few drops of 1N HCl.* Warm gently over a flame for 1 minute (but *do not boil*). Then blot off the excess HCl with a bit of clean tissue or paper toweling. Stain by adding a few drops of 0.5% aqueous solution of toluidine blue and warm gently for 1 minute (but *do not boil*); then blot off the excess stain.

Add a fresh drop of the stain and a coverslip. Cover the coverslip with a cork or a piece of folded paper and press down firmly with the thumb to squash the tissue (avoid any lateral movement of the coverslip).

Sealing the edges of the coverslip with melted paraffin or fingernail polish will prevent drying out during the laboratory period.

Acetocarmine stain† may be substituted for the toluidine blue.

---

*Dilute 83 ml of standard concentrated HCl (usually about 12N) by adding distilled water to make 1 liter.
†Add carmine to boiling acetic acid of 45% strength until no more carmine will dissolve. Filter the solution and cool.

SOME TYPICAL CELLS

Name _____

Date _____ Section _____

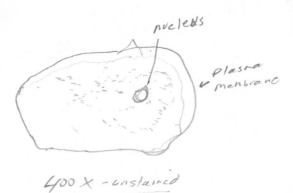

nucleus

plasma membrane

400 X - unstained

plasma membrane

Nucleus

400 X - stained

Squamous cells from cheek

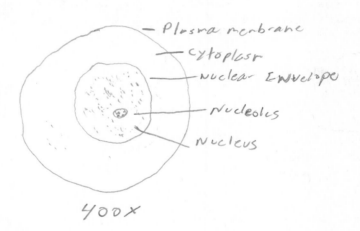

— Plasma membrane

— cytoplasm

— nuclear Envelope

— Nucleolus

— Nucleus

400X

Sea star egg

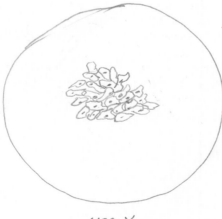

400 X

Liver cells

MITOSIS, as seen in

Interphase   **Early prophase**   **Late prophase**

Metaphase   **Early anaphase**   **Late anaphase**

Early telophase   **Late telophase**   **Daughter cells**

CHAPTER 3

CELL FUNCTION

The life substance of cells and tissues is not a single chemical substance but is made up of a mixture of many compounds in different states of matter. It is a mixture of visible granules and globules together with submicroscopic particles, the whole forming a colloidal system. Water is the most common of its constituents, but the organic constituents of proteins, carbohydrates, and fats also make up a considerable percentage of its weight. Inorganic salts are also a small but vital part of its makeup. The four chemical elements—oxygen, carbon, hydrogen, and nitrogen—make up about 97% to 98% of living matter.

Living matter exhibits several properties that distinguish it from the nonliving, such as irritability (the capacity to respond to stimuli), metabolic production of energy, movement, biosynthesis of cellular constituents and their maintenance and repair, exchange with the environment, and, especially, reproduction.* Several of the characteristics of living cells and tissues can be illustrated in the laboratory by simple experiments that you can perform and some by demonstrations set up by the instructor. They may help give you a better insight into the properties of living matter.

*See Hickman, C.P., Jr., Roberts, L.S., and Hickman, F.M., 1984. Integrated principles of zoology, ed. 7. St. Louis, The C.V. Mosby Co., Chapters 2 and 5.

EXERCISE 3A
Diffusion phenomena

MATERIALS

Note: Instructions for preparing many of the solutions are given in Appendix B.

Part 1: Brownian movement
- Tissue grinder
- Slides
- Coverslips
- Live tissue (leaf, insect, worm, and so on)
- Powdered carmine

Part 2: Diffusion and osmosis

Diffusion
1. Volatile oil (wintergreen, peppermint, other)
2. Small beakers
   Sugar cubes
3. Glass tubing
   Agar-agar solution
   Potassium hydroxide solution
   Neutral red
4. Screw-cap or stoppered tubes
   2% agar-agar solution
   Wax pencil
   0.02M methylene blue solution
   0.02M eosin solution

Osmosis
1. Osmometer (see Appendix B, p. 455, for assembly)
2. Skin of frog leg
   Glucose solution (such as white Karo syrup)
   Glass capillary tubes
   Beakers
   Support stands
3. Stomach and intestine
4. Potatoes
   Sugar
   Shallow dish
5. Clean slides and coverslips
   Distilled water
   0.9% NaCl
   3%-5% NaCl
   70% alcohol
   Sterile lancets

Part 3: Sol-gel transformation of colloids
- Gelatin
- Test tubes
- Alcohol lamps or water bath

Liquids in Mixtures

For background to the experiments and demonstrations you will perform in this exercise, let us review the different states in which living matter may exist in cells.

Crystalloid, or True, Solutions. A true solution is a homogeneous, one-phase mixture of two substances. The substance in excess is called the solvent (always water in biological systems), and the dissolved substance is called the solute. The molecular dimension of the solute is so small—less than 1 nm in diameter—that it will pass through an ultrafilter, such as collodion or cellophane. Solutions are completely stable; that is, solutes will not settle out of the solvent. Examples are solutions of sugars or salts.

Colloidal Systems. Colloidal systems are made up of molecules or aggregates of molecules having a molecular size of 1 to 100 nm in diameter. It is a condition in which one substance (such as a protein) is dispersed in another (water) to form many small phases suspended in one continuous phase. Colloids are stable systems and may look perfectly clear, as do true solutions. However, the particles are two large to pass through filters of collodion or cellophane. Examples of colloidal systems are proteins, polysaccharides, fine carbon particles, or soap in water; India ink; gelatin; and smoke, fine dust, or mist. The internal environment of living cells is colloidal in nature. A characteristic of many protein colloidal systems that we will study later is their capacity to exist in either of two states—sol or gel. The sol is fluid, and the gel is more rigid. Gelation of proteins is thought to occur when the proteins form numerous stabilizing linkages between molecules.

Suspensions. Suspensions comprise large particles of a micrometer or more in size, which are visible—at least with a microscope. Suspensions will not diffuse, and the particles in a suspension will settle out in time. Blood is a suspension of blood cells in plasma.

An emulsion is a suspension of fine particles or globules of one liquid in another liquid, such as fat in globules of milk.

WRITTEN REPORT

Record the results and observations of the following experiments on pp. 39-41.

PART 1: BROWNIAN MOVEMENT

Molecules of matter are in constant motion, caused by the kinetic energy of molecules. This molecular traffic is responsible for diffusion of solutes through a solvent and is an important factor in the passage of materials into and out of cells. Although we cannot see individual molecules, we can observe the results of their motion, as millions of water molecules collide with particles suspended in the water. This phenomenon is called brownian movement.

☞ 1. Thoroughly crush a bit of living tissue, such as a leaf fragment or a piece of a worm or insect. This may be done with a mortar and pestle or a tissue grinder or even by mashing the living fragment on a microscope slide with a solid glass rod that has been flame polished on one end. Add a very small amount of water to the tissue if necessary to assist grinding. After removing larger pieces of tissue, transfer some of the creamy residue to a drop of water on a clean slide. A toothpick is handy for this purpose. Cover with a cover glass and examine with high power, reducing the light as necessary. When all water currents have ceased, look for the smallest freely suspended particles you can see and watch their movement.

Even the smallest visible particles are far larger than molecules. Their movement is the result of recoil when they are struck by rapidly moving water molecules.

Do you think the motion you see is limited to living material? Let us look at some nonliving particles.

☞ 2. On a clean slide place a drop of water containing a suspension of powdered carmine or India ink and add a coverslip. Again try to follow a single particle.

What do you conclude about the presence of brownian movement in living and nonliving material? The movement will continue indefinitely or for as long as you can keep the water from evaporating.

PART 2: DIFFUSION AND OSMOSIS

In this exercise we will examine the mechanisms involved in the exchange of materials through the plasma membrane that separates a cell from its environment.

Diffusion

When two substances are placed in contact, the molecules of one substance will move randomly and spontaneously among the molecules of the other substance, resulting in the uniform dispersion or intermingling of the molecules of the two substances. This is free diffusion. It is the natural result of the continuous random motion of all molecules, witnessed in the preceding exercise. It occurs rapidly in gases, more slowly in liquids, and extremely slowly in solids.

1. The instructor will place a few drops of volatile oil, such as oil of wintergreen or oil of peppermint, on a tissue or blotter on the service desk. As the odor reaches each student, he or she will indicate by raising a hand. On a diagram of the room drawn on the blackboard someone can check the direction of the spread of the odor.

☞ Time how long the odor takes to reach the nearest student and how long to reach the farthest.

Note that the odorous molecules are moving from an area of greater concentration (the blotter on the service desk) to areas of lesser concentration (room air). This is a characteristic of any diffusing substance; net movement is "downhill" along a concentration gradient. If left undisturbed the diffusion would continue until the odorous molecules are equally distributed throughout the room.

2. Fill three small beakers with water—A and B with cool water and C with warm water.

☞ When the water is quiet, drop a cube of sugar into each one. Gently stir the water in A, but leave B and C motionless. Record the time required for each cube to dissolve completely. Explain the results.

3. A glass tube has been three-fourths filled with agar-agar colored with neutral red, and the remainder of the tube has been filled with a solution of potassium hydroxide (KOH). The ends are covered to prevent drying out. The rate of diffusion of the KOH may be determined by the change in color of the agar. Neutral red appears red in acid medium and yellow in alkaline medium.

☞ Measure daily for several days. Explain the results.

4. The rates of diffusion of a colloidal and noncolloidal solution in agar-agar may be compared in this demonstration.

☞ Fill two screw-cap or stoppered tubes about two-thirds full with 2% agar-agar. After it has hardened, mark the level with a wax pencil. Place 5 ml of 0.02M methylene blue solution (colloidal) in one tube and 5 ml of 0.02M eosin solution (noncolloidal) in the other. Screw on the caps to prevent evaporation. At 24-hour intervals for several days, record the distance diffused in each tube.

Osmosis

Osmotic pressure is one of the forces that determines the movement of water into and out of cells. Osmosis is the movement of solvent molecules (water) through a differentially permeable membrane. A differentially permeable membrane permits solvent molecules to pass through it freely but restricts the movement of solute molecules. For example, if water and a sugar solution are separated by such a membrane, the water molecules pass by diffusion through the membrane but the sugar molecules cannot pass through. The result is a net movement of water into the sugar solution.

Let us examine more closely what is happening. Suppose that a 10% sugar solution is enclosed in a bag composed of a differentially permeable membrane, and that the bag is placed in a container of distilled water. Knowing the molecular weights of water (18) and the sugar solution (342), we can calculate that there are approximately 169 water molecules in the sugar solution for every molecule of sugar. In an equal volume of distilled water outside there are approximately 170 molecules of water. Because of molecular motion, the membrane will be bombarded with molecules from both directions. Out of 170 hits on the inside, 169 are from water molecules and one is from a sugar molecule; on the outside however, all 170 hits are by water molecules. Since the membrane is permeable to water but not to sugar, slightly more water molecules enter the sugar solution than leave it. The result is a net flow of water inward; the sugar solution will increase in volume as the water volume outside decreases. Actually the passage of water inward is favored even more than our example suggests, because some of the water in the sugar solution becomes complexed with sugar molecules, leaving less free water to diffuse outward. Furthermore, free water molecules cannot pass out as readily as they can pass in because of interference by sugar molecules inside.

If we then fit a vertical tube to the bag of sugar solution, fluid will rise in the tube as water enters the solution across the membrane. This will continue until the hydrostatic pressure developed by the rise of fluid in the tube equals the osmotic pressure of the sugar solution. We can consider osmosis to be nothing more than the tendency of water molecules to move from a region of higher water concentration (distilled water in our example) to a region of lower water concentration (the sugar solution). Osmotic pressure is therefore a measure of the "escaping tendency" of water. Note that a differentially permeable membrane is essential for osmosis to occur. If the membrane is permeable *only* to water, it is said to be **semipermeable.** If the membrane allows one material (for example, sugar or salt) to pass through more readily then another, it is said to be **selectively permeable.** Both kinds of membranes are present in living systems.

As important as diffusion and osmosis are in life processes, they do not account for all movement of materials through living membranes. Many materials move *against* concentration gradients (in osmosis, water diffuses *down* a concentration gradient). Such materials are moved either by mediated transport involving membrane protein carriers, or by endocytosis (ingestion of fluids or solids by cells). In this exercise, however, we will demonstrate osmosis, a phenomenon of great importance in determining water flow between different compartments in living systems.

Following are a few of the simple means that can be used to demonstrate osmosis. Your instructor will indicate which one(s) will be used.

1. The instructor will have prepared ready for use an osmometer similar to that shown in Fig. 3-1. It consists of a tubular piece of unvarnished cellophane tied to a rubber stopper fitted to a long piece of glass tube. The cellophane tubing is filled with a concentrated sucrose solution. Gently lower the osmometer into a jar of water until the top of the rubber stopper is level with the water surface. Check the level of the sugar solution. If it is below the level of the top of the rubber stopper, wait until it rises to that level; this is your zero point. Mark the zero point with a wax pencil. At intervals of 10 to 15 minutes record the height of solution in the glass tube in millimeters and the time. Continue until the end of the period. Consult your

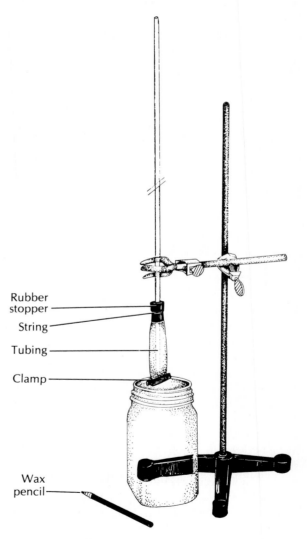

Rubber
stopper
String
Tubing
Clamp

Wax
pencil

Fig. 3-1. An osmometer.

instructor if the fluid column fails to rise or if it falls (osmometers occasionally develop leaks). Meanwhile continue with other exercises that may be assigned.

2. A frog skin "boot" or "glove" may be prepared by cutting the skin around the arm or leg of a frog and peeling the skin down and off the limb. Fill it with strong glucose solution, tie the open end very securely around a glass capillary tube, and suspend the skin in a beaker of water.

3. A setup similar to the one just mentioned can be made, using the stomach and intestine removed from a small vertebrate and thoroughly cleaned.

4. Prepare two potatoes by removing a slice from top and bottom of each potato. Remove the peeling 1.5 cm around the bottom of each potato. Scoop a small cup or hole about 2 to 3 cm deep in the top of each. Place in a shallow bowl of water. Add a pinch of sugar to the hole in one potato and use the other as a control. Check in 1 hour or so and explain what has occurred.

5. Osmosis in animal cells can be demonstrated by the use of red blood corpuscles. Clean and dry three slides and three coverslips.

On slide A place a drop of physiological salt solution (0.9% NaCl). Add a drop of blood (your own will do) and a coverslip. Examine under the high power of a microscope.

The solution is isotonic to the blood. What is the appearance of the blood cells?

On slide B mix a drop of distilled water (hypotonic) and a drop of blood. **Examine at once.** What occurs to the cells? What is this condition called?

On slide C mix a drop of blood with a drop of 5% saline solution (hypertonic). Examine and sketch. Wait a few minutes; then examine and sketch again. Has the appearance altered? Why? What term is used to describe this condition?

PART 3: SOL-GEL TRANSFORMATION OF COLLOIDS

Some proteins in living systems are able to undergo reversible transformation between a liquid sol condition and a semirigid gel state. You will observe in a future laboratory exercise how this transformation occurs in an ameba as the animal sends out a pseudopod. Flowing sol in the center of the pseudopod is converted rapidly into an external gel hyaline layer as the pseudopod extends. Sol-gel transformations also occur during cell division in higher organisms.

☞ Place a pinch of gelatin in a small test tube; then fill one-third full of water and shake. Heat gently over a flame or in a boiling water bath.

This is now called the **sol** state. Cool by placing the tube in cold running water or in a refrigerator. What change occurs? This is the **gel** state and the process is called **gelation.** Reheat and **solation** occurs again. Explain the results.

DIFFUSION PHENOMENA

Name _____

Date _____ Section _____

Brownian movement

What evidence of molecular movement did you see? _____

_____

_____

Was the movement more pronounced in living than in nonliving material? _____

What causes brownian movement? _____

_____

Diffusion

(1) How long did it take the diffusing odor to reach the nearest student? _____

The farthest? _____

(2) Compare the rate of diffusion of sugar: When stirred _____ In cool water _____

In warm water _____ Explain. _____

_____

_____

(3) What evidence of diffusion did you find in the tube? _____

_____

Rate of movement: 24 hr _____ 48 hr _____ 72 hr _____ 96 hr _____

(4) What was the rate of diffusion of colloidal and noncolloidal solution in agar-agar?

	24 hr	48 hr	72 hr	96 hr
Colloidal				
Noncolloidal				

What do these experiments show about the relative rates of diffusion in a gas, liquid, or solid medium? _____

_____

What are some of the factors that affect diffusion? _____

_____

Osmosis: use of a simple osmometer (p. 36)

Record the time and height of the column in millimeters for each reading of the osmometer.

Time	Height		Time	Height		Time	Height

Plot on graph paper (p. 41) the height of the column on the ordinate (Y-axis) and the time in minutes on the abscissa (X-axis). Plan your plot to avoid crowding. Label and title the graph and attach to your laboratory report.

Describe the shape of the curve on your graph and explain its significance. _____

_____

_____

Explain the forces involved in the rise of the solution in the tube. _____

_____

_____

What is the difference between simple diffusion and osmosis? _____

_____

_____

_____

Osmosis in red blood cells

Sketch the blood cells as you saw them in these solutions:

    (1) Isotonic   **(2) Hypotonic**   **(3) Hypertonic**

Which one represents hemolysis? _____ Define hemolysis. _____

_____

Which one demonstrates crenation? _____ Define crenation. _____

_____

Phase reversibility of colloids (p. 37)

In the sol state (warm gelatin) what is the external phase? _____

_____

The internal phase? _____

_____

In the gel state (cold or solidified gelatin) what is the external phase? _____

_____

The internal phase? _____

_____

EXERCISE 3B
Action of enzymes

MATERIALS
Two white spot plates per student pair
Test tubes and test tube rack
10 ml graduated cylinders
Ice
Warm water bath (such as a beaker of warm water)
Boiling water bath
Four solutions, prepared as directed in Appendix B, p. 455.

Enzymes are a special class of proteins that catalyze nearly all chemical reaction in cells. If enzymes were absent cellular reactions would proceed at a negligible rate. Enzymes speed reactions by reducing the amount of activation energy that substrates require to have strong chemical bonds disrupted. The enzyme combines with the substrate to form a precisely aligned enzyme-substrate complex. The substrate is split after passing through one or more intermediate steps, which require much less energy than would a single-step reaction. The shape of the enzyme molecule is crucial to the reaction, since an exact molecular fit between substrate and enzyme is required. Thus with few exceptions an enzyme will catalyze only one reaction and no other.

Although enzymes permit chemical reactions to proceed at the relatively low temperatures of animal bodies, they are sensitive to temperature, working faster at higher temperatures and more slowly at lower temperatures. Enzymes are also sensitive to pH (hydrogen ion concentration). Most cellular enzymes operate best at the near-neutral pH of the cellular environment, but many of the digestive enzymes work efficiently in slightly alkaline (salivary and intestinal enzymes) or strongly acid (stomach enzymes) media.

In this experiment the action of the enzyme α-amylase, which is present in the saliva and pancreatic juice of many vertebrates, will be studied. Amylase promotes the hydrolysis of starch to maltose, a disaccharide composed of two glucose units. Maltose, like glucose, is a reducing sugar and will give a positive test with Benedict's solution.

The class will be divided into two groups, with one half studying the effects of temperature on enzyme activity (Part 1) and the other half studying the effect of pH on enzyme activity (Part 2). At the end of the period experimental data will be exchanged between the two groups and laboratory reports prepared individually.

Preparation of Dilute Amylase for Parts 1 and 2. Each student pair, whether assigned to the temperature experiment (Part 1) or to the pH experiment (Part 2) will first prepare a diluted amylase solution.

Place 1.0 ml of 0.25% amylase in a test tube, and add 12 ml of water. This is your working diluted amylase solution used in all of the experiments that follow.

Part 1: Effect of Temperature on Enzyme Activity
In this four-part experiment, you will examine the effect of temperature on the *rate* of enzymatic activity. Since amylase breaks down starch (the substrate) to maltose (the product), a convenient assay of enzymatic activity is to measure the time required for all of a given quantity of starch to disappear from solution. This is the end point.

Preparations. Prepare for the test by labeling four test tubes *A1* through *A4*. In tubes A1, A2, and A3 place 2 ml of the diluted starch solution. Now add 2 ml of pH 7.0 buffer in each tube. Buffers are mixtures of substances that interact so that the pH of the mixture remains constant even though small quantities of acidic or basic substances are added. Tube A1 is to be left at room temperature. Place tube A2 in a beaker of hot tap water. Let it stand, keeping the water in the beaker quite warm to the touch by replacing it with hot tap water from time to time. The temperature of the water bath should be quite warm at all times but not scalding hot. Place tube A3 in a beaker of ice water, and let it stand. Some unmelted ice must always be present in the beaker. To tube A4, which is empty, add 2 ml of the diluted amylase solution, and put the tube in a boiling-water bath, and leave it until it is ready for testing. Do not allow the tube to boil dry. Add water as required to maintain the volume.

Testing for Amylase Activity. Prepare for starch tests by placing one drop of iodine-potassium iodide (I-KI) solution in each depression of your porcelain test plate. To tube A1 (room temperature) add 2 ml of unheated diluted amylase. Agitate to mix thoroughly, and immediately test for the presence of starch by adding a drop to the I-KI in one depression. The development of a blue color indicates starch is present. Continue to test for starch at short intervals, initially about 30 seconds, keeping track of the elapsed time. Use a different depression for each test. As the reaction proceeds, the blue color will be less apparent. The end point is reached when no blue color can be detected. Record the time required to reach the end point.

As a further test of enzymatic activity, you will then test for the appearance of maltose, a product of the reaction. To the mixture remaining in the test tube, add about one half as much Benedict's solution as the volume of remaining mixture. Test for the presence of reducing sugar (maltose) by placing the tube in a beaker of boiling water, and observe until no further change occurs—at least 5 minutes. If reducing sugar is present, the deep blue color of copper hydroxide will change to an orange-red as the cupric hydroxide $Cu(OH)_2$ is reduced to red cuprous oxide, Cu_2O.

Note that you have tested for both the disappearance of substrate (starch test) and the appearance of product (Benedict's test). Why cannot the Benedict's test alone be used as an assay of enzymatic rate?

Wash your porcelain test plate and proceed with tube A2 exactly as described for tube A1. Keep the tube in the warm water bath while the starch tests are being performed. It will be necessary to test for starch at short intervals, since the reaction should occur quite rapidly. Record the time required to reach the end point and the results of the test for reducing sugar.

Wash the test plate and proceed with tube A3 exactly as with tubes A1 and A2. Keep the tube in the ice bath while the tests are being made. Much longer intervals may be used in testing the contents of this tube. Record the time required to reach the end point and the results of the test for reducing sugar.

You are ready for the final test with test tube A4. This test differs from the preceding three because you will use the amylase that has been boiled. Remove the tube from the boiling bath, and if necessary restore the original volume with tap water. Add an equal volume of diluted starch solution and 2 ml of pH 6.8 buffer. The resulting mixture is similar to that of the previous tests, except that boiled amylase is present. The tube should be kept in the test tube rack while the tests are performed. Proceed as in the previous tests. Perform the starch tests at intervals until near the end of the period, at which time the solution remaining in the test tube should be tested for the presence of reducing sugar. Record the results. Directions for the written report follow Part 2.

Part 2: Effect of pH on Enzyme Activity

In this three-part experiment, you will examine the effect of pH on the rate of enzymatic activity of amylase. Since amylase hydrolyzes starch (substrate) to maltose (product), a convenient assay of enzyme activity is to measure the time required for all of a given quantity of starch to disappear from the solution. This is the end point.

Preparations. Place 2 ml of diluted starch solution (shake the stock bottle) in each of 3 test tubes. To one tube add 2 ml of pH 7.0 buffer, and mark this tube B1. To a second tube add 2 ml of pH 8.0 buffer, and mark this tube B2. To a third tube add 2 ml of pH

3.4 buffer, and mark this tube B3. Agitate all tubes to mix thoroughly. Keep the tubes in the test tube rack at room temperature during the tests.

Testing for Amylase Activity. Prepare to test for starch by placing one drop of I-KI solution into each depression of the porcelain test plate. To tube B1 add 2 ml of diluted 0.25% amylase and agitate to mix thoroughly. Immediately place 1 drop into a depression with I-KI solution. The development of a blue color indicates starch is present. Continue to test for starch at short intervals, initially about 30 seconds, keeping track of the elapsed time. Use a different depression for each test. As the reaction proceeds the color will be less apparent. The end point has been reached when no blue color can be detected. Record the time required to reach the end point.

To be assured that the enzymatic reaction you have performed has indeed produced maltose as a product, you will test for its presence with the Benedict's test for reducing sugar. To the mixture remaining in the test tube, add one half as much Benedict's solution as volume of remaining mixture. Test for the presence of reducing sugar (maltose) by placing the tube in a beaker of boiling water, and observe until no further change occurs—at least 5 minutes. If a reducing sugar is present, the deep blue color of copper hydroxide will change to an orange-red as cupric hydroxide $Cu(OH)_2$ is reduced to red cuprous oxide, Cu_2O.

Note that you have tested for both the disappearance of substrate (starch test) and for the appearance of product (Benedict's test). Why cannot the Benedict's test alone be used as an assay of enzymatic rate?

To tube B2 add 2 ml of diluted amylase and test for the disappearance of starch as just described. Record the end point time. Test the remaining mixture in tube B2 for the presence of reducing sugar with the Benedict's test.

To tube B3 add 2 ml of diluted amylase and test for the disappearance of starch as before. Record the end point time. Test the remaining mixture in tube B3 for the presence of reducing sugar with the Benedict's test.

WRITTEN REPORT

Part 1. Write a report on separate paper, giving a brief description of your hypothesis, technique, results, and conclusions concerning the effect of temperature on an enzyme-catalyzed reaction.

Part 2. Write a brief description of your hypothesis, technique, results, and conclusions concerning the effect of hydrogen ion concentration on the "active" site of an enzyme.

CHAPTER 4

GAMETOGENESIS
AND EMBRYOLOGY

EXERCISE 4A
Meiosis—maturation division of germ cells

MATERIALS
Prepared slides
   Grasshopper testis, showing stages of spermatogenesis
   Ascaris uterus, showing sperm entrance and maturation
     stages
Microscopes

In ordinary cell division—mitosis (exercise 2B)—each daughter cell receives identical copies of the genetic information. **Meiosis,** however, is a distinctive type of nuclear division that differs from mitosis in two very important ways. First of all, meiosis is a gamete-producing process in which primary germ cells (oogonia or spermatogonia) become mature gametes (ova or spermatozoa) for sexual reproduction. Second, in meiosis the chromosomes split once, but the cell splits twice, producing four cells—each with *half* the original number of chromosomes. The result is that mature eggs and sperm have only one member of each homologous chromosome pair, that is, a **haploid (n)** number of chromosomes. (In mitosis, it will be recalled, each cell ends up with a full complement of chromosomes, the **diploid [2n]** number.) The gametes then fuse during fertilization to form a zygote containing the full number of chromosomes.

The process of forming mature gametes from primary germ cells is called **gametogenesis,** and it involves both mitosis and meiosis. Early in the embryonic development of a sexually reproducing animal certain special cells called the **primordial germ cells** are set aside. These predestined cells migrate to the developing gonads, which later become the ovaries or the testes. The primordial germ cells are larger than ordinary somatic cells and are the future stock of gametes for the animal. Once in the gonad they begin to multiply by ordinary mitosis.

Thus both the body cells and the potential gametes contain an identical and complete, or diploid, set of chromosomes. The **maturation process** refers to the final divisions necessary to produce functional ova and spermatozoa. Gametogenesis, then, includes many divisions by mitosis during the early multiplication stages, followed finally by a reduction in the number of chromosomes by the process of meiosis.

When an animal approaches sexual maturity, the gonads begin to produce mature eggs or sperm by meiosis. Meiosis involves two divisions. The first division separates the homologous chromosomes (homologs), so that each of two resulting cells has one member of each pair. In the second division each haploid cell divides by a process similar to mitosis. The result is four cells, or gametes, each with a haploid number of chromosomes.

Gametogenesis in the testis is called **spermatogenesis,** and in the ovary it is called **oogenesis.** The same processes are involved in both, but there is an important difference in the end result. An oocyte undergoes meiosis to produce one large, functional egg (ovum) and three abortive cells (polar bodies), whereas a spermatocyte produces four functional spermatozoa.

SPERMATOGENESIS
Consider first the development of the spermatozoa (spermatogenesis). The early germ cells, which have migrated to the testis, are called primordial germ cells. They become **spermatogonia** as they multiply rapidly by mitosis. These early germ cells, like ordinary somatic cells, have the diploid number of chromosomes and divide by ordinary mitosis. (For a review of mitosis see pp. 26-29.)

As the organism reaches sexual maturity, some of the spermatogonia cease to divide; they enlarge, become **primary spermatocytes,** and begin the meiotic divisions. The first stage in the first meiotic division, prophase, is more complex than mitotic prophase because the chromosomes become rearranged by genetic recombination.

During early prophase the chromosomes become visible in the nucleus as faint threads (Fig. 4-1). Each

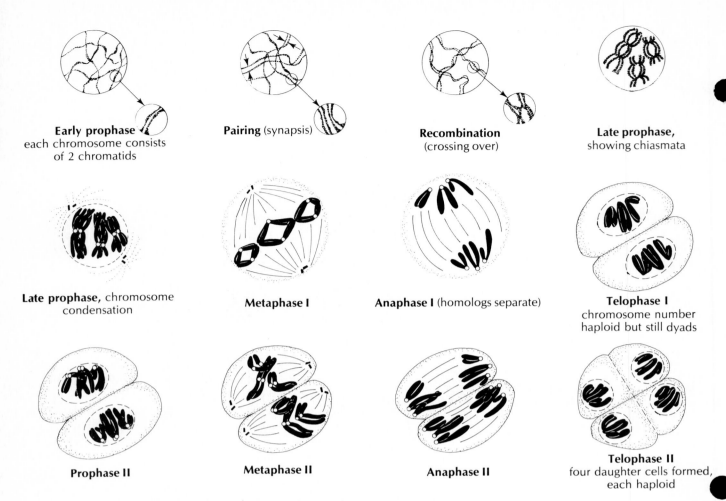

Early prophase
each chromosome consists of 2 chromatids

Pairing (synapsis)

Recombination
(crossing over)

Late prophase,
showing chiasmata

Late prophase, chromosome condensation

Metaphase I

Anaphase I (homologs separate)

Telophase I
chromosome number haploid but still dyads

Prophase II

Metaphase II

Anaphase II

Telophase II
four daughter cells formed, each haploid

Fig. 4-1. Stages of meiosis, showing division of cell with three pairs of chromosomes.

chromosome appears as a single thread when viewed with the light microscope, but in fact each has already doubled during premeiotic interphase and consists of two chromatids so closely aligned that they cannot be distinguished. Next, the two members of each pair of chromosomes, called **homologs,** become closely associated, lying side by side in **synapsis.** One member of each pair is derived from the male parent and the other from the female parent. The same genes of the homologs are now lined up in perfect order, although alternate forms (**alleles**) of the genes may be present in either homolog. The two homologs in synapsis are often referred to as a **bivalent.** However, *each* homolog is really composed of two chromatids and is now called a **dyad.** And because two homologous chromosomes (two dyads) are synapsed in each pair, the structure is called a **tetrad.**

At this stage **crossing over** occurs. In this process hereditary material is exchanged among the four chromatids in the tetrad. Later in prophase the crossing places become visible as **chiasmata,** the connecting

points between homologous chromosomes (Fig. 4-1). The chromosomes continue to condense, becoming very compact. This marks the end of prophase of meiosis I. At this time a spindle forms, and centrioles complete their replication, as in mitosis. Metaphase begins. The homologous pairs separate and move to opposite spindle poles (anaphase). Each chromosome is still double since it contains two chromatids (a dyad). Two daughter cells (**secondary spermatocytes**) have been produced as a result of meiosis I, each containing a haploid number of chromosomes. Each secondary spermatocyte thus has half as many chromosomes as any somatic cell.

The second meiotic division immediately follows. During meiosis II the dyads of each chromosome separate into individual chromatids (Fig. 4-1). One chromatid goes to one daughter cell and one to the other. Nuclear envelopes form around the chromatids, which now become full-fledged chromosomes. The daughter cells, which result from the division of the secondary spermatocyte, are called **spermatids.** Each spermatid

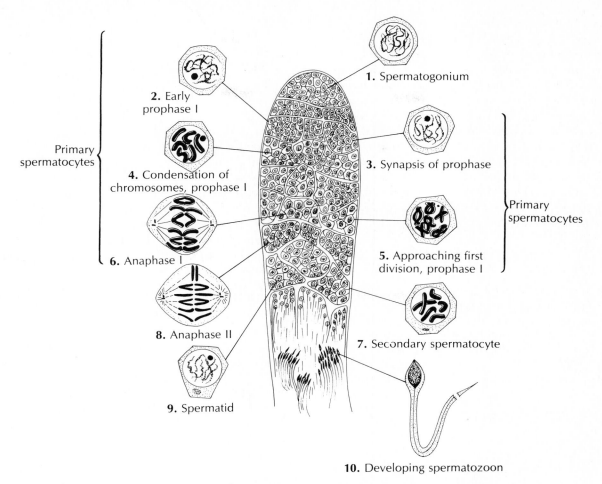

Fig. 4-2. Spermatogenesis as seen in one lobe of a grasshopper testis.

has the haploid number of chromosomes or just half as many as the spermatogonia or the body cells. Without further division each of the four spermatids transforms into a mature spermatozoan.

Spermatogenesis in the Grasshopper Testis

The maturation stages of sperm can be seen by studying cross sections of the testes of many kinds of animals. Fig. 4-2 shows a section through a lobe of the testis of the lubber grasshopper, *Romalea*. The testis consists of a number of lobes, whose pointed ends empty into the vas deferens. Each lobe contains a number of compartments (cysts) separated by tissue septa, and each cyst contains a number of cells all in the same stage of development. These cysts were formed at the blunt or apical end of the lobe. Here a group of primordial germ cells are continually and rapidly dividing by mitosis to form spermatogonia. As a new group of cells is formed from the primordial cells, it is pinched off to form a cyst. The spermatogonia in the new cyst begin to grow in volume, without dividing, and become known as primary spermatocytes.

☞ Try to find on your slide, with lower power, a longitudinal section through one of the lobes. Identify the various cysts, then study with high power.

In the section you should be able to identify cysts containing all of the following stages, with those containing primary spermatocytes at the apical end and those containing mature spermatozoa at the other end. Not all of the lobes will have been cut longitudinally, and so not all of them will show all of these stages. Some lobes will be cut transversely and may show only one or two stages. Search the slide for an ideal section or, if necessary, make a composite drawing from several sections.

1. **Spermatogonia.** Cells small and crowded at the apical (blunt) end.

2. **Primary Spermatocytes.** These cells are larger and are found in cysts nearest those containing spermatogonia (Fig. 4-2). They may be seen in several stages, with successive stages showing (a) chromatin threads, (b) chromatin threads broken into chromosomes that are pairing, (c) chromosomes thicker and

in pairs (dyads) and fused so that they seem to be haploid, and (d) split pairs (tetrads) formed, with chromosomes now showing curious shapes (coils, bars, rings, and others) and ready for the first division.

3. Secondary Spermatocytes. Cells that have undergone the first maturation division. They may appear smaller than the primary spermatocytes and with a reduced amount of chromatin.

4. Spermatids. Cells that have undergone the second maturation division. They are perfectly round, with a short filamentous tail.

5. Spermatozoa. Mature gametes, with long thin dark heads and filamentous tails seven to eight times longer than the heads.

DRAWINGS

On p. 51 sketch one of the lobes of the grasshopper testis, indicating the location of cysts containing the various stages. Or draw a section of some other testis or seminiferous tubule, as provided by your instructor, locating in it the various stages of spermatogenesis.

OOGENESIS AND FERTILIZATION IN ASCARIS

To illustrate oogenesis, we have chosen the eggs of *Ascaris*, which are especially good because they have only four large chromosomes. The slides you will use are sections through the long egg-filled uterus of a female *Ascaris* after copulation (because, in *Ascaris*, fertilization must occur before oogenesis is completed).

Since these eggs have been sectioned at random, only a part of them will present the typical views of division stages that are described here. Search through the sections for typical examples. Use high power and reduce the light, if necessary.

In the female the primordial germ cells become oogonia, which increase in number by mitosis (**multiplication period**). Some of the oogonia enlarge by taking in food (**growth period**) to become **primary oocytes**, which are ready for meiosis (**maturation period**). However, in *Ascaris* the egg does not mature until it has been entered (fertilized) and activated by a mature spermatozoan.

Sperm Entrance. Locate unfertilized primary oocytes, which characteristically have thin cell membranes, inconspicuous nuclei, and vacuolated cytoplasm. Scattered between the oocytes, find the heads of spermatozoa, appearing as small dark triangular bodies with a centriole at the base of the triangle. Find a primary oocyte with a sperm just entering (Fig. 4-3, *A*). After entrance the spermatozoon nucleus becomes more spherical, and the primary oocyte develops a shell that prevents penetration by other spermatozoa. While the egg is undergoing meiotic divisions, the spermatozoon nucleus (male pronucleus) will remain inactive at one side of the egg nucleus. It has already undergone meiosis in the male testis and now has the haploid number of chromosomes. The egg still has the diploid number, or four chromosomes.

 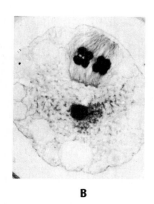 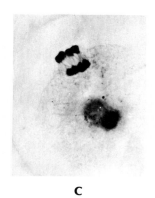

| A | B | C |

Fig. 4-3. Maturation of fertilized *Ascaris* egg. **A,** Fertilization—spermatozoon (lower left) entering egg. **B,** Primary oocyte with two tetrads in metaphase stage of first meiotic division. **C,** Tetrads dividing (telophase stage) to form secondary oocyte and first polar body. (Courtesy Carolina Biological Supply Co., Burlington, N.C.)

Formation of the First Polar Body (First Meiotic Division). Find a primary oocyte ready for its first meiotic division. It has a thick **shell**, with a **perivitelline space** between the shell and the cell membrane. Synapsis has occurred, bringing the maternal and paternal homologs together in pairs. The members of each pair have doubled so that each pair appears as four half chromosomes, or a **tetrad.** The *Ascaris* egg in Fig. 4-3, *B*, has two tetrads lined up on a spindle in the metaphase stage. Locate this stage on your slide. The male pronucleus is now spherical. In Fig. 4-3, *C*, the egg has reached the telophase stage, and division is almost complete. One **dyad** from each tetrad will go to each daughter cell. However, one daughter cell (**secondary oocyte**) will retain all the cytoplasm, while the dyads of the other are pushed out into the perivitelline space and become the **first polar body.** Find a secondary oocyte with two dyads in the cytoplasm and a polar body appearing as a dark spot in the perivitelline space. The spermatozoon nucleus still waits to one side (Fig. 4-4, *A*).

Formation of the Second Polar Body (Second Meiotic Division). In the secondary oocyte the two dyads divide, throwing off two chromatids in the second polar body, which, like the first, has no cytoplasm and is nonfunctional. The other two chromatids (the haploid number) remain in the egg, now a **mature ovum,** and take part in the formation of the **female pronucleus.** The first polar body may or may not divide, making two or three tiny nonfunctional polar bodies in the perivitelline space; but only one functional mature egg is produced from each oogonium.

How does this compare with spermatogenesis? Note that the nucleus of each mature ovum has now the haploid number of chromosomes, the same as the mature spermatozoan. Find an egg in which the second polar body is being formed.

Mature Egg. Look now for eggs that have completed their meiotic divisions. Note that the egg nucleus, or **female pronucleus,** is a round ball with a nuclear membrane and granular chromatin material very much like the nucleus of an interphase stage (Fig. 4-4, *B*). Note that the **male pronucleus** looks very much like the female pronucleus. They are called pronuclei because they help to form the **zygote nucleus.** How many chromosomes does the zygote have now? Although the two pronuclei come close to each other in the egg, only rarely do they fuse together in *Ascaris.* The nuclear membrane of each pronucleus disappears, and the two chromosomes from each nucleus move on to the spindle, which forms for the first cleavage division. Fig. 4-4, *C*, shows the chromosomes, **two maternal** and **two paternal,** lined up in metaphase (polar view). This will be the first of the series of mitotic (cleavage) divisions that will occur to produce the new embryo.

DRAWINGS

Prepare drawings as requested by the instructor. Or the student may wish to make drawings for future reference.

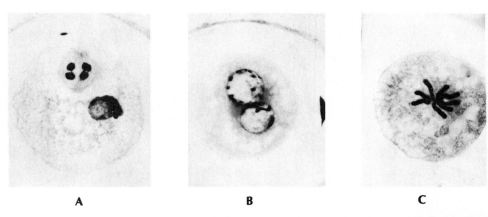

A   B   C

Fig. 4-4. **A,** Dyads of secondary oocyte dividing; will form mature ovum and second polar body; first polar body is seen above in perivitelline space; note male pronucleus (center right). **B,** Mature ovum with male and female pronuclei, each with haploid number of chromosomes. **C,** Diploid number of chromosomes (four) in zygote, now in metaphase stage (polar view) preparing for first cleavage division; this represents the beginning of a new embryo. (Courtesy Carolina Biological Supply Co., Burlington, N.C.)

STAGES OF SPERMATOGENESIS IN GRASSHOPPER TESTIS

Phylum _____

Class _____

STAGES OF OOGENESIS IN ASCARIS

EXERCISE 4B
Early embryology of the sea star

MATERIALS
Prepared slides of sea star development

The development of an animal from egg to adult is a process of almost unbelievable perfection that continues to excite and fascinate biologists. In this exercise you will study a sea star's early stages of growth and differentiation. You have studied gametogenesis, the formation of mature gametes (Exercise 4A). On fertilization of an egg by a spermatozoan, a **zygote** is formed. The paternal and maternal genomes are combined, restoring the diploid number of chromosomes, and the developmental process is activated. The zygote divides by mitosis into two cells (called **blastomeres**), those two divide into four, and so on, until the large, yolk-filled egg has been transformed into a bundle of small cells clustered together like a mass of soap bubbles. This process is called **cleavage.** Each successive division produces smaller and smaller cells. There is no growth during this period, only subdivision of mass. After several hundred or several thousand cells have been formed by cleavage, the cells rearrange themselves around a fluid-filled cavity. The embryo is now a **blastula.** Then cells begin to regroup to form a **gastrula,** in which important new cell associations are formed. Cells become increasingly committed to specific directions of development, which eventually lead to the **differentiation** of an adult animal.

SEA STAR DEVELOPMENT
The sea star belongs to phylum Echinodermata, which, like the chordates, belongs to the deuterostome group of animals.

The development of the sea star is rather easy to study because the sea star egg is of the **isolecithal** type; that is, the yolk material is evenly distributed throughout. Such a type during cleavage forms blastomeres about equal in size. This early division is called **total,** or **holoblastic, cleavage.** What are the other types of cleavage? See the text.

☞ On your slide you will find all stages of development from the unfertilized ovum to the late gastrula. Study your slide *with low power only* because these slides are thicker than average.

Unfertilized Ovum. You have already seen these cells in Exercise 2A (Fig. 2-2). There are several of these scattered on the slide. They can be identified by their spherical shape and their spherical nuclei. Note the small dark nucleolus in the nucleus.

Fertilized Undivided Ovum (Zygote). Find an ovum in which the nucleus and nucleolus are indis-tinct. This is a fertilized ovum. It has a light-colored membrane, the **vitelline membrane,** around it (Fig. 4-5, *A*). This membrane usually rises up from the cytoplasm shortly after the entrance of the spermatozoan. There is no shell, however, such as that found on the *Ascaris* egg. The space between this membrane and the cytoplasm is called the **perivitelline space.**

Cleavage Stages. Cleavage, or cell division, starts just after fertilization. The first mitotic division results in two **blastomeres,** which stay together. How does each blastomere compare in size and shape with the undivided ovum? Is the vitelline membrane still present? Now find a 4-cell stage. The cleavage planes run through the poles perpendicular to each other. Compare the size of the blastomeres with those in the 2-cell stage.

Find an 8-cell stage. This is the result of a third cleavage, the plane of which passed through the equator of the zygote. You may not be able to see more than 6 cells at one time, but by careful focusing you can identify the correct stage. Note that at each succeeding stage the total volume of the embryo does not exceed that of the early 1-cell stage.

Morula Stage. A mass of 16 or more cells with no large cavity in its center is called the **morula,** or "mulberry," stage (Fig. 4-5, *E*). How does its diameter compare with the diameter of the undivided zygote? What about the size of the blastomeres as compared with those in preceding stages?

Blastula Stage. As cleavage continues, the cells become smaller and are finally arranged in a single layer (**blastoderm**) around a central cavity, the **blastocoel,** or **segmentation cavity.** This is the blastula stage. You can distinguish this stage by the regular arrangement of the blastomeres and the difference between the dark outer zone of cells and the lighter central cavity (Fig. 4-5, *F* to *H*).

Gastrula Stage. The embryo is no longer a perfect sphere because one side of the blastula has been pushed in (invaginated) (Fig. 4-5, *I*). This process involves unequal cell division with some migration of cells. In the early gastrula stage there is only a slight indentation where invagination begins, but in later stages the invagination continues until the pushed-in side reaches the opposite wall, thus forming a two-walled sac. The new cavity is now called the **archenteron,** and its opening to the outside is the **blastopore.** The outer layer of cells of the gastrula is called the **ectoderm;** the inner layer, which forms the archenteron, is called the **endoderm.** These two cell layers are called the **germ layers.** The ectoderm will give rise

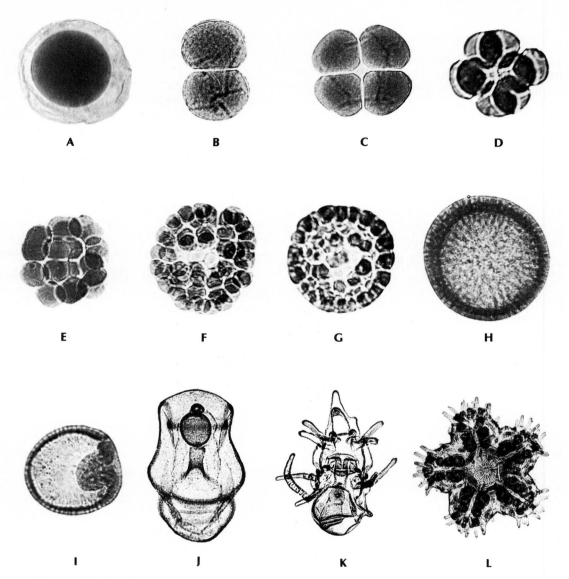

Fig. 4-5. Embryology of sea star. **A,** Fertilized egg. **B,** Two-cell stage. **C,** Four-cell stage. **D,** Eight-cell stage. **E,** Sixteen-cell stage (morula). **F,** Thirty-two-cell stage (morula). **G,** Sixty-four-cell stage (morula). **H,** Blastula. **I,** Gastrula. **J,** Bipinnaria larva. **K,** Brachiolaria larva. **L,** Young sea star. (Courtesy Carolina Biological Supply Co., Burlington, N.C.)

to the epithelium, which covers the body surface; the inner endoderm gives rise to the epithelial lining of the digestive tube. As the archenteron enlarges, the old blastocoel becomes smaller and finally disappears.

Later Stages of Development. In later stages a third germ layer, the **mesoderm,** is formed between the other two. All structures in the body of the animal will come from these three germ layers. The mesoderm gives rise to the great bulk of the body, such as the skeleton, muscles, blood vessels, and reproductive system; the ectoderm forms the outer layer of the skin, parts of the sense organs, and the nervous system; and the endoderm forms the lining of the alimentary canal and parts of other organs that are derived from the digestive tube. The archenteron be-

comes the cavity of the alimentary canal; the blastopore remains as the anus of the larva. The gastrula soon becomes a free-swimming **bipinnaria larva,** which is bilateral in symmetry (Fig. 4-5, *J*). The bipinnaria metamorphoses into a **brachiolaria larva** and finally into the radially symmetrical young sea star (Fig. 4-5, *K* and *L*).

In the deuterostomes we find (1) indeterminate cleavage—that is, any one of the blastomeres of the first cleavage stages has the capability, if it is separated from the others, of developing into a whole individual; (2) the mesoderm and coelom originate as pouches of the archenteron; and (3) the mouth is not derived from the blastopore. What are the characteristics of the protostomes?

PROJECT

Fertilization and cleavage in sea urchin eggs. Where fresh material is available, fertilization may be started prior to the class period so that several cleavages can be observed.

When in breeding condition, sea urchins can be induced to spawn by injecting through the peristome, with a hypodermic syringe and a 25-gauge needle, 0.5 ml of 0.5M KCl. Invert the animal over a small (250 ml) beaker of seawater to collect the shed gametes. Eggs will appear yellow or orange, sperm white. Stir the sperm suspension with a glass rod and add a small amount to the eggs (too many sperm may cause ab-

normal development). If several stages are desired at one time, divide the eggs into several bowls and fertilize at different times. A dozen or so eggs may be placed in a finger bowl of seawater and observed under a scanning lens. Evidence that cleavage is occurring is the appearance of a furrow, first on one side and then continuing around the fertilized egg. Because temperature and amount of oxygen affect the rate of development, only approximate times may be given. At room temperature we might find:

2-cell stage	$3/4$ hour
4-cell stage	$1\frac{1}{2}$ hours
8-cell stage	$1\frac{3}{4}$ hours
16-cell stage	3 hours
Blastula stage	6-10 hours
Free-swimming larvae	24-48 hours

Indirect light from a desk lamp may be preferable to substage lighting. Free-swimming stages tend to swim at the surface.

EXERCISE 4C
Frog development*

MATERIALS

Preserved frog eggs in development stages
Prepared stained slides
   Frog ovary, cross sections
   Frog development stages, cross sections
Models of frog development stages (optional)
Microscopes
Hand lenses or binocular dissecting microscopes

Frog eggs are much larger than the sea star eggs in which you studied development in Exercise 4B. The frog egg is a **telolecithal** type of ovum, in which the yolk is more or less concentrated in one hemisphere while the cytoplasm with the nucleus is found in the other hemisphere. Such an egg shows definite polarity. The **animal pole** is the region of the egg just above the nucleus in the darkly pigmented hemisphere; the **vegetal pole** is a similar point on the light-colored yolk hemisphere diametrically opposite the animal pole. In this type the entire ovum divides during cleavage, but the **blastomeres** are unequal in size and in rate of division.

You will study both the preserved specimens and the prepared slides of the developing stages of the frog. Containers with the various stages of frog egg development will be available. When you select a stage, be sure to keep it immersed in water while studying it; *do not allow it to dry*. Return your spec-

imen to the proper container when you have finished your study of it. If the specimens have been mounted in deep well slides, they are to be studied without removal from the containers.

STUDY OF EGGS IN FROG OVARY

☞ Obtain a stained slide showing a section through the ovary.

Note the **ova,** or eggs, of various sizes. The smaller ova contain little yolk and have conspicuous nuclei. Note the nature and position of the **chromatin granules** in these nuclei. Find connective tissue and muscle scattered among the groups of developing ova. Compare the location of the nuclei in small and large ova. Early germ cells have distinct **cell membranes,** but older ova have **follicular cell layers** surrounding them. Find an ovum with such a follicular layer. What is the function of this follicular layer?

Eggs in Cleavage Stages

☞ Use both the whole developing eggs and stained slides showing cross sections of the various stages of development.

One-Cell Stage. The 1-cell stage may be represented by an unfertilized ovum or, as is more commonly the case, by a fertilized egg (zygote) that has not yet started to develop. Note the thick jellylike layer surrounding the egg. There are three coats of

*See Hickman, C.P., Jr., L.S. Roberts, and F.M. Hickman. 1984 Integrated principles of zoology, ed. 7. St. Louis, The C.V. Mosby Co., pp. 126-132, 574-576.

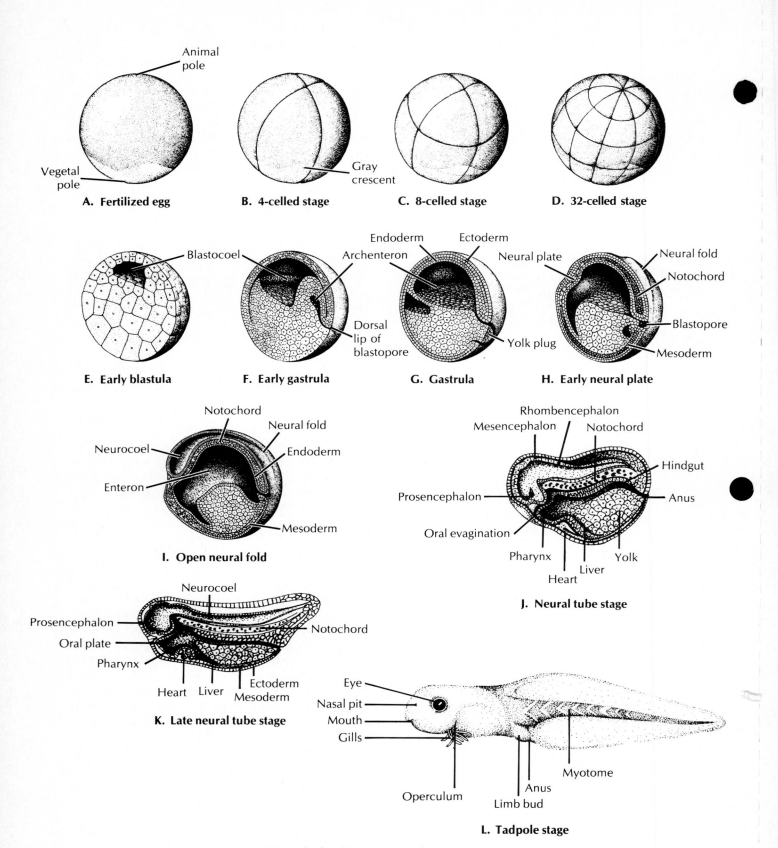

Fig. 4-6. Early development of frog to the tadpole stage.

jelly in this layer. What is the function of the jelly layer? What reproductive organ secretes this jelly layer? What is the shape of the ovum? The black or **animal hemisphere,** which is not as heavy as the **vegetal hemisphere,** always floats uppermost in the water, where it can absorb the sun's warmth, which is necessary for development (Fig. 4-6, *A*). Does fertilization occur internally or externally in the frog?

Soon after fertilization a **gray crescent** appears as a light indistinct crescent-shaped area on the margin of the pigmented zone (Fig. 4-6, *B*). It is formed on the side of the egg opposite that which the spermatozoan entered. It is produced by a migration of pigmented cytoplasm away from that region. Note that before fertilization the egg has radial symmetry; after fertilization it has bilateral symmetry.

Early Cleavage Stages. The frog egg undergoes cleavage just as do the eggs of other animals. The presence of yolk slows down cell division so that the cells in the animal hemisphere divide more rapidly than those in the vegetal hemisphere, and thus the cells of the two poles rapidly become unequal in size and number. Recall that *cleavage is ordinary mitosis:* therefore each cell has the diploid number of chromosomes.

The first cleavage plane, which results in the **2-cell stage,** is vertical (meridional), beginning at the animal pole and passing through the vegetal pole. The second cleavage plane, resulting in the **4-cell stage,** is also vertical but at right angles to the first cleavage plane. The third cleavage plane is equatorial or horizontal but passes closer to the animal pole. The four blastomeres (**micromeres**) of the animal pole of this **8-cell stage** are much smaller than the four **macromeres** at the vegetal pole (Fig. 4-6, *C*).

The **morula stage** contains 16 to 32 cells. There are two fourth cleavage planes; both are vertical, and they appear simultaneously. The cells of the animal pole complete their division before those of the vegetal pole, thus causing temporarily a condition in which there are 8 small cells at one end (animal pole) and 4 at the other. The fifth cleavage, which results in 32 cells, is also made up of two furrows, both of which are horizontal (Fig. 4-6, *D*). How are they placed with reference to the third cleavage plane of the 8-cell stage? Note that in these later stages the faster rate of cell division in the animal hemisphere results in a larger number of micromeres than macromeres. From now on the cells and cleavage planes are difficult to recognize.

Blastula Stage. The blastula stage begins at about 32 cells. A **segmentation cavity** or **blastocoel** has begun near the animal pole. The blastula stage is not easily recognized in the unsectioned embryo; look for it in a cross-section slide (Fig. 4-6, *E*).

Gastrula Stage. The gastrula stage is recognized by the crescent-shaped slit at the margin of the pigmented animal hemisphere. Gastrulation in the frog is more complicated than in the sea star. In the sea star eggs you studied earlier you saw that the gastrula stage was brought about by a simple folding in of one side. In the frog embryo the yolk cells are heavy, preventing such a simple arrangement. Instead, the pigmented cells of the animal hemisphere begin growing down over the vegetal cells (a process called **epiboly**), folding in at the equator along a crescent-shaped line (Fig. 4-6, *F*). The opening thus formed is called the **blastopore.** As the overgrowth continues, the lips of the blastopore draw closer together over the vegetal cells until only a small mass of vegetal cells is left showing, which is called the **yolk plug.** On whole mounts this will appear as a light circle on one side of the gastrula (Fig. 4-6, *G*). Look at the cross section of this stage and identify the yolk plug and blastopore. The blastopore represents the future posterior end of the embryo.

As gastrulation progresses, the inturning (invaginating) cells form an inner layer, the **endoderm,** surrounded by an overgrowth of animal cells called the **ectoderm.** A third layer, the **mesoderm,** forms between these (Fig. 4-6, *H*), derived from certain cells that were involuted at the lips of the blastopore. These three layers are called the **germ layers,** and from them the various tissues and organs of the body will be formed. The new cavity formed by gastrulation, of which the blastopore is the opening, is called the **archenteron** (gastrocoel or primitive gut). The blastocoel gradually becomes obliterated.

Neural Groove Stage. Examine preserved specimens and transverse sections (slide) of this stage (Fig. 4-6, *I*). Note that in the preserved specimen the embryo has assumed an elongated form. The embryo is still in its jelly covering and has not yet hatched. The posterior end is more pointed than the anterior end. Find along the dorsal side the **neural groove.** Later this groove becomes a tube.

In the prepared slide of a transverse section, note the neural groove with a **neural fold** on each side of the groove. The neural groove develops in a thick ectodermal plate, the **neural plate.** Just ventral to the neural groove is the **notochord.** The mesodermal layer is now well defined. What is the appearance of the mesoderm? Do you notice any yolk material?

Neural Tube Stage. In a preserved specimen, note that the neural groove has now closed (Fig. 4-6, *J*). A **neural tube** has been formed by the meeting of the neural folds in the midline. Find on each side of the head two ventral ridges. **Eyes** will develop from the first of these ridges, **gills** from the second. Turn the specimen over and look for **ventral suckers** by which the tadpole may hold fast to an object. Look for the **oral plate** at the anterior end and the **anus** at the posterior end (Fig. 4-6, *K*).

Now examine a transverse section through the embryo at this stage. Note the neural tube, notochord, and gastrocoele (archenteron). Observe the condition of the mesoderm. Masses of epimeric mesoderm lateral to the notochord give rise to the dermis, muscles, and skeleton; lateral to them is intermediate mesoderm, which forms most of the excretory system; and a thin hypomeric layer around the yolk mass is destined to split and give rise to the coelomic cavity.

Tadpole Stage. Examine specimens of tadpoles at various stages of development. What are the changes in the external body form? These have emerged from their jelly membranes (hatched). Find the **suckers** on the anterior ventral surface. In front of the suckers, find the oval **mouth** (Fig. 4-6, *L*). Look for two **external nares** on the anterior dorsal side. On each side of the head are fingerlike **external gills.** At the base of the tail is the **anus.** Note the condition of the **eyes.** Look for the **spiracle,** a tiny opening on the left side posterior to the eye. This is an exit for the water that bathes the gills.

Movement of the embryo within its jelly layers is accomplished by means of cilia on its ectoderm. The embryo is said to hatch when it frees itself from its jelly surroundings. Hatching time is not always correlated with development stages, depending somewhat on living conditions. Therefore the first tadpole stages are rather indefinite.

More advanced stages of the tadpole will show a skin fold (**operculum**) over the external gills, which disappear to be replaced by the **internal gills.** The transformation that the larva undergoes in becoming an adult frog is called **metamorphosis,** a process that takes place in all animals in which the larva differs greatly in form and structure from the adult.

Late tadpole stages may be studied with profit if specimens are available. What happens to the tail when the tadpole is transformed into an adult frog? Which pair of legs emerges first?

DRAWINGS (OPTIONAL)

On separate paper sketch such stages of frog development as you may wish to have for future reference.

PROJECTS AND DEMONSTRATIONS

1. Obtaining living amphibian eggs and embryos.

a. FRESHLY COLLECTED EGGS. Many species of frogs lay their eggs in the spring early enough to be studied before school closes. Among the species that lay their eggs early are the wood frog, the grass frog, and the many species of tree frogs. The places where they lay their eggs can usually be located during the breeding season, when frogs do a great deal of croaking. Place collected eggs in small aquariums having constant aeration. Avoid overcrowding. A few eggs may be kept in finger bowls if the water is changed frequently. Healthy eggs soon turn dark; dead eggs turn white and should be discarded. The rate of development is determined by the temperature of the water. At room temperature, many will hatch out within 2 days. Tadpoles in small numbers thrive in well-balanced aquariums. They will do well on the crumbled yolk of hard-boiled eggs, but they should also have access to algae and other debris.

b. INDUCTION OF OVULATION. Living eggs may be provided from September to spring by inducing ovulation. One adult female will provide about 2000 eggs. Most species can be used, but *Rana pipiens* is probably most satisfactory.

Obtain from collectors seven large females and two large males, freshly caught. Keep the largest female for ovulation and remove the pituitary glands from the other six. To do this, remove the cranium by inserting scissors in the angle of the jaw and cutting posteromedially behind the eyes; then cut across and through the occipital region to the other jaw. Invert the skull. Insert sharp scissors into the posterior brain cavity, lateral to the brain tissue, and cut forward through the floor of the cranium on each side of the brainstem, avoiding injury to the brain tissue. Cut anteriorly to the orbits. Deflect the cut floor of the cranium and find the kidney-shaped pinkish **pituitary gland** either in the cranial floor or attached to the optic chiasma. Remove the gland with forceps and place it in less than 2 ml of water. Place all six glands in the same water and use immediately.

Suck the whole glands up into the barrel of a 2 ml hypodermic syringe; add a large-bore hypodermic needle (no. 18). Hold the living female firmly on a table and inject into the posterolateral abdominal cavity, avoiding injury to the large veins and to the viscera. Place the animal into a battery jar in about 1 inch of water. Ovulation time can be controlled by the room temperature. At 25° C the eggs should be ready in 2 to 3 days; at 10° C, allow about 1 week.

Test for the presence of eggs in the uteri by bending the body forward so that the body and legs are at right angles; then apply pressure carefully by gentle squeezing of the abdomen from anterior to posterior with the closing hand. If eggs come readily and in large numbers, they are ready for insemination. When the female is ready, remove the testes from the males. With clean scissors, cut up the testes into 20 ml of pond or spring water. Let stand about 20 minutes and then divide into two clean finger bowls. When stripping the female, discard the first few eggs and divide the remainder into the two bowls of sperm suspension, distributing the eggs evenly. Rotate the bowls to encourage contact of the spermatozoa with every egg and then let stand for 3 minutes. Flood with pond or spring water and leave undisturbed for 2 hours. When the eggs have rotated so that they are dark side up, gently lift them from the bottom of the dish with a clean section lifter.

2. *Effect of thyroid on metamorphosis.* To speed up metamorphosis, separate a group of tadpoles into similar aquaria. Into one aquarium, dust small amounts of powdered thyroid. The tadpoles will eat some of the thyroid or, even if they do not eat it, will be affected by the hormone in solution. When thyroid is added, the water must be changed frequently—daily if necessary. Compare the development of the two groups.

When the tadpoles are transformed into small frogs, transfer them immediately to a moist terrarium or return them to their natural habitat.

3. *Demonstrations of chick embryo slides.*

4. *Demonstrations of pig embryo slides.*

5. *Sections through mammalian tests and ovaries.*

CHAPTER 5

TISSUE STRUCTURE AND FUNCTION*

MATERIALS

Compound microscopes
Prepared slides (the selection will depend on time and materials available)
 Frog skin, cross section
 Amphibian small intestine, cross section (*Amphiuma* is best)
 Artery, vein, and nerve, cross section
 Trachea, cross section
 Kidney cortex, cross section
 Ascaris, cross section through intestine
 Ground bone, cross section
 Cartilage (can be seen on the trachea slide)
 Skeletal muscle
 Cardiac muscle
 Smooth muscle (can be seen on intestine or artery)
 Human blood
 Amphibian or reptile blood
 Areolar connective tissue
 Others as desired

A **tissue** is an aggregation of cells and cell products of similar structure and embryonic origin that performs a common function. Tissues represent specializations of the properties that all protoplasm possesses, namely irritability, contractility, conductivity, absorption, excretion, and the like. The study of tissues, especially their structure and arrangement, is called **histology.**

However complex an animal may be, its cells fall into one of five major groups of tissues. These basic tissues are named **epithelial** tissue, **supporting** tissue (which includes connective tissue as well as the denser supporting tissues, such as bone and cartilage), **muscle** tissue, **nervous** tissue, and **vascular** tissue.

An **organ** is an aggregation of tissues organized into a larger functional unit, such as the heart or the kidney. Organs work together in teams called **systems.**

The following study is made on vertebrate tissue, but invertebrate tissues are built on essentially the same lines and may be substituted at the discretion of the instructor.

*See Hickman, C.P., Jr., L.S. Roberts, and F.M. Hickman. 1984. Integrated principles of zoology. ed. 7. St. Louis, The C.V. Mosby Co., pp. 133-138.

 Read the general description of the tissues (pp. 61-71) and familiarize yourself with the general types, their functions, and where they are found. As you do, look at slides showing examples of the various tissues to familiarize yourself with their appearance. Later, because tissues are usually found working together with other tissues in organs, you will study some sections through certain organs, each of which will contain several types of tissues.

DRAWINGS

On pp. 75 and 76 you will find places to sketch the various types of tissues that you study as assigned by your instructor. A number of the tissues can be identified by studying the sections of skin, intestine, trachea, artery, and nerve (pp. 71-73). The rest you will find on special slides.

Draw only tissues that you actually see. Do not copy the photographs or drawings. Where possible, show the shape of the cells and the size and location of the nuclei. Indicate under each drawing (1) what organ the tissue was seen in and (2) what animal the tissue was taken from, if that information is indicated on the slide.

GENERAL DESCRIPTION
Epithelial Tissue

An **epithelium** is a sheetlike layer of cells that covers an external or internal surface. This surface may be large (outer layer of the skin, lining of the digestive tract) or small (sebaceous glands, microscopic tubules). One surface of the epithelium is free, and the other rests usually on a bed of vascular connective tissue. Between the epithelium and the underlying connective tissue there is usually a thin basement membrane of intercellular substance (not easily seen in many preparations). Epithelial tissue itself lacks blood supply; its nourishment comes by diffusion from the blood supply of the connective tissue below it.

Epithelial tissues are derived from all three of the

embryonic layers but largely from ectoderm and endoderm.

The chief function of an epithelium is protection, but the cells are also variously specialized for secretion, excretion, absorption, lubrication, and sensory perception.

Epithelial tissue may be classified according to the number of layers of cells (simple or stratified) or according to the shape of the cells (squamous, cuboidal, or columnar).

Simple Epithelium. Simple epithelium is made up of a single layer of epithelial cells. It is found where there is not much wear and tear or where diffusion or absorption occurs through a membrane.

1. Simple squamous epithelium. Simple squamous epithelium is composed of thin, flat cells with round or oval nuclei (Figs. 5-1, *A*, and 5-2, *A*). The cells appear hexagonal in surface view; in side view or cut section they appear extremely thin, bulging a bit where the nucleus is located.

Simple squamous epithelium is specialized for diffusion. It lines the walls of blood vessels, the body cavity, Bowman's capsules of the kidney, and a few other places.

2. Simple cuboidal epithelium. Simple cuboidal cells are usually six sided, but they appear square in side view and polygonal or hexagonal in surface view (Figs. 5-1, *B*, and 5-2, *B*).

Simple cuboidal tissue is adapted for more wear and tear than is squamous tissue and for some diffusion and secretion. It is found in kidney tubules, salivary glands, mucous glands, thyroid follicles, and so on.

3. Simple columnar epithelium. Simple columnar cells are taller than they are wide and are closely packed like the cells of a honeycomb (Figs. 5-1, *C*, and 5-3). In surface view they appear hexagonal; in vertical section they appear as a row of rectangles, with the nuclei frequently all at the same level, usually in the lower part of the cell.

Columnar cells may be ciliated on the free surface

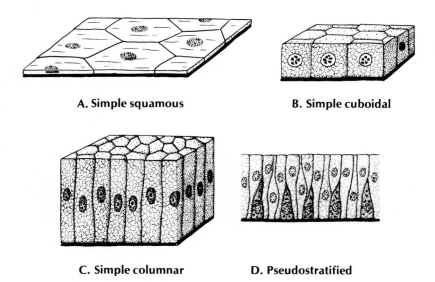

A. Simple squamous   B. Simple cuboidal

C. Simple columnar   D. Pseudostratified

Fig. 5-1. Types of simple and pseudostratified epithelium.

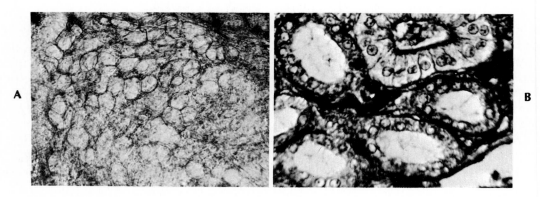

Fig. 5-2. Epithelial tissue. **A,** Simple squamous cells in endothelium, surface view. **B,** Simple cuboidal and simple columnar epithelium in kidney tubules.

or may be covered with a nonliving cuticle. The cells may be specialized for secretion in the form of single-celled glands that are called **goblet cells** (Fig. 5-3).

Columnar epithelium is found where the wear and tear is considerable and where there is some need for secretion; some absorption can occur through it. It lines a large part of the digestive tract, the oviducts, and many glands.

4. Pseudostratified columnar epithelium. Pseudostratified columnar cells are actually simple epithelium with all the cells resting on the basement membrane, but they have the appearance of stratified epithelium because they are not all the same height and their nuclei are located at different levels (Fig. 5-1, *D*). Pseudostratified columnar cells line the tra-chea (windpipe), bronchi, male urethra, and a few other places.

5. Stratified epithelium. Stratified epithelium is composed of two or more layers of cells (Figs. 5-4 and 5-5). It is found where the wear and tear is very great—often where surface cells are continually being sloughed off. It is not adapted for absorption or secretion.

Stratified epithelium is named according to the shape of the cells at the free surface—stratified **squamous** (Fig. 5-5) (outer layer of the skin, mouth, esophagus, anus, vagina, others); stratified **cuboidal** (urinary tract, ducts of sweat glands, testis tubules, others); and stratified **columnar** (parts of the pharynx, the larynx, the urethra, the salivary ducts, and so on).

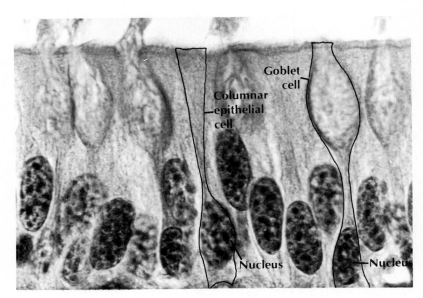

Fig. 5-3. Columnar epithelium containing goblet cells. Boundaries of two cells have been outlined. (Courtesy Carolina Biological Supply Co., Burlington, N.C.)

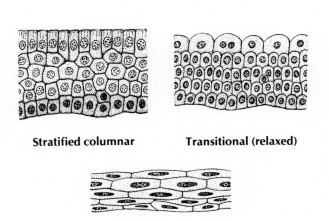

Stratified columnar   **Transitional (relaxed)**

Transitional (stretched)

Fig. 5-4. Stratified and transitional epithelium.

Some stratified epithelia that can change appearance by stretching or relaxing is called **transitional epithelium,** such as that in the bladder (Fig. 5-4).

Supporting Tissue

The connective, or supporting, tissues bind together and anchor other tissues and organs and give form and support to the body and its organs. They include both **loose connective tissue** (areolar, reticular, adipose) and **dense connective tissue** (sheaths, ligaments, tendons, cartilage, bone).

Supporting tissues are derived from embryonic mesoderm. They are characterized by various kinds of **cells,** varying amounts of nonliving **fibers** (which may be elastic or inelastic, branched or nonbranched), and an amorphous ground substance, or **matrix** (Fig. 5-6, *B*). The matrix and fibers are secreted by the cells.

The varying types, amounts, and arrangements of the cells, matrix, and fibers give rise to the many forms of connective and supporting tissues. For example, **adipose (fat)** tissue is composed mainly of cells; tendons and ligaments are mostly fibers; and cartilage is largely matrix.

The type and arrangement of fibers are also characteristic of certain tissues. In fasciae the fibers are interlaced and predominantly collagenous. In tendons and ligaments they are predominantly collagenous and arranged in a parallel fashion. In vocal cords they are predominantly elastic and parallel. Interlaced fibers supply resistance to tension from all directions. Parallel fibers are best oriented to withstand tension from one direction.

Areolar Connective Tissue. Areolar connective tissue is a loose fibroelastic tissue, the most widespread of all the connective tissues. It is found in every microscopic section of the body, fastening down the skin, membranes, vessels, and nerves, and binding the muscles and other parts together. In fresh tissue it appears whitish, translucent, soft, and stretchy. It consists of a clear jellylike matrix, various cells, and all three types of fibers (Fig. 5-6).

Adipose (Fat). Adipose connective tissue cells are specialized for fat storage and do not form ground substance or fibers. Note the clear bubblelike appearance of the tissue (Fig. 5-7). The cytoplasm and nucleus have been pushed to one side of the cell by the globule of fat that fills the cell.

Cartilage. The most common form of cartilage is **hyaline cartilage** (Fig. 5-8). It is found on the ends of long bones and in the nose, trachea, and other places. Its ground substance is firm but flexible. Scattered through it are **lacunae,** little cavities each containing at least one cell. Two or more cells together indicate recent cell division. The cells secrete the ground substance but no fibers. In the **fibrocartilage** of intervertebral discs, insertion of tendons, and other places the ground substance contains many inelastic fibers. **Elastic cartilage,** found for example in the external ear, epiglottis, and larynx, contains many elastic fibers.

A description of the trachea with its cartilage and other tissues is found on p. 73.

Bone. Bone is the most specialized of supporting connective tissues. It not only supports, but it also protects vital organs by means of bony frameworks,

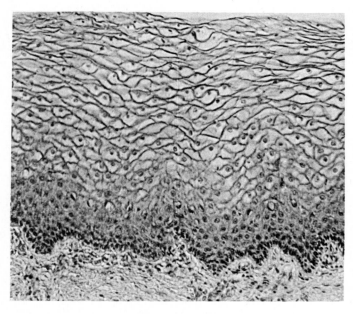

Fig. 5-5. Stratified squamous epithelium from outer layer of skin. Note that cells are flat at surface but more cuboid at base. A layer of connective tissue underlies the epithelium.

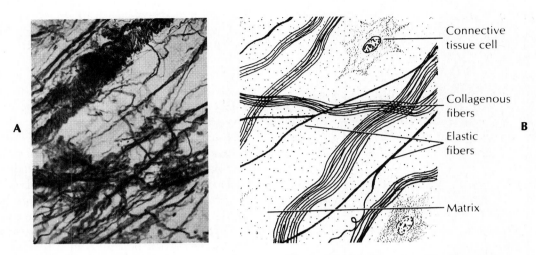

Fig. 5-6. Areolar connective tissue, the most widespread of all the connective tissues. **A,** Photomicrograph. **B,** Detail of structure.

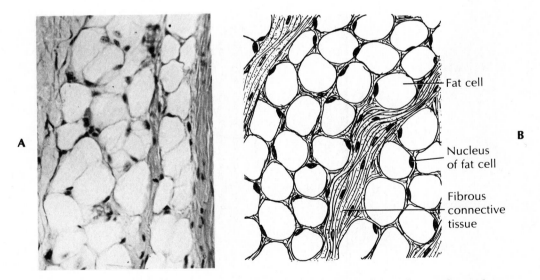

Fig. 5-7. Adipose tissue. In these cells globules of clear fat push the nucleus and cytoplasm to one side. **A,** Photomicrograph. **B,** Interpretive drawing.

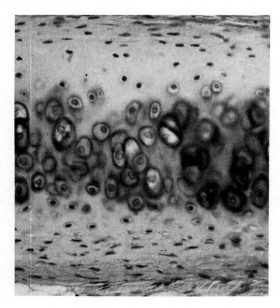

Fig. 5-8. Hyaline cartilage.

and it forms the red blood corpuscles and most of the white corpuscles. Bone also forms a complete system in its own right, the **skeleton.**

Bone matrix is heavily infiltrated with calcium and phosphate salts secreted by bone-forming cells called **osteoblasts.**

A long bone such as the femur or humerus consists of a shaft of **compact bone** surrounding a **bone marrow cavity** (Fig. 5-9). The enlarged ends of the bone are made up of **spongy bone,** the ends of which are padded with cartilage. The bone is covered with thin **periosteum.** Nerves and blood vessels penetrate and nourish the bone tissue.

Compact bone is made up of a series of **osteons (haversian systems)** (Figs. 5-9 and 5-10), each one built around a narrow canal containing blood vessels. In the formation of compact bone the bone-forming cells, osteoblasts, become arranged in the matrix around the blood vessels in layers called **lamellae.** Each osteoblast occupies a space called a **lacuna.** The cells have branching processes that are continuous with the processes of adjacent cells, thus forming tiny **canaliculi,** through which nourishment can reach the cells from the blood supply. Through the activity of the

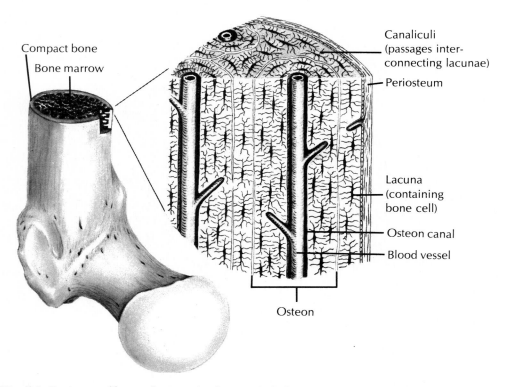

Compact bone

Bone marrow

Canaliculi (passages interconnecting lacunae)

Periosteum

Lacuna (containing bone cell)

Osteon canal

Blood vessel

Osteon

Fig. 5-9. Structure of bone, showing the dense calcified matrix and bone cells arranged into osteons (haversian systems). Bone cells are entrapped within the cell-like lacunae but receive nutrients from the circulatory system via tiny canaliculi that interlace the calcified matrix. Bone cells were known as osteoblasts when they were building bone, but in mature bone (shown here) they become resting osteocytes. Bone is covered with a compact connective tissue called "periosteum."

osteoblasts the area all around the cells becomes calcified, leaving only the narrow osteon canal containing the blood vessels. When the osteoblasts die, the little empty lacunae and their canaliculi are left in rows (lamellae) around the canals, giving the characteristic appearance to the osteons. Throughout life the osteons are continually being destroyed through the activity of bone-destroying cells, called **osteoclasts** and rebuilt by new osteoblasts.

☞ On a slide containing a piece of compact bone ground very thin, identify an **osteon** (haversian system) with its canal, lacunae, lamellae, and canaliculi. From the appearance of the slide do you think all the osteons are running parallel to each other? Do you see any incomplete osteons or lamellae? Can you explain their presence?

Vascular Tissue

Blood, lymph, and **tissue fluids** are the fluid tissues of the organism. Blood and lymph are composed of various types of cells (**corpuscles**) circulating in a fluid matrix (**plasma**) and flowing within a system of blood vessels. Lymph and tissue fluids are filtered from the blood to bathe all body cells. Vascular tissue transports nutritive substances, oxygen, and hormones to the tissues and carries away wastes.

The cells in blood include the following:

1. Erythrocytes (Red Corpuscles) (Fig. 5-11). In the human being and other mammals erythrocytes are round, flat biconcave discs that do not have nuclei. They average about 7 μm in diameter but are very thin. Amphibian (frog and salamander) blood has oval-shaped erythrocytes, with large granular nuclei. Red corpuscles carry oxygen and carbon dioxide between the lungs and the body tissues.

2. Leukocytes (White Corpuscles) (Fig. 5-11). There are several types of leukocytes, all with large, darkly staining nuclei. In human blood they are larger than the red cells; in frog blood they are smaller. By means of their ameboid movement they can pass through capillary walls and can surround and ingest foreign particles and invading organisms (phagocytosis). They also release special substances that help organize defense. They are active in fighting disease, inflammation, and allergies, performing most of their functions outside the blood vessels in connective tissue. The pus that forms in and around an infection consists largely of dead white blood cells.

Blood Platelets (Fig. 5-11). Platelets are tiny discs, 2 to 3 μm in diameter and very fragile. They are active in the clotting process. Platelets are found only in mammalian blood. Lower vertebrates have spindle-shaped cells called **thrombocytes,** which seem to have a similar function.

☞ Compare a stained slide of human blood with one of amphibian or reptile blood. Erythrocytes are usually stained red and the nuclei of leukocytes appear blue. In what other ways can you distinguish the erythrocytes from the leukocytes? How do mammalian erythrocytes compare with the nonmammalian erythrocytes in size, shape, and nucleation? How do the leukocytes compare?

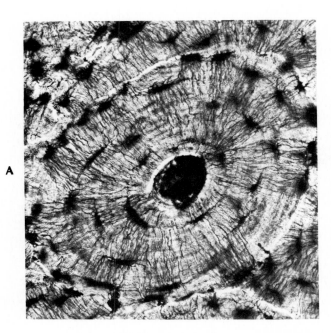

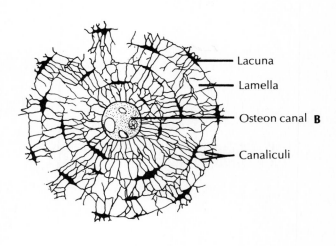

A
B

Lacuna
Lamella
Osteon canal
Canaliculi

Fig. 5-10. Section of ground bone. **A,** Structure of an osteon (haversian system) in cross section. **B,** Interpretive drawing. (Photograph by Carolina Biological Supply Co., Burlington, N.C.)

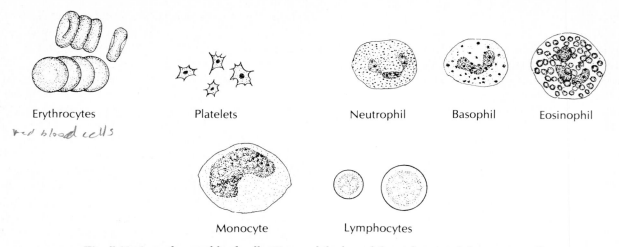

Erythrocytes

red blood cells

Platelets

Neutrophil

Basophil

Eosinophil

Monocyte

Lymphocytes

Fig. 5-11. Some human blood cells. Neutrophils, basophils, and eosinophils are types of leukocytes called granulocytes. Monocytes and lymphocytes are also leukocytes.

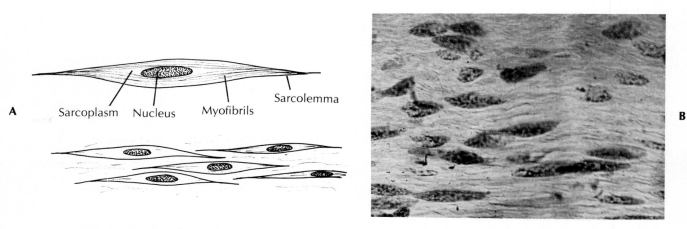

A

Sarcoplasm   Nucleus   Myofibrils   Sarcolemma

B

Fig. 5-12. Smooth muscle. **A,** Some smooth muscle fibers. **B,** Smooth muscle as seen in intestinal wall. Smooth muscle is involuntary in action. (Photograph by F.M. Hickman.)

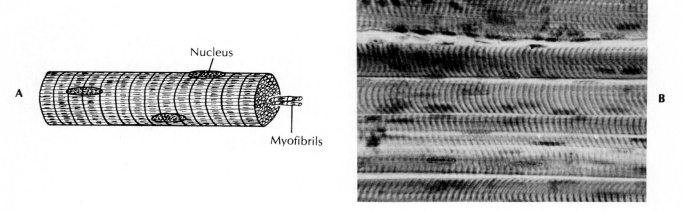

Nucleus

A

Myofibrils

B

Fig. 5-13. Skeletal muscle is striated and voluntary in action. **A,** Skeletal muscle fiber made up of many myofibrils. **B,** Several muscle cells. (**B** photograph by Joseph Bamberger.)

Muscle Tissue

Smooth Muscle. Smooth muscle is also called non-striated, visceral, or involuntary muscle (Fig. 5-12). In structure it is the simplest of the three types of muscle. It is found where slow sustained contractions are needed, such as in the digestive tract, uterus, and other visceral organs. It is entirely involuntary in its action.

Smooth muscle cells are long and spindle shaped, though rarely longer than 0.5 mm. The nucleus is located in the middle or thickest part of the cell, and the ends of the fiber diminish gradually to a fine point. Muscle fibers are composed of myofibrils, which in smooth muscle do not have cross striations, as in skeletal and cardiac muscle.

Smooth muscle will be studied later on a cross section of the small intestine (p. 72).

Skeletal Muscle. Skeletal muscle is also known as striated or voluntary muscle (Fig. 5-13). It makes up the bulk of the muscular system. Skeletal muscle contracts more rapidly than smooth muscle, but it fatigues more easily and is less sustained.

A muscle fiber, or cell, is long (1 to 40 mm) and cylindric and has blunt ends. Each fiber is made up of many myofibrils, all enclosed in a tough cell membrane. Each cell has a number of nuclei located just beneath the cell membrane. The myofibrils bear alternate light and dark bands and are so arranged that the entire muscle fiber has a striated appearance. The muscle fibers are grouped into bundles, and the bundles are grouped into functional skeletal muscles. Fibers, bundles, and muscles are embedded in connective tissue fascia. The connective tissue of the muscle merges with the fibrous sheet or tendon that holds the muscle in place. Skeletal muscles are richly supplied with blood vessels and nerves.

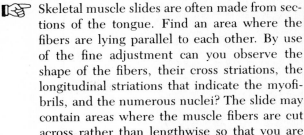 Skeletal muscle slides are often made from sections of the tongue. Find an area where the fibers are lying parallel to each other. By use of the fine adjustment can you observe the shape of the fibers, their cross striations, the longitudinal striations that indicate the myofibrils, and the numerous nuclei? The slide may contain areas where the muscle fibers are cut across rather than lengthwise so that you are observing the cut ends of the fibers. Note how the fibers are arranged into bundles held together by fasciae.

Cardiac Muscle. Cardiac muscle (Fig. 5-14) is much like skeletal muscle in structure in that it is striated; but it functions more like smooth muscle. It is involuntary and is found only in the heart and in the walls of the large vessels adjoining the heart. It is well adapted for rhythmic contractions; the fibers are arranged in branching columns interconnected in a fine meshwork that resembles a syncytium but is not. The myofibrils are transversely striated as in skeletal muscle, but the nuclei are centrally located as in smooth muscle.

On a slide mount of heart muscle note the branching nature of the striated fibers. Whereas the myofibrils are transversely striated, as in skeletal muscle, the nuclei are located centrally, as in smooth muscle. Nuclei lying between the fibers belong to connective tissue. Some preparations show dark bands called **intercalated discs,** which mark the boundaries between the ends of the cells.

Nervous Tissue

Nervous tissue is specialized for the reception of stimuli (perception) and for the conduction of nervous im-

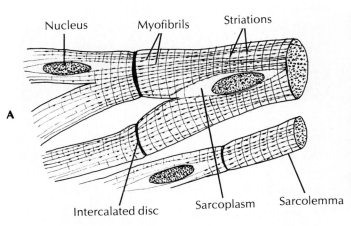
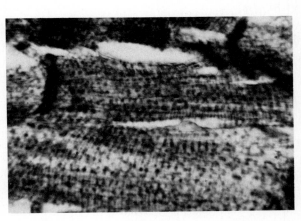

Fig. 5-14. Cardiac muscle is striated and involuntary. **A,** The cells are branching and separated by intercalated discs. **B,** Longitudinal section of cardiac muscle. (Photograph by Joseph Bamberger.)

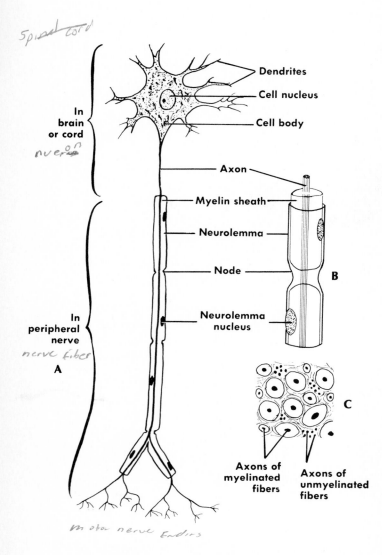

Spinal cord

nueron

nerve fiber

Dendrites

Cell nucleus

Cell body

In brain or cord

Axon

Myelin sheath

Neurolemma

Node

Neurolemma nucleus

B

In peripheral nerve

A

C

Axons of myelinated fibers

Axons of unmyelinated fibers

motor nerve Endins

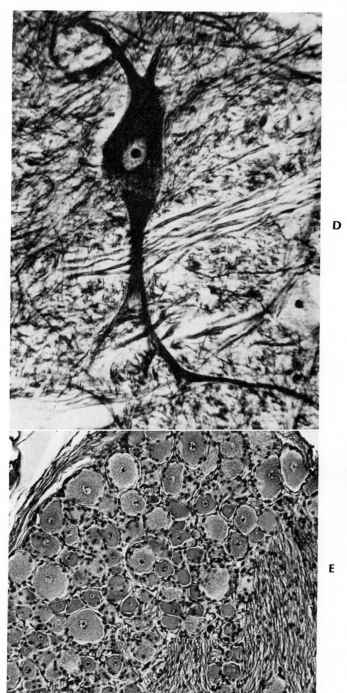

D

E

Fig. 5-15. **A,** Diagram of a neuron. **B,** Detail of a portion of an axon with myelin sheath. **C** and **E,** Portions of cross section of nerve trunk, showing the appearance of axons in cut section. **D,** A motor neuron from spinal cord. (Photographs by Joseph Bamberger.)

pulses to the muscles or glands that are to act on the impulses.

Nervous tissue is composed of nerve cells called **neurons.** A neuron is made up of a nucleated **nerve cell body** and several **processes** (Fig. 5-15). The processes that carry impulses into the nerve cell are called **dendrites;** those that carry impulses away are called **axons.** We often use the term "nerve cell" for the nucleated cell body and the term "nerve fibers" for

the processes. The nerve fibers are usually provided with a protective **myelin sheath** covered by a thin nucleated membrane, the **neurolemma.**

Nerve trunks (Fig. 5-15, *C* and *E*) and the white matter of the brain and spinal cord appear white because they are made up of many nerve fibers covered by their white myelin sheaths. The nucleated nerve cell bodies are all located in ganglia or in the gray matter of the brain or spinal cord (Fig. 5-16), and this

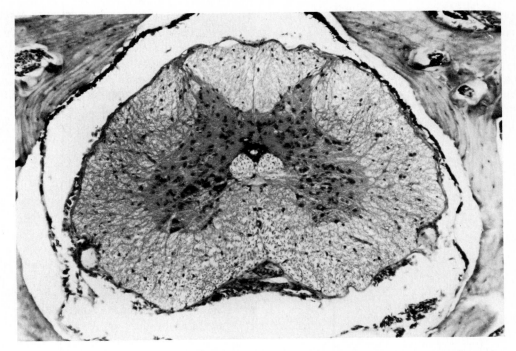

Fig. 5-16. Cross section of spinal cord of the lizard *Anolis*. (Photograph by F.M. Hickman.)

tissue in the natural state appears gray because of the cytoplasm of the cells.

☞ Examine a cross section of a spinal cord. This should show the gray matter in an **H**-shaped, or butterfly-shaped, area surrounded by the white matter (Fig. 5-16). Nerve cell bodies are easily distinguished in the gray matter. These are the cell bodies of motor and association neurons. In the white matter you will see the axons and dendrites that connect various parts of the nerve cord with each other and with the brain. Nerve fibers in cross section will appear as small dark dots, each surrounded by a clean circle, the myelin sheath. The natural color may not be evident in the slides, which are usually stained during preparation.

Tissues Combined into Organs

☞ In addition to the slides containing individual types of tissues that you have already seen, study the slides listed below, which contain sections through certain organs. These not only provide an elementary working knowledge of the appearance of most of the body tissues but also illustrate the manner in which tissues work together. While studying these slides, refer back freely to the descriptions and illustrations for help in identifying the various tissues.

The instructor may wish to vary the following list of slides according to the materials available in the laboratory.

Be prepared to recognize any of the various types of tissues that may later be set up by your instructor as "unknowns."

Section through the Skin of a Frog. The outer layer of the skin, the epidermis, is made up of **stratified squamous epithelium** (Fig. 5-17). Note the flat surface cells that give the epithelium its name. The columnar cells at the base divide to produce new cells that push out to the surface to replace the surface cells as they are worn off.

Beneath the epidermis is a thick layer of dermis, which is made up of **connective tissue** and contains glands and pigment. The connective tissue nearer the surface (spongy layer) contains loosely arranged fibers, whereas that in the deeper layer (compact layer) is much denser. Can you identify **elastic** or **collagenous fibers** in the dermis?

In the spongy layer you will find a number of mucous glands, each made up of a single layer of **cuboid epithelium**. The section of skin may also contain some very large **poison glands.** The glands open to the outside by small ducts, but, since the ducts are narrower than the glands, not all of the cut sections will include ducts.

Scattered through the dermis are small blood vessels. These may be capillaries, made up of a single layer of **squamous epithelium,** or small arteries or veins, containing layers of **smooth muscle.** The darkly stained, irregularly shaped bodies at the base of the epidermis and scattered through the dermis are pigment cells called **chromatophores.**

Cross Section of Amphibian Small Intestine (Preferably *Necturus* or *Amphiuma*). The intestine is a tube enclosing a cavity called the lumen. The lumen of the intestine is lined with a mucous membrane that

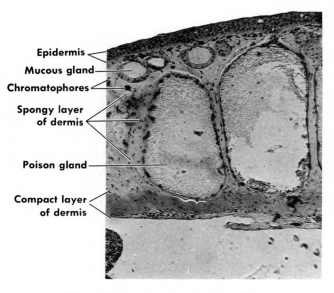

Epidermis

Mucous gland

Chromatophores

Spongy layer of dermis

Poison gland

Compact layer of dermis

Fig. 5-17. Section through skin of frog.

lies in many folds (Fig. 5-18). The mucous membrane is made up of **columnar epithelium** in which the nuclei are located near the base of the tall cells. (If the cells are cut at an angle, some of them may appear to have more than one nucleus. Many of the cells are specialized as secreting **goblet cells** and contain mucus-filled cavities opening into the lumen.

Surrounding the mucous membrane and largely conforming to its contours is a submucosal layer of **connective tissue,** containing mostly collagenous fibers. There are many blood vessels in this layer.

Outside the submucosa are layers of both circular and longitudinal **smooth muscle.** In the circular layer the long spindle-shaped cells can be seen fitting closely together. In the longitudinal layer only the cut ends of the fibers can be seen.

The outermost layer of the intestine is a thin layer of **squamous epithelium** that is a part of the peritoneum that covers all the visceral organs. The cells you see are in cut section, so they are very thin.

Section through an Artery, Vein, and Nerve. Most blood vessels are muscular organs. An artery will have a smaller diameter and thicker walls than will the vein that accompanies it, but otherwise their structure is similar. On your slide the blood vessels are probably collapsed so that the artery may appear flattened or ovoid in shape (Fig. 5-19), and the thinner

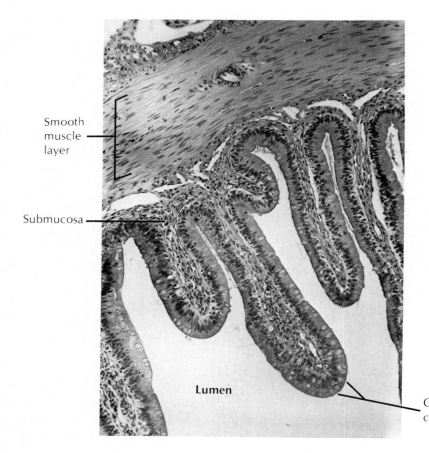

Smooth muscle layer

Submucosa

Lumen

Goblet cells in columnar epithelium

Fig. 5-18. Cross section through the small intestine of frog. (Photograph by F.M. Hickman.)

walls of the vein may be thrown into folds. There are three layers in an artery or vein. The innermost layer is an endothelium of **simple squamous epithelium.** Outside that is a layer of **smooth muscle** circularly arranged. There are often **elastic fibers** interspersed with the muscle. The outer layer is made up of elastic **fibrous connective tissue.**

This slide may contain one or several sections through nerves. Each nerve trunk is made up of many **nerve fibers,** each enclosed in its myelin sheath. Some of these fibers may be cut transversely and so appear circular; others may be cut diagonally or longitudinally and so appear ovoid or long.

There will probably be some **adipose tissue** scattered through the connective tissue that holds the vessels and nerves in place. This will appear as large clear cells, with the small nuclei pushed over to one side.

Cross Section through the Trachea. The trachea, or windpipe, is a tube that leads to the lungs. It is supported and held open by rings of **cartilage.**

The innermost layer of the trachea is a mucosal lining of ciliated **pseudostratified epithelium** containing many **goblet cells** (Fig. 5-20). Beneath this is a basement membrane of **connective tissue** containing a few fibers.

A submucosal layer contains many mucous glands composed of **cuboid epithelium.**

The cartilage bands in the trachea do not completely surround the trachea but are open dorsally.

The **cartilage** is easily recognizable by the large amount of clear ground substance interspersed with little lacunae, containing cartilage cells. In the space between the cartilage bands you may find some **smooth muscle** or **fibrous connective tissue.**

PROJECTS AND DEMONSTRATIONS

1. *Prepared slide of mesentery.* Examine the flat surface and hexagonal shape of **squamous epithelium.** A bit of the sloughed-off epidermis of a frog placed in a drop of water on a slide will also show this type of epithelium.

2. *Prepared slide of kidney.* Examine the **cuboidal epithelium** of kidney tubules. (The glandular tissue of thyroid or submaxillary gland may be used.)

3. *Prepared slide showing motor nerve endings on skeletal muscle.* Motor nerve endings are the ends of the motor nerves that stimulate the skeletal fibers to contract. Can you see the branching ends of a nerve fiber? Each terminal branch ends on some muscle fiber. Here it loses its myelin sheath and ends in a little crowfoot pattern that may look like a tiny disc on the muscle fiber. Each nerve fiber by means of these branching ends innervates many muscle fibers.

4. *Prepared slide showing decalcified bone.* Here you see bone tissue from which the minerals have been removed so that the remaining soft tissue could be cut and mounted on a slide.

5. *Long bone cut in longitudinal section.* Observe the compact bone, spongy bone, and marrow cavity.

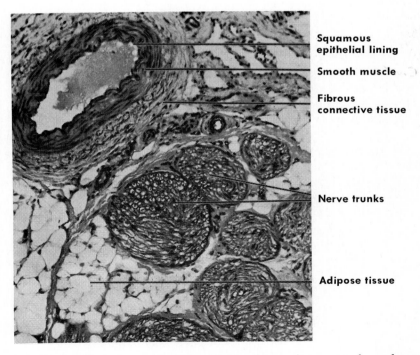

Squamous
epithelial lining

Smooth muscle

Fibrous
connective tissue

Nerve trunks

Adipose tissue

Fig. 5-19. Cross section through an artery *(upper left)* and several nerve trunks made up of many nerve fibers.

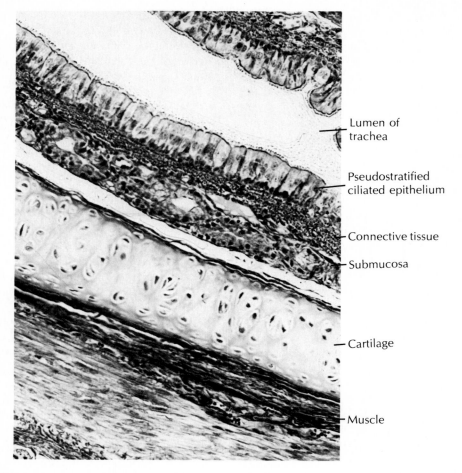

Lumen of trachea

Pseudostratified ciliated epithelium

Connective tissue

Submucosa

Cartilage

Muscle

Fig. 5-20. Cross section through portion of the trachea. (Courtesy Carolina Biological Supply Co., Burlington, N.C.)

6. *Preparation of motor end plates.* Following is a method for preparing motor end plates:

a. Remove skeletal muscle from the animal (rat, rabbit, lizard, snake, or other small vertebrate) that has been anesthetized with ether or Nembutal. Intercostal muscles are among the best. Cut muscles with sharp scissors into pieces about 5 mm long and 5 mm thick, following the long axis of the muscle fibers. Cut these pieces longitudinally into thin strips about 2 mm wide.

b. Soak the strips in freshly prepared filtered lemon juice for 10 minutes until they become clear. Now rinse in cold tap water four or five times. (Sometimes the next step is added without this washing in tap water.)

c. Place the muscle strips in 1% gold chloride in the dark for 10 to 20 minutes, depending on the kind of muscle. Intercostal muscles require about 10 min-

utes. While the strips are in gold chloride, stir frequently. The muscles should have a golden appearance.

d. Without washing, transfer the strips to a mixture of 1 part formic acid and 3 parts distilled water. Leave in the dark for 12 to 24 hours. Check to ascertain whether the tissue is too soft. Too long a time produces excessive softening.

e. Rinse quickly in tap water to remove formic acid on the surface of the muscles. Place in a mixture of equal parts 50% alcohol and pure glycerin.

f. Remove a bit of muscle, place on a drop of glycerin on a slide, and spread it out with sharp flat teasing needles. Add a clean coverslip. Lightly press down on the coverslip with a teasing needle, using a gentle lateral movement at right angles to the long axis of the muscle fibers. Check with a microscope. Seal with paraffin, gold size, or other sealant.

TISSUES

Name _____

Date _____ Section _____

Simple squamous epithelium

From _Frog skin_____

400 X

Simple cuboid epithelium

From _Human thyroid_____

Simple columnar epithelium

From _____

Pseudostratified epithelium

From _____

Stratified epithelium

From _____

Adipose
Loose ~~connective~~ tissue

From _____

Human Bone - Caniculi
Dense or ~~fibrous connective tissue~~

From _____

Blood cells

From _____Human_____

Skeletal muscle

From _____

Smooth muscle

From _____

Cardiac muscle

From _____

Moter Nerve Ending

~~Bone~~

From _____

Nerve Fibers

~~Cartilage~~

From _____

Nucrons

CHAPTER 6

INTRODUCTION TO ANIMAL CLASSIFICATION*

MATERIALS

Several species of amphibians, numbered and in trays (see pp. 78-80.)
or
Insects of several common orders (see p. 239)
or
Other animals to be classified, together with suitable simple keys
Suitable reference texts

Taxonomy involves the scientific naming of organisms and the grouping or classifying of them with reference to their exact position in the kingdoms of life. Scientific names are always Latinized and are recognized internationally. This tends to prevent confusion, for whereas one animal may be called by several different common names in different geographical areas, its scientific name is the same the world over.

Hundreds of thousands of different species of animals have been classified on the basis of morphology, size, relationships, and other factors. Most of these species tend to fall into certain large groups because of similarities in structural organization. These primary groups are known as **phyla** (singular, **phylum**). The members of a phylum share certain distinctive characteristics that set the members of that group apart from all other members of the animal kingdom. But there are differences within each phylum, and on the basis of these differences it is subdivided into smaller groups called **classes**; classes are further subdivided into **orders**; orders into **families**; families into **genera** (singular, **genus**); and genera into **species**. In large groups other categories, such as superclass, suborder, infraorder, subfamily, and so on, also exist.

A **species** is a distinctive kind of living thing. The species name consists of two parts, the genus name and the species epithet. We call this two-name system the Linnaean system of **binomial nomenclature**. The human species is *Homo sapiens; Homo* ("a man") is the genus and *sapiens* ("mighty" or "wise") is the species epithet, actually an adjective that modifies the genus name. The genus name can be used alone when one is referring to a group of species included in that genus, such as *Rana* (a large genus of frogs) or *Felis* (a genus of cats including wild and domestic species). The specific epithet, however, would be meaningless if used alone, since the same epithet may be used in combination with different genera. The domestic cat is designated *Felis domestica; domestica* used alone is without significance, since it is a commonly used epithet that identifies no particular organism. Therefore the species epithet must always be preceded by the genus name. It is permissible, however, to abbreviate the genus name when it is used in a context in which the genus is understood. *Felis domestica* might then be designated *F. domestica.*

In some instances three names may be used, in which case the last name indicates the **subspecies.** When three names are thus used, the method is called **trinomial nomenclature.** The long-tailed salamander *Eurycea longicauda longicauda,* listed in the key to common amphibians on p. 79, is classified as follows:

Phylum Chordata
  Subphylum Vertebrata
    Class Amphibia
      Order Urodela
        Family Plethodontidae
          Genus *Eurycea*
            Species *E. longicauda*
              Subspecies *E. longicauda longicauda*

Note that all except the species and subspecies names are capitalized; species and subspecies names begin with lowercase letters. Genus, species, and subspecies names are printed in italics or are underlined when written or typed.

Following is a brief exercise in classification that shows you how to use a taxonomic key to "run down" or "key out" the classification of an animal when neither its common nor its scientific name is known.

---

*See Hickman, C.P., Jr., L.S. Roberts, and F.M. Hickman, 1984. Integrated principles of zoology, ed. 7. St. Louis, The C.V. Mosby Co., pp. 147-153.

USE OF A KEY FOR ANIMAL IDENTIFICATION

This exercise will give you practical experience in working with a taxonomic key. A key is a systemic framework for biological classification. In this exercise, you will identify specimens of amphibians or insects (or other group of your instructor's choice). Below is a brief key to some common North American amphibians, and on p. 239 is a key to some common orders of insects. Your instructor may assign one of these keys or may substutite a different key based on forms common to your area. There are a number of excellent published keys available for a similar exercise on other groups.

The amphibians or other animals to be classified will be found in trays on the service table. All of the specimens in a single tray are of the same species. Each tray contains a different species. In the tray containing species A, each animal will bear a tag marked A with an individual number, for example, A1, A2, A3, and so on. Specimens of species B are marked B1, B2, B3, and so on.

Once you have made an identification, it is important to verify its accuracy by consulting one or more references containing a drawing or photograph of the species in question and a text description of its distinctive characters.

☞ Select a specimen and then, using the appropriate key and the following instructions, identify the specimen and record the species on pp. 81 and 82. Verify the identification in one or more of the reference books provided by the instructor. Identify as many of the species as requested by the instructor. Return all specimens to the proper trays on the service desk.

How to Use a Key

In the key two contrasting alternatives are offered at once, so that you can choose the one that fits your specimen. At the end of the choice you will find a reference number to the next set of alternatives to be considered. Again make a decision and proceed in the same manner until you arrive at the scientific name of the animal. Keep in mind the fact of individual variation; keys are based on the average or "typical" adult specimen, whereas your specimen may be immature or somewhat abnormal. It is often very helpful to examine more than one specimen of a species, if available, when a particular descriptive character proves troublesome. Remember too that preserved specimens tend to lose their original coloration with time.

Key to the Chief Phyla and Classes of the Animal Kingdom

Appendix A in the back of this book (p. 450) is a simple key to the more common phyla and classes of the animal kingdom. Since this key requires the previous knowledge of certain features, such as whether a body cavity is a true coelom or a pseudocoel, whether a symmetry is primary or secondary, and so forth, it is not suitable for the beginning student to use to identify unknown animals. It is, however, an excellent review device and should be kept in mind as such for use later in the course.

Taxonomic Chart

Folded inside the back cover of this manual you will find a useful taxonomic chart, Appendix C, which has been compiled for your benefit. Examine this chart. It will not only prove useful as a reference and as a review sheet, but it will also help you in forming a definite concept of the total organization of the major groups of animals and of the organ systems of animals. The chart can be attached to the inside back cover of the book where it can be unfolded for study, or you may prefer to attach it to your bulletin board where it will be ready to use. In any case use it frequently throughout the year.

TAXONOMIC KEY TO CERTAIN AMPHIBIANS
Glossary of terms used in key

Costal grooves Vertical grooves on the sides of salamanders lying between folds of skin.

Nasolabial groove Fine groove from nostril to upper lip (may require hand lens to see).

Plantar tubercles Wartlike projections on the soles of the feet.

Parotoid glands Glandular swollen regions on each side of the neck in toads.

Dorsolateral folds Prominent ridges extending from the head along each side of the back.

Cranial crests Bony ridges extending backward between the eyes.

Terminal discs Adhesive pads on the ends of the digits.

1. Amphibians with tails . . . . ***Order Uro-dela (Caudata)*** (salamanders)   2
   Amphibians without tails (in eastern United States) . . . . ***Order Salientia (Anura)*** (frogs and toads)   14
2. Adults with external gills . . . . ***Necturus maculosus*** (mudpuppy)
   Adults without external gills   3
3. Body eel-like; legs tiny and weak . . . . ***Amphiuma means*** (amphiuma)
   Body not eel-like; legs used in locomotion   4
4. Costal grooves very indistinct; hind legs flattened; red spots on sides . . . ***Diemictylus viridescens*** (newt)
   Costal grooves distinct; no red spots on sides   5
5. Nasolabial groove present; costal grooves numbering 14 or more   6
   Nasolabial groove absent; costal grooves numbering 10-12   12
6. Pale diagonal line present from eye to angle of jaw; hind legs larger and stouter than forelegs . . . . ***Desmognathus fuscus*** (dusky salamander)
   No diagonal line; all four limbs about same size   7
7. Tongue attached at the anterior margin   8
   Tongue free at anterior margin (attached in the middle)   9
8. Body with broad, dorsal band of reddish or brownish color; costal grooves 17-20 . . . . ***Plethodon cinereus*** (red-backed salamander)
   Body black or blue-black above, light slate color below; body peppered with silver-white spots . . . . ***Plethodon glutinosus*** (slimy salamander)
9. Pigment spots in more or less well-defined rows or stripes   10
   Pigment spots if present are scattered over the body   11
10. Bright orange or yellow with black spots, tail with vertical black herringbone pattern; tail usually at least 60% of body length . . . . ***Eurycea longicauda longicauda*** (long-tailed salamander)
    Body gray with two light dorsolateral lines edged in black . . . . ***Eurycea bislineata*** (two-lined salamander)
11. Ridge marked with light line running from eye to nostril, often bordered below by dark pigment . . . .

Gyrinophilus porphyriticus (spring salamander)
   No ridge or light line from eye to nostril; body red or reddish orange dotted on upper surface with irregular black spots . . . . ***Pseudotriton ruber*** (red salamander)
12. Dorsal light color forming whitish cross-bars on back or sides, often joined in ladderlike appearance . . . . ***Ambystoma opacum*** (marbled salamander)
    Dorsal light color distributed as spots or blotches   13
13. With two plantar tubercles on sole of the hind foot; dorsal yellow spots usually in dorsolateral series and along lower part of each side . . . . ***Ambystoma tigrinum*** (tiger salamander)
    With one or no plantar tubercles on sole of hind foot; dorsal yellow spots in dorsolateral series only . . . . ***Ambystoma maculatum*** (spotted salamander)
14. With parotoid glands and cranial crests; skin very warty . . . . ***Family Bufonidae*** (toads)   15
    Without parotoid glands and cranial crests; skin smooth or not very warty   16
15. Dark blotches, containing three or more warts, on dorsal surface; cranial crests usually parallel . . . . ***Bufo woodhousei fowleri*** (Fowler's toad)
    Dark blotches containing one or two warts; cranial crests usually divergent behind . . . . ***Bufo americanus*** (American toad)
16. Digits with terminal circular adhesive pads or discs; belly granulated or pebbled . . . . ***Family Hylidae*** (tree-frogs)   17
    Digits without terminal pads or discs . . . . ***Family Ranidae*** (true frogs)   20
17. Finger pads small, no wider than fingers   18
    Finger pads distinctly wider than fingers   19
18. Skin rough; hind toes mostly webbed; triangular dark mark between the eyes . . . ***Acris gryllus*** (cricket frog)
    Skin smooth; toes less than half webbed; three dark longitudinal bands on back . . . . ***Pseudacris nigrita*** (chorus frog)

19. Skin pale brown; a pale **X** on the back; small terminal discs on digits . . . . . . ***Hyla crucifer*** (spring peeper)

Skin green, brownish, or grayish; back may bear blotches but no **X**; terminal discs on digits large and conspicuous . . . . ***Hyla versicolor*** (gray treefrog)

20. Dorsolateral folds lacking or indistinct; large body size (12-18 cm); spots obscure or absent . . . . ***Rana catesbeiana*** (bullfrog)

Dorsolateral folds present; adult size less than 12 cm 21

21. Back with longitudinal folds between dorsolateral folds; black spots on back of elbow 22

Skin of back relatively smooth between dorsolateral folds; no spot on back of elbow 23

22. Spots rounded on trunk and legs; undersurfaces of legs white, or tinged with yellow in life . . . . ***Rana pipiens*** (leopard frog)

Spots square; undersurface of legs bright yellow in life . . . . ***Rana palustris*** (pickerel frog)

23. Body color brown; dark ear patch behind eye . . . . ***Rana sylvatica*** (wood frog)

Body color green, especially around jaws; no black ear patch . . . . ***Rana clamitans*** (green frog)

Use of a Key for Animal Identifications

Print the number of the specimen and the species (genus name and species epithet)—or order, in the case of insects—in the boxes. Enter the complete reference consulted to verify the identification.

Name _____

Date _____ Section _____

A _____ _____
Reference:

B _____ _____
Reference:

C _____ _____
Reference:

D _____ _____
Reference:

E _____ _____
Reference:

F _____ _____
Reference:

G _____ _____
Reference:

H _____ _____
Reference:

I _____ _____
Reference:

J _____ _____
Reference:

K _____ _____
Reference:

L _____ _____
Reference:

82

PART TWO

THE DIVERSITY OF
ANIMAL LIFE

CHAPTER 7

THE PROTOZOA*

PHYLA SARCOMASTIGOPHORA, APICOMPLEXA, AND CILIOPHORA

In the protozoa all the functions of life are performed within the limits of a single plasma membrane. There are no organs or tissues, but there is division of labor within the cytoplasm, where various complex **organelles** are specialized to carry out specific tasks. Each organelle is a small aggregate of macromolecules arranged in a definite manner that fits it for its specific function. Specialized organelles serve as skeletons, organs of locomotion, sensory systems, conduction mechanisms, defense mechanisms, contractile systems, and so on.

Among the different phyla there are various levels of organization, depending on the complexity of their structural plan. Since protozoa are not divided into cells, they are said to belong to the **"protoplasmic level of organization."**

Protozoa are often erroneously referred to as "simple" organisms. There are no simple organisms. Protozoa carry on all the functions of higher animals and are amazingly efficient in the performance of these functions. Each is extremely well adapted to its own environment. They are widespread ecologically, being found in fresh, marine, and brackish water and in moist soils. Some are free living, others live as parasites or in some other symbiotic relationship.

CLASSIFICATION

Traditionally four main groups of protozoa have been recognized: the flagellates, the amebas, the sporeformers, and the ciliates. More recent studies have shown that the classification based on these groups was unnatural, and the system that follows places these organisms in a more phylogenetic arrangement.

*See Hickman, C.P., Jr., L.S. Roberts, and F.M. Hickman. 1984. Integrated principles of zoology, ed. 7. St. Louis, The C.V. Mosby Co., pp. 167-200.

It is based on the 1980 revision adopted by the Society of Protozoologists, recognizing seven separate phyla of unicellular animals. We have listed only the five major phyla.

Phylum Sarcomastigophora (sar'ko-mas-ti-gof'o-ra). Locomotor organelles of flagella, pseudopodia, or both; usually only one type of nucleus and no spore formation.

  Subphylum Mastigophora (mas-ti-gof'o-ra). Having flagella for locomotion; autotrophic, heterotropic, or both; reproduction usually asexual by fission.

    Class Phytomastigophorea (fi'to-mas-ti-go-for'e-a). Plantlike flagellates with chromoplasts that contain chlorophyll. Examples: *Chilomonas, Euglena, Volvox, Ceratium, Peranema*.

    Class Zoomastigophorea (zo'o-mas-ti-go-for'e-a). Flagellates without chromoplasts; ameboid forms with or without flagella in some groups; predominately symbiotic. Examples: *Trypanosoma, Trichonympha, Leishmania*.

  Subphylum Opalinata (o'pa-lin-a'ta). Body with longitudinal, oblique rows of ciliumlike organelles; parasitic; cytostome (cell mouth) lacking; two to many nuclei of one type. Example: *Opalina*.

  Subphylum Sarcodina (sar-ko-di'na). Pseudopodia for feeding and locomotion; body naked or with external or internal skeletons; free living or parasitic.

    Superclass Actinopoda (ak'ti-nop'o-da). Pseudopodia with axial filaments; body often spherical; usually planktonic. Examples: *Actinosphaerium, Actinophrys, Thalassicolla*.

    Superclass Rhizopoda (ri-zop'o-da). Locomotion by lobopodia, filopodia, or reticulopodia or by cytoplasmic flow without discrete pseudopodia. Examples: *Amoeba, Arcella, Difflugia*, foraminiferans.

Phylum Apicomplexa (a'pi-com-plex'a). Characteristic set of organelles (apical complex) at anterior end in some stages; cilia and flagella usually absent; all species parasitic.

  Class Sporozoea (spor-o-zo'e-a). Spores or oocysts typically present, which contain infective sporozoites; locomotion of mature organisms by body flexion, gliding, or undulation of longitudinal ridges; flagella present only in microgametes of some species; pseudopodia usually absent; one- or two-host life cycles. Examples: gregarines, coccidians, malaria parasites.

    Class Perkinsea (per-kin'se-a). Small group parasitic in oysters. Example: *Perkinsus*.

Phylum Myxozoa (mix-o-zo'a). Spores enclosed by two or three (rarely one) valves; parasites of invertebrates and lower vertebrates, especially fishes. Example: *Myxosoma*.

Phylum Microspora (mi-cros'por-a). Parasites of invertebrates, especially arthropods, and lower vertebrates. Example: *Nosema*.

Phylum Ciliophora (sil-i-of'o-ra). Cilia or ciliary organelles present in at least one stage of life cycle; two types of nuclei usually; binary fission across rows of cilia; budding and multiple fission also occur; sexuality involving conjugation, autogamy, and cytogamy; heterotrophic nutrition; mostly free living; contractile vacuole typically present. (This is a very large group, now divided into three classes and numerous orders.) Examples: *Paramecium, Colpoda, Tetrahymena, Stentor, Blepharisma, Epidinium, Vorticella*.

EXERCISE 7A
The Sarcodina—*Amoeba* and others

MATERIALS
Amoeba culture
Other sarcodine cultures as available (see pp. 89-90)
Ringed slides (or plain)
Coverslips
Petrolatum
Thread
10% nigrosine
Radiolarian shells, preserved or on permanent slides
Foraminiferan shells, preserved or on permanent slides
Prepared slides
    Amoeba
    Entamoeba
    Other sarcodines as available
Microscopes

AMOEBA
Phylum Sarcomastigophora
    Subphylum Sarcodina
        Superclass Rhizopoda
            Class Lobosa
                Species *Amoeba proteus*

Where Found
The amebas* may be naked or enclosed in a shell. The naked amebas, which include the genera *Amoeba* and *Pelomyxa*, live in freshwater and seawater and in the soil. They are bottom dwellers and must have a substratum on which to glide. *A. proteus* is a freshwater species that is usually found in slow-moving or still-water ponds. They are often found on the underside of lily pads and other water plants. They live on algae, bacteria, protozoa, rotifers, and other microscopic organisms.

---

*The term *ameba* is used by many authors as a common name for any of the naked or shelled sarcodines. When referring to a specific genus the name is latinized and italicized, such as *Amoeba, Pelomyxa, Entamoeba*, and so forth.

Study of Live Specimens
☞ In the center of a clean ringed slide or slide ringed with petrolatum, the instructor will place a drop of culture drawn from the bottom of the culture bottle. Adjust the iris diaphragm of the microscope to provide *subdued light*. With *low power* explore the contents of the slide, at first *without a coverslip*.

You may see some masses of brownish or greenish plant matter, and there will probably be some small ciliates or flagellates moving about. The ameba, in contrast, is gray, rather transparent, irregularly shaped, and finely granular in appearance (Fig. 7-1). If it is not apparently moving, watch it a moment to see if the granules are in motion or if the shape is slowly changing. If you do not find a specimen after several minutes of careful examination, ask the instructor to check your slide.

☞ After you have found some specimens and observed their locomotion, carefully cover with a coverslip. If you are using a plain, unringed slide, you will need to support the coverslip to prevent crushing the amebas. You can do this by placing two short lengths of hair or thread or two pieces of a broken coverslip, one above and one below the drop of culture, before adding the coverslip. *Never use high power without using a coverslip!* From time to time you may need to add a drop of culture water to the slide at the edge of the coverslip to prevent the culture from drying out.

The heat from some types of substage microscope lamps may be injurious to small animals. If you are using such a lamp, turn it off occasionally to prevent overheating.

General Features
The outer cell membrane is the **plasmalemma**. Electron micrographs show the plasmalemma to be fringed

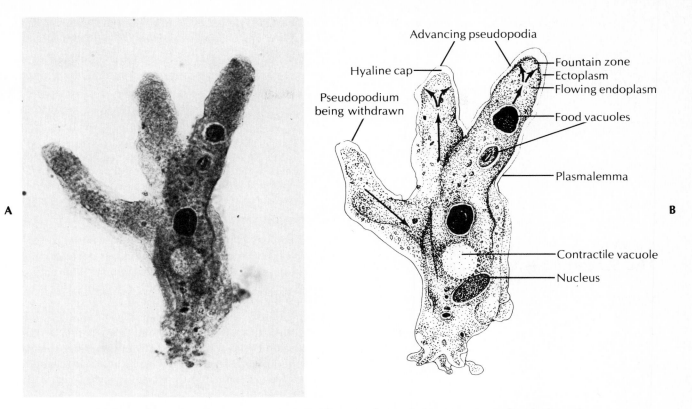

Fig. 7-1. Ameba. **A,** Whole mount. **B,** Interpretive drawing. (**A** courtesy Carolina Biological Supply Co., Burlington, N.C.)

with fine, hairlike projections that are thought to be involved in the adhesion of the cell surface to the substratum or to nutrient particles and to aid in the capture and intake of food (Fig. 7-2).

The enclosed cytoplasm is differentiated into a thin, clear peripheral layer of stiff **ectoplasm** and an inner portion that is fluid and finely granular in appearance, which is called the **endoplasm** (Fig. 7-1, *B*).

Locomotion. The ameba moves and changes shape by thrusting out **pseudopodia,** which are extensions of the cell body. In *A. proteus* and related amebas the large blunt pseudopodia are of the type called **lobopodia.** As the new lobopodium begins to form, a thickened extension of the ectoplasm called the **hyaline cap** appears, and the fluid endoplasm flows toward it. As the endoplasm flows into the hyaline cap, it fountains out to the periphery and is converted to ectoplasm, thus building up and extending the sides of the pseudopodium like a tube or sleeve. At some point the tube anchors itself to the substrate by means of its plasmalemma, and the animal moves forward. As the tube lengthens, the ectoplasm at the temporary "tail end" converts again to streaming endoplasm to replenish the forward flow. At any time the action can be reversed, the endoplasm streaming back, the tube shortening, and another pseudopodium forming elsewhere.

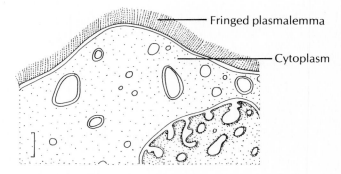

Fig. 7-2. Fringed plasmalemma of an ameba as it appears under electron microscopy.

A number of theories have been advanced to explain the force that causes the endoplasmic flow. Recent evidence seems to support the view that ameboid movement, like muscular movement, involves tiny filaments sliding past one another. Myosin filaments have been identified in *A. proteus*.

Observe the formation of pseudopodia. Does the animal have a permanent anterior and posterior end? Can you observe the change from endoplasm to ectoplasm, and vice versa? Does the animal move steadily in one direction? Do more than one pseudopodium ever start at once? How is a pseudopodium withdrawn? What happens when the animal meets an obstruction?

To observe the ameba's contact with the substrate, attach by means of petrolatum a coverslip on each side of a microscope slide, with edges extended as in Fig. 7-3. Place a drop of culture on the edge of the slide between the coverslips. Tilt the microscope 90 degrees and place the slide on the now vertical stage. Observe the ameba as it crawls on the edge of the slide.

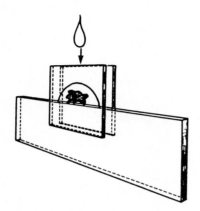

Fig. 7-3. Preparation for studying a side view of ameba.

Does the whole lower surface of the ameba contact the slide? Do pseudopodia ever extend vertically as well as laterally?

DRAWINGS

On p. 91 make a series of five or six outline sketches, each about 2 cm in diameter, to show the changes in shape that occur in an ameba in a period of 1 to 10 minutes. Sketch rapidly and use arrows freely to indicate the direction of flow of the cytoplasm. Indicate the magnification of the drawings.

Feeding. Note the **food vacuoles**, which are particles of food surrounded by water and enclosed in a membrane. Cells and unicellular animals that engulf foreign particles are called **phagocytes**, and this type of ingestion is known as **phagocytosis** (Fig. 7-4, A). Phagocytosis involves the encircling of the prey by pseudopodia to form a food cup. Subsequently the prey, along with a quantity of water, becomes completely enclosed by cytoplasm to form a food vacuole. Digestive enzymes diffuse into these vacuoles and digest the food.

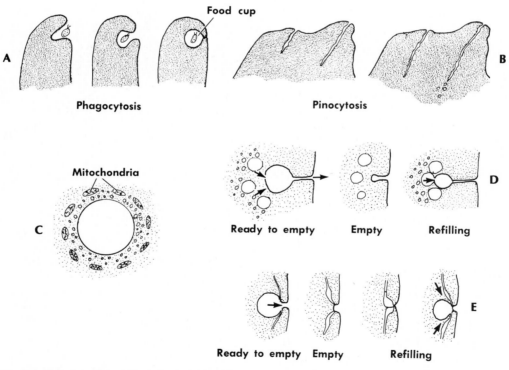

Fig. 7-4. Some protozoan functions. **A,** Successive stages in the engulfment of a ciliate by a pseudopodium. **B,** Stages in pinocytosis, or "cell drinking," by formation of channels and breaking off of fluid droplets in the cytoplasm. **C** to **E,** Osmoregulation by contractile vacuoles. **C,** Contractile vacuole of *Amoeba* surrounded by mitochondria and minute feeder vesicles. **D** and **E,** Stages in formation of ciliate vesicles by **D,** merging of smaller vesicles, and **E,** system of feeder canals; both types empty through a permanent pore.

Can you identify the contents of any of the food vacuoles? Watch the pseudopodia for signs of a food cup forming.

Undigested end products can be eliminated at any point through the plasmalemma.

Cells also take in fluid droplets and minute food particles by a process of channel formation called **pinocytosis** (Fig. 7-4, *B*). A demonstration method is given on p. 90.

Osmoregulation. Look for the **contractile vacuole** (also called the **water-expulsion vesicle**), a clear bubble containing no particles. Note that this bubble gradually increases in size by accumulation of fluid and then ruptures and disappears. This organelle rids the ameba of excess water that has been taken in along with food vacuoles or acquired by osmosis. The contractile vacuole is surrounded by smaller vesicles with a ring of mitochondria forming a circle around them all (Fig. 7-4, *C*). Apparently the larger vacuole fills from the smaller ones and then is pushed against the plasmalemma, where it ruptures and empties to the outside. Because it does not actually contract, the term "water expulsion vesicle" may be more suitable.

Time the appearance and disappearance of the contractile vacuole. Note that in a moving ameba the vesicle, as it enlarges, tends to be located in the temporary posterior end of the moving animal.

Nucleus. Locate the **nucleus.** It is disc shaped, often indented, finely granulated, and somewhat refractive to light. You can distinguish it from the contractile vacuole because the latter is perfectly spherical, increases in size, and finally disappears. The nucleus is usually carried along in the cytoplasm and is often found near the center of the animal. As it turns over and over, it sometimes appears oval and sometimes round.

Reproduction. It is possible, although not too likely, that you might find a specimen in division. If you do, call it to the attention of your instructor. The amebas reproduce asexually by a type of mitotic cell division, known as **binary fission.** The life cycles of some sarcodines are much more complex than that of *Amoeba*.

Study of a Stained Slide

☞ Check on a stained slide the structural features that you have studied in the living specimen. Check especially the **nucleus,** the **plasmalemma,** the **ectoplasm,** the **endoplasm,** the **contractile vacuole,** and the **food vacuoles.**

WRITTEN REPORT

Answer the questions on p. 92.

OTHER SARCODINES

 Examine stained slides of as many of the following Sarcodines as are available in your laboratory.

Parasitic Amebas

A number of species of *Entamoeba* (Fig. 7-5) are found in humans and other vertebrates. *E. gingivalis* lives in the mouth and feeds on bacteria around the base of the teeth. *E. histolytica* lives chiefly in the large intestine, where it causes dysentery by feeding on body tissues and red blood cells. These forms exist in two phases, the trophozoite, or active feeding phase, and the encysted stage in which nuclear divisions oc-

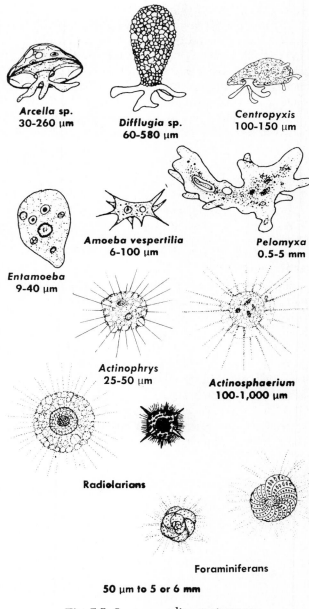

Arcella sp.
30-260 μm

Difflugia sp.
60-580 μm

Centropyxis
100-150 μm

Entamoeba
9-40 μm

Amoeba vespertilia
6-100 μm

Pelomyxa
0.5-5 mm

Actinophrys
25-50 μm

Actinosphaerium
100-1,000 μm

Radiolarians

Foraminiferans

50 μm to 5 or 6 mm

Fig. 7-5. Some sarcodine protozoans.

cur. The mature cyst usually contains four small nuclei. After ingestion by a suitable host the multinucleated ameba emerges from the cyst and undergoes a series of fissions resulting in uninucleate daughter amebas.

Endamoeba is a similar genus found in cockroaches and termites.

Some Shelled Amebas

Protozoa show great originality in the choice of building materials for their tests. *Arcella,* which occurs in bogs or swamps where there is much vegetation, has a hemispherical hat-shaped test (up to 260 μm in diameter) made up of siliceous or chitinous material set in a base of polymerized proteins. Small fingerlike pseudopodia extend through an opening in the flat underside of the test (Fig. 7-5).

Difflugia is often found in leaf-choked puddles or in delicate aquatic vegetation. This protozoan has an inverted flask-shaped test (up to 500 μm long) made up of sand grains cemented together with the polymerized protein base, sometimes with the addition of diatom shells or sponge spicules.

Foraminiferans are marine sarcodines that secrete a test of one or more chambers, usually calcareous, and sometimes incorporating silica, sand, or sponge spicules. Long, delicate, feeding pseudopodia extend through pores in the test. As the foraminiferan grows, it adds new chambers to the shell. On death their tests are added to the ooze of the ocean bottom. Many limestone deposits were formed by the tests of these animals.

Actinopod Amebas

Actinosphaerium and the smaller *Actinophrys* are more or less spherical, and their cytoplasm has a vacuolated, or frothy, appearance (Fig. 7-5). They do not have tests. Long, slender, feeding pseudopodia radiate in all directions. The animals are carried by currents, or roll slowly by means of contraction and expansion of the stiff pseudopodia, whose tips may become viscid for anchoring the organism. They feed on other protozoa and certain algae that, on contact, become attached to the pseudopodia and then paralyzed. These forms occur in ponds, usually floating among aquatic plants, and may be collected with a plankton net or on glass slides left submerged for 2 or 3 weeks. They may also be obtained from biological supply houses.

Radiolarians are marine sarcodines that secrete about themselves a transparent shell, or test, of silica, and then thrust out slender pseudopodia through pores in the shell. They float in surface plankton, and when they die, their tests become a part of the ocean bottom ooze. Examine a prepared slide showing a variety of shell types.

> **DRAWINGS**
>
> Sketch any of the above sarcodines or their tests on p. 91.

DEMONSTRATIONS

1. Demonstration of pinocytosis. Pinocytosis or "cell drinking" may be demonstrated in *Amoeba* by placing the amebas in a solution of 0.125M NaCl in 0.01M phosphate buffer at a pH of 6.5 to 7.0. Channels should begin to form in 2 or 3 minutes and continue for several minutes.*

2. Demonstrating the contractile vacuole (water-expulsion vesicle). Add a drop of 10% nigrosine to a drop of *Amoeba* culture on a slide, add a coverslip, and examine with high power.

3. Demonstration of osmoregulation. How would the osmoregulatory requirements of marine protozoans compare with those of freshwater forms? This might be demonstrated by placing a drop of rich *Amoeba* culture in each of four deepwell slides. Fill three of the wells with culture water that has been made up to 2%, 4%, and 6% saline (by adding 2, 4, or 6 g of salt per liter of water), and the fourth with plain culture water. After a few minutes place a drop from the bottom of each well on a slide and examine. Time the rate of discharge of the contractile vacuoles in specimens from each osmotic concentration. If there is a difference, can you explain it?

---

*Chapman-Andresen, C. 1964. Measurement of material uptake by cells: pinocytosis. In D.M. Prescott, (ed.). Methods in cell physiology. New York, Academic Press, Inc.

Phylum _____   Name _____

Class _____   Date _____ Section _____

Genus _____

LOCATION SKETCHES OF AMOEBA

OTHER AMEBAS

THE AMEBA

1. What is the average size of the specimens on your slide? _____

2. How can you distinguish between an ameba and a bit of stained debris? _____

    _____

3. Does the ameba react to stimuli? _____ What reactions have you observed? _____

    _____

4. What is the main function of the contractile vacuole? _____

5. How many times per minute did the vacuole empty? _____

6. Does the ameba have a respiratory organelle? _____ How does it breathe? _____

    _____

7. How does the ameba reproduce? _____

    _____

8. If you observed ingestion or egestion occurring, or if you could identify the contents of any of the food vacuoles, describe what you saw. _____

    _____

    _____

    _____

    _____

9. How would you describe and explain ameboid movement? _____

    _____

    _____

    _____

    _____

Subphylum Mastigophora—*Euglena* and *Volvox*

The mastigophorans, or flagellates, are borderline forms that are thought to include the most primitive of all the protozoan forms from the standpoint of phylogeny. Plants are characterized by the capacity to make their own food, and animals are characterized by their reliance on other organisms for food. Many flagellates combine both of these qualities to some extent and hence must be considered as a primitive type of plant-animal.

MATERIALS

Euglena culture
Volvox culture
Stained slides of *Euglena* and *Volvox*
Other flagellate cultures or slides as available (see pp. 96-97)
Methylcellulose (p. 462), polyvinyl alcohol, or Protoslo
Methyl violet or Noland's stain (see p. 462)
Microscopes
Slides and coverslips

EUGLENA

Phylum Sarcomastigophora
  Subphylum Mastigophora
    Class Phytomastigophorea
      Order Euglenida
        Genus *Euglena*

Where Found

Euglenoids are most common in still pools and ponds where they often give a greenish color to the water. Ornamental lily ponds are excellent sources. *E. viridis* and *E. gracilis* are commonly studied species.

Study of Live Specimens

☞ Place a drop of 10% methylcellulose or Protoslo on a clean slide, spread it thin, and add a drop of rich euglena culture and a cover glass. Study with high power.

Members of *E. gracilis* range from 40 to 50 μm in length, are spindle shaped, and are greenish. The color is caused by the presence of **chloroplasts,** which contain **chlorophyll** and are scattered through the cytoplasm (Fig. 7-6). A nucleus is located centrally, but is difficult to see in living specimens.

Locomotion. The blunt anterior end bears a little whiplike **flagellum** that you may be able to see with reduced light when a specimen has slowed down considerably.* The flagellum emerges from the **reservoir,**

---

*Dilute methyl violet or Noland's stain may help in making the flagella easier to see.

a clear flask-shaped space in the anterior end. In some flagellates, such as *Peranema,* the flagellum extends forward, but in *Euglena* it is directed backward along the side of the body. Movement involves the generation of waves originating at the base of the flagellum and transmitted along its length to the tip. The flagellum beats at the rate of about 12 beats per second and not only moves the organism forward but also rotates it and pushes it to one side, causing it to follow a corkscrew path, rotating about once every second as it goes. Electron micrographs show a second very short flagellum within the reservoir, which is not seen by the light microscope.

Euglenoid Movement. Watch how the animal changes its shape. These peculiar peristaltic contractions are referred to as "euglenoid movements," and authorities believe that the movements are made possible by microtubules, tiny hollow fibrils about 20 μm in diameter, lying just beneath the pellicle.

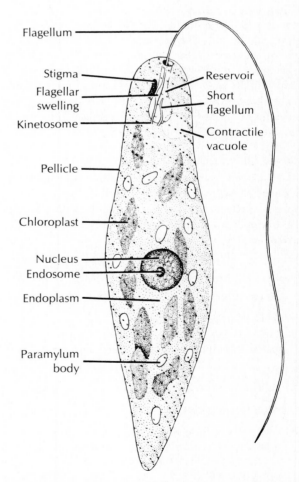

Flagellum
Stigma
Flagellar swelling
Kinetosome
Pellicle
Chloroplast
Nucleus
Endosome
Endoplasm
Paramylum body
Reservoir
Short flagellum
Contractile vacuole

Fig. 7-6. *Euglena.* Features shown are a combination of those visible in living and stained preparations.

Body Covering. The body is covered with a protective but flexible **pellicle** secreted by the clear **ectoplasm** that surrounds the **endoplasm.** The **stigma,** or eyespot, is a reddish pigment spot that shades a swollen basal area of the flagellum, which is thought to be light sensitive. Why is light sensitivity important to a euglena?

Osmoregulation. A large contractile vacuole empties into the reservoir. It is fed by smaller vesicles around it.

Feeding. The euglena takes in no solid food; the opening of the reservoir is merely for elimination of waste and excess water discharged by the contractile vacuole. Nutrition is **holophytic,** making use of photosynthesis—the manufacture of food from carbon dioxide and water with the aid of chlorophyll and sunlight. The chlorophyll is carried in oval **chloroplasts** in the cytoplasm. **Paramylum bodies** are scattered reddish particles in which starch reserves are stored.

Not all phytoflagellates are holophytic. Some are also saprozoic, some are phagotrophic, and some use a combination of methods.

Reproduction. The euglena reproduces by **longitudinal fission,** which may occur when the animal is free or when it is in the encysted state. The fission begins at the anterior end and proceeds posteriorly. During encystment the euglena rounds up and forms a gelatinous wall around itself. In this state it can withstand drought and other harsh conditions. A number of longitudinal divisions may occur during the encystment period, and many separate individuals may emerge later from a single cyst.

Stained Slide of Euglena

Select a good specimen on the slide and check the following structural features: **flagellum, stigma, chloroplasts, pellicle, reservoir,** and **contractile vacuole.** The **nucleus** is oval and near the center of the body. **Paramylum bodies** appear as lightly stained rods or ovals scattered through the endoplasm.

DRAWINGS

On p. 99 make a series of outline drawings, about 2 to 3 cm long, showing the various shapes your specimen assumes in its "euglenoid movement."

WRITTEN REPORT

On p. 99 describe and explain the manner in which your specimens move about (locomotion) and alter their shape (euglenoid movement). Could you see any of the flagella? Record any other interesting observations.

VOLVOX

Phylum Sarcomastigophora
  Subphylum Mastigophora
    Class Phytomastigophorea
      Order Volvocida
        Family Volvodicae
          Genus *Volvox*

Where Found

Volvox is often found with other protozoans in stagnant pools and ponds and slow-moving streams. It is also found occasionally in the blanket algae that cover ponds, especially in early spring and summer. Usually it is not common except in restricted areas.

General Features

Colonies of *Volvox* are large, green protozoan colonies that may reach a diameter of 2 to 3 mm (Fig. 7-7).

Colonies of *Volvox* are particularly interesting because they represent a transition between the protozoa and the many-celled metazoans. They are the most complex of the colonial flagellates. In *Volvox* we see the beginning of **cell differentiation,** resulting in a division of labor among cells—an important step toward the metazoans. This group also illustrates an important stage in the **development of sex.**

☞ Study living or preserved colonies and stained slides. For mounting the colonies, used ringed or depression slides, place the culture in the concavity, and cover with a cover glass. Study with *low power* and focus up and down to get all the details. Be especially careful if you use high power, for you may injure both the slides and the objective.

The colony is composed of a variable number of 1-celled individuals called **zooids.** There may be a few hundred or many thousands of zooids in a colony, arranged on the surface of a gelatinous ball and connected to each other by fine **protoplasmic strands.** The zooids are differentiated into **somatic cells** and a smaller number of **reproductive cells.** The somatic cells are quite similar to other flagellate animals and make up most of the colony. They handle the nutrition, locomotion, and response to stimuli for the entire colony. Each somatic cell contains **chloroplasts** (which give the colony its green color), a **stigma** for light sensitivity, and a pair of **flagella** for locomotion.

Locomotion. If you have living colonies, observe the locomotion. Do you notice that one end usually goes foremost? This is the anterior pole, and the anterior zooids are so highly specialized that they are unable to reproduce. In colonies of *Pleodorina* the anterior somatic cells are smaller than the rest and have larger stigmata, implying, perhaps, a greater sensitivity to light. Some of the colonies on your slide will contain smaller **daughter colonies** revolving about

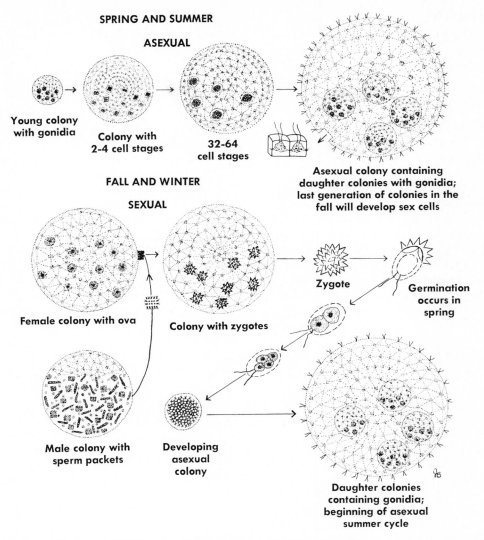

SPRING AND SUMMER

ASEXUAL

Young colony
with gonidia

Colony with
2-4 cell stages

32-64
cell stages

Asexual colony containing
daughter colonies with gonidia;
last generation of colonies in the
fall will develop sex cells

FALL AND WINTER

SEXUAL

Female colony with ova

Colony with zygotes

Zygote

Germination
occurs in
spring

Male colony with
sperm packets

Developing
asexual
colony

Daughter colonies
containing gonidia;
beginning of asexual
summer cycle

Fig. 7-7. *Volvox*, a colonial flagellate. *Above*, Asexual development of new daughter colonies from gonidia. *Below*, Development of colonies from fertilized eggs. Note the seasonal relationship of these types of reproduction. Some species are monoecious, and eggs and sperm are produced in the same colony, instead of in separate colonies as shown.

inside the gelatinous center of the mother colony (Fig. 7-7).

Asexual Reproduction. Certain of the cells are specialized for reproduction. *Volvox* reproduces asexually during the spring and summer months, producing four or five asexual generations. The reproductive cells, called **gonidia**, are enlarged cells located usually in the lower half of the colony. These cells divide, each going through a 2-cell, 4-cell, 8-cell stage and so on, finally producing hollow, spherical daughter colonies that remain protected within the mother colony for some time. **Gonidial colonies** are usually larger than sexual colonies and contain four to eight daughter colonies, which on the death of the mother colony escape and swim away.

Sexual Reproduction. In the fall the last generation of asexual colonies develops **sex cells.** These may be potentially male or female. Some species have both types of cells in a colony; other species have either male or female cells (Fig. 7-7).

Female sex cells enlarge without division to form **macrogametes,** or **ova.** These are larger and more numerous than asexual gonidia. They do not all mature at the same time. Male sex cells, by dividing repeatedly, form **packets** of tiny spindle-shaped biflagellated **microgametes,** or **sperm cells.** The sperm escape when mature and swim about. An egg fertilized by one of the sperm becomes a **zygote** and secretes a spiny cyst wall about itself. During the winter the somatic cells of the mother colony die, but the zygotes survive and in the spring give rise by cell division to new asexual colonies (Fig. 7-7). Compare the development of *Volvox* to the early development of sea stars (Chapter 4). To what embryologic stage might the colony be compared?

DRAWINGS

Sketch one or more of the colonies on p. 100.

PROJECTS AND DEMONSTRATIONS

1. Demonstration of living Trichonympha, a symbiont of the termite. The flagellate *Trichonympha* belongs to an order of zoomastigophorans that do not have chloroplasts or paramylum but have many flagella and live in the digestive tracts of roaches and termites. *T. campanula* (Fig. 7-8), which lives in the gut of American termites, is a good example of **symbiotic mutualism.** Neither the termite nor the flagellate can live without the other. The flagellates secrete an enzyme (cellulase) that digests the wood (cellulose) the termites eat. Termites cannot themselves digest cellulose.

Squeezing the posterior abdomen of the termite to extrude a drop of fluid from the anus may provide enough specimens on a slide for study. Or hold the termite with forceps, and with another pair of forceps pull off the posterior segment. Pull the intestine out of the body and tease it apart on a slide. Add a drop of 0.6% NaCl solution and cover.

2. Demonstration of a blood trypanosome. Members of the genus *Trypanosoma* are parasitic zoomastigophorans that live in the blood and tissue fluids of all classes of vertebrates (Fig. 7-9, *A*). They are usually transmitted by blood-feeding invertebrates. *Trypa-*

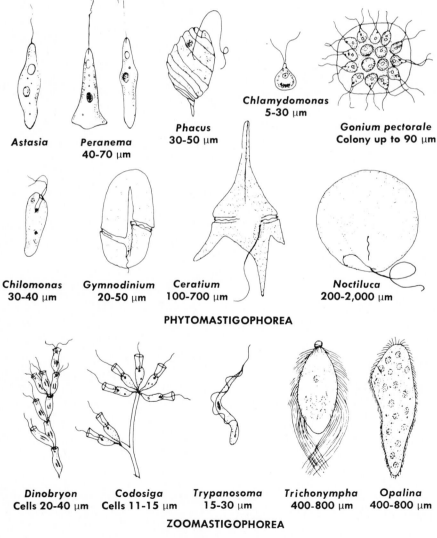

Astasia   Peranema
40-70 µm

Phacus
30-50 µm

Chlamydomonas
5-30 µm

Gonium pectorale
Colony up to 90 µm

Chilomonas
30-40 µm

Gymnodinium
20-50 µm

Ceratium
100-700 µm

Noctiluca
200-2,000 µm

PHYTOMASTIGOPHOREA

Dinobryon
Cells 20-40 µm

Codosiga
Cells 11-15 µm

Trypanosoma
15-30 µm

Trichonympha
400-800 µm

Opalina
400-800 µm

ZOOMASTIGOPHOREA

Fig. 7-8. Some flagellate protozoans.

nosoma brucei is a species that parasitizes native African antelopes and other ruminants. It is also very pathogenic in domestic livestock. Two subspecies, *T. b. gambiense* and *T. b. rhodesiense* cause African sleeping sickness in humans. Trypanosomes are transmitted by the tsetse fly.

Slides of stained blood smears from a rat infected with trypanosomes may be used to demonstrate these flagellate parasites. Or living trypanosomes may be obtained from the blood of a rat infected with the nonpathogenic *T. lewisi*, which is transmitted by the northern rat flea.

Snip off the tip of the rat's tail with sharp scissors and place a drop of the blood on a slide. Quickly apply a coverslip. Look for erythrocytes with erratic motion to locate the trypanosomes, and then observe the vigorous movements of the flagellates.

Note the fusiform, slightly twisted body of the organism (Fig. 7-9, *B*). Are the two ends alike? Note that there is an anterior flagellum and that along one side of the organism is what is known as an **undulating membrane**. The flagellum begins posteriorly at the kinetosome, runs along the surface of the body close to the pellicle, and continues anteriorly as a free whip. When the flagellum beats, the pellicle is pulled up into a fold; the fold and the flagellum together constitute the undulating membrane.

3. Dinoflagellates. Dinoflagellates, a group of phytomastigophorans found mostly in marine plankton, have a peculiar arrangement of two flagella; one extends backward to propel the animal; the other encircles the body in a transverse groove and causes it to rotate. Some dinoflagellates have chromatophores of various colors; others are colorless. Some, such as *Noctiluca*, possess luminescent granules in the cytoplasm, which are largely responsible for the phosphorescence observed in some coastal waters. Some occasionally appear in enormous numbers and cause the "red tides" off the coasts of Florida (*Gymnodinium*) and California (*Gonyaulax*). *Ceratium*, with long spines and green chromatophores, is found in both marine and fresh water (Fig. 7-8).

4. Colonial phytomonids. Several colonial forms of this group are common in fresh water and can be obtained in living cultures from biological supply houses. In *Gonium sociale*, with four biflagellated zooids, and *Gonium pectorale* (Fig. 7-8), with 16, all the cells can produce new colonies. In *Pleodorina* four of the 32 cells are sterile, but in *Eudorina* each of the 32 zooids can produce a new colony.

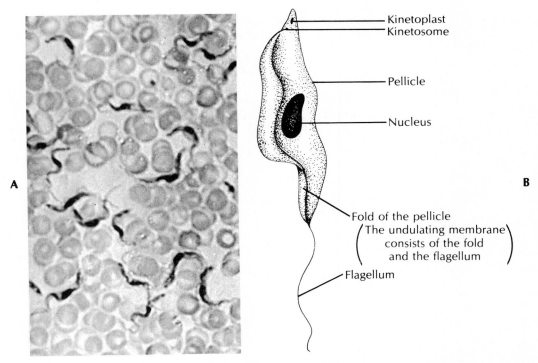

Fig. 7-9. A, Trypanosomes scattered among human red blood cells. Carried by the tsetse fly, they cause African sleeping sickness. **B,** Structure of trypanosome. The kinetoplast carries the mitochondrial DNA and gives rise to the mitochondria.

Phylum _____   Name _____
Class _____   Date _____ Section _____
Genus _____

SKETCHES OF EUGLENOID MOVEMENT

REPORT ON LOCOMOTION OF EUGLENA

EXERCISE 7C
Class Sporozoea—*Gregarina* and *Plasmodium*

Sporozoans are entirely endoparasitic. On the whole they lack special locomotor organelles, although some can move by gliding or by changes in body shape similar to euglenoid movement, and some species have flagellated gametes. Their means of distribution is by resistant spores or sporocysts, which contain sporozoites. The process of **sporogony**, or spore formation, in most sporozoans follows the formation of the zygote, so that sexual reproduction leads into the production of spores.

Many sporozoans also undergo asexual fission, or **schizogony,** a means of spreading the population within the host, and this too is followed by sporogony. Sporozoans have no mouth, and nutrition is osmotrophic (nutritives are absorbed from the immediate environment).

Common sporozoans are the gregarines (subclass Gregarinia), parasitic in such invertebrates as annelids, echinoderms and ascidians; and the coccidians (subclass Coccidia), which are endoparasites in both invertebrates and vertebrates. Among the Coccidia are *Eimeria*, which causes coccidiosis in domestic rabbits and chickens, and *Plasmodium*, which causes malaria in humans.

MATERIALS
Mealworms (*Tenebrio* larvae)
    or
Cockroaches (*Periplaneta, Blatta*)
Watch glasses
Invertebrate Ringer's solution, or 0.65% saline solution
Stained slides of gregarines
Slides and coverslips
Teasing needles
Pipettes
Stained human blood smears containing *Plasmodium* stages
Microscopes

GREGARINA
Phylum Apicomplexa
    Class Sporozoea
        Subclass Gregarinia
            Genus *Gregarina*

Various species of *Gregarina* can be found in the guts of cockroaches (such as *Blatta* and *Periplaneta*), mealworms (the larvae of *Tenebrio*), and grasshoppers.

☞ Using a live mealworm (or cockroach), cut off the head. Make two longitudinal cuts through the dorsal body wall; then carefully pull away the tail end with the gut attached. Place the gut in a watch glass and cover with a little invertebrate Ringer's solution. Tease into small bits; then transfer a little of the contents in a drop of the saline to a slide for examination. Add a coverslip.

If **trophozoites** are present, they will appear as large, distinctly shaped forms ranging from nearly round to elongate and wormlike, depending on the species (Fig. 7-10). In general the body is constricted into three unequal parts, an anterior epimerite modified as a holdfast organelle (which may have been detached), a middle protomerite, and a larger posterior deutomerite containing the nucleus.

Life Cycle. When the trophozoite is fully developed, its epimerite, which is attached to the epithelium of the host's gut, breaks off from the rest of the body. The remaining organism glides about in the gut and is ready to join end to end with another trophozoite, a process called **syzygy** (siz' i-ji) (Fig. 7-10).

The pair becomes encysted (gametocyst), and each member of the pair divides to produce numerous gametes. While still encysted, each gamete from one parent unites with a gamete from the other parent. Each resulting zygote secretes a wall about itself (oocyst) and then divides to form eight sporozoites. The gametocyst and its load of oocysts, each containing sporozoites, is shed from the host in the feces.

When ingested by a new host, the sporozoites are released into the gut, where they penetrate the epithelial cells. As the organism (now called a trophozoite) grows within the cell, it elongates and extends outside the cell, attached only by the anterior portion

Fig. 7-10. *Gregarina*, a parasitic sporozoan that lives in the gut of the cockroach. Two trophozoites are paired end to end (syzygy). The anterior portions, or epimerites, have been detached.

(epimerite). Here it remains until maturity, thus completing the life cycle.

If living material is not available, stained slides of gregarines showing trophozoites in syzygy may be substituted.

PLASMODIUM

Phylum Apicomplexa
  Class Sporozoea
    Subclass Coccidia
      Order Eucoccidia
        Genus *Plasmodium*

Plasmodium is the genus of sporozoan parasites that causes malaria. *Plasmodium* is a digenetic sporozoan; that is, it requires two hosts. It is transmitted to the human by a female *Anopheles* mosquito that has previously had a blood meal from a malaria-infected human. The mosquito, as it feeds, also injects into the human bloodstream some of its salivary juice, which contains infective **sporozoites** (Fig. 7-11).

In the human the sporozoites (Fig. 7-12, *A*) leave the bloodstream and penetrate into liver cells, where they undergo **schizogony** (asexual cleavage multiplication) and produce **merozoites.** These are released into the blood and may either infect other liver cells or enter red blood cells (erythrocytes), in either of which they undergo further multiplication.

In the red blood cells they become **trophozoites** (adult stage). The cytoplasm of the trophozoite becomes nucleated and takes on a ringlike appearance called the "ring stage" (Fig. 7-12, *B*).

☞ Examine a slide containing a stained blood smear with various infective stages. Find a red blood cell containing a ring stage. Slides stained with blood-differentiating dyes usually show the cytoplasm of the trophozoite blue and the nucleus reddish pink.

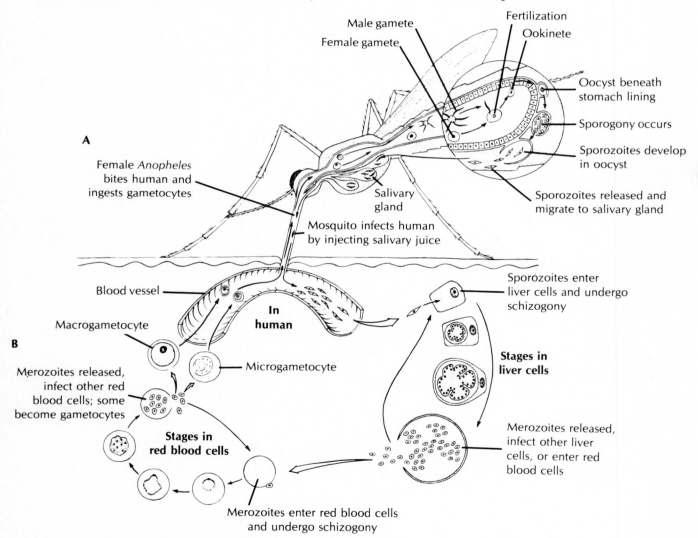

Fig. 7-11. *Plasmodium* life cycle. *Plasmodium vivax* causes malaria in humans. It is carried by the *Anopheles* mosquito.

In living material the trophozoites are motile, actively producing pseudopodia.

As the ring stage enlarges, the parasitized erythrocyte takes on a stippled appearance. The stipples are called Schüffner's dots, and they appear as deeper pink dots (Fig. 7-12, C).

The nucleus in the trophozoite divides one to four times (schizogony), producing more merozoites, which may infect more red cells. This dividing stage is called the schizont, and it appears multinucleated.

Some of the merozoites are destined to become gametocytes (sexual forms), which may be picked up by a feeding *Anopheles* female. The mosquito may take in both sexual and asexual stages when feeding, but for the asexual stages this is the end of the line. The gametocytes, however, give rise within the mosquito to gametes, which unite to become zygotes. The zygotes divide to produce sporozoites, which migrate to the salivary glands from which they may be injected along with the saliva into the blood of a human, perhaps to cause another malarial infection.

PROJECTS AND DEMONSTRATIONS

1. Demonstration of Monocystis. Monocystis, another gregarine, can often be demonstrated by removing the seminal vesicles from live earthworms and examining portions teased in 0.65% saline solution. If trophozoites are present, note their feeble movements. Stained slides of sections or smears of earthworm seminal vesicles showing developmental stages of *Monocystis* are available at biological supply houses.

2. Demonstration of coccidia slides. Stained slides of *Eimeria stiedae* from bile ducts of infected rabbit or *Eimeria tenella* from infected chicken intestine are available at biological supply houses. *Eimeria* causes coccidiosis, an important cause of death in domestic rabbits and chickens.

On a stained slide of infected rabbit bile duct, the developing trophozoites appear as spherical bodies, large and small, embedded in the outer ends of the columnar cells that line the ducts. You may find some of them undergoing multiple fission. The largest stages are the male gametocytes, in which large numbers of microgametes develop. The female gametocytes are somewhat smaller and have darkly staining granules around the periphery and a nucleus. The female gametocyte encysts (oocyst) and becomes fertilized. The encysted zygote escapes into the lumen of the bile duct and is shed with the feces. You may find a number of oocysts lying free in the lumen of the duct.

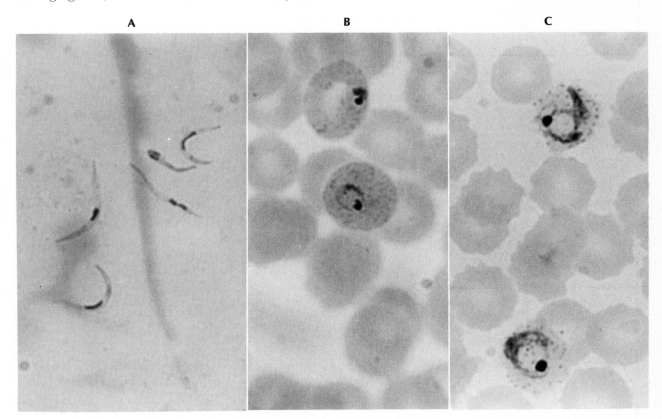

Fig. 7-12. *Plasmodium*, the malarial parasite. **A,** Sporozoites. **B,** The ring stage of the trophozoite. **C,** Developing trophozoites showing Shüffner's dots. (× 1500.) (From Farmer, J.N., 1980. The Protozoa. St. Louis, The C.V. Mosby Co.)

EXERCISE 7D
Phylum Ciliophora—*Paramecium* and other ciliates

MATERIALS
Stained slides of paramecia—normal, undergoing binary fission, and in conjugation
Paramecia cultures
Vorticella, Stentor, Spirostomum, or other ciliate cultures
10% methylcellulose (p. 462), polyvinyl alcohol, or Protoslo
Congo red–yeast mixture or Congo red–milk mixture (p. 461)
Acidified methyl green (p. 461) for staining nuclei
0.25% NaCl solution
Dilute picric acid
Weak acetic acid
Salt crystals
Cotton
Toothpicks or pins
Slides and coverslips
Microscopes

PARAMECIUM
Phylum Ciliophora
   Class Oligohymenophora
      Genus *Paramecium*

Where Found

Paramecium is an active protozoan common in most fresh water that contains vegetation and decayed organic matter. Pond scum and even cesspools are also good sources. Some of the most commonly studied forms are *P. multimicronucleatum, P. caudatum, P. aurelia,* and *P. bursaria.*

Study of Live Specimens
Locomotion and Behavior
 Place a drop of 10% methylcellulose, polyvinyl alcohol, or Protoslo on a slide and spread it thin. Add a drop of *Paramecium* culture but do not cover. Examine with a dissecting microscope, the scanning lens of a compound microscope, or a tripod magnifier.

A paramecium is slipper shaped, rather transparent and colorless, and very active. It can swim at the rate of 1 to 3 mm per second. Watch its swimming habits. Does it swim in a straight line, a circle, or zigzag? Does it keep one side uppermost or revolve on an axis? Does it have a definite anterior end? Can it reverse its direction? Does it seem to be contractile as *Euglena* is? Why is the term "spiral movements" used to describe its swimming habits?

> **WRITTEN REPORT**
>
> On p. 111 describe the locomotion of *Paramecium,* answering as many of the preceding questions as you can in the description.

☞ Now add *very few* fibers of absorbent cotton to the drop of culture on the slide, cover with a coverslip, and study with low power. Add water as necessary to prevent drying out.

What does a paramecium do when it encounters a barrier? How does it go about finding an opening? Is its body flexible enough to bend or to squeeze through tight places?

> **WRITTEN REPORT**
>
> On p. 111 describe the attempts of a paramecium to avoid or to pass under or around a cotton fiber barrier. Use diagrams and arrows if you wish.

General Structure and Function
☞ As a specimen slows down, study its structure with both low and high power.

Note the **oral groove** that extends obliquely from the anterior end to about the middle of the body. At the posterior end of the groove is the **mouth (cytostome).** From the mouth the oral groove extends into the body as a little canal, the **gullet (cytopharynx).** The groove and gullet are lined with strong cilia that are used in drawing in food (Fig. 7-13).

Pellicle. The cytoplasm is made up of two zones, the clear outer **ectoplasm** and the inner **endoplasm.** Are there any granules in the endoplasm? Outside of the ectoplasm is a complex living **pellicle.** A delicate **plasma membrane** lies just underneath the pellicle. The pellicle and plasma membrane are more easily seen in stained preparations.

Osmoregulation. A contractile vacuole (water-expulsion vesicle is usually located in each end of the body. *P. multimicronucleatum* may have more than two.

Observe the pulsations of the vacuoles caused by their alternate filling and emptying. After one empties, note the starlike **radiating canals** that appear (Fig. 7-14). These radiating, or nephridial, canals collect

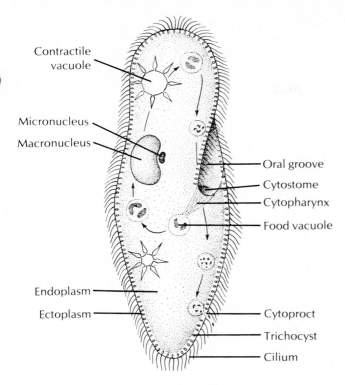

Fig. 7-13. *Paramecium*.

Labels for Fig. 7-13:
- Contractile vacuole
- Micronucleus
- Macronucleus
- Oral groove
- Cytostome
- Cytopharynx
- Food vacuole
- Endoplasm
- Ectoplasm
- Cytoproct
- Trichocyst
- Cilium

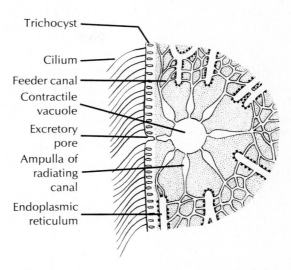

Fig. 7-14. Section of a contractile vacuole of *Paramecium*. Water is believed to be collected by endoplasmic reticulum and emptied into feeder canals and then into the vesicle. The vesicle contracts to empty its contents to the outside, thus serving as an osmoregulatory organelle.

Labels for Fig. 7-14:
- Trichocyst
- Cilium
- Feeder canal
- Contractile vacuole
- Excretory pore
- Ampulla of radiating canal
- Endoplasmic reticulum

liquid from a network of minute tubules and empty into the vacuoles which, when filled, rupture, thus expelling the fluid to the outside.

The vacuoles and radiating canals are most prominent in animals that have quieted down, when most of the culture water on the slide has evaporated.

Do the two vacuoles empty at the same time or alternately? Time the rate of pulsations. In normal pond water a vacuole may discharge once every 6 to 10 seconds. How does this animal's rate compare?

The pulsation period depends on the temperature and osmotic pressure of the medium. As the temperature increases, the pulsation period decreases. The pulsation rate can be slowed down by adding drops of 0.25% solution of sodium chloride to the slide. Knowing this, how do you think the rate in marine ciliates would compare with that of freshwater ciliates? Why?

Ciliary Action. These fine projections perform very much like the oars of a boat; that is, they have an effective stroke that propels the animal forward and a recovery stroke that offers little resistance. For a discussion of the infraciliature see your text.

☞ Find a quiet specimen, cut down the light, and look closely with high power at the margin of a paramecium. Study the action of the **cilia**.

Nuclei. The nuclei may be difficult to see in living paramecia. Your instructor may prefer you to use prepared stained slides, or you can stain the specimens on your slide by adding a drop of acidified methyl green to the slide at the edge of the coverslip so that the stain is drawn under by capillary action. This will kill the organisms but should give the cytoplasm a bluish tinge and stain the nuclei green.

Ciliates possess two types of nuclei. The macronucleus has several sets of chromosomes, divides amitotically, and is somatic in function; the micronucleus is diploid, divides mitotically, and has a genetic function. The number of nuclei in various ciliates may vary from one of each kind to several of either kind. *P. aurelia* has one large, kidney-shaped macronucleus and one small micronucleus. *P. multimicronucleatum* has one macronucleus and four or more micronuclei.

The form of the macronucleus is diverse among other ciliates. It is long and curved in *Vorticella*, *Euplotes*, and others, moniliform in *Spirostomum*, dumbbell shaped in *Blepharisma*, and resembles a string of beads in *Stentor*.

Trichocysts. If you reduce your light properly, you can, with very careful focusing, observe the many small spindle-shaped **trichocysts** lying in the ectoplasm just under the pellicle and perpendicular to the surface. Under certain kinds of stimulation the trichocysts explode, each one releasing a liquid that hardens in water to form a long slender threadlike filament. The tangle of these filaments is believed to have some protective function, although the function is still unclear despite much study.

In some ciliates, but not paramecia, the trichocysts are known to be toxic. In some, the discharged trichocysts may be used to anchor the animal while feeding. Trichocysts are absent in some ciliates.

Perhaps the acidified methyl green you just used to demonstrate the nuclei has caused the explosion of trichocysts on your slide. Alternate methods of bringing about their discharge in live paramecia are to add a drop of dilute picric acid or a drop of 1% aqueous solution of tannic acid or a drop of ink to the slide at the side of the coverslip and allow the fluid to be drawn under the slip. As the fluid reaches the paramecia, you may be able to observe the discharged trichocysts by use of subdued light and high-power lens. How does the length of the trichocysts compare with the length of cilia?

Report your results on p. 111.

Feeding. Nutrition in ciliates is holozoic. The paramecium is a particulate feeder; that is, it lives on small particles such as bacteria, which it moves toward the cytostome by the action of cilia in the oral groove.

☞ Spread a thin layer of methylcellulose or polyvinyl alcohol on a slide and add a drop of paramecium culture in milk colored with Congo red; cover. Or use a drop of regular culture, then dip the tip of a toothpick into a mixture of yeast and Congo red, and transfer a very small quantity to the culture on the slide. Mix gently with the toothpick. The mixture should be light pink. If too much yeast is added, the protozoans will be obscured. Carefully apply a coverslip.

Do not allow the culture to dry out during your observation. Using a hanging drop (see Appendix B, p. 455) will permit long observation without drying.

As a specimen slows down, note the currents created by the cilia in its **oral groove.** Watch the passage of the yeast particles into the groove and through the **cytostome,** or cell mouth, into a passageway called the **cytopharynx.** Watch the formation of a **food vacuole,** which is a membranous sac containing water and suspended food particles. When it reaches a certain size, the vacuole breaks away and another forms. Note the

direction in which the vacuoles are carried by streaming endoplasm, a movement called **cyclosis.**

Follow the course of a food vacuole. Does it vary any in size during its trip? Congo red, which is red in weak acid to alkaline solutions (pH 5.0 or above), turns blue in stronger acid solutions (pH 3.0 or below). Can you observe any changes in color in the vacuoles that might indicate a change in the condition of the vacuole contents? Is there any subsequent color change as the vacuoles near the anal pore? How might this be explained? Are the vacuoles any smaller as they complete their circuit than they were when first formed?

An **anal pore** (cytoproct) is found between the mouth and posterior end of the body. It is a temporary opening where indigestible food is discharged, and it is seen only at that time. Have you noticed such a discharge of material in one of your specimens?

WRITTEN REPORT

Report your observations on feeding and digestion on p. 112.

Response to Stimuli. The following simple experiments are designed to determine the type of response (**taxis**) paramecia make to selected types of stimuli. In these experiments the responses are **chemotactic.** Perhaps you can design and perform other experiments to determine responses to other types of stimuli, such as **phototaxis** (response to light rays), **thigmotaxis** (to contact), or **geotaxis** (to gravity).

☞ 1. Place a drop of culture on a clean slide with no coverslip. Use a hand lens, scanning lens, or dissecting microscope. (Some of you will be able to see paramecia with the naked eye.) Place a drop of weak acetic acid on the slide **near** the culture but **not touching it.** Locate the specimens; while observing them, draw a line with the point of a pin or toothpick from the acid to the culture.

Describe the reaction of the animals to the approach of the acid. Are they positively or negatively chemotactic to weak acid? Do they prefer the area where the acid is strongest, weakest, or in between?

☞ 2. Place a drop of culture on a clean slide. Place a few grains of salt on the slide near the culture but not touching it. While observing the animals, draw a grain or two of salt into the side of the culture drop.

Describe the reaction of the animals. Do they prefer the area nearest the salt, farthest from the salt, or in between?

Color in Ciliates. Although most ciliates are colorless, some ciliates are green, brown, blue, or pink. The color in some is caused by pigments in the ectoplasm; in others it is by the presence of green or brown algae. *Paramecium bursaria*, for example, is green because it contains symbiotic green algae called "zoochlorellae." *Stentor coeruleus*, the blue stentor, has rows of blue-green pigment, called stentorin, in the ectoplasm; however, there are other species of stentors that are green, brown, or rose, because of algae. *Blepharisma* (Fig. 7-15) is pink or rose because it contains the pigment purpurin.

Study of Stained Slides

With both low and high power, study a paramecium on a stained slide. Focus up and down on the body and note that its entire surface is covered with **cilia**. Examine the **pellicle**. Specially prepared slides will show the peculiar pattern of hexagonal areas. Look especially for features difficult to see in the living unstained specimens, such as the **macronucleus**, one or more **micronuclei, trichocysts, oral groove, cytostome, cytopharynx,** and **contractile vacuoles**.

Binary Fission. Study a stained slide of paramecia undergoing binary fission. Note the constriction across the middle of the body. What is happening to the macronucleus and micronucleus? At the end of the process the halves produced by the constriction will be separate daughter animals.

Conjugation. Study also a stained slide of conju-

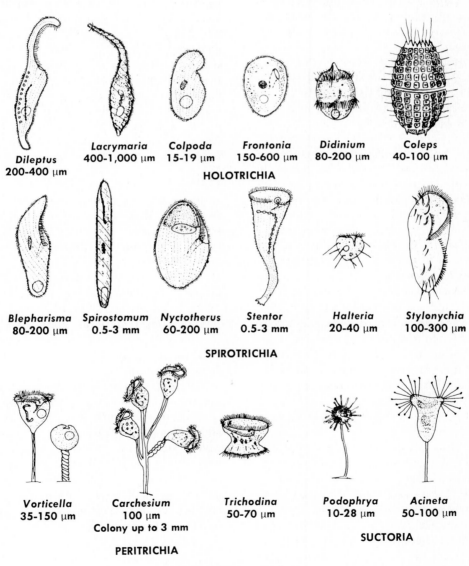

Dileptus
200-400 μm

Lacrymaria
400-1,000 μm

Colpoda
15-19 μm

Frontonia
150-600 μm

Didinium
80-200 μm

Coleps
40-100 μm

HOLOTRICHIA

Blepharisma
80-200 μm

Spirostomum
0.5-3 mm

Nyctotherus
60-200 μm

Stentor
0.5-3 mm

Halteria
20-40 μm

Stylonychia
100-300 μm

SPIROTRICHIA

Vorticella
35-150 μm

Carchesium
100 μm
Colony up to 3 mm

Trichodina
50-70 μm

Podophrya
10-28 μm

Acineta
50-100 μm

SUCTORIA

PERITRICHIA

Fig. 7-15. Some ciliates.

gation. Look for paired individuals lying *with oral grooves attached*. In this position they exchange micronuclear material. Consult your textbook for details of the process of conjugation.

DRAWING

On p. 111 sketch a paramecium in the process of binary fission and a pair of paramecia in the process of conjugation. If you can find a live specimen dividing, watch it over a period of time and make a series of sketches to illustrate the process.

OTHER CILIATES
Vorticella

Vorticella (Fig. 7-15) is a sessile ciliate. It clings to aquatic vegetation of stagnant ponds and streams. The blanket algae of ponds and small lakes is a favorite place for this animal.

Study both the living specimens and the stained slides. What is the color of the living animal? The vorticellid is attached by a long slender **stalk** that can contract into a spiral spring shape when it is disturbed. The **body** is bell shaped with a flaring rim, the **peristome,** at its distal end. Within the peristome is a circular **oral disc.** Note the **cilia** on the edges of the peristome and the oral disc. Note the beating of the cilia. What is their function? The **cytostome** (mouth) is found between the peristome and the oral disc. From the cytostome a short tube, the **cytopharynx,** leads into the interior. Food particles are swept into the cytopharynx by the action of the cilia, and **food vacuoles** are formed as they are in other protozoa. Note that the **nucleus** is made up of an elongated U-shaped body, the **macronucleus,** and a much smaller **micronucleus.** Does the animal have a **contractile vacuole?** *Vorticella* has a surrounding **pellicle,** which helps maintain the shape of the body. Underneath the pellicle is the **ectoplasm,** and beneath this is the **endoplasm,** which is granular. Reproduction is mostly by longitudinal binary fission, but **budding** also occurs. Sketch a specimen on p. 110.

Stentor

Spread methylcellulose or Protoslo thinly on the slide before adding *Stentor* culture and coverslip. *Stentor* (Fig. 7-15) is a large ciliate with many of the same characteristics as *Paramecium* and *Vorticella*. Estimate its length when expanded and when contracted. How would you describe its shape? See how many of the structures of this ciliate you can identify from your knowledge of paramecia. Can you locate the cytostome, cytopharynx, and contractile vacuole? Is there an oral groove? Can you see food vacuoles? Are the cilia uniform in length? Observe its striped appearance caused by longitudinal bands of pigmentation. The blue pigment stentorin causes the blue-green color of *S. coeruleus*.

Note the difference in shape when the animal is swimming freely and when it is anchored. Like many ciliates, but not paramecia, The stentor's ectoplasm contains many contractile **myonemes,** primitive forerunners of muscles, which make contraction possible. Note the large **macronucleus,** stretched out like a string of beads. Small dots nearby are the micronuclei. On p. 110 draw and label an extended specimen. Sketch its shape when the animal is swimming. Estimate its size.

Spirostomum

Spirostomum (Fig. 7-15) is one of the largest common freshwater protozoa. Estimate its length. Different species range from 50 μm to 3 mm long. Can you see why it might be mistaken for a worm? Locate the large contractile vacuole posteriorly. It is fed by a long canal. The macronucleus is long and beadlike. Locate the cytostome and long oral groove. Focus on the surface and see the arrangement of the cilia and the trichocysts. Describe its swimming or other movements. Do you think it has myonemes? On p. 110 draw and label an extended specimen. Outline a contracted one.

PROJECTS AND DEMONSTRATIONS

1. Mating reaction and conjugation. Many ciliates have been found to have mating types within each species or variety. Members of one mating type will mate with members of another mating type but not with their own type.

Pure lines of two mating types of paramecia, together with instructions on mating them, can be obtained from a biological supply house. These cultures should be used within 24 hours of delivery.

2. Use of nickel sulfate for quieting paramecia. Dissolve 0.1 g of nickel sulfate in 100 ml distilled water; then prepare a stock solution by adding 1 ml of nickel sulfate solution to 1000 ml of distilled water. For use, dilute the stock solution 1:5.

Fill a vial half full of dense *Paramecium* culture and slowly add an equal volume of the diluted nickel sulfate solution, shaking constantly to prevent exposure of the paramecia to too high a concentration of the nickel sulfate. The animals should be ready for observation in 15 to 20 minutes. Evaporation of fluid on the slide will result in a lethal concentration, but treated animals in a closed vial should remain alive for several days.

3. Concentrating Protozoa. Select a large-mouthed jar provided with a two-holed stopper (or lid in which holes can be cut [Fig. 7-16]). Into one hole

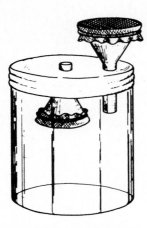

Fig. 7-16. Simple method of concentrating protozoans. Tie a piece of fine-meshed silk bolting cloth over the mouth of the inverted funnel inside the jar. Cheesecloth can be used on the outside funnel to strain out debris from pond water.

of the stopper insert the stem of a small funnel in an inverted position with its broad end covered by fine silk bolting cloth. Into the other hole of the stopper insert a larger funnel in an upright position. Put the stopper snugly in place in the jar and pour the culture water into the upright large funnel. As the jar fills, the water will filter out through the small funnel, and the organisms will be retained by the bolting cloth. An indefinite quantity of water may be run through this mechanism. The device may be used in the field to transport concentrated protozoans to the lab. If the water being filtered is full of algae or scum, it may be necessary to strain this out by tying a piece of cheesecloth over the entrance funnel.

4. Method of obtaining abundant dividing stages of Paramecium. During fission most paramecia tend to settle toward the bottom of the culture. Put a concentrated culture of *Paramecium* in a large funnel or other vessel provided with a stopcock at its lower end. After being provided with suitable food, three or more

generations are produced each day. By opening the stopcock at intervals and drawing off a few milliliters of culture, you can obtain large numbers of fission stages.

Early stages appear to be spindle shaped. The whole process of dividing usually requires about 25 minutes.

5. Microaquariums. A fascinating study of microorganisms and their relations to each other may be carried out in the laboratory at no expense and with only a few minutes of time at each laboratory period. Use a series of small clean empty jars such as have held baby food, jelly, mayonnaise or the like. Fill each jar two-thirds full with water from the tap, pond, ditch, or any other source. Add to each a teaspoonful of some source material. This material could be rich soil, plants (dry grass, leaves, hay, water plants, moss, or rotting leaf mold), pond scum, sludge from a sewage plant, or any other source you can think of. Label each jar. To retard evaporation, cover with a piece of glass, a plastic wrap, or the jar lid placed on loosely. Examine weekly, or oftener if you like, for several weeks or months. Each jar becomes a community that provides an interesting variety and an ever-changing cycle of life. In addition to protozoans you may find a variety of crustaceans, flatworms, rotifers, gastropods, annelids, hydras, algae, and diatoms. Keep a weekly record of what is found in each jar.

6. Electric response in paramecia. With a 6-volt dry-cell battery, a couple of wires, and a concavity slide, one can demonstrate the response of paramecia to an electric current. Remove the insulation from one end of each wire and tape the wires to the slide so that the bare ends almost reach each other across the bottom of the concavity. Fill the depression half full of culture and examine under a binocular dissecting microscope. Attach the free ends of the wires to the battery and watch the action of the paramecia. Before the paramecia reach the tip of the wire, detach one wire.

NOTES OR SKETCHES

REPORT ON PARAMECIUM LOCOMOTION

AVOIDING REACTION OF PARAMECIUM

TRICHOCYSTS

Binary fission   **Conjugation**

NOTES ON CHEMOTACTIC RESPONSE

CHAPTER 8

THE SPONGES

PHYLUM PORIFERA*

The members of the phylum Porifera are among the simplest forms of metazoans in the animal kingdom. Because sponges are little more than loose aggregations of cells, with little or no tissue organization, they are said to belong to the **cellular level of organization.** There is division of labor among the cells, but there are no organs, no systems, no mouth or digestive tract, and only very primitive nervous integration. There are no germ layers; therefore, sponges are neither diploblastic nor triploblastic. Adult sponges are all sessile in form. Some have no regular form or symmetry; others have a characteristic shape and radial symmetry. They may be either solitary or colonial.

Their chief characteristics are their **pores** and **canal systems,** the flagellate **choanocytes,** which line their cavities and create currents of water, and their peculiar internal skeletons of **spicules** or organic fibers (**spongin**). They also have some form of internal cavity (**spongocoel**) that opens to the outside by an **osculum.**

---

*See Hickman, C.P., Jr., L.S. Roberts, and F.M. Hickman. 1984. Integrated principles of zoology, ed. 7. St. Louis, The C.V. Mosby Co., pp. 202, 204-218.

Most sponges are marine, but there are a few freshwater species. The freshwater forms are found in small slimy masses attached to sticks, leaves, or other objects in quiet ponds and streams.

The phylum is divided into four classes:

Class Calcispongiae (= Calcarea) (cal′si-spun-je-e). Sponges with spicules of calcium carbonate, needle-shaped or three- or four-rayed; canal systems asconoid, syconoid, or leuconoid; all marine. Examples: *Scypha*, *Leucosolenia*.

Class Hyalospongiae (= Hexactinellida) (hy′a-lo-spun′je-e). Sponges with three-dimensional, six-rayed siliceous spicules; spicules often united to form network; body often cylindrical or funnel shaped; canal systems syconoid or leuconoid; all marine, mostly deep water. Examples: *Euplectella* ("Venus's flower basket"), *Hyalonema*.

Class Demospongiae (de′mo-spun′je-e). Sponges with siliceous spicules (not six-rayed), or spongin, or both; canal systems leuconoid; one family fresh water, all others marine. Examples: *Spongilla* (freshwater sponge), *Spongia* (commercial bath sponge), *Cliona* (a boring sponge).

Class Sclerospongiae (skle′ro-spun′je-e). A small group; leuconoid, with internal skeletons of siliceous spicules and spongin and an outer encasement of calcium carbonate; marine, associated with coral reefs. Examples: *Astrosclera*, *Calcifibrospongia*, *Merlia*.

EXERCISE 8A
Class Calcispongiae—*Scypha*

MATERIALS
Preserved or living *Scypha (Sycon, Grantia)*
Examples of asconoid and leuconoid sponges
Prepared slides
  Scypha, transverse sections
  Leucosolenia, transverse sections
  Spicule strew (see p. 119 for preparation)
Single-edged razor blade
Chlorine bleach (sodium hypochlorite)
Microslides
Coverslips
Watch glasses
Hand lenses or dissecting microscopes
Compound microscopes

SCYPHA, A SYCONOID SPONGE
Phylum Porifera
  Class Calcispongiae
    Order Heterocoela
      Genus *Scypha* (= *Sycon, Grantia*)

Where Found
Scypha is strictly a marine form, living in clusters in shallow water, usually attached to rocks, pilings, or shells. *Scypha* is chiefly a North Atlantic form. *Rhabdodermella* is a somewhat similar Pacific intertidal form, also belonging to class Calcispongiae.

Gross Structure
☞ Place a preserved specimen in a watch glass and cover with water. Examine with a hand lens or a dissecting microscope.

Scypha is a **syconoid** type of sponge (Fig. 8-1). In general it is vase shaped, with a body wall made up of a system of tiny, interconnected dead-end canals whose flagellated cells draw in water from the outside through minute pores, take from it the necessary food particles and oxygen, and then empty it into a large central cavity (spongocoel) for exit to the outside. All sponges have some variation of this general theme of canals and pores on which they depend for a constant flow of water.

External Structure. Note that the base of the sponge is closed. The other end has an opening, the **osculum,** surrounded by a fringe of stiff, rodlike **spicules.** The external surface appears bristly because of myriad tiny spicules that protrude through the body wall.

☞ Completely cover the sponge with water and examine under a dissecting microscope.

Note that the body wall seems to be made up of innumerable bristly, fingerlike processes pointing outward. Inside each of these "processes" is a **radial canal,** which is closed at the outer end but which opens into a central cavity called the **spongocoel.** The external

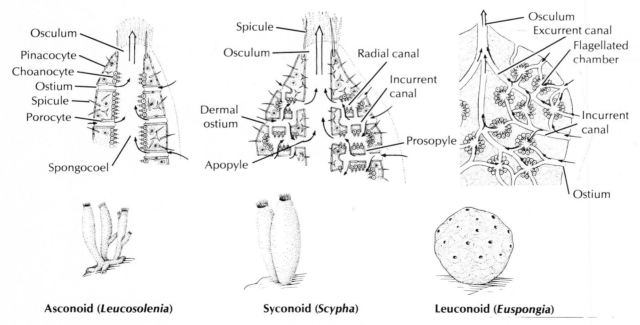

Asconoid (*Leucosolenia*)   **Syconoid (*Scypha*)**   **Leuconoid (*Euspongia*)**

Fig. 8-1. Three types of sponge structure. The degree of complexity from asconoid to complex leuconoid type involves mainly the water-canal and skeletal systems, accompanied by outfolding and branching of the collar-cell layer. The leuconoid type is considered the major plan for sponges because it permits greater size and more efficient water circulation.

spaces between these enclosed canals are **incurrent canals,** which open to the outside but end blindly at the inner end. The outside openings, or pores, are called **dermal ostia.**

Water enters the incurrent canals and passes through minute openings called **prosopyles** into the radial canals and then to the spongocoel and out through the osculum. There is no mouth, anus, or digestive system. What kind of symmetry does this sponge have?

Spongocoel. To study the spongocoel; do the following:

☞ Make a longitudinal cut through the midline of the body from osculum to base with a sharp razor blade. Place the two halves in a watch glass, and cover with water.

Find the small pores, called **apopyles,** that open from the radial canals into the spongocoel (Fig. 8-2). Can you distinguish the tiny canals in the cut edge of the sponge walls?

Cellular Structure

Transverse sections of the sponge are difficult slides to make because the spicules prevent the cutting of sections thin enough for the study of the cells. Therefore the spicules have been dissolved away during the preparation of the tissue for cutting.

☞ On a prepared slide of a cross section of *Scypha,* examine the entire transverse section with low power to get an idea of its general relations.

Note the **spongocoel** in the middle of the section (Fig. 8-3). Study the canal system. Find the **radial canals,** which open into the spongocoel by way of the **apopyles.** Since these openings are smaller in diameter than the radial canals, some of them will be lacking in this section, and some of the radial canals will appear closed at the inner end. Follow the radial canals outward and note that they end blindly. The radial canals may contain young larvae, called **amphiblastula larvae.** Identify the **incurrent canals,** which open to the exterior by the **dermal ostia.** Follow these canals inward and note that they also end blindly. Water passes from the incurrent canals into the radial canals through a number of tiny pores, or **prosopyles,** which will not be evident on the slides.

Sponge cells are loosely arranged in a gelatinous matrix called **mesohyl** (also called mesoglea or mesenchyme). It holds together the various types of ameboid cells, skeletal elements, and fibrils that make up the sponge body.

Choanocytes. With high power, observe the "collar cells," or choanocytes, that line the radial canals (Fig. 8-4). Although they are flagellated, you probably will not see the flagella. What is the function of the choanocytes?

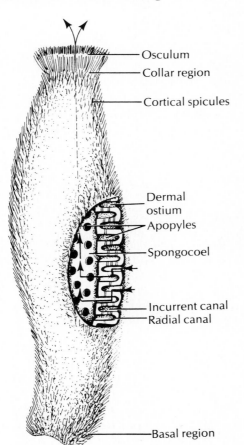

Osculum
Collar region
Cortical spicules
Dermal ostium
Apopyles
Spongocoel
Incurrent canal
Radial canal
Basal region

Fig. 8-2. Structure of the syconoid sponge, *Scypha*. The cutaway section shows the interior cavity, the spongocoel, with the apopyles that lead into it from the radial canals.

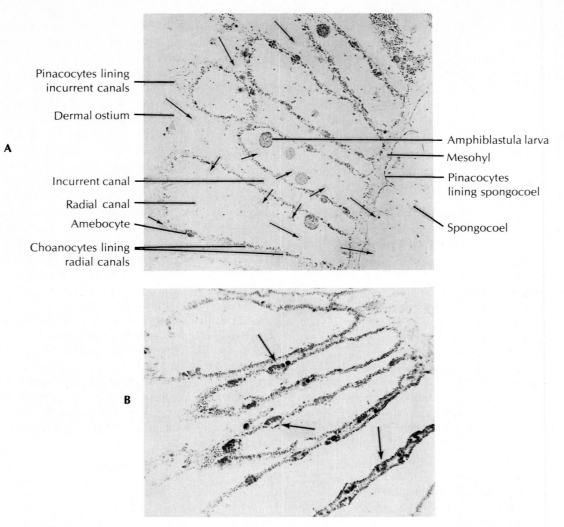

Pinacocytes lining
incurrent canals

Dermal ostium

A

Incurrent canal

Radial canal

Amebocyte

Choanocytes lining
radial canals

Amphiblastula larva

Mesohyl

Pinacocytes
lining spongocoel

Spongocoel

B

Fig. 8-3. A, Portion of section through wall of *Scypha*. Arrows indicate direction of water flow in canals. **B,** Another section, showing early cleavage stages of young embryos *(arrows)* developing in the mesohyl. At the amphiblastula stage (shown in **A**) the embryos break into the radial canals and are carried out to sea by flagellate currents.

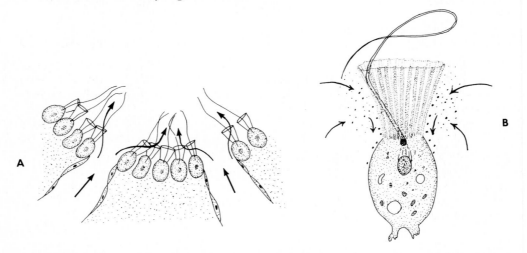

Fig. 8-4. A, Water current created by choanocytes in flagellated chambers. **B,** Choanocyte as a food-catching cell. The "collar" is a series of protoplasmic extensions that screen out larger food particles, letting them fall to the side of the cell for ameboid ingestion. Smaller particles flow through and are carried away in the current.

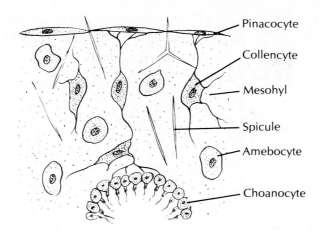

Fig. 8-5. Small section through sponge wall showing four types of sponge cells. Pinacocytes are protective and contractile; choanocytes create water currents and engulf food particles; amebocytes have a variety of functions; collencytes appear to have a contractile function.

Pinacocytes. Dermal amebocytes, called pinacocytes, may be seen as extremely thin (squamous) cells lining the incurrent canals and spongocoel and covering the outer surface (Fig. 8-5). What is their function?

Amebocytes. In the jellylike **mesohyl** that lies in the wall between the pinacocytes and choanocytes, look for large wandering amebocytes of various functions. Some are spicule-forming cells; some form sex cells; and others secrete spongin or spicules, serve as contractile cells, or aid in digestion (Figs. 8-3, *A*, and 8-5).

☞ If living sponges are available, tease a bit of tissue on a slide with a drop of seawater, and look for the various types of cells.

Reproduction

Sexual. *Scypha* sponges are monoecious, producing eggs and sperm in the mesohyl. The fertilized eggs develop in the mesohyl into little blastula-like ciliated embryos called **amphiblastula larvae,** which break through into the radial canals and finally leave the parent by way of the osculum. They soon settle down on a substratum and grow into sessile adults. Look for the embryos in the radial canals of the cross-section slide. Specially prepared slides that show early cleavage stages in the mesohyl may be available (Fig. 8-3, *B*). What is the advantage to a sessile animal of producing free-swimming larvae?

Not all sponges have amphiblastula larvae. In most Demospongiae and some of the calcareous sponges, the zygote develops into a **parenchymula larva** in which the flagellated cells invaginate to form a solid internal mass.

Asexual. Many sponges, including those of the genus *Scypha*, reproduce asexually by budding off new individuals from their base, thus forming sessile clusters. What would be the disadvantage if this were the sole means of reproduction? Is there a bud on your specimen?

The freshwater sponges and some marine Demospongiae reproduce asexually by means of **gemmules,** made up of clusters of amebocytes. Gemmules of the freshwater sponges are enclosed in hard shells and can withstand adverse conditions that would kill the adult sponge. In the spring they develop into young sponges. Marine gemmules give rise to flagellated larvae.

Skeleton

☞ Place a small bit of the sponge on a clean microscope slide and add a drop of commercial chlorine bleach such as Clorox (sodium hypochlorite) to dissolve the cellular matter. Break up the piece with dissecting needles, if necessary. Add a coverslip and examine under the microscope.

Look for **short monaxons** (short and pointed at both ends), **long monaxons** (long and pointed), **triradiates** (Y-shaped with three prongs), and **polyaxons** (T-shaped). These spicules of crystalline calcium carbonate ($CaCO_3$) form a sort of network in the walls of the animal (Fig. 8-5). What is the advantage of spicules to a loosely constructed animal such as *Scypha*?

Spicule types are used in the classification of sponges, along with the types of canal system. The Demospongiae have siliceous (mainly $H_2Si_3O_7$) spicules, or spongin fibers (composed of an insoluble scleroprotein that is resistant to protein-digesting enzymes) (Fig. 8-6), or a combination of both. Their spicules are either straight or curved monaxons or tetraxons, never triaxons. The glass sponges (Sclerospongiae) (Fig. 8-6) have siliceous triaxon (six-rayed) spicules.

117

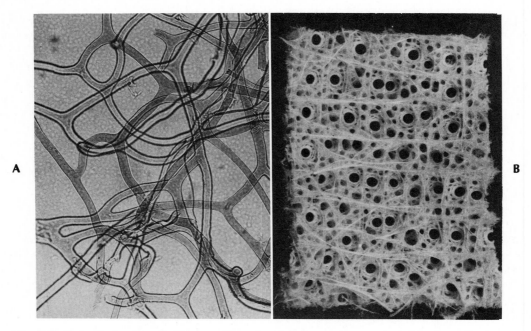

Fig. 8-6. Skeletal elements. **A,** Spongin fibers found in Demospongiae (greatly enlarged). **B,** Portion of wall of glass sponge *Euplectella* (Hyalospongiae) in which the spicules are arranged in a definite pattern (about natural size).

DRAWINGS

On p. 121 draw the following:
1. External view of *Scypha*, gross structure
2. Longitudinal section showing spongocoel and internal ostia
3. Types of spicules you have seen

On p. 122 draw a pie-shaped segment of a transverse section through *Scypha*, showing a few canals and some of the cellular details of their structure.

On p. 122 sketch any other sponges you have studied.

Label all drawings fully.

OTHER TYPES OF SPONGE STRUCTURE
Asconoid Type of Canal System

The asconoid canal system is best seen in *Leucosolenia*, another marine sponge of the class Calcispongiae. *Leucosolenia* grows in a cluster, or colony (Fig. 8-1, *A*), of tubular individuals in varying stages of growth. Large individuals may carry one or more buds.

☞ After observing the external structure of a submerged specimen, cut it in half longitudinally, place it on a slide with a little water, and cover. Study with low power. Or use a prepared slide.

The body wall is covered with pinacocytes on the outside and filled with mesohyl containing amebocytes and spicules. Incurrent pores extend from the external surface directly to the spongocoel, which is lined with flagellated choanocytes. The choanocytes produce the water current and collect food. An osculum serves as the excurrent outlet of the spongocoel. On living specimens you may be able to see some flagellar activity in the spongocoel.

Leuconoid Type of Canal System

The majority of sponges are of the leuconoid type and most leuconoids belong to the class Demospongiae (Fig. 8-1, *C*). In leuconoid sponges there are clusters of flagellated chambers lined with choanocytes, and the water enters and leaves the chambers by systems of incurrent and excurrent canals. Water from the excurrent canals is collected into spongocoels and emptied through oscula. In large sponges there may be many oscula. *Spongilla* and *Heteromeyenia*, which are freshwater sponges, and many marine sponges, such as *Halichondria*, *Microciona*, *Cliona*, and *Haliclona*—all belonging to Demospongiae—are of the leuconoid type. Examine any such sponges available, in both external view and cut sections, to see this type of canal system.

PROJECTS AND DEMONSTRATIONS

1. Freshwater sponges. Examine preserved specimens of the freshwater sponge (*Spongilla*). With low power, look for the **oscula**, which are found scat-

118

tered over the surface. What type of canal system does this sponge have? Look for **gemmules,** the winter bodies so characteristic of these forms. Prepared slides of gemmules may also be observed. Study also prepared slides of cross sections of the sponge.

2. Commercial sponges (bath sponges). The common genus of this leuconoid type is *Spongia.* Commercial sponges have complicated organizations. Study both preserved specimens and dry macerated specimens from which the soft parts have been removed, leaving only the skeleton. Can you observe the canals and oscula? Mount a very small piece in water, tease apart, and study the **spongin fibers** under the low power of the microscope.

3. Demonstrations. Examine a number of different types of sponge skeletons, including some of the glass sponges.

4. Prepared slides. Examine prepared slides of spicules, spongin, gemmules, and sections or whole mounts of various sponges.

5. Preparation of spicule strew. Permanent slides may easily be made of the spicules of any calcareous or siliceous sponge by boiling a fragment of the sponge in a 10% potassium hydroxide solution, or in 5% sodium hypochlorite solution, or in full-strength commercial bleach. Several different kinds of sponges can be boiled simultaneously in separate labeled test tubes in a beaker of boiling water. When the tissue has dissoved, let settle, pipette off the solution, add water to wash, and decant off the water; repeat the washing twice more. Add 90% alcohol, decant, and replace with fresh 90% alcohol. Allow to settle again; then transfer the spicules to a microscope slide and heat gently over a flame to evaporate the alcohol, or apply a match to burn it off. Now apply a drop of Canada balsam or other good mounting medium, add a coverslip, and label.

6. Permanent mounts of gemmules. Place gemmules in a test tube containing cold concentrated nitric acid and let stand 1 to 6 hours, or until they turn a translucent orange or yellow. Then wash several times with water, dehydrate, and mount as indicated above for sponge spicules.

7. Demonstration of water currents. Submerge a living sponge or cluster of sponges in a finger bowl or beaker of clean seawater (or pond water if freshwater sponge is used). Leave undisturbed for several hours or until natural water currents are restored. Fill a Pasteur pipette with a dilute suspension of carmine and gently expel it near the sponge. Note where the carmine crystals enter and leave the sponge. If the sponge does not respond, it may be dead or dying.

8. Sponge embryology. Gemmules of freshwater sponges such as *Spongilla* or *Heteromeyenia* can be collected in fall or winter months and stored in a refrigerator for several weeks. To start development, transfer each gemmule to a coverslip and place in a Petri dish of clean pond water. The sponges should hatch in approximately 10 days to 3 weeks, depending on the season and temperature. As the young sponge spreads out over the coverslip, details of its organization can be studied on the microscope.

9. Field trip to study sponges. In many inland communities there are ponds, lakes, or springs where freshwater sponges are known to occur. Examining sponges firsthand in their habitat is very satisfying, and a specimen can be brought back for further study. For schools along the coast many marine sponges, such as *Cliona, Microciona,* and many others are easily available at low tide. These can be kept for a while in marine tanks, although some sponges may prove toxic to other tank forms if kept there long.

10. Drying whole freshwater sponges. Remove the sponges from water and place in a warm shady place to dry out completely. Dried sponges are very fragile and must be handled carefully. Pack in soft, crushed tissue paper or toilet tissue (never in cotton) for shipping or storing. *Warning:* do not rub your eyes after handling a dry sponge!

Phylum _____   Name _____

Class _____   Date _____ Section _____

Genus _____

SCYPHA

External view   Longitudinal section
                                 Internal view

Calcareous sponge spicules

OTHER SPONGES

CHAPTER 9

T H E   R A D I A T E   A N I M A L S

PHYLUM CNIDARIA AND PHYLUM CTENOPHORA

In the two radiate phyla, Cnidaria (ni-dar'i-a) (=Coelenterata) and Ctenophora (ten-nof'-o-ra), we see many of the basic foundation structures on which all other metazoans build to some extent. There is a definite tendency for similar cells to be collected together to form tissues (the **tissue level of organization**). They have, however, only one really well-defined tissue, the nervous tissue.

Their advancement is shown by (1) their **radial** (or **biradial**) **symmetry** constructed around an oral-aboral (mouth-to-base) axis, (2) embryological primary **germ layers** that are homologous to those of higher metazoans, and (3) a **gastrovascular cavity** with a mouth opening. Some of the cniderians have a skeleton; but in most of the radiates, fluid in the body cavity serves as a simple form of **hydrostatic skeleton.**

Each phylum has, however, some unique characteristics. The cnidarians have characteristic stinging organelles (nematocysts), usually lacking in the ctenophores. Polymorphism is also common in cnidarians, but absent in ctenophores. The ctenophores have distinctive adhesive cells, called "colloblasts," on their tentacles and unique rows of ciliated comb plates not found in other phyla.

PHYLUM CNIDARIA*

There are two main types of body form in the cnidarians—the **polyp** (hydroid) form, often sessile, and the **medusa** (jellyfish) form, which is free swimming. In some groups of cnidarians both polyp and medusa stages are found in the life cycle; in others, such as the sea anemones and corals, there is no medusa; and

*See Hickman, C.P., Jr., L.S. Roberts, and F.M. Hickman. 1984. Integrated principles of zoology, ed. 7. St. Louis, The C.V. Mosby Co., pp. 219-252.

in still others, such as the scyphozoans, or "true jellyfish," the polyp stage is reduced or lacking. In life cycles in which both polyps and medusae are found, the juvenile polyp stage gives rise asexually to the medusa, which reproduces sexually. Both the polyp and medusa have the diploid number of chromosomes, but the gametes are haploid.

Classification

Phylum Cnidaria (=Coelenterata) (ni-dar'i-a)

Class Hydrozoa (hy-dro-zo'a). Both polyp and medusa stages represented, although one type may be suppressed; medusa with a velum; found in fresh and marine water. Examples: *Hydra, Obelia, Gonionemus, Tubularia, Physalia.*

Class Scyphozoa (sy-fo-zo'a). Solitary; medusa stage emphasized; polyp reduced; enlarged mesoglea; medusa without a velum. The true jellyfish. Examples: *Aurelia, Rhizostoma, Cassiopeia.*

Class Anthozoa (an-tho-zo'a). All polyps, no medusae; coelenteron subdivided by mesenteries (septa).

  Subclass Alcyonaria (=Octocorallia) (al'cy-o-na're-a) Polyp with eight pinnate tentacles; septal arrangement octamerous. The "soft corals." Examples: *Gorgonia, Renilla, Alcyonium.*

  Subclass Zoantharia (zo'an-tha'ri-a). Polyp with simple unbranched tentacles; septal arrangement hexamerous; skeleton, when present, external. Sea anemones and stony corals. Examples: *Metridium, Taelia, Astrangia.*

  Subclass Ceriantipatharia (se-ri-ant'i-pa-tha'ri-a). With simple unbranched tentacles; mesenteries unpaired. Tube anemones and black or thorny corals. Examples: *Cerianthus, Antipathes.*

Class Cubozoa (ku'bo-zo'a). Solitary; polyp stage reduced; bell-shaped medusae square in cross section, with a tentacle or group of tentacles at each corner; margin without velum but with velarium; all marine. Examples: *Carybdea, Chironex.*

Class Hydrozoa—*Hydra, Obelia, Gonionemus*

Three hydrozoan forms are traditionally used as examples of this class. The freshwater hydras are easily available, are conveniently large (2 to 25 mm in length), and can be studied alive. The hydras are solitary polyp forms, but they are somewhat atypical of the class because they have no medusa stage.

Obelia is a marine colonial hydroid that is more plantlike than animal-like in appearance. Its hydroid colonies are 2 to 20 cm tall, depending on the species, but its medusae are minute (1 to 2 mm in diameter). *Gonionemus*, another marine form, has a minute, solitary polyp stage but produces a beautiful little medusa about 2 cm in diameter. For convenience we combine the *Obelia* hydroid and the *Gonionemus* medusa for a life-history study.

The hydroid stage, considered to be the juvenile stage, produces, by asexual budding, the medusa, which is the sexual adult.

MATERIALS
Living material
 Hydras
 Artemia larvae,* *Daphnia*, or enchytreid worms
 Marine hydroids, if available
Preserved material
 Obelia
 Gonionemus
Prepared slides
 Stained hydras
 Budding hydras
 Male and female hydras
 Cross sections of hydras
 Obelia colonies
 Obelia medusae
Clean microscope slides
Coarse thread
Watch glasses
Glutathione
Bouin's fluid (see p. 461 for preparation)
Compound microscopes
Dissecting microscopes or hand lenses
Coverslips

HYDRA, A SOLITARY HYDROID
Phylum Cnidaria
 Class Hydrozoa
  Order Hydroida
   Suborder Anthomedusa
    Genus *Pelmatohydra (Hydra)* or *Chlorohydra*

---

*For instructions on rearing brine shrimp (*Artemia*), see pp. 457-458.

Where Found
The hydra is found in pools, quiet streams, and spring ponds, usually on the underside of the leaves of aquatic vegetation, especially lily pads. To collect hydras, bring in some aquatic plants with plenty of pond water and place in a clean battery jar or aquarium. In a day or so, the hydras, if present, may be seen attached to the plants or to the sides of the jar.

Behavior
☞ Place a live hydra in a drop of culture water on a ringed slide or watch glass. Examine with a hand lens or dissecting microscope.

 The hydra may be contracted at first. Watch it as it recovers from the shock of transfer. Does it have great powers of contraction and expansion? How long is it when fully extended? How many **tentacles** does it have? Compare with other specimens at your table. Do the tentacles move about? Touch one with the tip of a dissecting needle and watch its reaction. What parts of the animal are most sensitive to touch? After a quiet period, when the animal is extended, tap the watch glass and note what happens. The elevated area surrounded by tentacles is called the **hypostome** and bears the **mouth**. Does your hydra have any **buds** or **gonads**? Does it attach itself to the glass by its **basal disc**? Does it ever travel about in the dish? How?

 The warty appearance of the body, especially the tentacles, is caused by clusters of special cells called **cnidocytes**, which contain the stinging organoids called **nematocysts**.

> Record your animal's reactions on p. 131.

 Symbionts of Hydra. Small ciliated protozoans such as *Kerona* or *Trichodina* are sometimes seen gliding over the body and tentacles of the hydra, where they live as ectoparasites but apparently do little harm.

 The common green hydra *Chorohydra viridissima* takes its color from symbiotic algae, called zoochlorellae, that live in its cells.

 Feeding and Digestion. To observe the feeding reaction of the hydra; do the following:
☞ Add to the hydra culture on your slide a drop of water containing thoroughly washed *Artemia* larvae or other suitable food organisms, such as *Daphnia* or enchytreid worms.

 How does the hydra react to the presence of food? How does it capture its prey? Does the prey struggle to escape? When does the prey stop moving—before it is eaten or after? What is the reaction of the hypostome? How long does the feeding reaction take? How does the hydra act when its appetite has been satisfied?

If food is not available, or if the hydras will not eat, your instructor may want to demonstrate the feeding reaction by adding a little of the chemical **glutathione** to a watch glass containing some hydras. Glutathione is found in living cells. It is released, in certain of the hydra's prey, from the wound made by nematocysts. It stimulates the hydra to open its mouth and secrete mucus to aid in the swallowing process.

Some digestion occurs within the **gastrovascular** cavity into which gland cells secrete digestive enzymes (**extracellular digestion**). Food particles are then engulfed by cells of the gastrodermis in which digestion is completed (**intracellular digestion**). Indigestible materials must be regurgitated, since there is no anus.

On p. 131 record any feeding reactions you have observed, using notes and sketches.

External Structure

☞ Place a live hydra on a slide, add short (1 cm) pieces of coarse thread above and below it, and cover with a coverslip. Or use a ringed slide, if you prefer.

Note the cylindrical body, at the oral end of which is the conical **hypostome**, which bears the **mouth** and is surrounded by the **tentacles** (Fig. 9-1). The basal, or aboral, end secretes a sticky substance for attachment. Can you make out the outline of the **gastrovascular cavity** by focusing up and down? Is there any anal opening? What kind of symmetry does the hydra have?

Cnidocytes. Examine the tentacles closely on the live specimen and note the little wartlike elevations. These are areas containing groups, or batteries, of specialized cells called **cnidocytes,** each of which encloses a stinging organoid called a **nematocyst** (ne-mat'o-cyst). The nematocyst is a tiny capsule containing a coiled, threadlike, tubular filament that can be everted. Focus on a battery of cnidocytes on the edge of a tentacle. Do you see tiny projecting hairlike "triggers" called **cnidocils**? These are modified flagella and

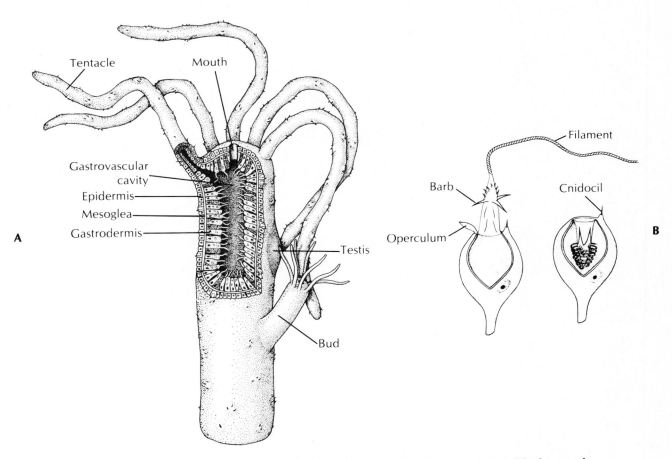

Fig. 9-1. A, Structure of hydra. Both bud and developing gonad are shown, but in life they rarely develop simultaneously. **B,** Cnidocytes with discharged (*left*) and undischarged (*right*) "stinging cells" called nematocysts.

seem to be involved in the discharge of the nemato-cysts. However, a chemical stimulation (as from food) is necessary to lower the threshold before the cnidocil can be stimulated by contact.

Are there as many cnidocytes on the body as on the tentacles? Why? Are there any on the basal disc?

☞ Now add a drop of Bouin's fluid to the slide at the edge of the coverslip. This will cause the nematocysts to be discharged and will lightly stain the cells.* Examine the discharged ne-

---

*A quantity can be prepared by adding Bouin's fluid to a watch glass of specimens. A drop or two of acetocarmine on the preparation is also effective. The specimens may then be stored in 80% alcohol for future use.

matocysts under high power with subdued light.

There are three different kinds of nematocysts in the hydra. The largest and most striking type is equipped with **barbs** near the base of a long hollow **thread** (Fig. 9-1, *B*). **Hypnotoxin,** a poison that penetrates the prey and paralyzes it, is discharged through the hollow thread. The other kinds of nematocysts are smaller and are specialized for adhesion.

Cross Section, Stained Slide

☞ Study a prepared stained slide of a cross section of the body. Examine under both low and high power.

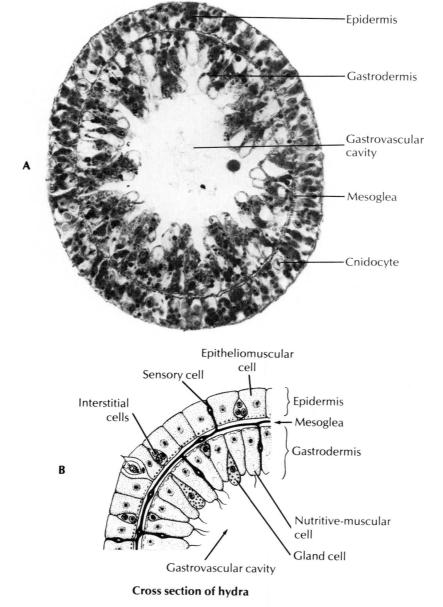

Cross section of hydra

Fig. 9-2. **A,** Cross section of a hydra. **B,** Diagrammatic cross section of a portion of the body wall (**A** courtesy of Carolina Biological Supply Co., Burlington, N.C.)

Note that the body wall is made up of two layers of cells, the outer **epidermis** (from ectoderm) and the inner **gastrodermis** (from endoderm), separated by a thin noncellular layer, the **mesoglea** (mes′o-glee′a) (Fig. 9-2). The hydra is **diploblastic**, being derived from two germ layers, the ectoderm and the endoderm.

Epidermis. Most of the cells in this layer are **epitheliomuscular cells** (Fig. 9-2, *B*). These are medium-sized cuboid cells with darkly stained nuclei. Their inner ends are drawn out into slender contractile fibers that run longitudinally in the mesoglea and make possible the rapid contraction of the animal. The contractile fibers are closely associated with the **nerve net,** which also lies just beneath the epidermal layer. Now look for occasional cnidocytes containing the spindlelike **nematocysts.** At the bases of the epitheliomuscular cells may be found some small dark **interstitial** cells. These are embryonic cells that can transform into the other kinds of cells when needed. **Gland cells** secrete mucus onto the body surface, particularly around the mouth and basal disc.

Mesoglea. The mesoglea layer appears very thin, and the contractile or nerve fibers in it may not be visible.

Gastrodermis. The **gastrodermis** layer is made up principally of flagellated columnar **nutritive-muscular** cells (Fig. 9-2). The flagella create currents in the gastrovascular cavity. Contractile fibers at the base of the cells run in a circular direction in the mesoglea. When these contract, the hydra becomes longer and thinner. The gastrodermis may give the appearance of two layers of cells because the outer part of the cells contain large fluid-filled vacuoles, and the free ends of the cells contain food vacuoles that have been formed by food engulfed by pseudopodia (phagocytosis). Which protozoa and which sponge cells did you find using this method? Intracellular digestion occurs in the nutritive-muscular cells. **Gland cells** (which may be very difficult to see with student microscopes) discharge enzymes into the gastrovascular cavity where extracellular digestion occurs. There are small **interstitial cells** at the bases of the other cells. What types of cells are common to both layers of the body wall?

How does digestion in the hydra compare with that in the sponge? Does the sponge have intracellular or extracellular digestion?

Sensory cells would be found in both layers but would be difficult to identify.

Reproduction

Asexual. Budding is the asexual method of reproduction. A part of the body wall grows out as a hollow outgrowth, or bud, that lengthens and develops tentacles and a mouth at its distal end. Eventually the bud constricts at the basal end and breaks off from the parent. Buds may be found on some of the live hydras.

☞ On a stained slide showing a budding hydra study the relation of the bud to the parent. Note that the gastrovascular cavities of the two are continuous and that both layers of the parent wall extend into the bud.

Sexual. Some species are **monoecious** and have both testes and ovaries; other species are **dioecious.** Sex organs, or gonads, develop in the epidermis (from the interstitial cells) of a localized region of the body column.

☞ Examine stained slides showing testes (spermaries) and ovaries.

The **testes,** small outgrowths containing many spermatozoa, are found toward the oral end; the single **ovary** is a large, rounded elevation nearer the basal end. The ovary produces a large ripe **egg,** which breaks out and lies free on the surface. **Spermatozoa** break out of the testis wall, pass to the egg, and fertilize it in position. The **zygote** so formed undergoes several stages of development before dropping off the parent.

DRAWINGS

On p. 132 sketch a budding hydra and specimens that show testes (spermaries) and ovaries.

OBELIA, A COLONIAL HYDROID
Phylum Cnidaria
  Class Hydrozoa
    Order Hydroida
      Suborder Leptomedusae
        Genus *Obelia*

Where Found

Obelia is one of the many colonial hydroids found in marine waters and attached to seaweeds, rocks, shells, and other objects. Its minute medusae make up a part of the marine plankton.

Life History

The life history of *Obelia* includes both the polyp (hydroid) and the jellyfish (medusa) stages (Fig. 9-3). The hydroid colony arises from a free-swimming **planula** larva, which settles down and attaches itself to a substratum. Then by a process of budding, a colony is formed, which includes two kinds of polyps, the nutritive polyps called **hydranths** and the reproductive polyps called **gonangia.** Medusa buds produced in the gonangia break away to become free-swimming **medusae,** or jellyfish. The medusae are dioecious. When each sex is mature, it discharges its gametes into the

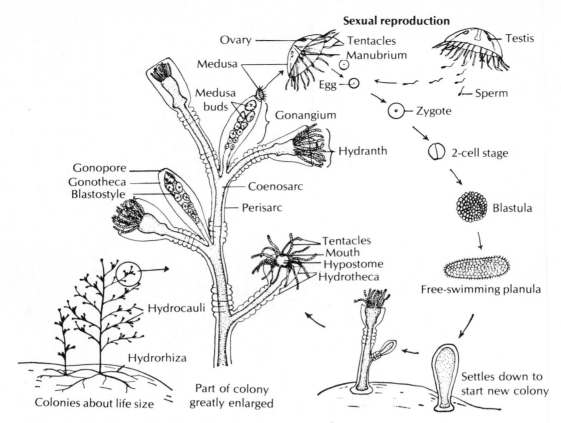

Sexual reproduction

Ovary — Tentacles
Medusa — Manubrium
Medusa buds
Egg
Gonangium
Hydranth
Gonopore
Gonotheca
Blastostyle
Coenosarc
Perisarc
Testis
Sperm
Zygote
2-cell stage
Blastula
Tentacles
Mouth
Hypostome
Hydrotheca
Free-swimming planula
Hydrocauli
Hydrorhiza
Colonies about life size
Part of colony greatly enlarged
Settles down to start new colony

Fig. 9-3. *Obelia* life cycle, showing alternation of polyp (asexual) and medusa (sexual) stages. In *Obelia* both its polyps and stems are protected by continuations of the perisarc. In some hydroids only the stems are so protected.

surrounding water, where fertilization occurs. The zygote develops into the planula larva, which attaches to a substratum, and the cycle is repeated. Thus the medusae give rise sexually to the asexual hydroid colonies, which in turn produce medusae.

Obelia has a macroscopic hydroid stage, but its medusa stage is microscopic. Another hydrozoan form, *Gonionemus*, has a similar life history, but its medusa is fairly large and its hydroid colony inconspicuous. To illustrate the life cycle, it is convenient to use the polyps of *Obelia* and the medusa of *Gonionemus*.

The Hydroid Colony—Behavior

☞ If living colonies of any hydroid species are available, examine them in fresh seawater under low power.

Allow time for some of the tentacled hydranths to extend their tentacles. Determine their reaction to gentle localized touch. They may be fed *Artemia* larvae, and the feeding reaction may be observed as in the hydra.

Make notes or sketches on separate paper.

Study of a Preserved Colony

☞ Place a small piece of seaweed with attached *Obelia* colonies in a watch glass of water and observe with hand lens or dissecting microscope.

The colonies resemble tiny plants attached to the seaweed by rootlike **hydrorhiza**. From the hydrorhiza arises the main stem, or stolon (**hydrocaulus**), which gives rise to many lateral branches. On these branches are found two kinds of individuals, or zooids, the nutritive **hydranths** and the reproductive **gonangia**. The hydranths can be recognized by their vase shape and the tentacles at their free ends, the gonangia by their elongated club shape and lack of tentacles. Note where the different individuals occur on the hydrocaulus. Budding polyps may also be found on your specimen. Buds are usually small and lack tentacles. Note that the entire colony is encased in a thin transparent protective **perisarc**, secreted by the epidermis.

Study of a Stained Slide

☞ Study a stained slide of *Obelia* and compare with the preserved colony and with any living colony you may have examined. Use low power of the compound microscope.

Examine the hydrocaulus. Its inner protoplasmic

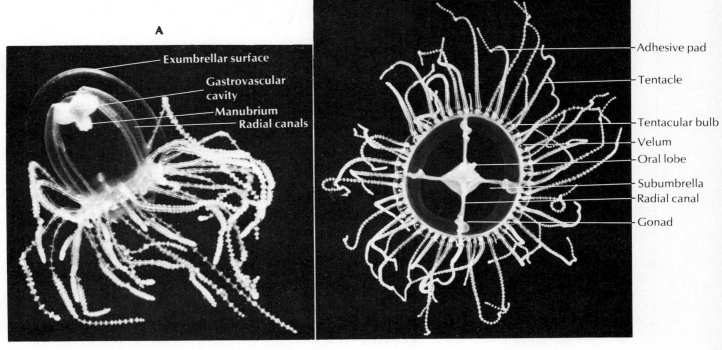

Fig. 9-4. *Gonionemus nurbachi*. **A**, Swimming. **B**, Oral view. (Photographs by D.P. Wilson, Plymouth, England.)

part is the **coenosarc**, a hollow tube composed, like the hydra, of **epidermis**, **mesoglea**, and **gastrodermis**. It encloses a **gastrovascular cavity** that is continuous throughout the colony. Surrounding this living part is the transparent, nonliving perisarc (Fig. 9-3).

Each nutritive polyp, or hydranth, is continuous with the coenosarc. The **hydrotheca** is a transparent extension of the perisarc that forms a protective cup around the hydranth. Each hydranth has an elevated **hypostome** terminating in the **mouth** and bearing a circle of **tentacles** around the base. Each tentacle has in its epidermis rings of swellings caused by clusters of **cnidocytes**, which bear the **nematocysts**. Trace the continuous **gastrovascular cavity** from the hydranth through the branches and stem of the hydrocaulus. The food taken by the hydranths can thus pass to every part of the colony.

Reproductive **gonangia** are club-shaped and arise at the junction of the hydranth and coenosarc. Each gonangium is made up of a hollow continuation of the coenosarc called the **blastostyle**. A number of saucer-shaped **medusa buds** arise from the blastostyle by transverse budding and develop into mature **medusae**. Where are the most mature buds located? The transparent **gonotheca** around the blastostyle is a modification of the perisarc. Young medusae escape through the opening, or **gonopore**, at its distal end.

GONIONEMUS, A HYDROMEDUSA
Phylum Cnidaria
  Class Hydrozoa
    Order Hydroida
      Suborder Limnomedusae
        Genus *Gonionemus*

Where Found
Gonionemus is a marine medusa found mainly in shallow protected coastal and bay areas along both coasts of the United States. Hydrozoan medusae are often called "hydromedusae," as distinguished from the usually larger scyphomedusae, or jellyfish of the class Scyphozoa. *Gonionemus* has a minute polyp stage in its life history.

Behavior
☞ If living hydromedusae, either freshwater *Craspedacusta* or marine *Gonionemus*, or others, are available, watch their movements.

Note the pulsating contractions that force water from the subumbrella and so propel the animal by a feeble "jet propulsion" (Fig. 9-4, *A*). Can they change direction? How? Does the animal swim continuously, or does it float sometimes? What happens if a tentacle is touched? Feeding reactions differ with different species. Some apparently depend on chance contact with

129

food; others, such as *Gonionemus*, swim to the surface and then turn over and float downward with tentacles spread in search of food.

Record your observations on separate paper.

General Structure

☞ Place a preserved *Gonionemus* in a watch glass filled with water. Examine with a hand lens. It is fragile; so do not grasp it with the forceps. Orient it by lifting or pushing with a blunt instrument.

The convex, outer (aboral) surface is called the **exumbrella;** the concave (oral) surface is the **subumbrella.** Each of the **tentacles** around the margin of the bell bears rings of clustered stinging cells, an **adhesive pad** near its distal end, and a pigmented **tentacular bulb** at its base (Fig. 9-4, *B*). The tentacular bulbs make and store nematocysts, help in intracellular digestion, and act as sensory organs. Between the bases of the tentacles are tiny **statocysts,** which are considered to be organs of equilibrium. They are little sacs containing concretions. You will need the low power of the compound microscope to see them.

Now look at the subumbrellar surface. Around the margin find a circular shelflike membrane, the **velum.** The medusa moves by a form of jet propulsion, forcing water out of the subumbrellar cavity by muscular contractions, thus propelling the animal in the opposite direction.

Note the **manubrium** suspended from the central surface of the subumbrellar cavity. At its distal end is the **mouth** with four liplike **oral lobes** around it.

The **gastrovascular cavity** includes the **gullet,** the **stomach** at the base of the manubrium, four **radial canals** extending to the margin, and a **ring canal** around the margin.

Note the convoluted **gonads** suspended under each of the radial canals. The sexes look alike and can be determined only by microscopic inspection of the gonadal contents.

The medusa has the same cell layers as the hydroid or polyp form. All surface areas are covered with **epidermis;** the **gastrodermis** lines the entire gastrovascular cavity and between these two layers is the jel-

lylike **mesoglea,** which is a great deal thicker in the medusa than in the hydroid form.

DRAWING

On p. 132 label the diagrammatic drawing of an oral-aboral section through *Gonionemus*.

PROJECTS AND DEMONSTRATIONS

1. Freshwater jellyfish (Craspedacusta sowerbyi). Freshwater jellyfish are found occasionally in various parts of the United States, chiefly in the eastern half of the country. They have both polyps and medusae in the life cycle, but the polyp is a small (2 mm), nontentacled simple tube with a mouth at one end and the other end attached. It is found in small colonies. The medusa, or jellyfish, is about 15 to 20 mm in diameter and is provided with more than 200 marginal **tentacles.** Note that these tentacles are of different lengths and do not have adhesive discs. There are at least three sets of tentacles. Four of these tentacles are much longer than the others. Are the tentacles attached on the margin of the bell? Do you find sense organs? Note the length of the **manubrium.** The **gonads** are saclike and hang down in the subumbrella. Does this jellyfish have a **velum**? Although it has separate sexes, most individuals living in the same locality are of the same sex.

2. Portuguese man-of-war (Physalia pelagica). The Portuguese man-of-war illustrates the highest degree of **polymorphism** among the cnidarians, for several types of individuals are found in the same colony. Study a preserved specimen. Note the **pneumatophore,** or bladder. What is its function? Suspended from the pneumatophore are many zooids that have budded from it. These are a complex of hydroid individuals, which include nutritive **gastrozooids,** with long stinging tentacles, and medusoid individuals such as the reproductive **gonophores,** the swimming bells (**nectophores**), and the **jelly polyps.**

3. Obelia. Examine some medusae of *Obelia*.

4. Various colonial hydroids. Examine such forms of colonial hydroids as *Eudendrium, Tubularia, Pennaria, Sertularia,* or *Bougainvillia,* either preserved or on slides.

OBSERVATIONS ON HYDRA BEHAVIOR

FEEDING REACTIONS OF THE HYDRA

Budding hydra   Hydra with testes   Hydra with ovary

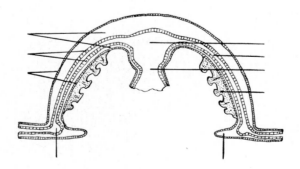

GONIONEMUS,
Diagrammatic oral-aboral
section—to be labeled

EXERCISE 9B

Class Scyphozoa—*Aurelia*, a "true" jellyfish

MATERIALS
Preserved *Aurelia*
Living jellyfish, if available
Hand lenses or dissecting microscopes
Finger bowls

AURELIA

Phylum Cnidaria
    Class Scyphozoa (= Scyphomedusae)
      Order Semaeostomae
        Genus *Aurelia*

Where Found

Aurelia is common along both coasts of North America. It is mainly confined to the coastal waters of temperate and tropical latitudes, although some may be found in the polar regions.

Scyphozoan medusae (often referred to as scyphomedusae) are generally larger than hydrozoan medusae (hydromedusae), most of them ranging from 2 to 40 cm, but some reaching as much as 2 m or more in diameter. The jelly layers (mesoglea) are thicker and contain cellular materials, giving the medusae a firmer consistency than the hydromedusae. Scyphozoans are often called the "true" jellyfish. Scyphomedusae are constructed along a plan similar to that of hydromedusae, but they lack a velum. Their parts are arranged symmetrically around the oral-aboral axis, usually in fours or multiples of four, so that they are said to have **tetramerous** radial symmetry. Their gastrovascular systems have more canals and more modifications than those of hydrozoans.

Behavior

If living *Aurelia* or other scyphozoan genera are available, observe their swimming movements (Fig. 9-5, A). How is movement achieved? Are they strong swimmers? How many times per minute does the medusa pulsate? Does it swim horizontally or vertically? Use a gentle touch with a small camel's hair brush to test reaction to touch. To test response to food chemicals, the brush may be dipped into glucose solution, clam or oyster juice, or other food substances; or small crustaceans or other small food organisms may be placed near their tentacles.

Record your observations on separate paper.

General Structure

☞ Using a ladle (the medusa is too fragile to be handled with a forceps), transfer a preserved specimen of *Aurelia* to a finger bowl of water, and spread out flat.

Note that *Aurelia* is more discoidal and less cup shaped than *Gonionemus*. When spread flat, the jel-

A   **B**

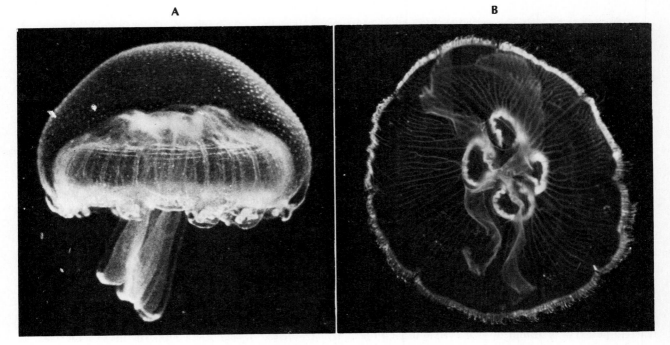

Fig. 9-5. *Aurelia*. **A**, Swimming. **B**, Oral view. (Photographs by D.P. Wilson.)

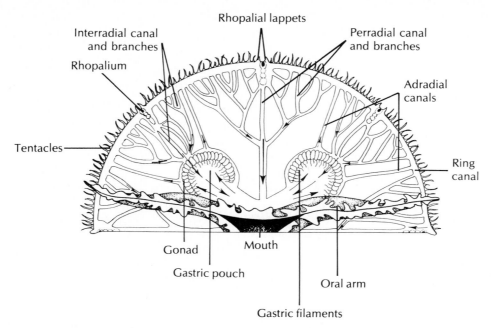

Fig. 9-6. Structure of *Aurelia*. Arrows show direction of circulation in the canal system of the gastrovascular cavity.

lyfish shows a circular shape broken at eight regular intervals by marginal notches (Fig. 9-5, *B*). Each marginal notch contains a **rhopalium** (ro-pay′li-um), a sense organ consisting of a statocyst and an ocellus. This is flanked on each side by a protective marginal **lappet** (Fig. 9-6). What is the function of a statocyst? An ocellus?

☞ Snip out a rhopalium with scissors and examine under the higher power of a dissecting microscope.

In living medusae the excision of all rhopalia would interfere with swimming, either slowing down the contractions or stopping them altogether.

Note the short **tentacles** that form a fringe around the margin of the animal. Compare these tentacles with those of *Gonionemus* medusae. How do they differ?

Gastrovascular System. In the center of the oral side are four long troughlike **oral arms.** These are modifications of the manubrium. Note that the oral arms converge toward the center of the animal, where the square **mouth** is located. The mouth opens into a short **gullet,** which leads to the **stomach.** From the stomach four **gastric pouches** extend. They can be identified by the horseshoe-shaped **gonads** that lie within them. Near the inner edges of the gonads are numerous thin processes, the **gastric filaments,** which are provided with **nematocysts.** What would be the use of stinging cells here? On the oral side of each gastric pouch is a round aperture that leads into a blind depression, the **subgenital** pit. These pits have

no connection with the gonads or gastric pouches and may be respiratory in function. A complicated system of **radial canals** runs from the gastric pouches to the **ring canal** that follows the outer margin (Fig. 9-6). This system of stomach and canals, arranged to resemble the hub, spokes, and rim of a wheel, forms the medusoid gut, or gastrovascular cavity. Contrast this with the simple, saclike gut of polypoid individuals.

Aurelia medusae feed on plankton. The organisms are caught in mucus secreted on the exumbrella and licked off the margin by the oral arms. Water currents created by flagella carry the particles along grooves in the arms to the stomach. The gastric filaments with their nematocysts help pull in and subdue the larger organisms and also secrete digestive enzymes. Food particles are carried by flagella to the gastric pouches, then through unbranching adradial canals to the ring canal (Fig. 9-6). Fluid returns from the ring canal by way of the branching perradial and interradial canals. Digestion is both extracellular and intracellular.

Reproduction. Sex cells are shed from the gonads into the gastrovascular cavity and are discharged through the mouth for external fertilization. Within the folds of the oral arms the young embryos develop into free-swimming **planula larvae.** These escape from the parent, attach to a substratum, and develop into tiny polyps called **scyphistomae.** These later begin to strobilate, or bud off young medusae (**ephyrae**), in layers resembling a stack of saucers.

DEMONSTRATIONS

1. Slides. Examine slides that show stages in the life cycle of *Aurelia*—scyphistoma, strobila, ephyra.

2. Scyphozoan jellyfish. Examine various species of preserved scyphozoans.

3. Cassiopeia. Cassiopeia is a genus of jellyfishes (order Rhizostomeae) common in the shallow waters around Florida. They can usually be obtained from Florida marine supply houses and can be kept for several days or weeks in a marine aquarium. This interesting jellyfish has eight thick gelatinous oral lobes that are fused in such a manner as to obliterate the central mouth and form numerous canals with small openings in the oral lobes. The oral lobes also bear many small tentacles with nematocysts. There are 16 rhopalia, but no tentacles around the scalloped margin.

Although *Cassiopeia* can swim by rhythmic pul-sations of the bell, it spends much of its time lying oral side up on the bottom of lagoons and tidal pools, anchored by a suckerlike action of its aboral surface. (Its common name is the "upside-down jellyfish.") Here, as it pulsates, it draws water over its oral lobes, bringing in food and oxygen. Small organisms are paralyzed by nematocysts, entangled by mucus, and swept into the canals by flagellar action.

The young scyphistoma that forms from a planula larva buds off a single ephyra (young medusa) at a time (monodisc formation), instead of the strobila formation found in *Aurelia*.

Cassiopeia can be cultured in marine tanks of artificial seawater. The scyphistomae reproduce readily by two asexual methods, one by budding off young medusae, the other by budding off small planuloid larvae that detach and swim about and finally settle down and develop into scyphistomae.

EXERCISE 9C

Class Anthozoa—*Metridium*, a sea anemone, and *Astrangia*, a stony coral

MATERIALS

Metridium, preserved
Living sea anemones and corals, if available
Finger bowls and/or dissecting pans
Dried corals
Astrangia, preserved

METRIDIUM

Phylum Cnidaria
    Class Anthozoa
        Subclass Zoantharia
            Genus *Metridium*

Where Found

Metridium is a genus of sea anemones commonly found on wharf piles or stones and in rocky crevices from shallow water to a depth of 50 to 75 m along the coast. Most sea anemones are solitary sessile animals and do not live in colonies. The members of the class Anthozoa are all polyps in form; there are no medusae. There is a great variety in size, structure, and color among the sea anemones (Fig. 9-7). All are marine.

Behavior

If living anemones are available, allow one to relax completely and then touch a tentacle lightly with a glass rod or glass needle. What is the response? Stroke the tentacle more firmly. Do other tentacles also respond? Can you explain this on the basis of what you know of cnidarian neuromuscular anatomy?

Using another relaxed anemone, drop bits of clean filter paper on the tentacles and time the type and speed of the response. Now test with bits of filter paper-soaked in clam or mussel juice and compare the reactions. Test again with bits of clam or other sea food. Compare.

Is the response to food similar to the response to touch? Is the reaction of both animals a part of the normal feeding reaction? Use the glass rod to probe the pedal disc. What happens? If live sea stars are available, try touching an anemone with the arm of a star. What happens? Is this a feeding response or a defense reaction? Some anemones react to certain predatory stars by detaching their pedal discs and moving away from the star.

> Record your observations on separate paper.

Fig. 9-7. Sea anemones, the "flowers of the sea." **A,** *Corynactis viridis*, or jewel anemones. These are small red or pink anemones with white-tipped tentacles, often found on the underside of overhanging rocks. **B,** *Anemonia sulcata* has unusually long tentacles. **C,** *Cerianthus* is a burrowing anemone that builds a tube in the sand, through which it can ride up to feed or down for protection. (**A,** photograph by D.P. Wilson, Plymouth, England; **B** and **C,** photographs by C.P. Hickman, Jr.)

External Structure

☞ Place a preserved specimen in a dissecting pan. Note the sturdy nature of its body structures compared with those of other cnidarians you have studied.

The **body** is cylindrical in shape (Fig. 9-8), but in preserved specimens it may be somewhat wrinkled. Note that the body of the animal can be divided into three main regions: (1) the **oral disc** or free end, with numerous conical **tentacles** and the **mouth;** (2) the cylindrical **column,** forming the main body of the organism; and (3) the **basal disc** (aboral end), by which during life the animal attaches itself to some solid object by means of its glandular secretions. Although it is called a sessile animal, the sea anemone can glide slowly on its basal disc.

Is there more than one row of tentacles? Note that the inner surface of the mouth is lined with ridges and that a smooth-surfaced ciliated groove, the **siphonoglyph** (sy'fun-o-glif), is found at one side of the mouth. (In some specimens there may be two of these grooves.) What is the function of the siphonoglyph? The mouth is separated from the nearest tentacles by a smooth space, the **peristome.**

Note the tough outer covering (**epidermis**) of the specimen. Small pores on tiny papillae are scattered over the epidermis, but they are hard to find.

Internal Structure

☞ Study of the internal anatomy is best made by a comparison of two sections, one cut longitudinally through the animal and a second cut transversely. Study first one and then the other of these two sections to get the general relations of the animal. Determine through what part of the animal the transverse section was made.

Look at the longitudinal sections and note that the **mouth** opens into a **pharynx,** which extends down only partway in the body where it opens into the large **gastrovascular cavity.** Thus the upper half of the body appears as a tube within a tube; the outer tube is the **body wall,** and the inner tube is the pharynx. Look at the cross section and determine these relations. Notice that the gastrovascular cavity is not only the space below the gullet but it also extends upward to surround the pharynx.

The gastrovascular cavity is subdivided into six **radial chambers** by six pairs of **primary** (complete) **mesenteries** (septa), which run from the oral to the aboral end. In the gullet region these primary mesenteries extend from the body wall to the gullet; below or aboral to the gullet their inner edges are free in the gastrovascular cavity (Fig. 9-8). Are they the same

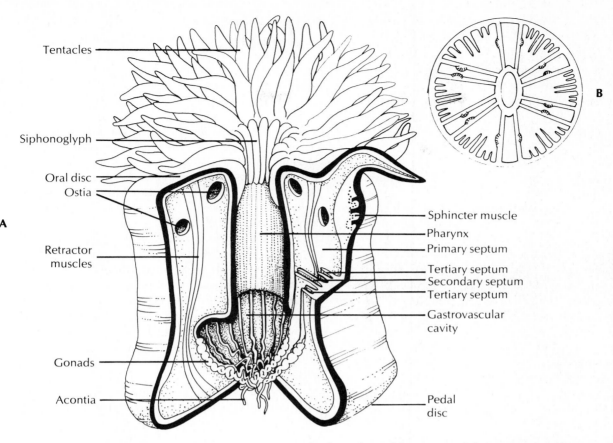

Tentacles

Siphonoglyph

Oral disc
Ostia

A

Retractor
muscles

Gonads

Acontia

Sphincter muscle
Pharynx
Primary septum
Tertiary septum
Secondary septum
Tertiary septum
Gastrovascular
cavity

Pedal
disc

B

Fig. 9-8. A, Structure of a sea anemone. The free edges of the septa and the acontia threads are equipped with nematocysts to complete the paralyzation of prey begun by the tentacles. **B,** Diagrammatic transverse section through the pharynx showing arrangement of the septa.

width throughout their length? Examine transverse sections to determine this. Note that the chambers formed by the primary mesenteries in the gullet region communicate with each other by means of small oval holes, the **ostia.** Find these ostia next to the gullet near the oral end.

Now notice that these six larger chambers are partially subdivided by smaller pairs of **incomplete mesenteries** (septa), which extend varying distances from the body wall into the gastrovascular cavity. Note that all these incomplete mesenteries are free at their inner edges, although they extend from the oral to the aboral ends of the animal. Each mesentery is composed of a double sheet of gastrodermis.

The free edges of the mesenteries are trilobed and are called **septal filaments.** They bear nematocysts and secrete digestive enzymes. Aboral to these filaments are attached long delicate threads called **acontia** (Fig. 9-8). Each acontium is provided with stinging cells and may be protruded through the mouth or body

pores for defense. The **gonads** are thickened bands on the mesenteries, lying next to the digestive filaments. Is *Metridium* monoecious or dioecious? In what ways does this polyp form resemble the hydra? In what ways is it different?

ASTRANGIA, A STONY CORAL
Phylum Cnidaria
   Class Anthozoa
      Subclass Zoantharia
         Genus *Astrangia*

The stony corals resemble small anemones but are usually colonial. Each polyp secretes a protective calcareous cup into which the polyp partly withdraws when disturbed. Some corals form colonies consisting of millions of individuals, each new individual building its skeleton upon the skeletons of dead ones, thus forming, over many years, great coral reefs. Reef-building corals build only in tropical or subtropical

137

waters, where the water temperature stays at or above 21° C (70° F).

Astrangia lives in temperate waters along both the North Atlantic and Gulf coasts. It does not build reefs but forms small colonies of 5 to 30 individuals encrusted upon rocks or shells. Its food consists of small organisms, such as protozoans, hydroids, worms, crustaceans, and various larval forms.

Behavior

The same sort of touching and feeding experiments as suggested for *Metridium* are applicable to corals, making allowance for the smaller size.

Structure

☞ Examine living or preserved coral polyps.

Note the delicate, transparent polyps expanded from the circular skeletal cups (Fig. 9-9). The polyps resemble those of anemones, with a column, oral disc, and crown of tentacles. Two dozen or more simple tentacles, supplied with nematocysts, are arranged in three rings around the mouth. Siphonoglyphs are absent. The edges of the mesenteries can usually be seen through the transparent polyp walls. They form a hexamerous pattern, as in the sea anemones. Digestion in the gastrovascular cavity is similar to that of anemones.

Colonies usually arise by budding or division from a single polyp that has been sexually produced. The surface of the colony between the thecae is covered by a sheet of living tissue, which is an extension of the polyp walls. This tissue connects all members of the colony and contains a gastrovascular cavity that is continuous with the gastrovascular cavity of the polyps.

☞ Now study a piece of skeleton from which the polyps have been removed.

Each cup, called a **corallite** (Fig. 9-9, *B*), was the home of a polyp, which secreted it. The cups are located on a mass of calcium carbonate known as the **corallum.** The rim of the cup is called the **theca,** and the base is the **basal plate.** The **sclerosepta,** or radial partitions within the thecae, form the same hexamerous pattern, as do the mesenteries. They are laid down by the folds of the epidermis at the base of the polyp. The theca is formed by the epidermis of the lower part of the column. As the coral polyp grows, the cup tends to fill up with calcium carbonate, and the theca and sclerosepta are continually extended upward. Note how the complete sclerosepta fuse at the center to form the **columella.** Between them are the incomplete sclerosepta with free ends. The entire skeleton is outside the body of the polyp.

PROJECTS AND DEMONSTRATIONS

1. Various hard coral skeletons. Examine such skeletons as the white corals; the red coral, *Corallium;* the staghorn coral, *Acropora;* the organpipe coral, *Tubipora;* or the brain coral, *Meandrina.*

2. Preserved soft corals. Examine forms such as the sea plume, *Gorgonia,* or the sea pansy, *Renilla. Gorgonia* is a horny coral that has a wide distribution.

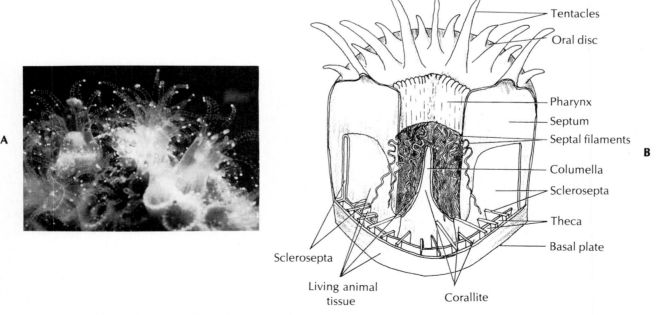

Fig. 9-9. Stony corals. **A,** Living polyps of *Astrangia danae*. **B,** Structure of a coral polyp, diagrammatic.

Its polyps are embedded in a layer of mesoglea that contains calcareous spicules. The central core of the branch is an axial rod. The polyps are connected by tiny canals called solenia (or stolons) in the mesoglea layer.

The sea pansy has a leaf-shaped body on which the polyps are scattered, and a stem or peduncle, one end of which is embedded in the soft sea floor. As in *Gorgonia*, the gastrovascular cavities of the polyps are all connected by the network of solenia.

3. *Spicules of soft corals.* Spicules of various shapes can be extracted by boiling a small piece of the coral in a strong aqueous solution of potassium hydroxide. When the cells have dissolved, let the spic-ules settle, and pour off the fluid. Add water, let settle, and pour off again. Do this several times; then pipette some of the spicules to a slide and add a cover glass.

4. *Nematocysts.* Nematocysts may be collected from various live cnidarians. Zoantharian corals (soft corals) usually have fairly large nematocysts, and the little red colonial anemone, *Corynactis*, is also good for this purpose. Wiping a clean coverslip across the oral disc is likely to attract both discharged and undischarged nematocysts, which can be studied by inverting the coverslip over a drop of seawater on a slide. A good quality microscope is necessary.

Collect and compare the nematocysts of various hydrozoans, scyphozoans, and anthozoans.

EXERCISE 9D
Phylum Ctenophora—*Pleurobrachia,* a comb jelly*

MATERIALS
Living ctenophorans, if available
Preserved *Pleurobrachia*
Finger bowls
Dissecting microscopes or hand lenses

PLEUROBRACHIA
Phylum Ctenophora
  Class Tentaculata
    Order Cydippida
      Genus *Pleurobrachia*

Where Found
Ctenophores, or comb jellies, live in both shallow and deep waters in all seas. *Pleurobrachia pileus* is common along our New England coast and *P. bachei* is more abundant on the western coast. *Mnemiopsis* is a larger form found along the Atlantic and Gulf coasts. *Pleurobrachia* are often called "sea walnuts" or "sea gooseberries" (Fig. 9-10). Ctenophores are easily collected by towing a plankton net.

Behavior
Comb jellies are fragile, transparent animals. Live specimens do not ship well, but if they are available, you will find them beautiful to watch. Note the eight ciliated **comb rows** (Fig. 9-11), which are the means of locomotion. The "combs," or plates of cilia, beat in succession, with those at the aboral end beating first, causing the action to pass down the rows in waves to propel the animal aboral end first. Note the activity of the long trailing **tentacles** on which small prey that chance to come in contact are captured and held by specially adapted glue cells (colloblasts or collocytes). Food is ingested by drawing the tentacles across the mouth. Ctenophores are carnivorous, feeding largely on planktonic larvae, small copepods, and fish eggs.

Ctenophores are bioluminescent. This is easily demonstrated if the laboratory can be darkened. A touch or a disturbance of the water will usually cause a flash of light in animals accustomed to the dark.

WRITTEN REPORT

Report your observations on the living specimens on separate paper.

Structure
☞ Study a preserved specimen of *Pleurobrachia* submerged in water.

Locate the **sense organ** at the aboral pole and the **mouth** at the oral pole. Note the two lobes, one on each side of the mouth. Ctenophores have biradial symmetry. How does this differ from the radial symmetry of the cnidarians? Find near the aboral pole the two openings of the **tentacle sheaths.** The **tentacles** in preserved specimens may be wholly retracted into the sheaths. Look at the eight **comb rows** that extend as meridional lines from near the aboral pole toward the oral pole.

☞ Cut out a piece of the body wall containing a comb row, place in a watch glass of water or mount on a slide, and examine under the mi-

---

*See Hickman, C.P., Jr., L.S. Roberts, and F.M. Hickman. 1984. Integrated principles of zoology, ed. 7, St. Louis, The C.V. Mosby Co., pp. 245-249.

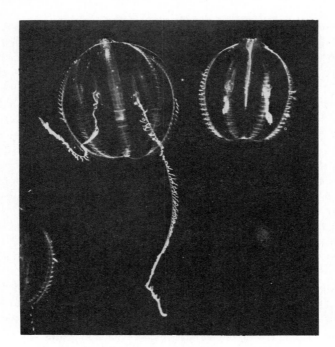

Fig. 9-10. Living sea walnuts, *Pleurobrachia*. (Photograph by D.P. Wilson, Plymouth, England.)

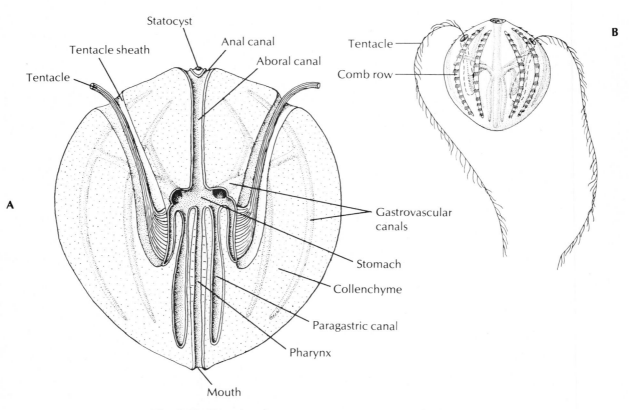

Fig. 9-11. *Pleurobrachia*. **A,** Hemisection. **B,** External view.

croscope. Note that each row is made up of many small plates, or ctenes (combs), each consisting of a row of long cilia.

By enlarging the cut opening of the body wall you can observe the tentacle sheaths and the branched tubular **gastrovascular cavity,** consisting of pharynx, stomach, two pairs of branched gastrovascular canals, and an aboral canal that ends near the sense organ. Carmine or India ink suspensions may be injected into the mouth to aid in following this system.

Between the **epidermis** and **gastrodermis** lies the **collenchyme,** a watery jelly that makes up the bulk of the animal and serves as a simple type of hydrostatic skeleton.

☞ Cut out the aboral cap of epidermis, mount on a slide, and examine the sense organ under the microscope.

This is a **statocyst,** or balancing organ, consisting of a small concretion, or statolith. The statolith is balanced on the ends of the four tufts of fused cilia, called balancers, that send impulses to the comb rows.

Ctenophores are monoecious (hermaphroditic).

Sex cells are produced in the gastrodermis and, when ripe, are discharged through the mouth. Fertilization occurs in the seawater.

Use p. 142 for sketches of corals you have studied or for notes on the behavior of living corals.

PROJECTS AND DEMONSTRATIONS

Experiments with living ctenophores might include (1) making cuts at various levels in the comb-plate rows and recording any changes in patterns of activity, (2) severing one or more of the ciliated bands that run from the statocyst to the comb rows, (3) excising the sense organ and noting the effects on ciliary coordination, equilibrium, and locomotion, (4) determining the location of muscle fibers by electrical stimulation, (5) timing the phases of digestion by feeding vitally stained plankton, or (6) determining the amount of stimulation necessary to produce flashes of luminescence in ctenophores that have been kept for some time in the dark.

CHAPTER 10

THE ACOELOMATE ANIMALS

PHYLUM PLATYHELMINTHES

THE ACOELOMATE PHYLA

The acoelomate animals, are animals that have no coelom (body cavity). They include the phylum Platyhelminthes (Gr., *platys*, flat, + *helmins*, worm) and the phylum Rhynchocoela (Gr., *rhynchos*, beak, + *koilos*, hollow). In acoelomate animals the space between the body wall and the digestive tract is not a cavity, as in higher animals, but is filled with muscle fibers and a loose tissue of mesenchymal origin, called **parenchyma,** both derived from mesoderm.

The platyhelminths, or **flatworms,** are a large and economically important group because they include not only the free-living **planarians** but also the parasitic **tapeworms** and **flukes.**

The Rhynchocoela (formerly called Nemertina) are the **ribbon worms,** often called nemertine or nemertean worms. They are all marine and are characterized by an eversible **proboscis** that can be thrown out with great speed to capture food. Because of the general unavailability of the nemertines, that phylum will not be included in the laboratory exercises.

Acoelomates are advanced over the radiate animals in several ways. (1) Their **bilateral symmetry** adapts them for forward movement. (2) Their tissues are well defined and organized into complex functional **organs.** (3) The nervous system is more highly organized, with a concentration of nervous tissue and sense organs in the anterior end (**cephalization**). (4) They no longer depend on simple diffusion for elimination of nitrogenous wastes but have an **excretory system** of specialized flame cells and tubules. (5) The platyhelminths have a gastrovascular system, but the ribbon worms have separated the two functions and have a complete mouth-to-anus digestive tract and a circulatory system.

Flatworms are said to have a tissue-organ level of organization.

PHYLUM PLATYHELMINTHES*

The phylum consists of the following classes:

Class Turbellaria (tur′bel-lar′e-a). Mostly free living, with a ciliated epidermis. Example: *Dugesia tigrina,* the brown planarian.

Class Monogenea (mon′o-gen′e-a) The monogenetic flukes. Body covered with a syncytial tegument without cilia; leaflike to cylindrical in shape; posterior attachment organ with hooks, suckers, or clamps, usually in combination; all parasitic, mostly on skin or gills of fishes; single host; monoecious; usually free-swimming ciliated larva. Examples: *Polystoma, Gyrodactylus.*

Class Trematoda (trem′a-to′da) The digenetic flukes. Body covered with nonciliated syncytial tegument; leaflike or cylindrical in shape; usually with oral and ventral suckers, no hooks; development indirect, first host a mollusc, final host usually a vertebrate; parasitic in all classes of vertebrates. Examples: *Fasciola, Clonorchis, Schistosoma.*

Class Cestoda (ses-to′da) The tapeworms. Body covered with nonciliated, syncytial tegument; scolex with suckers or hooks, sometimes both, for attachment; long ribbonlike body usually divided into series of proglottids; no digestive organs; parasitic in digestive tract of all classes of vertebrates; first host may be invertebrate or vertebrate. Examples: *Taenia, Diphyllobothrium.*

---

*See Hickman, C.P., Jr., L.S. Roberts, and F.M. Hickman. 1984. Integrated Principles of zoology, ed. 7, St. Louis, 1984, The C.V. Mosby Co., pp. 255-270.

EXERCISE 10A
Class Turbellaria—the planarians

MATERIALS
Live planarians
Prepared slides
  Planaria, stained whole mounts
  Bdelloura, stained whole mount
  Planaria, cross sections
Powdered carmine or talc
Black paper
Finger bowls
Camel's hair brushes
Small mirrors
Small flashlights (pocket size)
Modeling clay
Petri dishes
Spring, pond, or well water or dechlorinated tap water (*not* distilled or demineralized water)
Raw liver (or cut-up mealworms)
MS-222,* 1:1500 solution (0.066%)
Watch glasses or depression slides
Hand lenses or dissecting microscopes

---

*Tricaine methane sulphonate, available as "Finquil" from Ayerst Laboratories.

DUGESIA
Phylum Platyhelminthes
  Class Turbellaria
    Order Tricladida
      Family Planariidae
        Genus *Dugesia*

Where Found
Freshwater triclads, or planarians, are found on the underside of stones or submerged leaves or sticks in freshwater springs, ponds, and streams. Common forms are the brown planarian, *Dugesia tigrina*, adapted to warm sluggish water, and the black planarians, *D. dorotocephala* and *Dendrocoelopsis vaginatus*, found in cool spring waters. Other suitable genera for study are *Dendrocoelum* and *Polycelis*. Planarians can be collected on pieces of raw beef tied to stones or plants near the water's edge and can be kept in the laboratory if fed small amounts of fresh meat and if the water is changed frequently to prevent pollution. Fig. 10-1 shows a few of the freshwater turbellarians.

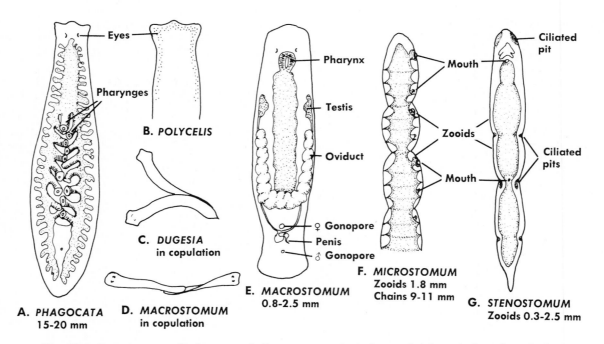

Fig. 10-1. Some common freshwater turbellarians. **A,** to **C,** Order Tricladida with three-branched gastrovascular cavity. *Phagocata* differs from most in having many pharynges. **D** to **G,** Order Rhabdocoela, with straight, unbranched gastrovascular cavity. They are small species, rarely more than 4 mm. *Macrostomum* is a single individual, but *Microstomum* and *Stenostomum* usually forms chains of zooids. Note repetition of mouth and ciliated pits as new zooids form.

Observation of Live Planarians

☞ Using a small camel's hair brush, place a live planarian on a ringed slide, depression slide, or Syracuse watch glass in a drop of culture water. Replace the water as it evaporates. Keep the surrounding glass dry to prevent the animal from wandering out of range. By holding the slide above a mirror you can also observe its ventral side.

External Structure. Observe the animal and decide which are its **anterior, posterior, dorsal,** and **ventral** aspects. Note the triangular **head.** Its earlike **au-**ricles (Fig. 10-2, A) bear many sensory cells, but they are tactile and olfactory, not auditory in function. Are the **eyes** movable? Do they have lenses? Note the pigmented **skin.** Is it the same color on the underside? Is its coloring protective? Are the length and breadth of the worm constant? Holding the slide for a moment over a microscope lamp, can you locate the muscular **pharynx** along the midline? When the animal feeds, the pharynx can be protruded through a ventral **mouth** opening. Can you verify this by use of the mirror?

Use a hand lens, dissecting microscope, or low

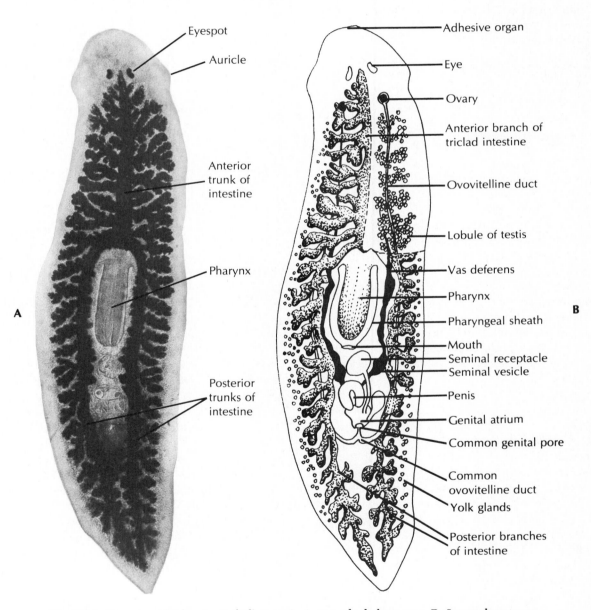

A **B**

Fig. 10-2. Planaria, a freshwater turbellarian. **A,** A stained whole mount. **B,** Internal anatomy; one half the anterior trunk of the intestine is removed to reveal the testes. (**A** courtesy Carolina Biological Supply Co., Burlington, N.C.)

145

power of a compound microscope for further examination of the eyes and body surface.

Locomotion. Observe the animal's gliding movement. Glands in its ciliated epidermis secrete a path of mucus on which the planarian propels itself by means of its cilia. Do you think the cilia alone are responsible for its movement? What do you think causes the waves of contractions along its body? Does the animal ever leave the drop of water and travel on the dry glass? Why? How does it use the head and auricles? Does it ever move backward?

Reactions to Stimuli. Observe the responses of planarians to touch (**thigmotaxis**), food (**chemotaxis**), and light (**phototaxis**) by doing some of the following simple experiments.

1. Response to touch. *Very gently* touch the outer edges of the worm with a piece of lens paper or a soft brush. What parts of its body are most sensitive to touch? Are its reactions more localized or less localized than those of the hydra?

2. Response to food. To observe the pharynx, smear a bit of fresh beef liver (or cut-up mealworms) over a slide, spreading it thin with a scalpel or the edge of another slide. Invert the slide over two supports (such as matchsticks or modeling clay) in a Petri dish, add culture water or pond water to barely cover the slide, and introduce into the dish a couple of planarians that have not been fed for several days. Examine under a dissecting microscope. The planarians will soon find the meat, and you will be able to watch the pharynx.

3. Response to light. Make a cover of black paper or cardboard to fit over and around a finger bowl containing several planarians. Leave a small opening in one side of the cover and place the bowl and cover in a well-lighted position. Check at 5- to 10-minute intervals to see whether the animals are attracted to or repelled by the light.

Are they positively or negatively phototaxic? Study in your textbook the structure and function of the eyespots.

4. Response to directional illumination. Direct a beam of light (a small flashlight will do) at the planarians from one side of the dish and observe their movements. Move the light 90 degrees around the dish and direct it again at the planarians. Do you think their movements indicate a "trial and error" response or a directed response? Try using two beams of light crossing at right angles. How do the animals react?

5. Response to light and dark backgrounds in nondirectional illumination. Prepare the lower half of a Petri dish by covering its sides and one half of its bottom on the outside with black tape (black paint or black paper will do). Set the dish on a white surface so that half of the bottom is black and the other half is white. Cut a circle of black paper a little larger in diameter than the top of the Petri dish. Cut out the center of the circle, leaving a ring of paper about 3 cm wide, or wide enough to extend inward from the edge of the dish about 1.5 cm to shade the sides of the dish from overhead illumination. Place the dish in a dark room or box, with a light source several feet above the dish. Place a few planarians in the center of the dish and leave for a while. When they have ceased moving about, count the animals on the dark surface and those on the white surface and compare the number. Now remove the animals from the dish, add a suspension of talc or carmine to the water, and rotate the dish gently. The movements of the planarians can be seen where the talc particles adhere to the mucous trails left on the bottom of the dish. Do you note any difference in the length of the trails on the black and white surfaces? Is there a directional response?

WRITTEN REPORT

On pp. 151 and 152, report your observations and the conclusions you drew from observing the live planarians.

Regeneration in Planarians

Planarians have remarkable powers of regeneration and are easy to work with.

Before starting to operate on a worm, have ready the materials needed and decide exactly how you wish to cut the worm. You may want to cut it in three pieces, one cut anterior to the pharynx and the other posterior to it. Or you may want to bisect it longitudinally or divide the head or the tail longitudinally (Fig. 10-3). In the latter case, be sure to hold the blade in the cut for a moment to prevent the pieces from coming together again and healing in the original position.

Have ready a razor blade and as many watch glasses as you will need for the pieces, each dish half filled with clean culture water or aged tap water. Do not use distilled water. Prepare an "operating table" by folding a piece of lens or toilet tissue over a glass slide. Transfer the planarian to the wet paper by means of a camel's hair brush or a pipette. Let the animal relax and extend fully before cutting but *do not let it become dry*. When it is fully extended, cut quickly and cleanly. When the cut has been made, dip the paper into the watch glass and rinse the pieces off into the water with a gentle stream of water from a pipette. Never scrape them off with a hard instrument.

Another method sometimes used is to place the planarian on a cube of ice. Let the worm extend fully; then cut it right on the ice and rinse off the pieces.

For delicate operations, such as the removal of an

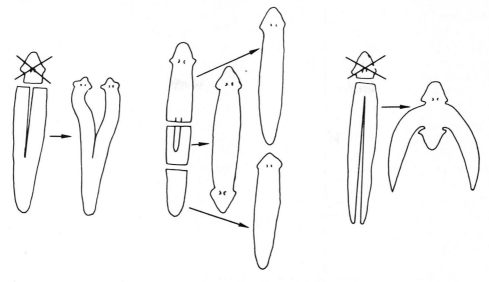

Fig. 10-3. Some suggestions for regeneration experiments.

eyespot, the animals may be anesthetized in a 1:1500 (0.066%) solution of the commercial preparation MS-222 (tricaine methanesulfonate) for 15 minutes. The animals will remain quiescent during the operation and will recover in half an hour or so when returned to the culture water.

Number or label the dishes for your record. A piece of adhesive tape will hold your name, date, and dish number and may be attached to the side of the watch glass. The dishes may be stacked and the top one covered with an empty dish. Two or three times a week, carefully pour off the water, rinse the bottom of the dish, and refill with fresh water. The worms will cling to the bottom of the dish. It is important that the dishes be kept covered and not allowed to dry up. Do not feed the worms. Remove any dead pieces (they will appear grayish or fuzzy) at once.

As the tissue begins to heal, a blastema will form on the cut edge. Is growth faster at the posterior end of a cephalic piece or at the anterior end of a caudal piece? On decapitated pieces, when do new eyes appear? When can you distinguish new auricles? Which is regenerated first, eyes or pharynx? Do you find evidence of polarity in the manner of growth of these pieces?

DRAWINGS

On the record sheet, p. 153, make a sketch of the shape of each piece. Examine the specimens twice a week if possible, each time recording the date and sketching the appearance of each regenerating piece. Summarize the results on p. 154.

Observation of Stained Whole Mounts

The stained whole mount of the planarian shows an animal that was fed food mixed with India ink, carmine, or some other suitable stain before killing and fixing, resulting in a darkly stained gastrovascular tract.

☞ Using the dissecting microscope or the low power of a compound microscope, study stained whole mounts of the freshwater planarian and the marine *Bdelloura*, identifying the following structures:

The Digestive System. As in the cnidarians the digestive tract of the turbellarian is a **gastrovascular cavity,** the branches of which fill most of the body (Fig. 10-2, *B*). Since there is no anus, undigested food is ejected through the mouth.

The muscular **pharynx** is enclosed in a **pharyngeal sheath,** but its free end can be extended through the ventral **mouth.** Ingestion occurs by the muscular sucking action of the pharynx. The pharynx opens into the intestine, which has one **anterior trunk** and two **posterior trunks,** one on each side of the pharynx. Branching **diverticula** increase its capacity and its digestive surface. Some digestion occurs within the lumen of the digestive cavity by means of enzymes secreted by intestinal gland cells (**extracellular digestion**). As in the cnidarians, digestion is completed within the phagocytic cells of the gastrodermis (**intracellular digestion**).

Reproduction. Flatworms are monoecious, and their reproductive system is complex (Fig. 10-2, *B*). Because of the large digestive tract most of the reproductive organs of turbellarians are obscured on the stained whole mounts. However, the penis and genital pore may be seen on the *Bdelloura* slides. See your text for further description of the reproductive system and Fig. 10-1, *C* and *D*, for planarian copulation. The reproductive system will be more easily studied in the liver flukes and tapeworms.

Planarians and other freshwater turbellarians also reproduce asexually by transverse fission; in some species chains of zooids are formed asexually (Fig. 10-1, *F* and *G*).

Excretion and Osmoregulation. The excretory system, consisting of **excretory canals** and **protonephridia (ciliated flame bulbs),** cannot be seen in the whole mounts; see your text for a discussion of this system. The main function of the protonephridial system may be the regulation of the internal fluid content of the animal (osmoregulation). The system is often absent in marine turbellarians.

Nervous System. *Bdelloura* is a marine turbellarian that lives as an ectocommensal on the horseshoe crab, *Limulus*. Properly stained slides show the **ladder type of nervous system,** as well as the triclad digestive tract.

On a stained whole mount of *Bdelloura* find the **cerebral ganglia** at the anterior ends of the two lateral **nerve cords. Transverse nerves** connecting the cords and **lateral nerves** extending outward from the cords form the "rungs" of the ladder.

Sense Organs. A pair of **eyespots,** or **ocelli,** are light-sensitive pigment cups (Fig. 10-2, *A*). Chemoreceptive and tactile cells are abundant over the body surface, especially on the auricles.

DRAWINGS

1. On p. 155 complete the drawing of the external morphology of the planarian and label fully.
2. The outline of *Bdelloura* on p. 155 contains an outline of the digestive system. Draw in and label the cerebral ganglia, the nerve cords, and the transverse and lateral nerves.

Transverse Sections of Planarian

The appearance of a cross section will depend on whether it is cut from the anterior, middle, or posterior part of the planarian. Sections cut anterior to the pharynx will contain a centrally located section through the anterior trunk of the intestine; those cut posterior to the pharynx will contain laterally located sections of the two posterior trunks of the intestine; those cut through the pharynx will show the conspicuous round muscular pharynx, with branches of intestine on each side (Fig. 10-4).

The **epidermis** of ciliated cuboidal epithelial cells (derived from ectoderm) contains many dark rodlike **rhabdites,** which, when discharged in water, swell and form a protective mucous sheath around the body.

Inside the epidermis is a layer of **circular muscles;** then there is a layer of **longitudinal muscles** (cut transversely and appearing as dark dots). **Dorsoventral muscle fibers** are also visible, particularly at the sides of the animal. Do these muscles explain the waves of contractions you saw in the living animal? Note that there is *no body cavity*. The flatworms are called **acoelomate** animals. **Parenchyma** (largely mesodermal) is the loose tissue filling up space between the organs.

Several hollow sections of the intestine and its diverticula (derived from endoderm) may be seen, depending on the location of the section. What kind of epithelial cells makes up the intestinal walls? In a middle section, note the thick circular **pharynx,** covered and lined with epithelium and containing layers of circular and longitudinal muscle similar to those in the body wall. The **pharyngeal chamber** is also lined with epithelium.

Nerve cords, reproductive and excretory ducts, testes, and ovaries are found in the parenchyma of adult animals, but they are difficult to identify.

DRAWING

Make a drawing on p. 155 of such section(s) as your instructor directs. Label completely.

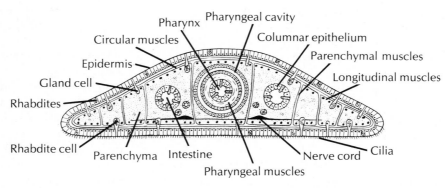

Fig. 10-4. Diagrammatic cross section of a planarian through the region of the pharynx.

PROJECTS AND DEMONSTRATIONS

1. Gradients of planarians. Section each of several live planarians into three parts—cephalic, body, and caudal pieces—and place the parts in three dishes labeled appropriately and containing fresh pond water. In 3 to 5 days colorless blastemas will protrude from the cut edges. Submerge the pieces in a 1% methylene blue solution for 5 to 8 hours or until specimens are dyed. Place each piece on a depression slide under anaerobic conditions by sealing the cover glass with petrolatum. When the oxygen is used up, the color disappears. Which end of the pieces shows the color response first? On exposure to air again, at which end do the blastemas regain color more rapidly? If color change is correlated with oxygen consumption, what might this tell you about the rate of metabolism in various parts of the body? What is meant by "polarity"?* How does this experiment demonstrate the theory?

2. Flame cells. Flame cells of the excretory system are difficult to see in the intact planarian because they are usually obscured by pigment in the integument. To demonstrate flame cells, compress the body (or sectioned portion of the body) of a planarian between a slide and a coverslip to crush and partially disperse the tissues. Search the tissue debris with high power of the microscope. Flame cells are located by the rapid flickering movement of the ciliated tuft. When you have located one (often several flame cells are found close together), switch to oil immersion for closer study. The acoelomate excretory system is a **protonephridium.** The flame cells (or flame bulbs) are cup-shaped terminations containing tufts of cilia that act to propel fluids through the excretory canals. They may also help to filter materials from the body fluids surrounding the flame cells.

3. Microstomum and Stenostomum. Microstomum and *Stenostomum* are common microscopic freshwater varieties that reproduce primarily by transverse fission. On living or mounted specimens, note the constrictions dividing the animal into chains of **zooids,** which will become the new individuals (Fig. 10-1, *F* and *G*).

Note the mouth, pharynx, and simple intestine. Some species have eyespots; some do not. Some have ciliated pits. *Microstomum* is interesting in that it has nematocysts in the epidermis apparently obtained by feeding on hydras. According to some authors the nematocysts may actually be discharged when the flatworm is bumped sufficiently by its prey.

---

*See Hickman, C.P., and others: Integrated Principles of zoology, ed. 7, St. Louis, 1984, The C.V. Mosby Co., p. 142.

REACTIONS OF PLANARIANS

Name _____

Date _____ Section _____

Locomotion _____

_____

_____

_____

_____

_____

_____

_____

Response to touch (thigmotaxis) _____

_____

_____

_____

_____

_____

_____

Response to food (chemotaxis) _____

_____

_____

_____

_____

_____

_____

_____

Response to light (phototaxis) _____

_____

_____

_____

_____

_____

Response to directional illumination _____

_____

_____

_____

_____

_____

_____

_____

Response to light and dark background in nondirectional illumination _____

_____

_____

_____

_____

_____

_____

_____

Name _____

Date _____ Section _____

Sketches of original cuts of specimens:					
Date	1	2	3	4	Control

Record of growth:					
Dates	1	2	3	4	Control

Method used

Results and conclusions

Phylum _____   Name _____

Class _____   Date _____ Section _____

Genus _____

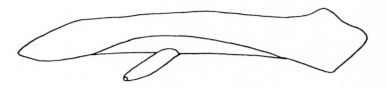

A. External morphology of planarian, shown
with pharynx extended

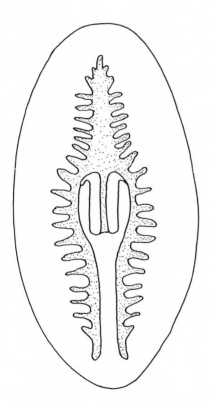

B. Nervous system of *Bdelloura*
(digestive system is already shown)

C. Transverse section(s) of planarian

Phylum _____

Class _____

Genus _____

TAPEWORM

Scolex and neck

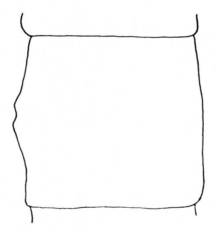

Mature proglottid

Gravid proglottid

EXERCISE 10B
Class Trematoda—the digenetic flukes

MATERIALS
Stained slides
  Clonorchis
  Miracidia
  Cercariae
Pithed leopard frogs (Rana pipiens)
Live snails (Physa, Cerithidea, other)
Dissecting tools
Normal saline solution (0.65%)
Syracuse watch glasses
Microscope slides
Microscopes, compound and dissecting

CLONORCHIS, THE LIVER FLUKE OF HUMANS

Phylum Platyhelminthes
  Class Trematoda
    Order Digenea
      Species: Clonorchis sinensis (human liver fluke)

Where Found

All trematodes are parasitic, harbored in or on a great variety of animals. Many of them have three different hosts in their life cycle.

The adult, or sexual, stage of Clonorchis is found in the human bile duct, chiefly in persons in Southeast Asia and the Far East. The asexual, or larval, stages are found in aquatic snails and fishes.

Study of a Stained Whole Mount

☞ Study a stained slide of an adult fluke, first with a hand lens, then with the low power of the microscope.

How long is the specimen? Compare it with the planarian in size and shape. Note the **oral sucker** at the anterior end and the **ventral sucker** (acetabulum) on the ventral surface. What is their function?

Body Wall. The body covering of both flukes and tapeworms was formerly believed to be a nonliving cuticle, but the electron microscope reveals it to be a living syncytial **tegument** consisting of the protoplasmic processes of cells that dip down into the parenchyma. Electron photographs also show the outer surface of the tegument to be adapted for parasitic life by having minute fingerlike processes, called **microtriches,** that can interlock with the microvilli of the host's intestinal mucosa, thus helping to anchor the worm in the host's intestine. The body surface is not ciliated.

The circular and longitudinal **muscle layers** of the body wall are similar to those of turbellarians, and body spaces are filled with **parenchyma.**

Digestive System. The **mouth** lies anteriorly in the oral sucker (Fig. 10-5). It leads to a muscular **pharynx** and short **esophagus** that divides into two lateral branches of the digestive tract. There is no anus. The fluke feeds on the body tissues and fluids of the host, sucked in by the pharynx. In view of the life habits of the planarians and the flukes, why do you think the digestive system is so much more extensive in the planarian than in the fluke?

Protonephridial System. The **excretory pore** at the posterior end is the outlet of the **excretory vesicle.** Follow the vesicle forward to see where it divides into two long tubules. The tubules collect from flame bulbs, which you will not be able to see.

Reproduction and Life Cycle. Flukes are monoecious; each animal has both male and female systems (Fig. 10-5). The male **testes** are conspicuous branched organs located one in front of the other in the posterior half of the body. From each testis a slender duct (**vas efferens**) extends forward and then unites with the one on the other side to form a single

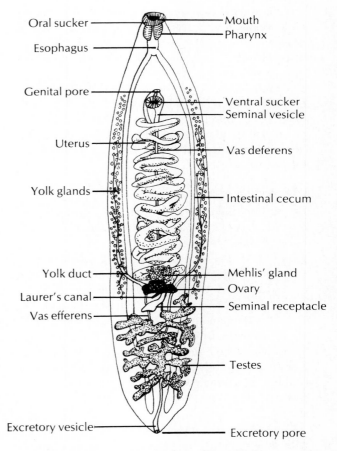

Fig. 10-5. Clonorchis, the liver fluke of humans.

sperm duct (**vas deferens**). This duct enlarges anteriorly to form a swollen **seminal vesicle** that empties through a small **genital pore** just anterior to the ventral sucker. A penis and cirrus sac are found in many trematodes but not in this species.

The female system consists of a single trilobed **ovary** anterior to the testes, a short **oviduct**, a small **Mehlis' gland** (formerly called the shell gland) near the ovary, large granular-looking **yolk (vitelline) glands** along the sides (each with a **yolk duct** to the region of the Mehlis' gland), an oval **seminal receptacle** connected to the oviduct, and a large coiled **uterus**, usually full of eggs, extending up to the **genital pore.** You will probably not be able to identify the oviduct and Mehlis' gland.

Eggs from the ovary are discharged into the oviduct where they are fertilized by spermatozoa that have been introduced through the **Laurer's canal** during copulation and stored in the seminal receptacle. The Laurer's canal is a copulatory canal extending from the seminal receptacle to the dorsal body surface. The zygotes acquire yolk and shells as they travel to the uterus. Cleavage stages have already begun by the time the eggs are discharged into the bile or feces of the human host.

When shed into the water, the ciliated embryos, or **miracidia,** are ingested by aquatic snails. Here they develop into baglike **sporocysts** that produce, asexually, a generation of larval forms called **rediae**, which in turn produce, asexually, a number of tadpole-like **cercariae.** These leave the snail and, if they are successful in finding a suitable fish, bore under the scales or into the muscle. Here they encyst. If the fish is eaten raw or improperly cooked by humans, young flukes emerge, travel to the bile ducts, and mature into sexual adults.

Observation of Living Flukes

Living lung flukes and bladder flukes can usually be obtained from pithed leopard frogs *(Rana pipiens)* that have not been too long in captivity.

☞ Make a ventral incision into the body cavity of the frog and remove the lungs and bladder. Place the organs in individual Syracuse watch glasses of normal saline solution (0.65%). Tease each organ apart with teasing needles, looking for living flukes.

Lung Flukes. You may find either lung flukes or nematodes or both in the frog lungs. The nematodes (Phylum Nematoda) are slender, transparent roundworms that thrash about by bending in S and C shapes.

The flukes are mottled dark and light flatworms with slow writhing movements. More than 40 species of *Haematoloechus* (Fig. 10-6, *A*) have been found in the lungs of amphibians over the world. How does their movement compare with that of planarians? Do they have eyes? The uterus is probably so filled with thousands of eggs that most internal organs are obscured.

Place a specimen on a slide with some tap water and a cover glass. The tap water and the pressure of the glass may cause the animal to discharge most of its small brown eggs and thus leave the animal transparent enough to study.

You may now be able to see the intestinal tract filled with ingested blood, and the uterus with its remaining eggs. The eggs of *Haematoloechus* are ingested by a certain water snail in which the cercariae develop. The free-swimming cercariae encyst in the gills of the aquatic nymph of a dragonfly. The frog becomes infected by ingesting either the nymph or the adult dragonfly.

Bladder Flukes. In the urinary bladder, look for small flukes with large suckers, or acetabula. Study the details of the flukes on a slide with a drop of saline solution. Several species of *Gorgodera* and *Gorgoderina* occur in the bladder (Fig. 10-6, *B* and *C*). These have a very large ventral sucker at the end of the anterior third of the body. In their life cycle miracidia enter small bivalves (family Sphaeriidae), and the cercariae that develop there are usually ingested by a damselfly nymph or other aquatic insect larva that serves as food for the frog or tadpole.

Polystoma (Fig. 10-6, *F*) is a fluke that has at its posterior end a large holdfast, or **opisthaptor**, consisting of six suckers and two hooks located on a muscular disc. Reproduction in *Polystoma* is associated with the frog's breeding season, and there is only one host. Its larval stage is parasitic on the gills of the tadpole.

Miracidia. If bladder flukes were found in the frog bladder, examine the saline solution in which the bladder was teased. It may contain eggs or the ciliated miracidia. If not present, this stage may be seen on prepared slides.

Living Cercariae. Cercariae may usually be found by crushing the common pond snail *Physa* in a watch glass and removing the broken shell. Just visible to the naked eye, the cercariae show up well under the dissecting microscope, characterized by their erratic swimming movements.

Or you may use the common mud snail (for example, *Cerithidea californica*) by cracking the tip of the shell, removing the liver to a watch glass, and examining under a dissecting microscope. Uninfested livers are usually brown or green, whereas infested livers appear orange or tan or mottled.

Place some of the cercariae on a microscope slide, cover, and examine with the compound microscope. Do you find any miracidia in the snail? Why?

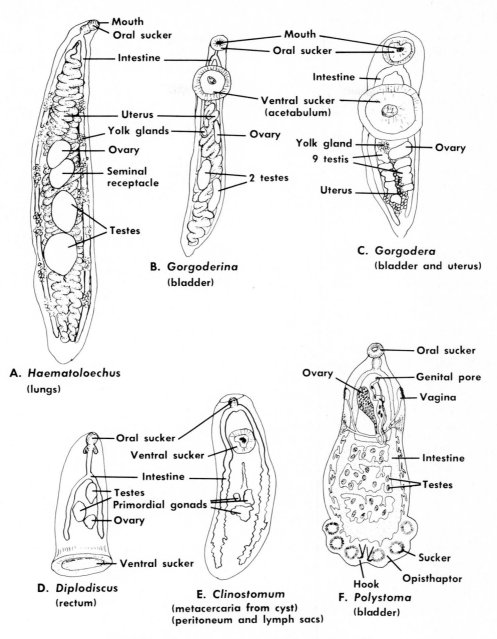

Fig. 10-6. Some common trematode parasites of frogs.

DRAWINGS (OPTIONAL)

Sketch on p. 162 any living flukes, miracidia, or cercariae you have found. Did you find any nematodes?

PROJECTS AND DEMONSTRATIONS

1. Prepared slides. Study prepared slides of (a) *Schistosoma, Paragonimus,* or other parasite and (b) the ova, miracidia, rediae, and cercariae of flukes.

2. Making permanent mounts of frog parasites. Remove the parasites from the lungs or bladder to a dish of 15% alcohol. When they have ceased moving, place one in the center of a clean glass slide, blot with a bit of paper towel, and cover with 3 or 4 drops of stain mountant. Drop a coverslip onto the preparation (straight down, not obliquely), allowing the medium to flow out to the edge of the slip. Keep level until firm.

EXERCISE 10C
Class Cestoda—the tapeworms

MATERIALS
Preserved tapeworms
Prepared slides of tapeworms showing scolex and immature, mature, and gravid proglottids
Microscopes
Hand lens

TAENIA
Phylum Platyhelminthes
  Class Cestoda
    Order Cyclophyllidea
      Genus *Taenia*
        Species *T. pisiformis* (dog tapeworm)

Tapeworms are all endoparasitic and show such extreme adaptations to their parasitic existence as the total lack of a digestive system and great overemphasis on reproduction. Most of them require two hosts of different species, with the adult tapeworm always living in the digestive tract of a vertebrate.

As their name implies, tapeworms are ribbonlike and their long bodies are usually made up of segments called **proglottids** (not to be confused with metamerism in higher forms) that are formed by a continuous process of **budding** from the head end, or **scolex**. As new proglottids are formed anteriorly, the older ones are pushed backward so that the oldest, or most mature, proglottids are always at the posterior end. The scolex, which serves as a holdfast, is usually equipped with **suckers** or **hooks** or both. The body covering, or **tegument,** is similar to that of the flukes.

The form described here is *Taenia pisiformis*, a dog or cat tapeworm whose larval stage is found in the liver of rabbits. However, other forms may be substituted in the laboratory. Among those that parasitize humans are *T. solium*, the larval stage of which encysts in the muscle of pigs and which, like *T. pisiformis*, possesses both hooks and suckers; *Taeniarhynchus saginata*, which has cattle as the alternate host and has no hooks; and *Diphyllobothrium latum*, whose larval stage is found in crustaceans or fishes and whose scolex is modified to bear long, sucking grooves, called bothria. The small dog tapeworm, *Dipylidium caninum*, is picked up by dogs or cats from its alternate hosts, which are lice or fleas. Its scolex bears a retractile rostellum armed with circlets of thornlike hooks.

General Structure

☞ Study a preserved whole specimen or one embedded in plastic.

How long is the specimen? Examine the **scolex** under a dissecting microscope and identify the **hooks** and **suckers.** Note the **neck** from which new proglottids are budded off. As older proglottids are pushed back by the addition of new ones, they mature and become filled with reproductive organs. Can you distinguish maturing from very young proglottids by the presence of a **genital pore** at one side? As eggs develop and become fertilized, the proglottid becomes distended (gravid) with the uterus filled with embryos. Where do you find **gravid proglottids**? Such proglottids soon break off and are shed in the feces of the host. Ingested by a suitable alternate host, the young embryos of *T. pisiformis* migrate to the liver and encyst there as cysticerci (bladder worms). How might a rabbit become infected? How would a dog acquire the adult tapeworm?

Microscopic Study

☞ Examine with low power a prepared slide of a tapeworm whole mount containing (1) the scolex with the neck and a few immature proglottids, (2) a mature proglottid, and (3) a gravid proglottid.

The Scolex. In *T. pisiformis* the scolex has both **suckers** (how many?) and a **rostellum** bearing a circle of **hooks.** What is the function of hooks and suckers? Is there a mouth? Is the **neck** segmented? Two lightly stained tubes, one on each side, are **excretory canals,** which extend the entire length of the animal. Some of the **immature proglottids** have developed **genital pores.**

The Mature Proglottid. The mature proglottid is a little narrower at its anterior end (Fig. 10-7). From the lateral **genital pore** two tubes extend medially. The more anterior of these, the **sperm duct,** is convoluted and branches from the middle of the proglottid into many small efferent ducts that collect sperm from many small **testes** scattered through the proglottid. The more posterior duct is the slender **vagina,** which curves posteriorly between the branched **ovaries** to connect with short **oviducts.** The oviducts also connect with the long central **uterus.** **Mehlis' gland** (function unknown) and a larger **vitelline gland** (yolk gland) are found posterior to the uterus.

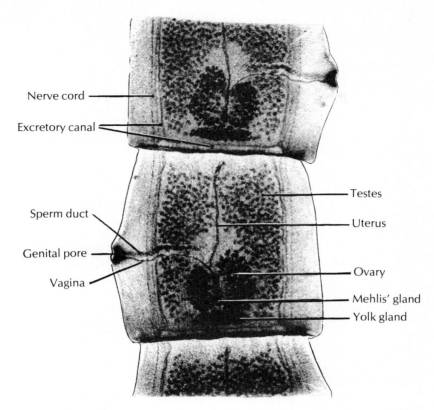

Nerve cord

Excretory canal

Testes

Uterus

Sperm duct

Genital pore

Vagina

Ovary

Mehlis' gland

Yolk gland

Fig. 10-7. Photomicrograph of mature proglottids of *Taenia pisiformis*, the dog tapeworm.

Note the **excretory canals** on each side. They are part of the protonephridial system and empty to the outside at the posterior end of the worm. They are connected by a **transverse canal** across the posterior margin of each segment.

The **nerve cords** run just lateral to the excretory ducts. The tapeworm nervous system resembles that of the planarian and fluke, although it is less well developed.

Do you find any digestive organs in the tapeworm? Why? What is the tapeworm's food? Where is it digested? How does the worm obtain it? In what ways is the structure of tapeworms adaptive? What control methods can be used against pork and beef tapeworms?

The Gravid Proglottid. Look at one of the **gravid proglottids.** What structures can you identify besides the distended **uterus?** Study the **eggs** with high power. (Be very careful when you change to high power, since these are whole mounts and therefore thicker than average.) Can you distinguish in any of the eggs a six-hooked **onchosphere** or embryo?

PROJECTS AND DEMONSTRATIONS

1. Measly meat. Study a piece of "measly" meat infected with tapeworm cysts or bladder worms.

2. Slides of cross sections of tapeworm. Identify the tegument, parenchyma, excretory ducts, nerve cords, uterus, and other reproductive organs that appear in section.

3. Preserved specimens or prepared slides. Examine preserved specimens or prepared slides of (a) various species of tapeworms and (b) prepared slides of ova and cysts.

161

CHAPTER 11

THE PSEUDOCOELOMATE ANIMALS*

PHYLUM NEMATODA
PHYLUM ROTIFERA
PHYLUM GASTROTRICHA
PHYLUM NEMATOMORPHA
PHYLUM ACANTHOCEPHALA

All the bilateral animal phyla except the acoelomates possess a **body cavity** belonging to one of two types: (1) a **true coelom** in which a peritoneum (an epithelium of mesodermal origin) covers both the inner surface of the body wall and the outer surface of the visceral organs contained in the cavity; or (2) a **pseudocoel,** or body cavity not entirely lined with peritoneum.

A body cavity of either type is an advantage because it provides room for organ development and for storage and allows some freedom of movement within

the body. Since the cavity is often fluid filled, it also provides for a hydrostatic skeleton in those forms lacking a true skeleton.

Several of the pseudocoelomate phyla were formerly grouped together as classes in a phylum called Aschelminthes (Gr., *askos*, wineskin, + *helmins*, worm), but the recent tendency is to separate these groups. There are some eight pseudocoelomate phyla, of which phylum Nematoda is by far the largest.

In general the pseudocoelomates tend to be cylindric in body form and unsegmented and have a complete (mouth-to-anus) digestive tract. The epidermis is usually covered with a cuticle. There are both aquatic and terrestrial members, and parasitism is fairly common.

*See Hickman, C.P., Jr., L.S. Roberts, and F.M. Hickman. 1984. Integrated principles of zoology, ed. 7. St. Louis, The C.V. Mosby Co., pp. 279-301.

EXERCISE 11A
Phylum Nematoda—*Ascaris* and others

The nematodes are an extensive group with worldwide distribution. They include terrestrial, freshwater, marine, and parasitic forms. They are elongated roundworms covered with a flexible, nonliving cuticle. Circular muscles are lacking in the body wall, and the longitudinal muscles are arranged in four groups separated by epidermal thickenings.

MATERIALS

Preserved *Ascaris* specimens
Prepared slides
  Stained *Ascaris* cross sections
  Trichinella cysts in pork muscle
  Trichinella, male and female
  Necator americanus
  Enterobius vermicularis
  Turbatrix
Living *Turbatrix*
Soil samples, if desired (see Soil Nematodes, p. 167)
Slides and coverslips
Dissecting pans
Microscopes

ASCARIS, THE INTESTINAL ROUNDWORM

Phylum Nematoda
  Class Phasmidia (Secernentea)
    Order Ascaridata
      Genus *Ascaris*
        Species *A. suum*

Where Found

Ascaris lumbricoides is a common intestinal parasite of humans. *A. suum*, which parasitizes pigs, is so similar to the human parasite that it was long considered merely a different strain of *A. lumbricoides*. *A. megalocephala* is common in horses.

Intestinal roundworms are often present in their hosts (human or otherwise) in such great numbers as to cause serious disorders.

General Features

☞ Place a preserved *Ascaris* in a dissecting pan and cover with water.

Females, which run 20 to 49 cm in length, are more numerous and are larger than males, which av-

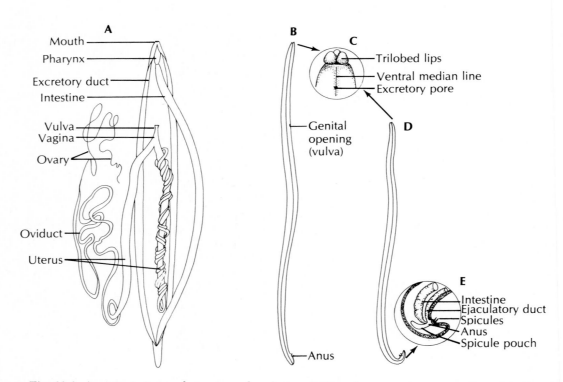

Fig. 11-1. *Ascaris.* **A,** Internal structure, dorsal view. **B,** Female, external structure, lateral view. **C,** Anterior end, ventral view. **D,** Male, external structure, lateral view. **E,** posterior end of the male, internal view.

erage 15 to 31 cm in length. The males have a curved posterior end and two chitinous **spicules** projecting from the anal region. The spicules are used to hold the female's vulva open during copulation. How long is your specimen? Is the body segmented? Compare your specimen with one of the other sex.

With a hand lens, find the **mouth** with three **lips**, one dorsal and two ventral (Fig. 11-1, *C*). Find the ventral **anus** at the posterior end. The anus in the male not only discharges feces from the rectum but also serves as a genital opening. The female genital opening (**vulva**) is located on the ventral side about one-third the length of the body from the anterior end. It may be hard to distinguish from scars. Use the hand lens.

Note the shiny **cuticle** that covers the body wall. It is nonliving and consists primarily of **collagen**, which is also found in vertebrate connective tissue.

Four **longitudinal lines** run almost the entire length of the body—the **dorsal** and **ventral median lines** and two **lateral lines.** The dorsal and ventral lines, which indicate the location of bundles of nerve fibers, are very difficult to see on preserved specimens. However, along the lateral lines the body wall is thinner, and the lines usually appear darker or somewhat transparent. Excretory tubes are located inside the lateral lines.

Internal Structure

☞ You will probably have a female specimen. Place the worm in a dissecting pan and cover it with water. Locate the lateral lines, where the body wall seems somewhat thinner. Now find the anus and vulva on the ventral side. This should help you identify the opposite, or middorsal line. Now slit open the body wall *along the middorsal line,* keeping the points of the scissors close up against the wall to avoid injuring the internal structures. Pin back the body wall to expose the viscera, *slanting the pins outward* to allow room for dissection.

Body Wall and Pseudocoel. Note the body cavity. Why is it called a **pseudocoel?** How does it differ from a true coelom? Note the fluffy masses lining the body wall. These are the large nucleated protoplasmic portions of the **longitudinal muscle cells** whose fibers extend longitudinally in the body wall. With your teasing needle, tease out some of the fibers from the cut edge of the wall. Examine fibers and cells under the microscope. Absence of circular muscles accounts for the thrashing movements of these animals. Note the absence of muscle cells along the **lateral lines.**

Excretory System. Excretory canals located in the lateral lines unite just back of the mouth to empty ventrally through an **excretory pore.** The canals are largely osmoregulatory in function. Excretion also occurs through the cuticle. Flame cells are lacking in *Ascaris* and other nematodes, although they are found in some other pseudocoelomate phyla.

Digestive System. The mouth empties into a short muscular **pharynx,** which sucks food into the ribbonlike **intestine** (Fig. 11-1, *A*). The intestine is thin walled for absorption of digested food into the pseudocoel. Trace it to the **anus.** What is meant by "tube within a tube" construction? Does *Ascaris* fit this description? Does the planarian?

Digestion is begun extracellularly in the lumen of the intestine and is completed intracellularly in the cells of the intestinal wall.

There are no respiratory or circulatory organs. Oxygen is obtained mainly from the breakdown of glycogen within the body, and distribution is handled by the pseudocoelomic fluid.

Reproductive System. The female reproductive system fills most of the pseudocoel. It is a Y-shaped set of long, convoluted tubes. Unravel them carefully with a probe. The short base of the Y, the **vagina,** opens to the outside at the **vulva.** The long arms of the inverted Y are the **uteri.** These extend posteriorly and then double back as slender, much-coiled **oviducts,** which connect the uteri with the threadlike terminal **ovaries.** Eggs pass from the ovaries through the oviducts to the uteri, where fertilization occurs and shells are secreted. Then they pass through the vagina and vulva to the outside. The uteri of an ascaris may contain up to 27 million eggs at a time, with as many as 200,000 eggs being laid per day. Study the life history of *Ascaris* from your text.

The male reproductive system is essentially a single, long tube made up of a threadlike **testis,** which continues as a thicker **vas deferens.** Both are much coiled. The vas deferens connects with the wider **seminal vesicle,** which empties by a short, muscular **ejaculatory duct** into the anus. Thus the male anus serves as an outlet for both the digestive system and the reproductive system and is often called a **cloaca.** **Spicules** secreted by and contained in spicule pouches may be extended through the anus. In copulation the male inserts the penial spicules into the vulva of the female and discharges spermatozoa through the ejaculatory duct into the vagina.

What is meant by "sexual dimorphism"? How does *Ascaris* illustrate this? How is *Ascaris* transmitted from host to host? How can infestation be prevented? In what ways is *Ascaris* structurally and functionally adapted to life as a parasite in the intestine?

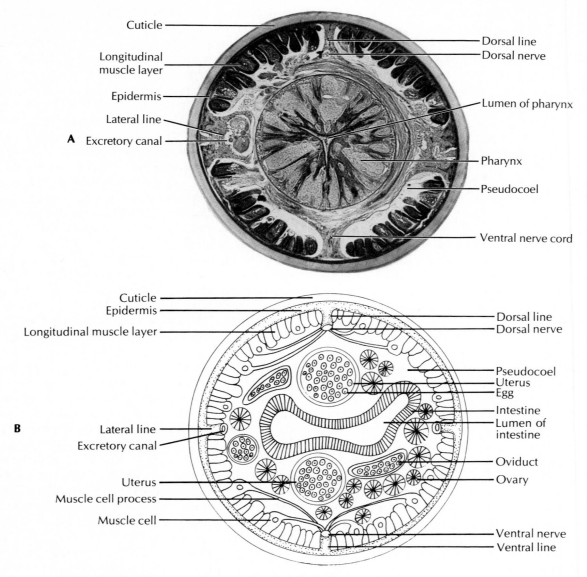

Cuticle — Dorsal line
Dorsal nerve
Longitudinal muscle layer
Epidermis
Lateral line — Lumen of pharynx
A Excretory canal
Pharynx
Pseudocoel
Ventral nerve cord

Cuticle
Epidermis — Dorsal line
Longitudinal muscle layer — Dorsal nerve
Pseudocoel
Uterus
Egg
Intestine
Lumen of intestine
Lateral line
Excretory canal
Oviduct
Uterus — Ovary
Muscle cell process
Muscle cell
Ventral nerve
Ventral line
B

Fig. 11-2. Transverse sections through an ascaris worm. **A,** Through the trilobed pharynx. **B,** Through the intestine. (**A,** Photograph by Carolina Biological Supply Co., Burlington, N.C.)

Transverse Section of Female Ascaris

☞ Study a prepared stained slide, at first under low power.

Note the thick noncellular **cuticle** on the outside of the body wall (Fig. 11-2). Below the cuticle is the thinner syncytial **epidermis,** which contains nuclei but few cell walls. The **longitudinal muscles** making up most of the body wall appear as fluffy, irregular masses dipping into the **pseudocoel,** with the tips of the cells directed toward the nearest nerve cord. Muscle continuity is interrupted by the **longitudinal lines.** Look for **excretory canals** in the lateral lines and look for the **dorsal** and **ventral nerve cords** in the dorsal and ventral lines. The lateral lines appear free of muscle cells. In the pseudocoel the large **uteri** (Fig. 11-2, *B*) are filled with eggs enclosed in shells and in cleavage stages. The thin-walled **oviducts** also contain eggs, whereas the wheel-shaped **ovaries** are composed of tall epithelial cells and have small lumens. The **intestine** is composed of a single layer of tall columnar cells (endodermal). The pharynx and the rectal region of the intestine are lined with cuticle. Why? Why is *Ascaris* not digested in the human intestine?

Cross sections of the male are similar in all particulars, except for the reproductive system, with its sections of **testis, vas deferens,** and seminal vesicle.

SOME FREE-LIVING NEMATODES
Vinegar Eels

Turbatrix aceti, the vinegar eel, is often found in fermented fruit juices, particularly the sediment of non-pastuerized vinegar, where it feeds on the yeasts and bacteria found there. As in *Ascaris* the females are longer than the males. Notice their swimming movements with the body bending in S's and C's. What does its movement tell you about the direction of its muscle fibers?

☞ Study quieted specimens or a prepared stained slide for body detail. Movement may be slowed down by adding a drop of dilute hydrochloric acid or by warming the slide very gently over a desk or alcohol lamp.

Note the blunt anterior end, the **mouth**, the **esophagus** with its posterior **esophageal bulb** and valve, the long straight **intestine,** and the ventral **anus** a short distance from the pointed tail end.

In a female specimen, note on the ventral side the **vulva,** which receives sperm in copulation and through which the young worms are born. *Turbatrix* is **ovoviviparous,** producing up to 45 young during its average life span of 10 months. The **seminal receptacle** lies posterior to the vulva, and the tubular **uterus,** lies anterior to the vulva. Eggs in various cleavage stages may often be found. The **ovary** extends anteriorly from the uterus and then doubles back dorsally.

In the male the tubular **testis** extending from about the middle of the body empties into a **genital tube,** which in turn empties into the **intestine** just before the anus. **Copulatory spicules** enclosed in a sac may be protruded through the anal opening for holding during copulation.

Soil Nematodes

Nematodes are abundant in almost every imaginable habitat. The main limiting factor is the presence of water because nematodes are aquatic animals in the strictest sense. They are capable of activity only when immersed in fluid, even if it is only the microscopically thin film of water that normally covers soil particles. If no water film is present, the nematodes either die or pass into a quiescent resting stage.

In the soil most nematodes are most numerous in the upper few inches. They are richer among the roots of plants than in open soil, so the best collecting source would probably be the top two inches of a long-established meadow turf.

Some nematodes are herbivorous, feeding on algae, fungi, or higher plants; many feed on plant roots. Examples are species of *Tylenchus, Heterodera, Dorylaimus,* and *Monhystera*. Carnivorous nematodes feed on other nematodes, rotifers, small oligochaetes, and so on. These include species of *Dorylaimus, Diplogaster,* and *Monochus*. Some nematodes are "saprophagous," such as *Rhabditis, Cephalobus,* and *Plectus,* which probably live on bacteria or other microorganisms. Some species are fairly omnivorous. Some soil nematodes are parasitic or ectoparasitic on plant or animal life.

Soil samples may be brought in from a wide variety of places. By using one of the methods listed on p. 169 you may collect the nematodes as well as many annelids and arthropods and study them under the microscope. Methods are also given for obtaining *Rhabditis* from the nephridia of earthworms.

SOME PARASITIC NEMATODES
Hookworm, *Necator americanus*

Hookworms live in the intestines of their vertebrate hosts. They attach themselves to the mucosa and suck up the blood and tissue fluids from it. The species most important to humans are *Necator americanus* and *Ancylostoma duodenale*. Hookworm is common in the southern United States, where 95% of the cases are *Necator*. *Ancylostoma caninum* is the common hookworm of domestic dogs and cats.

Hookworms mature and mate in the small intestine of the host. The developing eggs are passed out in the feces. On the ground they require warmth (preferably 20° to 30° C), shade, and moisture for continued development. They hatch in 24 to 48 hours into young juveniles, which feed on the fecal matter, molt their cuticles twice, and in a week or so are ready to infect a new host.

If the ground surface is dry, they migrate into the soil, but after a rain or morning dew they move to the surface, extend their bodies in a snakelike fashion, and wave back and forth. Thousands may group together waving rhythmically in unison. Under ideal conditions they may live for several weeks.

Infection occurs when the juveniles contact the skin of the host and burrow into it. Those that reach blood vessels are carried to the heart and then to the lungs. Here they are carried by ciliary action up the respiratory passages to the glottis and swallowed. In the small intestine they grow, molt, mature, and mate. In 5 weeks after entry they are producing eggs.

Whether hookworm disease results from infection depends on the number of worms present and the

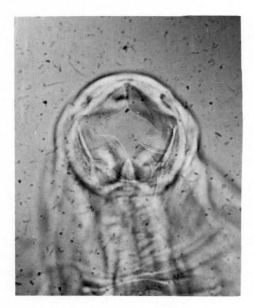

Fig. 11-3. Dorsal view of the mouth of *Necator americanus*. Note the broad cutting plates in the ventrolateral margin *(top)*. (From Schmidt, G.D., and L.S. Roberts, 1981. Foundations of parasitology, ed. 2. St. Louis, The C.V. Mosby Co.

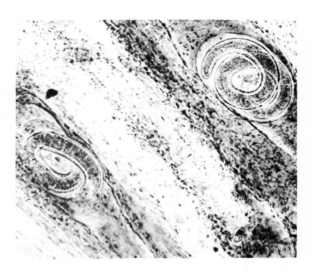

Fig. 11-4. Larvae of *Trichinella spiralis* encysted in pig muscle. Larvae may live 10 to 20 years in these cysts. If eaten by another host, they are liberated in the intestine, where they mature and release more larvae into the host's blood.

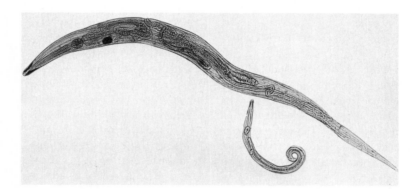

Fig. 11-5. Pinworms, *Enterobius vermicularis*. The male is smaller and has the curled posterior end. (Courtesy Indiana University School of Medicine, Indianapolis.)

nutritional condition of the infected person. Massive infections in the lungs may cause coughing, sore throat, and lung infection. In the intestinal phase moderate infections cause an iron-deficiency anemia. Severe infections may result in severe protein deficiency. When accompanied by chronic malnutrition, as in many tropical countries, there may be irreversible damage, resulting in stunted growth and below-average intelligence.

☞ Examine prepared slides of hookworms.

Adult males of *Necator* are typically 7 to 9 mm long, and females 9 to 11 mm. Specimens of *Ancylostoma* are slightly longer. The anterior end curves dorsally, giving the worm a hooklike appearance. Note the large buccal capsule, which bears a pair of dorsal and a pair of ventral cutting plates surrounding its margin (Fig. 11-3). A stout muscular esophagus serves as a powerful pump.

Note on the male the conspicuous copulatory bursa consisting of two lateral lobes and a smaller dorsal lobe, all supported by fleshy rays. Needle-like spicules are present, which in *Necator* are fused at the distal ends to form a characteristic hook. In the female the vulva is located in about the middle of the body.

Trichina Worm

The trichina worm, *Trichinella spiralis*, is a nematode parasite in humans, hogs, rats, and other carnivorous mammals. Study a slide of the larvae encysted in pork muscle (Fig. 11-4). How many cysts does your slide show? Would you be able to see the cysts in meat with the naked eye? The cyst wall is made of fibrous tissue that becomes calcified, thus protecting the host from the parasite. How many worms are coiled in a cyst? Study the life cycle of *Trichinella* in your text. Are they oviparous or ovoviviparous? How can you prevent trichinosis? Study a slide showing male and female adults. Where would the adults live in the human host?

Pinworm, *Enterobius vermicularis*

Pinworms (Fig. 11-5), the most common helminth parasite of humans in the United States, live in the large intestine and cecum. The females, up to 12 mm in length, lay their eggs at night around the anal region of their host. A single female may lay 4600 to 16,000 eggs. Scratching contaminates the hands and bedding of the host. The eggs, when swallowed, hatch in the duodenum and mature in the intestine.

PROJECTS AND DEMONSTRATIONS

1. Prepared slide of Wuchereria bancrofti. Examine prepared slides of the filarial worm, *Wuchereria bancrofti*. What disease is caused by this nematode? In what climate is infestation common? What is the alternate host? What control methods might be used?

2. Prepared slides of Dracunculus medinensis. Examine prepared slides of the guinea worm, *Dracunculus medinensis*. In what part of the world is this nematode common? Where do the larvae develop? How is this parasite acquired? What control methods would you use to prevent infestation?

3. Prepared slides of eggs, larvae, or cysts of any of the nematode parasites.

4. Soil-sampling methods.

a. BOILED POTATO. Leave some pieces of boiled potato for several days in various places—under a rock, plank, or bit of soil or in a garden, meadow, marsh, or empty lot—and then place them in sterile test tubes, plug with cotton, and set in a warm room for several days before examining.

b. BAERMANN APPARATUS. Wrap a sample of soil in several layers of cheesecloth and suspend from the arm of a ring stand (Fig. 11-6). Fit a short rubber tube with a petcock clamp to the spout of a funnel and attach the funnel to an arm of the ring stand below the bag of soil. Pour warm water (20° to 26° C) into the funnel so that 4 to 6 cm of the soil in the bag is immersed. After 1 to several hours, drain off a little of the water from the tube into a small test tube. Let the nematodes settle, and then pipette off the supernatant fluid. Examine the nematodes under the microscope.

5. Rhabditis, nematode parasite of earthworms. Fill some Petri dishes half full of agar jelly (made by boiling 2 g of agar per 100 ml of water) and allow to cool. Into each dish, cut 6-mm lengths from the posterior thirds of several fresh earthworms. Cover and leave for 2 to 5 days or until nematodes are seen on the putrefying pieces of earthworm. Adding small bits of raw meat from time to time and subculturing will maintain cultures indefinitely.

Another method is to slit freshly killed *Lumbricus* down the middorsal line posterior to the fifteenth segment. Pin back the walls, cover with water, and examine the nephridia. Immature nematodes may be found in or on some of the nephridia or brown bodies. If parasites are found, place the earthworm in a jar on wet paper toweling and cover. Keep the preparation wet for 2 to 4 days, depending on the temperature, while the earthworm decomposes and the nematodes reproduce. Examine some of the culture in a few drops of clear water. There should be much of the life cycle present—fresh eggs, eggs containing larvae, immature worms, and mature worms—in the culture. To preserve the worms for storage, put the culture into a test tube and let settle. Pour off most of the water and add an equal amount of boiling 10% formalin to kill and straighten the worms.

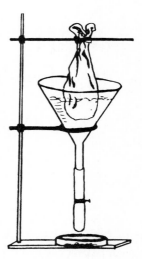

Fig. 11-6. Baermann apparatus.

EXERCISE 11B
A brief look at some other pseudocoelomates

MATERIALS
Living materials
   Philodina, or mixed rotifers
   Chaetonotus, or mixed gastrotrichs
   Gordius, or other nematomorphs, if available
Preserved or plastic-mounted material
   "Horsehair worms" (such as *Gordius*)
   Spiny-headed worms (such as *Macracanthorhynchus*)
Prepared slides of any of the above types
Slides and coverslips
Microscopes

PHYLUM ROTIFERA
Philodina or Others

☞ Place a drop of rotifer culture on a depression slide, cover, and examine with subdued light under low power.

How does the animal attach itself? Is it free swimming? Does it have a definite head end? Note the anterior discs of cilia (**corona**) that give the impression of wheels turning. Are they retractile? The cilia function both in swimming and in feeding.

The tail end (or **foot**) bears slender toes. How many? The toes have cement glands for clinging to objects. How does the rotifer use the toes?

Locate the **mastax**, a grinding organ. This is conspicuous in living rotifers because of its rhythmic contractions (Fig. 11-7, *C*). Rotifers feed on small plankton organisms swept in by cilia. Can you identify the digestive tract?

In *Philodina* (Fig. 11-7, *C*) the **cuticle** is ringed (annulated) so that it appears segmented. From watching its movements would you say it had circular muscles? Longitudinal? Oblique? Estimate the size of the rotifers. Is there more than one variety in the culture?

In many rotifers, the cuticle is thickened and rigid and is called a **lorica**, *Monostyla* and *Platyias* (Fig. 11-7, *B*) are examples. Some rotifers, such as *Floscularia*, live in a secreted tube. Most rotifers live in fresh water.

PHYLUM GASTROTRICHA
Chaetonotus or Others

Gastrotrichs include both freshwater and marine organisms. They are similar in size and general habits to the rotifers and are often found in the same cultures.

☞ Place a drop of culture on a slide, cover, and study first with low and then with high power.

Observe the manner of locomotion of the gastrotrichs. They slide along on a substratum by means of ventral cilia. *Chaetonotus*, a common genus, is covered with short curved dorsal spines (Fig. 11-7, *D*). The rounded head bears cilia and little tufts of sensory bristles. The tail end is forked and contains cement

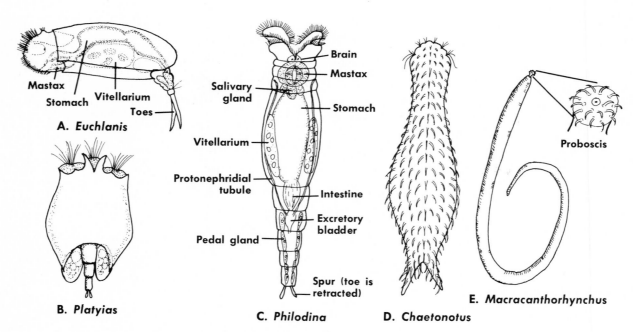

Fig. 11-7. Some representative pseudocoelomates. **A** to **C**, Rotifera. **D**, Gastrotricha. **E**, Acanthocephala. The rotifers and gastrotrichs shown here range from 120 to 500 μm in length. The spiny-headed worm *Macracanthorhynchus* grows up to 65 cm long.

170

glands similar to those of rotifers. Do gastrotrichs use the forked tail in the same manner as rotifers use the toes?

Gastrotrichs have a syncytial epidermis covered with a cuticle. In feeding they use the head cilia to sweep algae, detritus, and protozoans into the mouth. How long are the specimens?

Most marine gastrotrichs are hermaphroditic, but in freshwater species only parthenogenetic females are known.

PHYLUM NEMATOMORPHA
The Threadworms, or "Horsehair Worms"

The threadworms, or horsehair worms, such as *Paragordius* and *Gordius*, are long cylindrical hairlike worms often found wriggling in watering troughs, puddles, ponds, and quiet streams. Most of them range between 0.5 and 3 mm in diameter and from 10 to 300 mm in length, but some may reach a length of 1 m.

How long are your specimens? Is the diameter uniform throughout? If you have live specimens, what would you conclude about their muscular makeup, judging from their movements?

Nematomorphs have no lateral lines or excretory system, and in the adults the digestive system is degenerate. They differ from nematodes in having a cloaca in both sexes. The females lay long gelatinous strings of eggs on water plants. Larvae, which are encysted on plants, are sometimes eaten by arthropods, in whom the juvenile stage is parasitic for a while.

PHYLUM ACANTHOCEPHALA
The Spiny-Headed Worms,
Macracanthorhynchus

The adult *Macracanthorhynchus hirudinaceus* (Fig. 11-7, *E*) is parasitic in the small intestine of the pig, where it attaches to the intestinal lining and absorbs the digested food of its host. Like the tapeworm it has no digestive system at all.

The body is cylindrical and widest near the anterior end. A small spiny proboscis on the anterior end bears six rows of recurved hooks for attachment to the intestinal wall. The proboscis is hollow and can be partially retracted into a proboscis sheath.

The worms are dioecious. The male is much smaller than the female and has a genital bursa at the posterior end that may be partly evaginated through the genital pore and is used in copulation. Eggs discharged by the female into the host feces may be eaten by white grubs (larvae of the beetle family *Scarabeidae*) in whom they develop. Pigs are infected by eating the grubs or the adult beetles.

COMPARING REPRESENTATIVES
OF THREE PHYLA

Name _____

Date _____ Section _____

Features	Cnidaria Hydra	Platyhelminthes Planaria	Nematoda Ascaris
Symmetry			
Shape			
Germ layers			
Body covering			
Cephalization			
Body cavity			
Musculature			

Features	Cnidaria Hydra	Platyhelminthes Planaria	Nematoda Ascaris
Digestive tract and digestion			
Excretion			
Nervous system			
Sense organs			
Reproduction, sexual			
Reproduction, asexual			

CHAPTER 12

THE MOLLUSCS

PHYLUM MOLLUSCA*
A Protostome Eucoelomate Group

The molluscs, which rank next to the arthropods in number of named species, include the chitons, snails, slugs, clams, oysters, squids, octopuses, cuttlefish, and some others. They have retained the basic features introduced by the preceding phyla, such as triploblastic structure, bilateral symmetry, cephalization, and a body cavity, but the body cavity, though small, is now a **true coelom,** a characteristic shared by all remaining phyla. All of the organ systems are present.

Molluscs have a specialized muscular **foot,** generally used in locomotion. A fold of the dorsal wall, called the **mantle,** or **pallium,** encloses a **mantle cavity,** usually contains the **gills,** and secretes the **exoskeleton,** or shell. There is an open **circulatory system** with a pumping **heart** and a complete mouth-to-anus digestive system. Most molluscs have within the mouth a unique rasping organ, the **radula,** used for scraping off food materials. Most molluscs have a well-developed head.

Molluscs range in structural pattern from simple primitive forms to the very complex and highly developed cephalopods. Molluscs have left an extensive fossil record, indicating that their evolution has been a long one. Although many of them are among the most sluggish of animals, they occupy numerous ecological niches and are found in the sea, in fresh water, and on land.

The phylum consists of the following classes:

Class Monoplacophora (mon'o-pla-kof'o-ra). Body bilaterally symmetrical, with broad flat foot; a single dome-shaped shell; five or six pairs of gills in shallow mantle cavity; radula present; separate sexes. Example: *Neopilina*.

*See Hickman, C.P., Jr., L.S. Roberts, and F.M. Hickman. 1984. Integrated principles of zoology, ed. 7. St. Louis, The C.V. Mosby Co., pp. 302-337.

Class Polyplacophora (pol'y-pla-kof'o-ra). Chitons. Elongated, dorsally flattened body with reduced head; bilaterally symmetrical; radula present; shell of eight dorsal plates; foot broad and flat; gills multiple, along sides of body between foot and mantle edge; sexes usually separate. Examples: *Mopalia, Chaetopleura*.

Class Caudofoveata (kaw'do-fo-ve-at'a). Wormlike; shell, head, and excretory organs absent; radula usually present; mantle with chitinous cuticle and calcareous scales; oral pedal shield near anterior mouth; mantle cavity at posterior end with pair of gills; sexes separate; formerly united with solenogasters in class Aplacophora. Examples: *Chaetoderma, Limifossor*.

Class Solenogastres (so-len'o-gas'trez). Solenogasters. Wormlike; shell, head, and excretory organs absent; radula usually absent; mantle usually covered with scales or spicules; mantle cavity posterior, without true gills but sometimes with secondary respiratory structures; foot represented by long, narrow, ventral pedal groove; hermaphroditic. Example: *Neomenia*.

Class Scaphopoda (ska-fop'o-da). Tooth shells, elephant tusk shells. Body enclosed in a one-piece, tubular shell open at both ends; conical foot; mouth with radula and tentacles; head absent; mantle for respiration; sexes separate. Example: *Dentalium*.

Class Gastropoda (gas-trop'o-da). Snails, slugs, conchs, whelks, and others. Body asymmetrical, usually in a coiled shell (shell uncoiled or absent in some); head well developed, with radula; foot large and flat; one or two gills, or with mantle modified into secondary gills or lung; dioecious or monoecious. Examples: *Busycon, Physa, Helix, Aplysia*.

Class Bivalvia (bi-val'vi-a). Bivalves. Body enclosed in a two-lobed mantle; shell of two lateral valves of variable size and form, with dorsal hinge; head reduced; no radula; foot usually wedge-shaped; gills platelike; sexes usually separate. Examples: *Anodonta, Venus, Tagelus, Teredo*.

Class Cephalopoda (sef'a-lop'a-da). Squids and octopuses. Shell often reduced or absent; head well developed with eyes and radula; foot modified into arms or tentacles; siphon present; sexes separate. Examples: *Loligo, Octopus, Sepia*.

EXERCISE 12A
Class Bivalvia (Pelecypoda)—the freshwater mussel

MATERIALS
Living bivalves in an aquarium
Preserved or freshly killed mussels for dissection (see Appendix B, p. 464)
Clean empty bivalve shells
Dissecting pans
Pasteur pipette
Glass rod
Carmine suspension

FRESHWATER MUSSEL
Phylum Mollusca
   Class Bivalvia
      Subclass Palaeoheterodonta
         Order Unionoida
            Genus *Anodonta*, others

Where Found
Bivalves are found in both fresh water and salt water. Many of them spend most of their existence partly or wholly buried in mud or sand. Freshwater mussels are found in rivers, lakes, and streams and are particularly abundant in the Mississippi River valley. Some common freshwater genera are *Anodonta*, *Lampsilis*, *Elliptio*, and *Quadrula*.

Traditionally the freshwater mussels are used in the general laboratories because of their availability, but the sea clam, *Spisula* is quite similar to the freshwater clams and makes a good substitute. The sea mussel, *Mytilus*, or the quahog, *Mercenaria*, can also be substituted.

Behavior and General Features
If living bivalves in an aquarium are available, you can learn a good deal about them if you have patience. Freshwater mussels lie half buried in the sand. They are sluggish, and their reactions are slow.

What is the natural position of the clam at rest? When moving? Note that it leaves a furrow in the sand when it moves. The soft body is protected by a hard **exoskeleton** composed of a pair of **valves,** or shells, hinged on the dorsal side. When the animal is at rest, the valves are slightly agape ventrally, and you can see at the posterior end the fringed edges of the **mantle,** which lines the valves. The posterior edges of the mantle are shaped so as to form two openings (**apertures**) to the inside of the mantle cavity (Fig. 12-1).

☞ With a Pasteur pipette introduce a small amount of carmine dye into the water near the apertures and see what happens to it.

Which of the apertures has an incurrent flow, and which has an excurrent flow? A steady flow of water through these apertures is necessary to bring oxygen and food to the animal and to carry away wastes. Most bivalves are filter feeders that filter minute food particles from the water, trap them in mucus, and carry them by ciliary action to the mouth.

Some marine clams have the mantle drawn out into long muscular **siphons.** When the animal burrows deeply into the mud or sand, the siphons extend up to the surface to bring clear water into the mantle cavity.

☞ Gently touch the mantle edge with a glass rod.

What happens? The mantle around the apertures is highly sensitive not only to touch but to chemical stimuli, a necessity if the animal is to close its valves to exclude water containing unpleasant or harmful substances.

If the supply of live clams is plentiful, the instructor may want to remove one of the valves from a clam and sprinkle a few carmine granules on the gills, the labial palps, and the inside of the mantle to let you observe the ciliary action that maintains and controls this water flow.

☞ Lift a clam out of the sand to observe the hasty withdrawal of the foot. Then lay it on its side on the sand to see if it will right itself.

The foot is as soft, as agile, and as sensitive as the human tongue. Mucous glands keep the foot well protected with mucus.

If there is a marine tank containing scallops, compare the method of locomotion of the scallops with that of the mussels.

Fig. 12-1. Freshwater clam showing apertures in mantle. The dorsal, or upper, aperture is excurrent; the larger, lower aperture is incurrent.

Report your observations on p. 184.

External Structure
The Shell

☞ Procure a preserved or freshly killed mussel and an empty shell or valve.

The two valves of the mussel are attached by a **hinge ligament** on the dorsal side; the ventral side is free for the protrusion of the foot. A swollen hump, the **umbo** (pl. **umbones**), near the anterior end of the hinge is the oldest part of the shell. Concentric **lines of growth** around the umbo indicate growth periods. The youngest part of the shell is the edge.

The outer horny layer of the valve is the **periostracum,** which is secreted by a fold at the edge of the mantle. Has any of this layer been eroded away? Where is it thinnest? Why? The inner, iridescent mother-of-pearl surface is the **nacreous layer,** which lies next to the mantle and is secreted continuously by the mantle surface. Break off a piece of the empty valve and note the thick middle **prismatic layer** made up of crystalline calcium carbonate. It is secreted by glands in the edge of the mantle.

With the aid of Fig. 12-2, locate on the valve the **scars** of the **anterior** and **posterior adductor muscles,** which close and hold the valves together; the **protractor muscle,** which helps extend the foot; and the **anterior** and **posterior retractor muscles,** which pull in the foot. Find the **pallial line** where the **pallial muscle** of the mantle was attached to the valve.

Identify near the dorsal edge the calcerous **teeth**—the long ridges are **lateral teeth** and the pointed projections are **pseudocardinal teeth.** The shape and prominence of the teeth will vary a great deal among the genera.

Opening the Shell

☞ If your specimen has not already been pegged, pry the valves apart with the scalpel and insert a wooden block between them. Before opening the clam, become oriented by locating its dorsal and ventral sides, its anterior and posterior ends, and its right and left valves.

Decide where the **adductor** and **pallial muscles** are located. A fleshy **mantle** lines the inner surface of each valve. Insert the scalpel between the *left valve* and *left mantle* and gently separate the mantle from the valve along the pallial line. Now, keeping the blade close to the shell, locate and cut through each of the adductor muscles. Making sure that the mantle is completely separated from the left valve, lift the loosened left valve, letting the body remain in the right valve. Cover the clam with water.

Examine the elastic **hinge ligament.** It acts as a spring, forcing the valves apart when the adductors relax. Identify the **teeth.** Note that their tongue-and-groove arrangement keeps the valves in position. Is the arrangement the same on each valve?

The mantle. Now that the left valve is removed note the **mantle** that lined the valve and protects the gills and visceral mass. Examine the posterior edge of the mantle and see how the two mantle edges come together to form two posterior openings—a ventral

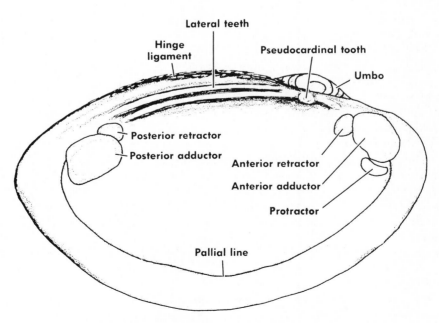

Fig. 12-2. Left valve of freshwater clam, showing the muscle scars.

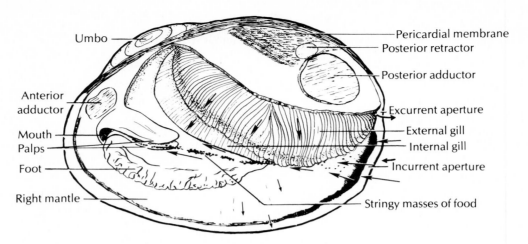

Fig. 12-3. Feeding mechanism of a freshwater clam.

incurrent aperture and a dorsal **excurrent aperture**—that permit a continuous flow of water through the mantle cavity (Fig. 12-3). Both the mantle and the gills serve as respiratory organs for the exchange of oxygen and carbon dioxide.

Note the thick **pallial muscle** along the free outer edge of the mantle and the **pallial line** where it was attached to the valve. The entire outer surface of the mantle secretes the nacreous layer of shell. This surface is not ciliated, so the animal is unable to rid itself of foreign objects that might get between the shell and the mantle. Instead the mantle secretes nacre around them. Do you find any pearls or other evidence of such irritation in the valve or in the mantle?

Look at the free edge of the mantle and note that it is lobed. The edge of the outer lobe secretes the prismatic layer, and cells on the medial side of the outer lobe secrete the periostracum. The middle lobe is sensory and on some bivalves, such as the scallops, may bear eyes or tentacles. The inner lobe is muscular.

Muscles. Locate the large **adductor muscles**, which close the valves (Fig. 12-3). Slightly dorsal to them are the **retractors**, which withdraw the foot. Find their origin and insertion. The **protractor muscle** arises from the anterior end of the visceral mass and compresses the visceral mass to extend the foot.

Pericardium. On the dorsal side of the animal, above the posterior adductor, locate the thin, almost transparent **pericardial membrane** (Fig. 12-3), which covers the pericardial cavity and the heart. *Do not puncture this membrane until instructed to do so later*.

The Mantle Cavity

☞ Lift up the mantle to expose the body mass to which the mantle is attached dorsally.

The space between the mantle and the body is called the **mantle cavity**, or **pallial cavity**. In the posterior region, suspended from the dorsal wall, is one of the two pairs of large **gills**. These are made up of vertical tubes, each with many minute openings (ostia) to the outside. Cilia on the mantle and gills keep water flowing into the tubes to bring fresh oxygen. If the outer gill is much thicker than the inner gill, the animal is probably a female in which the gill is serving as a **brood chamber** for developing eggs.

The soft portion of the body is the **visceral mass**. The muscular **foot** lies ventral to the visceral mass and in preserved animals feels hard to the touch. In life, however, the foot is as soft and muscular as the human tongue. The foot operates by a combination of muscular movement and hydraulic mechanisms. The clam can extend or enlarge the foot hydraulically by engorgement with blood and uses the extended foot for anchorage or for drawing the body forward.

Attached to the anterior end of the visceral mass are two pairs of **labial palps**, a pair on each side of the body. The left and right outer palps join anteriorly to form a protective lip for the slitlike **mouth**. The palps secrete a great amount of mucus and are ciliated to guide toward the mouth food particles trapped in the mucus.

Internal Structure and Function

Respiratory System. Gaseous exchange occurs in both the mantle and the gills. Both mantle and gills are ciliated to ensure water flow.

Each gill is made up of two walls (**lamellae**) connected to each other by a series of thin partitions that divide the gill into many vertical **water tubes**. Water enters the tubes through innumerable small holes, or **ostia**, in the lamellae. The water tubes empty into a **suprabranchial chamber**, which lies dorsal to the gills and empties to the outside through the excurrent aperture.

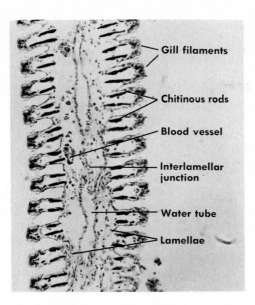

Fig. 12-4. Cross section through a portion of the gill of a mussel. (Photomicrograph by F.M. Hickman.)

The labels in the figure are:
- Gill filaments
- Chitinous rods
- Blood vessel
- Interlamellar junction
- Water tube
- Lamellae

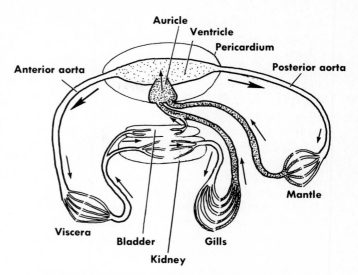

Fig. 12-5. Scheme of circulation of a freshwater clam. Note that all of the blood is carried either to the gills or to the mantle for oxygenation. Note also that in this "open" system, blood sinuses in the tissues replace the capillary beds found in animals with "closed systems."

The labels in the figure are:
- Auricle
- Ventricle
- Pericardium
- Anterior aorta
- Posterior aorta
- Mantle
- Viscera
- Bladder
- Kidney
- Gills

☞ Probe into the suprabranchial chamber from the excurrent aperture. Slit open the chamber and see the tops of the water tubes.

Now cut off a piece from the lower part of the gill and examine the cut edge of the tubes and lamellae with the hand lens. Then with a razor blade, cut a very thin transverse section of gill, mount cut side up with a little water on a slide, and study with the dissecting microscope or scanning lens.

The lamellae are made up of fine vertical ridges, called gill filaments, that are supported by chitinous rods (Fig. 12-4). Do you see any blood vessels in the walls or partitions? Why is a rich blood supply necessary to the gills?

For ventilation of the mantle cavity a continuous water current is necessary. This is effected by the cilia on the surface of the gills and mantle. A sedentary animal, such as a clam, which lives half buried in mud or sand, must have some means of clearing the water that goes through the gills of sediment, detritus, and fecal matter. This is accomplished partially by the small size of the gill pores and partially by mucus secreted by glands in the roof of the mantle cavity, which traps particles too large for the gill pores (Fig. 12-3). Larger debris drops off the gills and smaller food particles are carried on toward the mouth.

The water tubes in the female serve as brood pouches for eggs or larvae during the breeding season. If the outer gill is much thicker than the inner one, it may be full of eggs or larvae.

Circulatory System

☞ To open the thin-walled **pericardial sac,** slit the pericardium close to the dorsal edge with the tip of the dissecting needle and lift the thin tissue carefully so as not to injure the delicate **auricle** of the heart, which lies just beneath it.

The three-chambered **heart** is composed of a muscular **ventricle,** running along the dorsal side of the cavity, and a pair of thin-walled **auricles.** You will see the fragile, almost transparent left auricle extending fanlike from the ventricle to the floor of the pericardium. If it has been torn loose, you will see a slit, or **ostium,** in the side of the ventricle, through which blood passes from auricle to ventricle. The **anterior aorta** and the **posterior aorta** leave the anterior and posterior ends of the heart to run along the dorsal side of the rectum. Slip the blunt probe through the ostium of the heart and probe each aorta.

The pericardial space around the heart is a part of the **coelomic cavity,** which is greatly reduced in molluscs.

Fig. 12-5 shows the general plan of the **open system of circulation** found in the freshwater clam. From the ventricle the aortae carry blood to sinuses in the body tissues. From the visceral organs blood is carried first to the kidney for removal of wastes, then to the gills for gaseous exchange, and then back to the auricles and ventricle. Blood from the mantle, rich in oxygen, returns directly to the auricles. The blood of molluscs is colorless but contains either hemoglobin or hemocyanin for oxygen transport. It contains nucleated, ameboid corpuscles.

Coelom. Although molluscs have a true coelom, it is small. The pericardial cavity is part of the coelom, as is the small space around the gonads. A true coelom, you recall, is distinguished from other cavities by being lined with epithelium that arises from the mesoderm.

Excretory System. A pair of dark kidneys lies under the floor of the pericardial sinus. They are roughly U-shaped tubes. The kidneys pick up waste from the blood vessels, with which they are richly supplied, and from the pericardial sinus, with which they connect. The waste is discharged into the suprabranchial chamber and carried away with the exhalant current.

Digestive System. Locate again the **labial palps** and the **mouth.** The **anus** can be seen posteriorly where it empties at the excurrent aperture.

☞ Now make a sagittal section of the entire visceral mass. To do this, remove the clam from the shell and remove the mantle. Find the mid-dorsal line of the foot and visceral mass and cut a sagittal section with a scalpel or razor blade, bisecting the entire animal into right and left halves, cutting through the center of the heart, the mouth, and the foot. Examine both halves.

Find the **esophagus,** the **stomach** surrounded by greenish **digestive glands,** and portions of the **intestine** surrounded by the light brown tissue of the **gonad.**

The intestine makes several loops before it ascends above the first loop to enter the pericardial sac. Here the **rectum** passes through the length of the ventricle, emerges above the posterior aorta, and runs posteriorly to the anus. Fecal products are carried away by the exhalant current. The rectum has an inner longitudinal fold, the **typhlosole,** which increases its surface area.

☞ Cut the ventricle in two transversely and look inside it to see the rectum and its typhlosole.

Filter feeding. The clam depends on water currents created by cilia on the gills, and mantle and labial palps to bring in a food supply, mostly phytoplankton. As respiratory water enters the gills through the many tiny pores, food particles, sand, and other debris carried by the incoming current are filtered out by the small size of the pores. Gland cells on the gills and palps secrete copious amounts of mucus, which entangles the particles that are too large to enter the gills (Fig. 12-3). As the heavier particles of sediment drop off as a result of gravitational pull, the smaller food particles travel toward a ciliated food groove on the ventral edge of each gill and thence to the labial palps. The palps, being also grooved and ciliated, direct the mass of mucus, which contains trapped food, to the mouth.

Digestion. The floor of the stomach is folded into ciliary tracts for sorting particles. A rotating gelatinous rod called the **crystalline style** projects into the stomach and is kept rotating by means of cilia. It is composed of mucoproteins and digestive enzymes. The mucous food mass becomes caught on the rotating style and spins about too. As it does, enzymes are thrown off the style into the mass. Food particles in the spinning mass detach and land on the ciliated tracts on the stomach floor where they are sorted. Large, unsuitable particles go to the intestine for elimination. Smaller or partially digested particles are carried to the digestive gland, where digestion is completed intracellularly. As the crystalline style dissolves at the free end, it is continually being replaced at the other end. The crystalline style cannot be seen in preserved clams because it dissolves away after death.

Reproductive System. The sexes are separate but are difficult to distinguish, except by the swollen gills of the pregnant female. The **gonads** (**ovaries** or **testes**) are a brownish mass of minute tubes filling the space between the coils of the intestine.

☞ Make a wet mount of gonadal tissue and determine whether there are eggs or sperm in it.

The gonads discharge their products into the suprabranchial chamber. Spermatozoa are carried into the surrounding water. They enter a female with the inhalant current and fertilize the eggs in the suprabranchial chamber. The **zygotes** settle into the water tubes of the outer gill (brood pouch), where each zygote develops into a tiny bivalved larval form known as a **glochidium.** The glochidia, about the size of dust particles, escape through the excurrent siphon. The glochidia are provided with valves bearing hooks, by which they fasten themselves to the gills, fins, or skin of a passing fish. Here they encyst and live as parasites for several weeks. After a growth period the young clams break loose and sink to the bottom sand, where they develop as free-living adults. What advantages do the young clams derive from this relationship with the fishes?

☞ If a female with swollen brood pouch is available, make a wet mount of some of the gill contents and examine with the microscope to determine whether eggs or glochidia are present.

Nervous System. The nervous system of the clam is not very highly centralized. Dissection of the nervous system is difficult and often impractical in preserved specimens. Three pairs of **ganglia** (small groups of nerve cells) are connected to each other by nerves. The **cerebropleural ganglia** are found one on each side of the esophagus on the posterior surface of the anterior adductor muscle.

The **pedal ganglia** are fused and are found in the anterior part of the foot. You may have exposed them when you bisected the foot; they are located in the median line in the soft spongy tissue (mostly gonad) just above the muscular foot tissue. They are connected to the cerebropleural ganglia by a pair of nerves.

The **visceral ganglia** are fused into a star-shaped

body just ventral to the posterior adductor muscle. They are covered by a yellowish membrane and are connected to the cerebropleural ganglia by nerves.

How is the placement of the ganglia and their connectives related to the other organs of the body and to the important receptors and effectors?

Sense organs are poorly developed in the clam. They are involved with touch, chemical sensitivity, balance, and light sensitivity. They are most numerous on the edge of the mantle, particularly around the incurrent aperture, but you will not be able to see them. However, in the scallop, *Pecten*, the ocelli are large and numerous, forming a distinctive row of steel blue "eyes" along the edges of each mantle.

ORAL REPORT

Be prepared to demonstrate your dissection and explain the clam's structures and their functions.

DRAWINGS

1. Interpret a transverse section through the region of the heart and complete the diagram on p. 183, adding the structures you would find in such a section. Label fully.
2. Draw a cut section through a gill. Use p. 183.
3. If a female with distended brood pouch is available, sketch some of the eggs or glochidia on p. 183.

PROJECTS AND DEMONSTRATIONS

1. Cross section of entire clam after removal from the shell. This study is made from prepared slides. What you will see in it depends to some extent on the region through which the body was cut. Identify the **mantle, mantle cavity, gills, lamellae, water tubes, intestine, foot, suprabranchial space, gonad,** and other structures that show in the cross section.

2. Shells of oysters, scallops, and sea clams.

3. Shipworm, Teredo. Examine a piece of wood into which these wormlike molluscs have bored. Examine a preserved specimen of the shipworm. Note the small valves, the small body, and the prolonged siphon that makes up the bulk of the shipworm.

4. Prepared stained slides of glochidia larvae. Note especially the valves and the muscles. Do they have **larvae threads** between the valves? If so, what are the functions of these threads?

5. Female clam with glochidia in the gills. Preserved specimens of clams with brood pouches (marsupia) may be studied. Which of the gills serve as brood pouches? How numerous are the glochidia?

6. Dissection of the live bivalve. Examine a live bivalve to see the beating heart. (See Appendix B for narcotizing methods.)

7. Demonstration of a living gill. Examine under the microscope a gill mounted in water to show the action of the cilia.

8. Effect of temperature on ciliary action of the clam gill. Remove a piece of bivalve gill and pin out on a layer of wax in a glass dish. Support the gill on two small glass rods approximately 1 cm apart. Place a 1 mm disc of aluminum foil on the surface of the gill and time its progress across a given distance. Starting with ice-cold water, record the rate of transport. As the water warms to room temperature, record the rate at intervals of 5° C. Judicial additions of slightly warmer water may also be tried. Make several determinations at each temperature and average. Make a diagram plotting the mean rates of transport (millimeters per minute) against the degrees of temperature.

Diagrammatic transverse section through a clam
in the region of the heart

Cut section of clam gill   Eggs or glochidia of clam

183

NOTES ON BEHAVIOR OF BIVALVE

NOTES ON BEHAVIOR OF SNAIL

184

EXERCISE 12B
Class Gastropoda—the land snail

MATERIALS

Living snails (*Helix*, others)
Preserved or freshly killed snails (see Appendix B, p. 464)
Assortment of snail shells
Squares of glass plate
Finger bowls
Dissecting microscopes

LAND SNAIL

Phylum Mollusca
  Class Gastropoda
    Subclass Pulmonata
      Superorder Stylommatophora
        Genus *Helix, Polygyra*, others

Where Found

Most land snails (Fig. 12-6) prefer fairly moist habitats. They are common in wooded areas, where they spend their days in the damp leaf mold on the ground, coming out to feed on the vegetation at night. *Helix* (Fig. 12-6, *A*) is a large snail from southern Europe; *Polygyra* is an American land snail.

Behavior

☞ Place a living snail in a finger bowl and, using the following suggestions, observe its behavior.

Studying a snail will take patience, for snails are unhurried. Can you detect muscular waves along its **foot** as it moves? Does it leave a trail on the glass as it travels? What causes this? Place a piece of lettuce leaf near a snail and see whether it will eat. How long does it take the snail to find the food? Note the **head** with its **tentacles.** How do the two pairs of tentacles differ? Where are the **eyes** located? Touch an anterior tentacle gently and describe what happens. Touch some other parts of the body and note results. Pick up a snail very carefully from the glass on which it is moving and observe exactly what it does.

☞ Place a land snail on a moistened glass plate and invert the plate over a dish so that you can watch the ventral side of the foot under a dissecting microscope.

Do you see the waves of motion? Can you see evidence of mucus? Of ciliary action? If you are fortunate you may see the action of the **radula** in the mouth. The radula is a series of tiny teeth attached to a ribbonlike organ that moves rapidly back and forth with an action like that of a rasp or file (Fig. 12-7). (For further study of the radula see items 10 and 11 on p. 187).

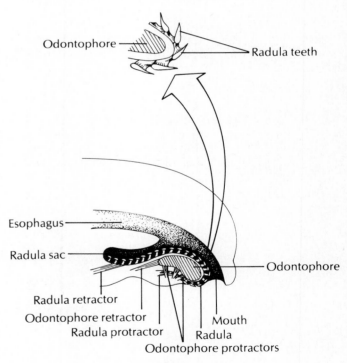

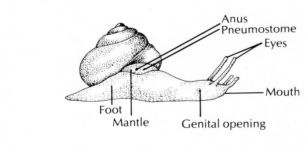

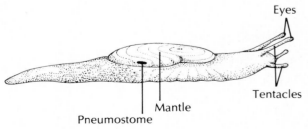

Fig. 12-6. Common pulmonates. *Top, Helix,* a common land snail. *Bottom, Limax,* a common garden slug.

Fig. 12-7. Diagrammatic longitudinal section of a gastropod head showing the radula and radula sac. The radula moves back and forth over the odontophore cartilage. As the animal grazes, the mouth opens, the odontophore is thrust forward against the substratum, the teeth scrape food into the pharynx, the odontophore retracts, and the mouth closes. The sequence is repeated rhythmically.

Prop the plate up in a vertical position. Which direction does the snail move? Now rotate the plate 90 degrees so that the snail is at right angles to its former position. When it resumes its travels, in which direction does it move? Now rotate the plate again and observe.

Do you think the snail is influenced in its movements by the forces of gravity (geotaxis)? Try several snails. Do they respond in a similar manner?

> Report your observations on the behavior of the snail on p. 184.

The Shell

Examine the shell of a preserved specimen. Note the nature of the spiral shell. Is it symmetrical? To what part of the clam shell does the **apex** correspond? The body of the snail extends through the **aperture**. A **whorl** is one complete spiral turn of the shell. On the whorls are fine **lines of growth** running parallel to the edge of the aperture. Holding the apex of the shell upward and the aperture toward you, note whether the aperture is at the right or left. **Dextral**, or right-handed, shells will have the aperture toward the right and **sinistral**, or left-handed, shells toward the left (Fig. 12-8). Examine a piece of broken shell for the characteristic three layers—the outer **periostracum**, the middle **prismatic layer**, and the shiny inner **nacre**.

Surface Anatomy

If the snail has not contracted enough during killing and preservation to allow you to slip it carefully from the shell, you may use scissors to carefully cut around the spiral between the

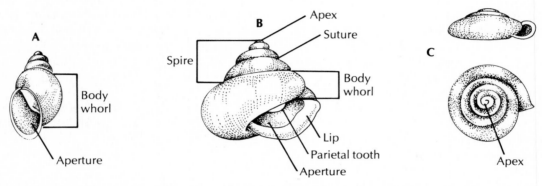

Fig. 12-8. Structure of a snail shell. **A,** *Physa,* a sinistral, or left-handed, freshwater snail with lymnaeiform shell (height exceeds width). **B,** *Mesodon* (= *Polygyra*), a dextral, (right-handed) snail with heliciform shell (width exceeds height). **C,** *Helicodiscus,* a land snail with planorbiform, or flattened, spire (two views).

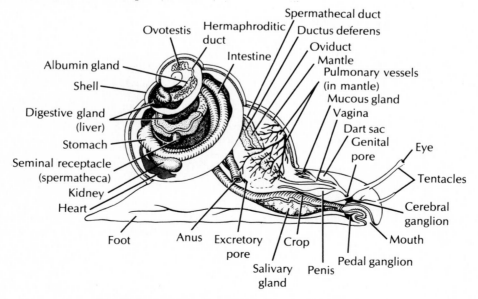

Fig. 12-9. Anatomy of a pulmonate snail (diagrammatic).

whorls, removing pieces of shell as you go and leaving only the central parts, or **columella.** Try not to damage the coiled part of the visceral mass.

The body is made up of the **head,** the muscular **foot,** and the coiled **visceral hump** (Fig. 12-9). Identify the **tentacles, eyes,** ventral **mouth** with three lips, and the **genital aperture** just above and behind the right side of the mouth. The foot bears a **mucous gland** just below the mouth. The mucous secretions aid in locomotion.

Note the thin **mantle** that covers the visceral hump and forms the roof of the **mantle cavity.** It is thickened anteriorly to form the **collar** that secretes the shell.

Find a small opening, the **pneumostome,** under the edge of the collar (Fig. 12-6). It opens into a highly vascular portion of the mantle cavity, located in the first half-turn of the spiral, that serves as a respiratory chamber (**lung**) in pulmonates. Here diffusion of gases occurs between the air and the blood. Oxygen is carried by the pigment hemocyanin. Most aquatic gastropods possess gills. The mantle cavity in the second half-turn contains the heart and a large kidney.

The rest of the coiled visceral mass contains the dark lobes of the digestive gland, the intestine, the lighter colored albumen gland (part of the reproductive system), and the ovotestis (Fig. 12-9).

PROJECTS AND DEMONSTRATIONS

1. Example of a pond snail. Watch a pond snail attached to the glass side of an aquarium. Note the broad foot by which it clings to the glass. Can you see the motion of the **radula** as the animal eats algae that has settled on the glass?

2. Examples of other pulmonate snail shells.

3. Shell-less pulmonate. Examine a shell-less pulmonate such as the common garden slug, *Limax* (Fig. 12-6, *B*).

4. Examples of shells of prosobranch snails. Examine some prosobranchs such as limpets, periwin-

kles, slipper shells, abalone, oyster drills, conchs, or whelks.

5. Examples of nudibranchs. Examine some nudibranchs such as sea slugs, sea hares, and sea butterflies.

6. Dissection of a prosobranch snail. Do a dissection of the marine whelk *Busycon.*

7. Masses of fresh snail eggs. Eggs of freshwater snails are frequently found on vegetation in aquaria where snails are kept or on leaves or stones in streams or ponds. Examine at intervals to follow the development of the embryo and young snails.

8. Egg cases of marine snails. Examine some egg cases such as those of *Busycon* or *Fasciolaria.*

9. Ciliary action in the intestine of snail. Obtain a live aquatic snail (*Physa*). Cut open the shell and remove the viscera. Carefully slit open the body and remove the intestine. Slit the intestine and place a small portion, the intestinal surface up, on a slide in a drop of saline solution. Examine with the high power of your microscope. Note the progressive undulations of the cilia over the surface. The ciliary action is best seen some time after you have made the preparation. Are the cilia independent of nerve action? Place a drop of warm (45° C) saline solution on the preparation and note what happens. What is the function of cilia in this region?

10. Isolation of the radula of a snail. Cut off the head of a snail and soak it in a 10% solution of KOH for 2 or 3 days or until the soft tissues are destroyed and only the radula remains. Transfer the radula to water and wash for 1 to 2 hours in running water. Pieces of attached tissues may be removed by gentle teasing with a needle. With a piece of paper on each side, the radula should be placed between two glass slides bound together by strong rubber bands. In this position, dehydrate it in 50%, 95%, and 100% alcohols, clear in xylol, and mount in balsam or euparal.

11. Study of whole mounts of various snail radulae. Study the patterns of the numerous teeth on the surface of the radula with the low power.

EXERCISE 12C
Class Cephalopoda—*Loligo,* the squid

MATERIALS
Preserved or freshly killed squids (see Appendix B, p. 464)
Dissecting tools
Dissecting pans

LOLIGO
Phylum Mollusca
   Class Cephalopoda
      Subclass Coleoidea
         Order Teuthoidea
            Genus *Loligo*

Where Found
The squids and octopuses are all marine animals. Whereas octopuses are often found in the shallow tidal waters, the active squids are free swimming and are found in offshore waters at various depths.

Squids range in size from 2 cm up. The giant squid *Architeuthis* may measure 15 m from tentacle tip to posterior end. *Loligo* averages about 30 cm.

Behavior
Since squids do not ordinarily survive long in the average aquarium, they may not be available for observation. If they are, notice the swimming movements. A squid can move swiftly, either forward or backward, by using its fins and ejecting water through its **funnel.** When water is ejected forward, the squid moves backward; when water is ejected backward, the animal moves forward—movement by jet propulsion! Even when it is resting there is a gentle rhythmical movement of the fins, a muscular movement that also aids in bringing water to the gills in the mantle cavity.

Notice the color changes in the integument caused by contraction and expansion of many **chromatophores** (pigment cells) in the skin. Each chromatophore is a minute sac of pigment to which radiating muscle fibers are attached. The spreading of the pigment throughout the cells causes darkening of the skin; the concentrating of the pigment lightens the skin color. The squid can change from almost white through shades of purple to almost black, with a system of elaborate changes of pattern as well as color. This serves as a complicated and highly developed means of communication.

When the animal is attacked, it can emit through its funnel a cloud of ink from its ink sac.

Food is caught by a pair of retractile tentacles extended with lightning speed, then held and manipulated by the eight arms, and killed by a poison injection. Small fishes, shrimps, and crabs are favorite foods.

External Structure
The squids are called "head-footed" because the end of the head, which bears tentacles, is really homologous to the ventral side of the clam, and the tentacles represent a modification of the foot. Squids are highly organized, with a well-developed nervous system and eyes that form images as do vertebrate eyes.

Notice the streamlined body with its **head** and **arms** at one end and a pair of **lateral fins** at the other (Fig. 12-10). The visceral mass, dorsal to the head, is covered with a thick **mantle,** the free end of which forms a loosely fitting **collar** about the neck. Near the collar are three articulating ridges, the **pallial cartilages,** which fit into the grooves of three **infundibular cartilages** at the base of the head and funnel.

The head bears a pair of complex, highly advanced **eyes** that can form images. Just ventral to each eye is a small **aquiferous pore** leading into the anterior channel of the eye. It admits water to regulate the pressure within the eye chamber. Dorsal to the eye is a fold of tissue called the **olfactory crest.**

The head is drawn out into 10 appendages: four pairs of **arms,** each with two rows of stalked suckers, and one pair of long retractile **tentacles,** with two rows of stalked suckers at the ends. The arms of males are longer and thicker than those of females, and in the mature male the left fourth arm becomes slightly modified for the transfer of spermatophores to the female. This is called **hectocotyly.** On the hectocotylized arm some of the suckers are smaller and form an adhesion area for carrying the spermatophore. The long tentacles can be shot out quickly to catch prey.

The **mouth** lies within the circle of arms. It is surrounded by a **peristomial membrane,** around which is a **buccal membrane** with seven projections, each with suckers on the inner surface. Probe in the mouth to find two horny beaklike **jaws.**

A muscular **funnel (siphon)** usually projects under the collar on the posterior side, but it may be partially withdrawn. Water forced through the funnel by muscular contraction of the mantle furnishes the power for "jet propulsion" locomotion. Wastes, sexual products, and ink are carried out by the current of water that enters through the collar and leaves through the funnel. The siphon, or funnel, of the squid is not homologous to the siphon of the clam; the clam siphon is a modification of the mantle, whereas the squid siphon, along with the arms and tentacles, is a modification of the foot.

The mottled appearance of the skin is caused by **chromatophores,** the pigment cells.

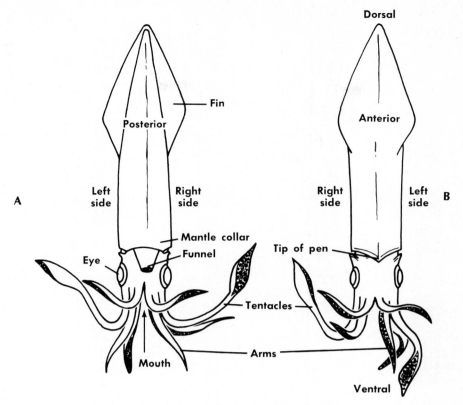

Fig. 12-10. External structure of a squid. **A,** Posterior view. **B,** Anterior view.

Mantle Cavity

☞ Beginning near the funnel, make a longitudinal incision through the mantle from the collar to the tip. Pin out the mantle and cover with water.

The space between the mantle and the visceral mass is the **mantle cavity.** The mantle itself is made up largely by circular muscles covered with integument. The funnel contains both circular and longitudinal muscles.

Locate again the interlocking cartilages. They help support the funnel and close the space between the neck and the mantle so that water inhaled around the collar can be expelled only by way of the funnel. Lateral to the funnel, find large saclike valves that prevent outflow of water by way of the collar.

☞ Slit open the funnel to see the muscular tonguelike valve that prevents inflow of water through the funnel.

This valve allows a buildup of hydrostatic pressure in the mantle cavity before a jet stream of water is ejected through the funnel.

Locate a large pair of **funnel retractor muscles** and beneath them the large **head retractor muscles.** Locate the free end of the **rectum** with its **anus** near the inner opening of the funnel. Between it and the visceral mass is the **ink sac.** Do not puncture it. When it is endangered, the squid can send out a cloud of black ink through the funnel as it darts off in another direction.

A pair of long **gills** are attached at one end to the visceral mass and at the other to the mantle (Fig. 12-11). They are located where water entering the mantle cavity passes directly over them. The gills are not ciliated, as in the bivalves. The mantle cavity is ventilated by action of the mantle itself. Contraction of the radial muscles in the body wall causes the wall to become thinner and the capacity of the mantle cavity to become greater, so that water flows in around the collar. The water is then expelled through the funnel when the mantle muscles contract. This movement, also used in locomotion, permits very efficient ventilation of the gills.

A round structure at the base of each gill is the **branchial heart.** The circulatory system of the cephalopod is a closed one. The branchial hearts receive blood returning from all over the body and pass it on through branchial arteries to the gills. The blood returns through veins to a large **systemic heart** lying between and somewhat anterior to the branchial hearts, which pumps it out to the body.

The skeletal system of the squid consists of several cartilages and the **pen.** Besides the articulating cartilages of the mantle and body, there are cartilages

189

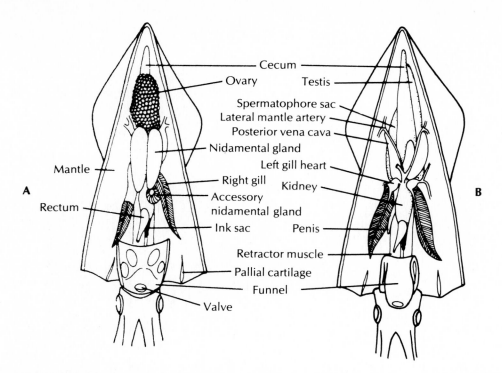

Fig. 12-11. Posterior view of mantle cavity of a squid. **A,** Female with funnel (siphon) opened to expose the valve. **B,** Male.

protecting the eye and ganglia. The pen, which represents the shell, is a thin, chitinous, feather-shaped plate that stiffens the anterior body wall and is concealed by the mantle (Fig. 12-10).

Since few introductory courses include the dissection of the squid's internal organs, that dissection will not be covered here.

DEMONSTRATIONS

1. *Microslides showing spermatophores of Loligo.*
2. *Preserved octopuses and cuttlefish (Sepia).*
3. *Shells of Nautilus.*
4. *Dried cuttlebone of Sepia.*
5. *Dissection of an injected cephalopod to show circulatory system.*
6. *Dissection of a cephalopod brain.*
7. *Living cephalopod, if available.*

CHAPTER 13

THE ANNELIDS

PHYLUM ANNELIDA*
A Protostome Eucoelomate Group

The annelids include a large variety of earthworms, leeches, and marine polychaetes. Their various adaptations fit them for freshwater, marine, terrestrial, and parasitic living. They are typically elongate wormlike animals, cylindrical in cross section, and have muscular body walls. The most distinguishing characteristic that sets them apart from other wormlike creatures is their segmentation. They are often referred to collectively as the "segmented worms." This repetition of body parts (**metamerism**) is not only external but is also seen internally in the serial repetition of body organs. The development of metamerism is of significance in the general evolutionary trend toward specialization, for along with segmentation comes the opportunity for the various segments, or metameres, to become specialized for certain functions. Such specialization is not as noticeable in the annelids as in the arthropods, but the introduction of metamerism made possible the rapid evolution of advanced organization seen in the arthropods and in the chordates, the only other phyla emphasizing metamerism.

The division of the coelomic cavity into fluid-filled compartments has also increased the usefulness of hydrostatic pressure in the locomotion of annelids. By shifting coelomic fluid from one compartment to another through perforations in the dividing septa, dif-

*See Hickman, C.P., Jr., L.S. Roberts, and F.M. Hickman. 1984. Integrated principles of zoology, ed. 7. St. Louis, The C.V. Mosby Co., pp. 338-361.

ferential turgor can be effected, permitting a preciseness of body movement not possible in the lower forms. The coordination between their well-developed neuromuscular system and this more efficient type of hydrostatic skeleton makes the annelids proficient in swimming, creeping, and burrowing.

Annelids have a complete mouth-to-anus digestive tract with muscular walls so that its movements are independent of bodily movements. There is a well-developed closed circulatory system with a pumping vessel, a high degree of cephalization, and an excretory system of nephridia. Some annelids have respiratory organs.

There are three main groups of annelids, classified chiefly on the basis of the presence or absence of a clitellum, parapodia, setae, annuli, and other features.

Class Polychaeta (pol'e-ke'ta). Segmented inside and out; parapodia with many setae; distinct head with eyes, palps, and tentacles; no clitellum; separate sexes; trochophore larva usually; mostly marine. Example: clamworm (*Nereis*).

Class Oligochaeta (ol'i-go-ke'ta). Body segmented inside and out; number of segments variable; clitellum present; few setae; no parapodia; head poorly developed; coelom spacious and usually divided by intersegmental septa; direct development; chiefly terrestrial and freshwater. Examples: *Lumbricus, Tubifex*.

Class Hirudinea (hir'u-din'ea). Segments 33 or 34 in number, with many annuli; clitellum present; anterior and posterior suckers; setae absent (except *Acanthobdella*); parapodia absent; coelom closely packed with a connective tissue and muscle; terrestrial, freshwater, and marine. Examples: *Hirudo, Placobdella*.

EXERCISE 13A

Class Polychaeta—the clamworm

MATERIALS

Preserved clamworms (*Nereis*)
Living nereids and/or other polychaetes as available
Dissecting tools
Dissecting pans
Clean slides
Hand lenses or dissecting microscopes
Compound microscopes

NEREIS

Phylum Annelida
  Class Polychaeta
    Subclass Errantia
      Genus *Nereis*

Where Found

Clamworms (also called sandworms, or ragworms) are strictly marine. They live in the mud and debris of shallow coastal waters, often in burrows lined with mucus. Largely nocturnal in habit, they are usually concealed by day under stones, in coral crevices, or in their burrows. The common clamworm *Nereis virens* may reach a length of half a meter.

Behavior

If living nereid worms are available, study their patterns of locomotion. When the animal is quiescent or moving slowly, note that the **parapodia** (lateral appendages) undergo a circular motion that involves an effective stroke and a recovery stroke, each parapodium describing an ellipse during each two-stroke cycle. In the effective stroke the parapodium makes contact with the substratum, lifting the body slightly off the ground. The two parapodia of each segment act alternately, and successive waves of parapodial activity pass along the worm.

For more rapid locomotion, undulatory movements of the body produced by muscular contraction and relaxation are used in addition to the parapodial action. As the parapodia on one side move forward in the recovery stroke, the longitudinal muscles on that side contract; as the parapodia sweep backward in their effective stroke, the muscles relax. Watch a worm in action and note these waves of undulatory movements. Can the worm move swiftly? Does it seek cover? Does it maintain contact with substratum, or does it swim freely?

Place near the worm a glass tube with an opening a little wider than the worm. Does the worm enter it willingly? Why? Place a bit of fresh mollusc or fish meat near the entrance of the tube. You may be able to observe the feeding reactions.

> Record your observations on p. 195 or on a separate sheet of paper.

External Features

☞ Place a preserved clamworm in a dissecting pan and cover with water.

Note the body with its specialized **head**, variable number of segments bearing **parapodia**, and the caudal segment bearing the **anus** and a pair of feelers, or **cirri**. Compare the length and number of segments of your specimen with those of other specimens at your table. Do they vary? The posterior segments are the smallest because they are the youngest. As the animal grows, new segments are added just anterior to the caudal segment.

Examine the **head.** It is made up of a small **prostomium** surrounded by the first segment, called the **peristomium** (Fig. 13-1). The prostomium bears a pair of small median **tentacles**, a pair of lateral fleshy **palps**, and four small dark **eyes**. The peristomium bears four pairs of **peristomial tentacles.** The palps and tentacles are sensory organs of touch and taste and the eyes are photoreceptors. The **mouth** is a slit on the ventral side of the peristomium through which the **pharynx** can be protruded. If the pharynx is everted on your specimen, do not confuse it with the head structures. The pharynx is large and muscular, bearing a number of small horny teeth and a pair of dark, pincerlike chitinous **jaws.** If the pharynx is fully everted, the jaws will be exposed. If not, you may be able to probe into the pharynx to find them.

☞ Cut off, close to the body, a parapodium from the posterior third of the body. Mount it in water on a slide, cover with a coverslip, and examine with hand lens or dissecting microscope.

The **parapodia** are used for respiration as well as for locomotion. Are they all identical? Each parapodium has a dorsal lobe called the **notopodium** and a ventral lobe called the **neuropodium** (Fig. 13-2). Each lobe bears a bundle of bristles called **setae.** A small process called the **dorsal cirrus** projects from the dorsal base of the notopodium, and a **ventral cirrus** is borne by the neuropodium. Each lobe has a long chitinous, deeply embedded spine called an **aciculum.**

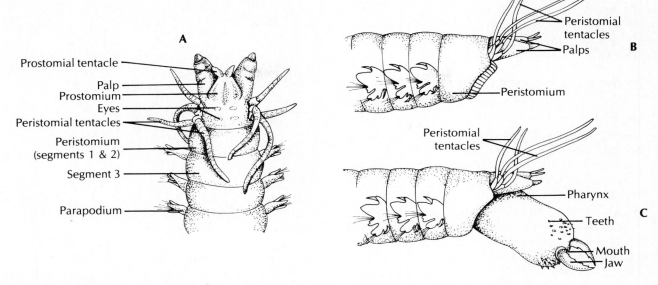

Fig. 13-1. Anterior view of the clamworm *Nereis*. **A,** Dorsal view. **B,** Side view with pharynx retracted. **C,** Side view with pharynx extended.

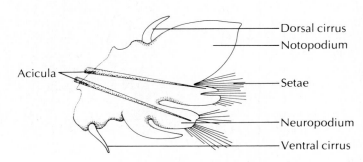

Fig. 13-2. A parapodium of the clamworm *Nereis*.

The acicula are the supporting structures of the parapodium (they are more conspicuous in the posterior parapodia). Each aciculum is attached by muscles that can protrude it as the parapodium goes into its effective stroke and retract it during the recovery stroke. The spines then aid in moving the body forward as the parapodium makes contact with the substratum.

☞ Peel off a piece of the thin cuticle that covers the animal and study it in a wet mount under the microscope.

It is fibrous, and its iridescence is caused by its cross striations. It is full of small pores through which the gland cells of the underlying epidermis discharge their products.

DRAWING

In this space sketch a small piece of the cuticle.

Internal Structure

Because the internal structure of the polychaete bears many general resemblances to that of the oligochaete, the study of the internal anatomy will be limited to that of the earthworm.

OTHER POLYCHAETES

There are over 10,000 species of polychaetes, most of them marine. They include some unusual and fascinating animals. Besides the errant, or free-moving, worms such as *Nereis* and its relatives (subclass Errantia), there are many sedentary species (subclass Sedentaria), including the burrowing and tube-building forms (Fig. 13-3).

☞ Examine the marine aquarium for living polychaetes and also the preserved material on the demonstration table.

Chaetopterus is a filter feeder often found on mud flats. It secretes a U-shaped parchment tube through which it pumps and filters a continuous stream of seawater. For further description and methods, see Projects and Demonstrations.

Diopatra is an errant tubeworm that camouflages the exposed portion of its tube with bits of shell, seaweed, and sand. Note its gills and long sensitive antennae. The fanworms (Fig. 13-3) are ciliary feeders, trapping minute food particles on mucus on the radioles of their feathery crowns and then moving the particles down toward the mouth in ciliated grooves.

Amphitrite is a deposit feeder. It runs long, extensible tentacles over the mud or sand, picking up food particles. These are carried along each tentacle in a ciliated groove to the mouth. Note the cluster of gills just below the tentacles.

Examine some of the tubes of various worms. What kinds of materials do you find? Examine an old seashell or rock bearing calcareous worm tubes. The variety of secreted tubes and the materials used to supplement them seem endless.

PROJECTS AND DEMONSTRATIONS

1. *Preserved polychaetes (Amphitrite, Arenicola, others).*

2. *Prepared slides.* For example, use whole mounts, or cross sections of polychaetes.

3. *Living feather-duster worms in marine aquarium.*

4. *Chaetopterus, a filter feeder in a glass tube.*

Chaetopterus secretes a long U-shaped parchmentlike tube in which it spends its life filtering food particles from the water. The worm is highly modified and specialized so that it creates a continuous current through the tube that keeps water and food particles moving over the body.

Fig. 13-3. Annelid tubeworms secrete the tubes in which they live. As sessile suspension feeders, their feathered crowns are well adapted for the capture of food. **A,** *Protula.* **B,** *Spirographis.* (Photographs by C.P. Hickman, Jr.)

Cut off the narrow end of its tube and slit along the side of the tube with scissors, being careful not to damage the occupant of the tube. Place the worm in a U-shaped glass or plastic tube of approximately the same length and diameter as the original tube. Place the tube, open ends up, in an aquarium for observation. Watch the rhythmic parapodial movements as the animal pumps the water through the glass tube.

A pair of modified parapodia near the anterior end secretes a mucous bag and suspends it across the tube so that the water passing through the tube is filtered through the mucous bag. Try placing some fine carbon particles at the incurrent end of the tube, then see if they become trapped in the bag. The mucous bag, when full of food, is detached, rolled into a ball, and sent along a ciliary tract anteriorly toward the mouth and swallowed. Can you observe this process?

Check the tube frequently for several days or weeks to see whether the worm will secrete a new mucous lining for the tube. What shape is the new lining?

5. *Bioluminescence in Chaetopterus*. In *Chaetopterus* photogenic cells are found in the hypodermis over most of the body, particularly in the parapodia and tentacles. These cells actively secrete a luminous material into the water.

Several worms should be removed from their tubes, placed in a bowl of seawater, and left in a very dark room for an hour or more before the experiment. While still in the dark, stimulate with the lowest direct-current (DC) electric shock that will evoke luminescence (or if equipment is unavailable prod with a probe). Find out the following: (a) Are some parts of the body more sensitive than others? (b) Does the light spread from the point of stimulus or does it occur simultaneously over the whole body? (c) How long does the luminescence last after a stimulus? (d) Is there a time lag between the stimulus and the response? (e) Does repeated stimulation cause fatigue? Why?

6. *Observation of burrowing polychaetes*. An observation chamber can be constructed of two pieces of glass, clear plastic, or Lucite (each about 30 to 40 cm square), a length of plastic or rubber tubing 10 to 15 mm in diameter, and some large rubber bands or small clamps (not metal). Arrange the tubing in a U shape on one glass, cover with the other glass, and hold in place with clamps or rubber bands. Fill the U space nearly full with mud or sand. Stand up in an aquarium and place one or more burrowing polychaetes on the mud surface. The lugworm **Arenicola** will burrow into sand, then form a tube with mucous lining. You might try *Clymenella*, which needs sand for its tube, or *Amphitrite*, or other burrowing species. You should be able to observe the burrowing and tube-building processes.

7. *Observation of circulation in polychaetes*. Many of the polychaetes are suitable for this, for example, the bloodworm, *Glycera;* the clamworm, **Nereis;** and many of the tube worms. The worm may be anesthetized or may be pinned by a few of the parapodia to a wax-bottomed pan and observed under the dissecting microscope. Note the pulsations of the dorsal vessel. Often the parapodia or the gills are transparent enough for you to observe the route of the blood cells. This is most interesting to watch.

NOTES ON THE BEHAVIOR OF POLYCHAETES

EXERCISE 13B
Class Oligochaeta—the earthworm*

MATERIALS
Living materials
  Earthworms
  Leeches
  Freshwater oligochaetes
Preserved or living materials for dissection
  Lumbricus
Tricaine methanesulfonate (MS222), 0.08%, for anesthetizing living worms for dissection (immerse worms in anesthetic $1\frac{1}{2}$ to 2 hours before class use)
Stained cross sections of earthworms
Paper toweling
Glass plates
Hand lenses

LUMBRICUS, THE COMMON EARTHWORM

Phylum Annelida
  Class Oligochaeta
    Order Haplotaxida
      Genus *Lumbricus*
        Species *L. terrestris*

Where Found
Earthworms prefer moist rich soil where it is not too dry or sandy. They are found all over the earth. They are chiefly nocturnal and come out of their burrows at night to forage. A good way to find them is to search with a flashlight around the rich soil of lawn shrubbery. The large night crawler is easily found this way, especially during warm moist nights of spring and early summer.

Behavior
☞ Place a sheet of paper toweling on your work area and wet the center of the paper, leaving the rest dry. Place a live earthworm on the moist area. Using the following suggestions, observe its behavior.

Is the skin of the worm dry or moist? Do you find any obvious respiratory organs? Where do you think exchange of gases occurs? Would this necessitate a dry or damp environment? What happens when the worm contacts the dry toweling? Is it positive or negative to moisture?

Notice the mechanics of crawling. Its body wall contains well-developed layers of circular and longitudinal muscles. As it crawls, notice the progressing peristaltic waves of contraction. Do these waves move forward or posteriorly? Run a finger along the side of

*See Hickman, C.P., Jr., L.S. Roberts, and F.M. Hickman. 1984. Integrated principles of zoology, ed. 7. St. Louis, The C.V. Mosby Co., pp. 340-349.

the worm. Do you detect the presence of small setae (bristles)? Notice that areas of the body in which the longitudinal muscles are contracting are in temporary contact with the substratum, thus forming the "feet" of the moving animal. The setae aid in providing traction or holding power when the animal is burrowing (or resisting a robin's pull).

How does the animal respond when you gently touch its anterior end? Its posterior end? Draw the towel to the edge of the desk and see what happens when the worm's head projects over the edge of the table. Is it positively or negatively thigmotactic? Turn the worm over and see if and how it can right itself.

Can you devise a means of determining whether the earthworm is positively or negatively geotactic (responsive to gravity)?

☞ Place the earthworm on a piece of wet glass. Does this difference in substratum affect its locomotion? Is friction important in earthworm locomotion?

Does the earthworm have eyes or other obvious sensory organs? Can you devise a means of determining whether it responds positively or negatively to light?

Record the responses of the earthworm on p. 206.

External Structure
☞ Examine a preserved or anesthetized earthworm under water, using a hand lens as necessary.

What are the most obvious differences between the earthworm and the clamworm?

The first four segments make up the head region. The first segment is the **peristomium.** It bears the **mouth,** which is overhung by a lobe, the **prostomium.** The head of the earthworm, lacking in specialized sense organs, is considered degenerate and is not a truly typical annelid head.

Find the **anus** in the last segment. Observe the saddlelike **clitellum,** which in mature worms secretes the egg capsules into which the eggs are laid. In what segments does it occur?

Feel the little bristles or **setae** used by the worm to prevent slipping. They are manipulated by small muscles at their bases. How many pairs are on each segment, and where are they located? Use the hand lens or binocular microscope to determine this. What does the name "Oligochaeta" mean? "Polychaeta"? Are these names well chosen?

☞ Strip off a small piece of the iridescent non-cellular **cuticle.** Float it out in water on a slide, cover, and examine with the microscope for pores through which mucus is discharged from gland cells in the epidermis. Reduce the microscope light if necessary.

Note the fine striations in the cuticle that give it its iridescent appearance. How does it compare with the cuticle of *Nereis?*

DRAWING

Sketch a bit of the cuticle in this space.

DRAWINGS

Complete the external ventral view in Fig. 13-4. Draw in and label prostomium, peristomium, mouth, setae, male pores, female pores, seminal grooves, clitellum, and anus.

There are many external openings other than the mouth and anus. The **male pores** on the ventral surface of somite 15 are conspicuous openings of the sperm ducts from which spermatozoa are discharged. Note the two long **seminal grooves** extending between the male pores and the clitellum. These guide the flow of spermatozoa during copulation. The small **female pores** on the ventral side of segment 14 will require a hand lens. Here the oviducts discharge eggs. You may not be able to see the openings of two pairs of **seminal receptacles** in the grooves between segments 9-10 and 10-11 or the paired excretory openings, **nephridiopores,** located on the lateroventral surface of each segment (except the first three and the last).

A **dorsal pore** from the coelomic cavity is located at the anterior edge of the middorsal line on each segment from 8 or 9 to the last. Many earthworms eject a malodorous coelomic fluid through the dorsal pores in response to mechanical or chemical irritation or when subjected to extremes of heat or cold. The dorsal pores may also help regulate the turgidity of the animal. How would the loss of coelomic fluid affect the animal's escape mechanism (quick withdrawal into its burrow)? How does the lack of dorsal pores in its anterior segments protect its burrowing ability?

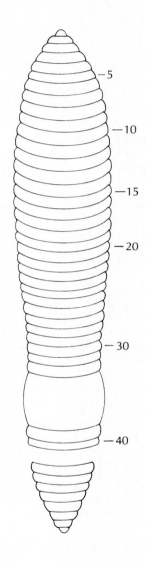

Fig. 13-4. External ventral structure of the earthworm.

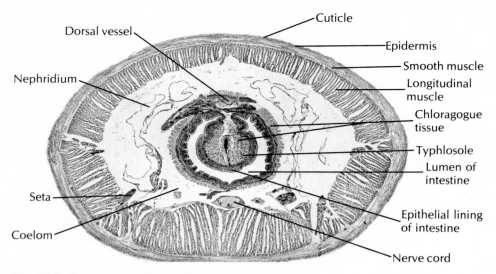

Fig. 13-5. Cross section of an earthworm through the intestinal region. (Courtesy Carolina Biological Supply Co., Burlington, N.C.)

Internal Structure and Function*

☞ Cut off the anterior end of the animal 1 or 2 inches back of the clitellum and save this end for dissection later. Cut off one or two segments from the posterior piece and examine the cut ends with a hand lens or dissecting microscope.

Note the outer **epidermis**, the thick middle layer of **muscles**, and the thin inner **peritoneum**. The intestine has a dorsal infolding, the **typhlosole**. Note the "tube-within-a-tube" arrangement—a tubular digestive tract within a tubular body wall, with the coelomic cavity between them (Fig. 13-5).

☞ Cut another piece of the worm from the posterior piece about 2.5 cm long and make a longitudinal section by cutting it in half dorsoventrally.

Note how the **coelom** is divided into a series of compartments by **septa** that coincide with the external segmentation. In earthworms, these septa are incomplete ventrally where the nerve cord passes through,

---

*Although this description applies to the study and dissection of preserved earthworms, living worms anesthetized with tricaine methanesulfonate (0.08% solution) may be substituted. The advantages of living material are that the body organs are their natural color (and thus are more easily identified), the beating of the "hearts" and peristaltic movements of the dorsal vessels are vividly displayed, and the gut, especially the gizzard, reveals its natural peristaltic contractions. Living, anesthetized worms should be opened middorsally by cutting carefully through the body wall with a razor blade or a new scalpel blade. Pin the animal as described for the preserved worm, but keep it moist with isotonic saline solution (amphibian Ringer's solution works well) applied with a pipette. Do not use water.

Do not cut the animal into anterior and posterior portions until *after* you have opened and pinned the anterior half and studied its internal anatomy. Then cut off the posterior portion for study of the transverse section.

but these openings can be closed by sphincter muscles so that there is little movement of fluid between segments during locomotion.

☞ Now, holding the *anterior portion* of the worm dorsal side up, insert the scissors point into the cut posterior end just beside the dark middorsal line made by the **dorsal blood vessel.** Keep the lower blade up to prevent damaging the viscera. Cut **close** to one side of the vessel, extending the cut forward to about the fourth somite. Loosen the septa on each side to spread the body wall apart. Pin down each side of segment 15 (identified by the male pores). Then count the segments carefully and pin down segments 5, 10, 20, 30, and so on, thus creating definite landmarks with the pins. Slant the pins out at a wide angle from the body (**Fig. 13-6**). Cover completely with water and use the hand lens freely for examination.

Identify the **digestive tract** (is its diameter uniform?); the three pairs of large whitish **seminal vesicles** in somites 9 to 12; the two pairs of small spherical **seminal receptacles** in somites 9 and 10; a pair of **nephridia** (thin coiled tubes) in the coelomic cavity of each segment (loosen some of these with a dissecting needle so that they float upward in the water); the **dorsal vessel** riding on the digestive tract; and the five pairs of **aortic arches**, formerly called "hearts," around the esophagus in somites 7 to 11 (some of these are covered by the seminal vesicles).

Reproductive System. The earthworm is monoecious; it has both male and female organs in the same individual, but cross fertilization occurs during copulation. First, consider the **male organs.** Three pairs of **seminal vesicles** (sperm sacs in which spermatozoa

198

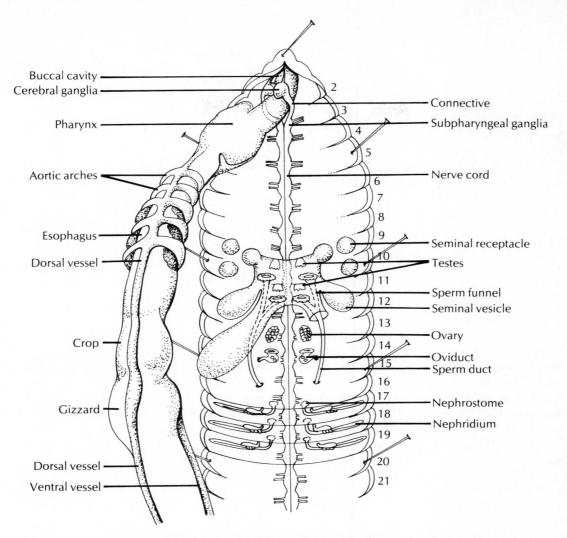

Buccal cavity
Cerebral ganglia
Pharynx
Aortic arches
Esophagus
Dorsal vessel
Crop
Gizzard
Dorsal vessel
Ventral vessel

2
3
4
5
6
7
8
9
10
11
12
13
14
15
16
17
18
19
20
21

Connective
Subpharyngeal ganglia
Nerve cord
Seminal receptacle
Testes
Sperm funnel
Seminal vesicle
Ovary
Oviduct
Sperm duct
Nephrostome
Nephridium

Fig. 13-6. Internal structure of the earthworm *Lumbricus,* dorsal view.

mature and are stored before copulation) are attached in somites 9, 11, and 12; they lie close to the esophagus. Two pairs of small branched **testes** are housed in special reservoirs in the seminal vesicles, and two small sperm ducts connect the testes with the **male pores** in somite 15; but both testes and ducts are too small to be found easily. The **female organs** are also small. Two pairs of small round **seminal receptacles,** easily seen in somites 9 and 10, store spermatozoa after copulation. You will probably not find the paired ovaries that lie ventral to the third pair of seminal vesicles or the paired **oviducts** with ciliated funnels that carry eggs to the female pores in the next segment.

Earthworm Copulation. When mating, two earthworms, attracted to each other by glandular secretions, extend their anterior ends from their burrows and, with their heads pointing in opposite directions, join their ventral surfaces in such a way that the sem-

inal receptacle openings of one worm lie in opposition to the clitellum of the other (Fig. 13-7). Each worm secretes quantities of mucus so that each is enveloped in a slime tube extending from segment 9 to the posterior end of the clitellum. Seminal fluid discharged from the sperm ducts of each worm is carried along the seminal grooves by contraction of longitudinal muscles and enters the seminal receptacles of the mate. After copulation the worms separate, and each clitellum produces a secretion that finally hardens over its outer surface. The worm moves backward, drawing the hardened tube over its head. The tube, as it is moved forward, receives eggs from the oviducts, sperm from the seminal receptacles, and a nutritive albuminous fluid from skin glands. Fertilization occurs in the cocoon. As the worm withdraws, the cocoon closes and is deposited on the ground. Young worms hatch in 2 to 3 weeks.

Digestive System. Identify the **mouth;** the mus-

199

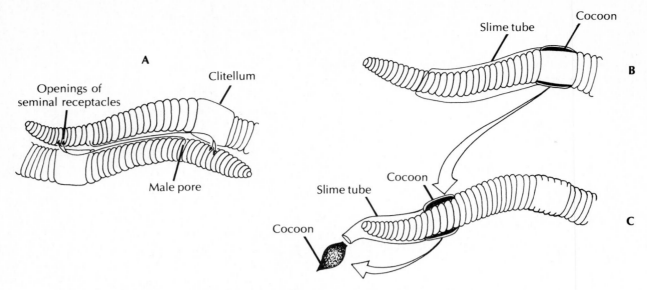

Fig. 13-7. **A**, Earthworm copulation. **B** and **C**, Formation and deposit of the cocoon.

cular **pharynx** attached to the body by **dilator muscles** for sucking action (the muscles, torn by the dissection, give the pharynx a hairy appearance); the slender **esophagus** in somites 6 to 13, which is hidden by the aortic arches and seminal vesicles; the large thin-walled **crop** (15, 16) for food storage; the muscular **gizzard** (17, 18) for food grinding; the **intestine** for digestion and absorption; and the **anus**. Two or three paris of white calciferous glands are seen on each side of the esophagus; they are thought to secrete calcium carbonate to neutralize acid foods. Bright yellow **chlor-agogue** cells, covering the intestine and extending along the dorsal vessel, are conspicuous in the living earthworm. They are known to store glycogen and lipids, but probably have other functions as well, similar to those of the vertebrate liver. Which wall of the intestine folds in to form the **typhlosole** so as to increase absorptive surface? Refer to the cut sections you made earlier. What is the relation of the septa to the intestine?

Circulatory System. The well-developed blood system of the earthworm is a **closed system**; that is, the blood flows in a continuous circuit of vessels rather than opening out into body spaces. Consult your text for the direction of flow and complete circulation.*
Identify the **dorsal vessel** along the dorsal side of the digestive tract, five pairs of **aortic arches** encircling the esophagus in segments 7 to 11, and the **vessel** ventral to the digestive tract. Both dorsal vessel and arches are contractile, with the dorsal vessel being the chief pumping organ and the arches maintaining a

*See Hickman, C.P., Jr., L.S. Roberts, and F.M. Hickman. 1984. Integrated principles of zoology, ed. 7. St. Louis, The C.V. Mosby Co., pp. 344-345.

steady flow of blood into the ventral vessel. Lift up the intestine at the cut posterior end of the worm and see the ventral vessel. Now lift up the white nerve cord in the ventral wall and use the hand lens to see the **subneural vessel** clinging to its lower surface and the pair of **lateroneural vessels**, one on each side of the nerve cord.

☞ Be able to trace blood flow from the dorsal vessel to the intestinal wall and back, to the epidermis and back, and to the nerve cord and back.

See Projects and Demonstrations, nos. 2 to 4, for demonstrations on circulation.

Excretory System. A pair of tubular **nephridia** lie in each somite except the first three and the last. Each nephridium has a ciliated funnel-shaped **nephrostome** in the segment anterior to it. The nephrostome draws wastes from one somite into the ciliated tubular portion of the nephridium in the next somite, which empties the waste to the outside through a **nephridiopore** near the ventral setae.

☞ Use a dissection microscope to study the nephridia. They are largest in the region just posterior to the clitellum. Be sure the worm is completely covered with clean water, which will float the nephridia up into view.

See Projects and Demonstrations, nos. 6 and 7.

Nervous System. To examine the nervous system, do the following.

☞ Very carefully extend the dorsal incision to the first somite.

Find the small pair of white **cerebral ganglia** (the brain), lying on the anterior end of the pharynx and partially hidden by dilator muscles; the small white nerves from the ganglia to the prostomium; a pair of **circumpharyngeal connectives,** extending from the ganglia and encircling the pharynx to reach the **subpharyngeal ganglia** under the pharynx; and a **ventral nerve cord,** extending posteriorly from the subpharyngeal ganglia for the entire length of the animal. Remove or lay aside the digestive tract and examine the nerve cord with a hand lens to see in each body segment a slightly enlarged **ganglion** and **lateral nerves.**

ORAL REPORT

Be prepared to (1) demonstrate your dissection to the instructor, (2) point out both the external and internal structures you have studied, and (3) explain their functions.

DRAWING

Sketch in the nervous system in the outline in Fig. 13-8.

Histology of Cross Section

☞ Examine a stained slide with low power. Note the tube-within-a-tube arrangement of intestine and body wall. Identify the following:

Cuticle. Thin, noncellular, and secreted by the epidermis (Fig. 13-5).

Epidermis (Ectodermal). Columnar epithelium containing mucous gland cells. Mucus prevents the skin from drying out.

Circular Muscle Layer. Smooth muscle fibers running around the circumference of the body. How does their contraction affect the body shape?

Longitudinal Muscle Layer. Thick layer of featherlike fibers that run longitudinally. What is their function? The muscle layers may be interrupted by the setae and dorsal pore.

Peritoneum (Mesodermal). Thin, epithelial layer lining the body wall and covering the visceral organs (Fig. 13-9).

Setae. If present, they are brownish spines in a sheath secreted by epidermis. They are moved by tiny muscles (Fig. 13-9). Why are they not present on all slides?

Coelom. Space between the **parietal peritoneum,** which lines the body wall, and the **visceral peritoneum,** which covers the intestine and other organs.

Alimentary Canal. Probably the intestine; surrounded by chloragogue tissue (Fig. 13-5), which

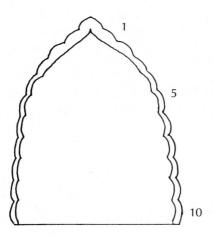

Fig. 13-8. Dorsal view of earthworm's dissected nervous system.

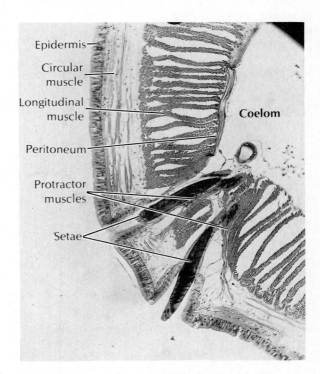

Fig. 13-9. Portion of cross section of an earthworm showing one set of setae with their protractor muscles. (Courtesy Carolina Biological Supply Co., Burlington, N.C.)

plays a role in intermediary metabolism similar to that of the liver in vertebrates. Inside the chloragogue layer is a layer of **longitudinal muscle.** Why does it appear as a circle of dots? Next is a **circular muscle layer,** followed by a layer of ciliated columnar epithelium (endodermal), which lines the intestine. Are these layers continuous in the **typhlosole?** Intestinal contents are moved along by peristaltic movement. Is such movement possible without both longitudinal and circular muscles? Is peristalsis possible in the intestine of the flatworm or *Ascaris?* What does force the food along in those animals?

Ventral Nerve Cord. Use high power to identify the three **giant fibers** in the dorsal side of the nerve cord and the nerve cells and fibers in the rest of the cord.

Blood Vessels. Identify the **dorsal** vessel above the typhlosole, **ventral** vessel below the intestine, **subneural** vessel below the nerve cord, and **lateral neurals** beside the nerve cord.

Some slides may also reveal parts of **nephridia, septa, mesenteries,** and other structures.

DRAWINGS

1. Sketch and label a cross section of the earthworm as it appears on your slide. Use p. 205.

2. Visualize a transverse section through segment 11 and one through segment 18 on p. 205. Make outline diagrams of these sections, putting in such structures as you think you would see. Do not include cellular structure. Label.

FRESHWATER OLIGOCHAETES

Some species of freshwater oligochaetes are available at biological supply houses. They may also be obtained from the mud and debris of streams, lakes, stagnant pools, and ponds. Some are common in filamentous algae. Some are found in deep lake bottoms, but most live in shallow water. Most of them range from 1 to 30 mm in length.

☞ If specimens are available, either living or preserved, examine with low power and compare with the terrestrial earthworms.

Identify the **somites, setae,** and **alimentary canal.** In the living specimen the pulsating of the blood, peristalsis of the alimentary canal, and ciliary action in the intestine may be observed.

Most of the freshwater oligochaetes feed by ingesting quantities of substratum, as earthworms do, though some feed on algae. *Chaetogaster* lives on small crustaceans, insect larvae, and the like. *Aeolosoma* (Fig. 13-10, *B*) uses cilia around the mouth to sweep in food particles. In *Stylaria* (Fig. 13-10, *A*) the prostomium is drawn out into a long proboscis.

Tubifex (Fig. 13-10, *E*) is a reddish oligochaete 2 to 3 cm long that builds burrows in the mud, where it lives head down and waves its extended tail back and forth to stir up water currents. It is especially common in sluggish and polluted streams and lakes.

Most gaseous exchange occurs through the thin body wall, but *Dero* and *Aulophorus* (Fig. 13-10, *C* and *D*) have ciliated anal gills. Gilled species lie quietly hidden in the substratum or in a tube with the posterior end projecting into the water.

Reproduction by budding is common, such that a new individual forms from one of the posterior segments. They also reproduce sexually, depositing cocoons on rocks, vegetation, and debris.

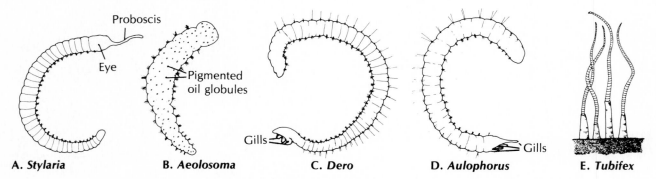

Fig. 13-10. Some common freshwater oligochaetes. These worms reproduce chiefly by budding off new individuals from posterior segments.

PROJECTS AND DEMONSTRATIONS

1. Cocoons of earthworms. Earthworm cocoons can be obtained from a biological supply house. How large are they? Explain from your textbook descriptions how they are formed. Is there more than one worm in a cocoon? Where do earthworms deposit their cocoons?

2. Blood vessels of earthworm. A good way to show up the blood vessels of earthworms is to open up the body cavities of fresh specimens and immerse them for 3 hours in a solution of 4 parts nitric acid and 2 parts hydrochloric acid. This method often reveals the blood vessels as black on a yellow background.

3. Demonstration of blood flow. The following experiment demonstrates the flow of blood and the pulsating of the hearts. Procure a fresh specimen of the small manure worm *Eisenia foetida*. Cut a cardboard of medium thickness to the same size as an ordinary glass slide (7.5 by 2.5 cm). In the center of this cardboard, cut a rectangular window about 5.5 by 1.8 cm. Put the cardboard on the glass slide and in the center cavity so formed place a live worm. Imprison it by taping another glass slide on top of the cardboard. By exerting some pressure on the top slide you can see the red blood in the dorsal blood vessel. What is the direction of blood flow? Observe also the pulsating of the paired aortic arches.

4. Invertebrate blood smears.

a. Draw out blood from a living mussel (heart), an arthropod, or an earthworm or other oligochaete by means of a syringe.

b. Smear on a clean glass slide and dry in the air.

c. Cover the preparation with a measured quantity of staining fluid (Wright's blood stain) for 1 minute.

d. Now add to the staining fluid on the slide an equal quantity of distilled water. Allow this mixture to remain for 2 to 3 minutes.

e. Wash in distilled water until the film has a pinkish tint. Examine under the microscope.

f. Dry between filter paper and mount in balsam. Look for formed elements (corpuscles).

5. Demonstration of nephridium. A method of demonstrating the nephridium, especially the difficult **nephrostome,** is to anesthetize a living earthworm by immersion in tricaine methanesulfonate (0.08% solution) or a chloretone solution (0.2% aqueous solution), make an incision in its middorsal region behind the clitellum, cut away the septa, and remove the intestine, being careful not to disturb the nephridia. Then squirt onto the preparation with a pipette several drops of a methylene blue solution (1 part of methylene blue to 2000 parts of a 0.6% salt solution). Leave the stain in place for a few minutes and wash it off with a 0.6% salt solution. Keep the preparation immersed with the salt solution and observe the nephrostomes and other parts of the nephridia. The nephridium may also be removed and studied under the microscope.

6. Ciliary movement. Ciliary movement may be observed in the nephridium of a freshly killed worm. Remove a nephridium, place on a slide with water, and cover with a cover glass. Observe under high power of the microscope.

7. Isolation of annelid setae. Boil pieces of the worm in a 5% potassium hydroxide solution until the tissue has dissolved. Allow the setae to settle, and decant the fluid. Wash by adding water, allowing to settle, and then decanting; repeat several times. There are several types of setae—capilliform, straight, and curved and those having the ends bifurcated, hooked, pectinate, and the like. The type, number, and location of the setae are frequently of taxonomic significance in the classification of annelids.

EXERCISE 13C
Class Hirudinea—the leech

MATERIALS
Living leeches
Preserved *Hirudo*

HIRUDO

Phylum Annelida
  Class Hirudinea
    Order Gnathobdellida
      Genus *Hirudo*

Where Found

Leeches are predaceous and mostly fluid feeders. Some are true blood suckers, but they are not truly parasitic since they usually attach themselves to a host only during feeding periods.

Hirudo is one of the freshwater leeches found in lakes, ponds, streams, and marshes. Most marine leeches feed upon turtles, fishes, dolphins, and other species. *Hirudo medicinalis* is often referred to as the "medicinal leech" because it was formerly used in blood-letting. It feeds on the blood of vertebrates, to which it attaches itself periodically. Living leeches are readily available from biological supply houses.

Behavior

Note the use of the ventral **suckers**—the smaller one at the anterior end and the larger one at the posterior end. Attempt to pull an attached leech free from its substratum. Are the suckers powerful? The leech com- bines the use of its suckers with muscular body contractions in creeping. The anterior sucker attaches and the body contracts, then the posterior sucker attaches and the body extends forward before the anterior sucker attaches again.

Drop a leech into the water to observe its undulating free-swimming motions.

External Features

Study a preserved specimen of *Hirudo*. How does its shape compare with that of the earthworm? Can you distinguish between the dorsal and ventral surfaces? How? The leech is segmented inside and outside, but externally the segments are also marked off into one to five **annuli** each, the larger number being found in segments 8 to 23 inclusive. Each true segment bears a pair of **nephridiopores*** and, on one of its annuli, a row of **sensory papillae** or (at the anterior end) eyespots.

The anterior sucker contains the **mouth.** The **anus** is located in the middorsal line at the junction with the posterior sucker. The **male genital pore** is located ventrally on segment 11, and the **female pore** is on segment 12. Segments 9, 10, and 11 serve as the functional **clitellum,** secreting the capsule in which the eggs develop. Leeches are monoecious.

---

*Nephridiopores can be demonstrated by gently squeezing a freshly narcotized specimen, thus causing a little fluid to be exuded from the nephridial bladders.

EARTHWORM

Transverse section through the body of an earthworm

Hypothetical section through segment 11   Hypothetical section through segment 18

OBSERVATIONS ON THE BEHAVIORAL RESPONSES
OF THE EARTHWORM

Hydrotaxis _____

_____

_____

_____

Locomotion _____

_____

_____

_____

Thigmotaxis _____

_____

_____

_____

Phototaxis _____

_____

_____

_____

Importance of friction _____

_____

_____

_____

CHAPTER 14

T H E   A R T H R O P O D S

PHYLUM ARTHROPODA*
A Protosome Eucoelomate Group

There are approximately a million named species of arthropods. These include the chelicerates (comprising the spiders and scorpions and their allies) and the mandibulates (comprising the crustaceans, millipedes, centipedes, and insects).

Arthropods have **jointed appendages,** a fact that has given them their name (Gr., *arthron*, joint, + *pous*, *podos*, foot). Like the annelids, with whom they are probably closely related, the arthropods have **segmented bodies,** but whereas the somites of the annelid are more or less similar, in the arthropod some somites may be modified for certain functions and others may be fused or even lost altogether. There is a tendency for the somites to be combined or fused into functional groups called **tagmata.** Specialization of the somites has made possible a rapid evolution with great complexity of form and a broad adaptive radiation.

In addition to features shared by previous phyla, such as **triploblastic development,** a **true coelom, bilateral symmetry, cephalization,** and **all organ systems,** they have developed **striated muscle** for rapid movement, an **exoskeleton** of cuticle containing **chitin** for support and protection, **gills** and a very efficient **tracheal system** for gaseous exchange, and **greater specialization** of body organs, especially a great specialization of form and function among the appendages. The coelom is reduced, consisting largely of hemocoel, or spaces in the tissues filled with blood. The circulating fluid, which is a mixture of blood and lymph, is not restricted to blood vessels.

The arthropods include the following subphyla and classes:

Subphylum Trilobita (tri′lo-bi′ta) **trilobites.** All extinct forms; Cambrian to Carboniferous; body divided by two longitudinal furrows into three lobes; distinct head, thorax, and abdomen; biramous (two-branched) appendages.
Subphylum Chelicerata (ke-lis′e-ra′ta) **eurypterids, horseshoe crabs, spiders, ticks.** First pair of appendages modified to form chelicerae; pair of pedipalps and four pairs of legs;

no antennae, no mandibles; cephalothorax and abdomen usually unsegmented.
  Class Merostomata (mer′o-sto′ma-ta). Aquatic chelicerates that include the horseshoe crabs *(Limulus)* and the extinct Eurypterida.
  Class Pycnogonida (pik′no-gon′i-da). Sea spiders.
  Class Arachnida (ar-ack′ni-da). Spiders and scorpions. Somites fused into cephalothorax; head with paired chelicerae and pedipalps; four pairs of legs; abdomen segmented or unsegmented, with or without appendages; respiration by gills, tracheae, or book lungs. Example: *Argiope.*
Subphylum Mandibulata* (man-dib′u-la′ta). Head appendages consisting of one or two pairs of antennae, one pair of mandibles, and one or two pairs of maxillae.
  Class Crustacea (crus-ta′she-a). Crustaceans. With gills; body covered with carapace; exoskeleton with limy salts; appendages biramous and variously modified for different functions; head with two pairs of antennae. Examples: *Cambarus, Homarus.*
  Class Diplopoda (di-plop′o-da). Millipedes. Subcylindric body elongated and wormlike; variable number of somites; usually two pairs of legs to a somite. Example: *Spirobolus.*
  Class Chilopoda (ki-lop′o-da). Centipedes. Elongated with dorsoventrally flattened body; variable number of somites, each with pair of legs; tracheae present. Example: *Lithobius.*
  Class Pauropoda (pau-rop′o-da). Pauropods. Minute, soft-bodied forms with 12 segments and 9 or 10 pairs of legs. Example: *Pauropus.*
  Class Symphyla (sym′fy-la). Garden centipedes. Centipede-like bodies of 15 to 22 segments and usually 12 pairs of legs. Example: the garden centipede, *Scutigerella.*
  Class Insecta (in-sek′ta). Insects. Body with distinct head, thorax, and abdomen; thorax usually with two pairs of wings; three pairs of jointed legs. Example: *Romalea* (lubber grasshopper).

---

*Traditionally the subphylum Mandibulata has contained those arthropods possessing mandibles and antennae, that is, the crustaceans, myriapods, and insects. However, some authorities now believe that crustacean mandibles and insect-myriapod mandibles have resulted from convergent evolution rather than a natural, close relationship. They consider the Crustacea and Uniramia (insects and myriapods) as subphyla. For the present we will retain the traditional arrangement.

---

*See Hickman, C.P., Jr., L.S. Roberts, and F.M. Hickman. 1984. Integrated principles of zoology, ed. 7. St. Louis, The C.V. Mosby Co., pp. 367-378.

The chelicerate arthropods—the horseshoe crab and garden spider

The chelicerates include the horseshoe crabs, sea spiders, spiders, scorpions, mites, ticks, and some others. They are the arthropods that do not possess mandibles (jaws) for chewing. Instead, the first pair of appendages, which are called chelicerae (kuh-liss'er-e), are feeding appendages that are adapted for seizing and tearing. The body of the chelicerate is made up of a cephalothorax and an abdomen. There are no antennae.

MATERIALS

Living materials, if available
  Limulus, in marine aquarium
  Spiders, in terrarium
Preserved materials
  Limulus
  Garden spiders
Hand lenses or dissecting microscopes

HORSESHOE CRAB

Phylum Arthropoda
Subphylum Chelicerata
  Class Merostomata
    Subclass Xiphosurida
      Genus *Limulus*
        Species *L. polyphemus*

Where Found

The horseshoe crabs are not really crabs at all but members of an ancient group of chelicerates, the Merostomata, most of which are long extinct. They are marine bottom dwellers, scavengers that search the muddy bottom for molluscs, worms, algae, and the like. In the spring they come up onto beaches to spawn. *Limulus polyphemus* lives along the North American Atlantic and gulf coastlines.

External Features

The entire body of the horseshoe crab is covered with a tough, leathery **exoskeleton** that contains a good deal of scleroprotein. As the animal grows it must shed (molt) the exoskeleton, a process called **ecdysis.**

Cephalothorax. Covering the cephalothorax dorsally and laterally is a hard, horseshoe-shaped **carapace,** concave below and convex above. A pair of lateral **compound eyes** and a pair of median **simple eyes** are on the dorsal side.

On the ventral side are six pairs of appendages, located around the **mouth** (Fig. 14-1). The first pair, called the **chelicerae,** are small and are used in food manipulation. The second pair are called **pedipalps.** The next three pairs are similar to the pedipalps. All five pairs are chelate (bear pincers, or **chelae**), and all except the chelicerae have spiny masticatory processes called **gnathobases** on the basal segments. Move the legs and see how these processes would serve to tear up food and move it toward the mouth. These appendages are also used in walking, and their chelae pick up food and pass it up to the gnathobases. Note that the sixth pair of legs have no chelae, but each has instead four movable spines. These legs are used to push against the sand to help in forward movement and in burrowing. Between the sixth pair of appendages is a small rudimentary pair called the **chilaria.**

Abdomen (opisthoma). The abdomen bears six pairs of spines along the sides and, on its ventral side, six pairs of flat platelike appendages. The first of these forms the **genital operculum,** on the underside of which are two **genital pores.**

The other five abdominal appendages are modified as **gills.** Lift up one of these flaps to see the many (100 to 150) leaflike folds called **lamellae.** Because of this leaflike arrangement, such gills are called **book gills** (Fig. 14-1). The exchange of gases between the blood and the surrounding water takes place in the lamellae. The movement of the gills not only circulates water over them but also pumps blood in and out of the lamellae. The blood contains hemocyanin, a respiratory pigment used in oxygen transport, as well as a type of amebocyte that aids in blood clotting.

The beating of the abdominal flaps can also be used in swimming and may aid the animal in burrowing by creating a water current that washes out the mud or sand posteriorly.

Telson. The long slender **telson,** or tailpiece, is used for anchoring when the animal is burrowing or plowing through the sand. The **anus** is located under the proximal end of the telson.

Life Cycle

Horseshoe crabs are dioecious. In the spring the females swim up on the sandy beaches at the highest tides, followed closely by the males, whose pedipalps have become modified for clasping. As the female burrows into the sand to lay her eggs, the male discharges his sperm over them and they are then quickly covered with sand (Fig. 14-2). As the tide recedes, the 200 to 300 eggs in the sandy nest are warmed by the sun and in about 2 weeks hatch into active larvae about a centimeter long. The larva of the horseshoe crab bears a striking superficial resemblance to the ancient trilobites, so it has come to be called the "trilobite larva." After several molts the larva assumes the adult form. Sexual maturity is reached in the third year.

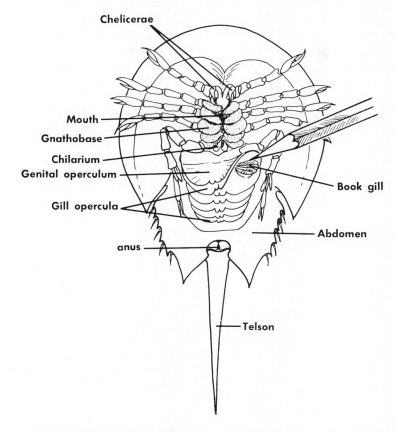

Fig. 14-1. Ventral view of *Limulus*. One gill operculum is raised to reveal location of book gill.

Fig. 14-2. Mating of horseshoe crabs. (Photograph by F.M. Hickman.)

Behavior

If live horseshoe crabs are available in a marine aquarium, you will enjoy making these observations. First, where do you find the resting animals? Swimming about, resting on the sand, or covered with sand? Before disturbing them drop some bits of fresh shrimp, oyster, or fish meat near them. Do they respond? How? (Normally they do most of their feeding at night.)

To observe their respiratory gill movements, lift up a crab in the water so that you can watch the beating of the gills. The gill movement is probably faster than when the animal is resting. Why? Each time the flaps move forward, blood flows into the gill lamellae; and as the flaps move backward, the blood flows out.

Free the animal near the surface of the water and see how it swims. Does it turn over? Which appendages does it use in swimming? As it settles to the sandy bottom, watch its reactions. Does it try to burrow? How? Can you see the use of the last pair of legs? Does the animal arch its back? Does it use the telson?

GARDEN SPIDER

Phylum Arthropoda
　Subphylum Chelicerata
　　Class Arachnida
　　　Order Araneae
　　　　Genus *Argiope*

Where Found

Spiders are found in most places where animals can live. They are distributed in all kinds of habitats such as forests, deserts, mountains, swamps, land, and water. The garden spider (*Argiope*), which builds its orb webs in sunny places in gardens and tall grass, is found throughout the United States. Two common species are *A. aurantia*, with a mottled back and yellow abdomen (Fig. 14-3), and *A. trifasciata*, with black and yellow bands on the abdomen. The males are about one-fourth the size of the females (from 4 to 8 mm) and are rarely seen. They are said to spin smaller webs near those of the female.

Behavior

The garden spider will spin its orb web in captivity, and if kept in a large glass container provided with tall grass or twigs and covered with screen or cheesecloth, the process can be watched. The spider spins a symmetric orb web and likes to hang head downward in the center, holding close together its forelegs and also its hind legs. If the web building can be observed, note the process—how it makes a supporting framework, then the radial spokes, and finally the spiral—first a temporary spiral beginning outside and then a permanent spiral beginning at the center. If possible examine bits of the silk under a microscope and note the different kinds.

Adding insects to the terrarium containing the spider may allow you to witness a capture in the net, the biting of the prey to paralyze it, and then the securing of the prey with the silken thread.

Spiders secrete enzymes to begin the digestive process outside the body. *Argiope* can apparently crush and tear the prey as well as suck out the liquid parts. How long does it take the spider to complete a meal?

Can you see the eyes? Most spiders have poor vision but are covered with sensory hairs and are very sensitive to touch and vibration.

External Features

☞ Study a preserved specimen and handle gently. There is always danger of breaking off appendages or the abdomen from the rest of the body. Use the hand lens or a dissecting microscope to examine the parts. Keeping the specimen moist will help prevent its becoming so brittle.

The chitinous **exoskeleton** is hard, thin, and somewhat flexible. **Sensory hairs** project on all parts of the body. The tagmata of the arachnid include the anterior

Fig. 14-3. Black and yellow garden spider, *Argiope aurantia*, which builds its orb web in gardens or tall grasses and then hangs there head down, awaiting an unwary insect. (Photograph by L. L. Rue III.)

cephalothorax and the posterior **abdomen** joined by a slender waist or **pedicel.**

Cephalothorax. Most spiders have six to eight **eyes** on the anterior dorsal surface but some have fewer. Spiders do not have compound eyes; all are simple ocelli. How many eyes does your specimen have? Use the hand lens to find them.

Identify the paired, two-jointed chelicerae, which are vertically oriented on the front of the face. One of the joints of a chelicera is a terminal **fang** by which the spider ejects poison from its poison gland (Fig. 14-4, *B*). Does the spider have true jaws (mandibles)? The pedipalps are six-jointed and used for gripping the prey. In the male the pedipalp is modified as an intromittant organ to transfer sperm to the female. The basal parts (coxal endites) of the pedipalps are used to squeeze and chew the food. Find the mouth between the pedipalps. The spider can ingest only liquid food.

There are four pairs of **walking legs.** Each leg is made up of seven segments as follows (from base to distal end): **coxa, trochanter, femur, patella, tibia, metatarsus,** and **tarsus.** The tarsus has claws and a tuft of hair at its terminal end.

Abdomen. On the ventral surface at the anterior end find two lateral triangular spots that mark the location of the **book lungs.** Book lungs are fashioned very much like the book gills of the horseshoe crab, except that they are enclosed internally in pockets.

The inner walls of these pockets are folded into long, thin plates (the leaves of the "book") held apart by bars so that there are always air spaces between them. Gas exchange occurs between the blood circulating inside the lamellae and the air flowing in the spaces between the lamellae. These air spaces connect with a small air chamber in each lung that opens to the outside through a slitlike opening, or spiracle, just posterior to each book lung (Fig. 14-4). Some spiders, the tarantula for example, have two pairs of book lungs. By peeling forward the cover of the lung and examining with the dissecting microscope you should be able to see the many thin respiratory leaves.

Locate, between the spiracles, the **epigynum** (e-pij'in-um), which conceals the female genital pore. The preserved specimens are probably all females.

Posteriorly on the abdomen, just in front of the spinnerets, is a small **tracheal spiracle.** This is an opening into a small chamber from which tracheal tubes extend into the body. Tracheal tubes carry air to the blood in the body tissues. Arachnid tracheal systems are similar to those of insects but less extensive. The garden spider has both book lungs and tracheae, but not all spiders do; some have only one type of respiratory organ.

There are three pairs of **spinnerets** on a raised surface. The middle pair is quite small, but the other two pairs are rather large and conical and in life are readily movable. The ends of the spinnerets have a

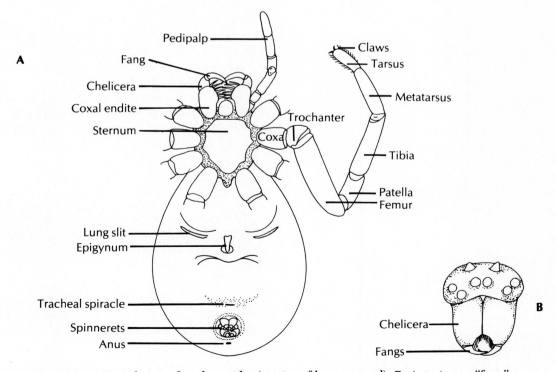

Fig. 14-4. A, Ventral view of garden spider (portion of legs removed). **B,** Anterior, or "face," view.

variety of tiny silk spouts, each producing a particular type of silk. The silk is secreted as a fluid by the silk glands and hardens on exposure to the air. Examine the spinnerets with the dissecting microscope.

A small papilla just posterior to the spinnerets bears the **anus**.

PROJECTS AND DEMONSTRATIONS

1. The tarantula, trap-door spider, black widow spider, and other spiders, alive or preserved.

2. Various mites and ticks.

3. Scorpions, alive or preserved.

4. Fossil trilobites and/or eurypterids.

5. Mounted stained preparations of the "trilobite larva."

6. Collection and preservation of arachnids. Besides the obvious methods of hand-picking and looking under stones or bark, other valuable aids are sweeping the grass with a net; beating or shaking bushes, hedges, or low branches over newspapers, cloth, or an upturned umbrella; and sifting leaves and ground litter. A simple sifter can be constructed from a square

Fig. 14-5. Simple device for gathering up tiny arachnids.

of wire netting folded in half and joined at the ends. The sifter can be carried flat, then opened into a boat-shaped sifter for use.

The difficulty of picking up tiny arachnids without injuring them is often a problem. Forceps or a camel's hair brush may be used, or a simple "pooter" can be made for sucking them up into a glass or plastic tube (Fig. 14-5). The tube is closed at each end by a one-hole stopper. A short piece of plastic tubing inserted in the bottom stopper accepts the organisms. A longer piece at the top becomes the mouth piece. A piece of gauze under the top stopper prevents the animals from being sucked into the collector's mouth.

A variation of the Berlese funnel (Fig. 14-18) can be made by setting a tin can, with top and bottom removed, into a glass or metal funnel, which is placed in the neck of a bottle of preservative. Fill the can with leaf litter and place over it a small electric light bulb (25 watts or less) or drop a mothball onto the litter. The arachnids, to avoid the heat, light, or fumes, will drop into the preservative.

A pitfall trap can be made by sinking a tin can, glass jar, or plastic container into the ground with the rim even with the surface. Leaves or pebbles in the bottom provide the captives some protection from the sun. If the trap is not to be visited daily, it is well to use a plastic container with drainage holes punched in the bottom in case of rain, or else to use a vessel containing some 2% formalin.

If spiders are to be kept alive, they should, because of their cannibalistic habits, be kept in separate containers. All but the largest can be kept in glass or disposable plastic tumblers stoppered with cotton wrapped in cheesecloth. A drop or two of water should be added daily. If they are to be kept for long periods, insects should be supplied as food.

Arachnids can be killed and fixed in 70% alcohol. They can be stored in propylene phenoxytol or in 70% alcohol. Another recommended storage fluid for spiders and daddy longlegs (harvestmen) is:

70% alcohol	85 parts
Glycerol	5 parts
Glacial acetic acid	5 parts

If a specimen is allowed to dry out completely while studying, it becomes rigid and is ruined. Keep the specimens moist while studying. Small specimens can be studied under water.

Labels on figure:
Buccal cavity
Cerebral ganglia
Pharynx
Aortic arches
Esophagus
Dorsal vessel
Crop
Gizzard
Dorsal vessel
Ventral vessel

Chelicerae

Mouth

Gnathobase

Chilarium

Genital operculum

Gill opercula

Book gill

Abdomen

anus

Telson

Fig. 14-1. Ventral view of *Limulus*. One gill operculum is raised to reveal location of book gill.

Fig. 14-2. Mating of horseshoe crabs. (Photograph by F.M. Hickman.)

EXERCISE 14B
The aquatic mandibulates
CLASS CRUSTACEA*—THE CRAYFISH (OR LOBSTER) AND OTHER CRUSTACEANS

The mandibulate arthropods make up a large assortment of crustaceans, insects, and myriapods, all of which possess **mandibles,** or jawlike appendages, instead of chelicerae. All mandibulates have, in addition to the mandibles, at least one pair of **antennae** and a pair of **maxillae** on the head.

Since nearly all crustaceans are aquatic and most insects and myriapods are terrestrial, they are usually designated as the aquatic mandibulates and the terrestrial mandibulates, although some investigators contend that the mandibles of the crustaceans and the mandibles of the insects and myriapods are not homologous but result from convergent evolution. There does not at this time, however, seem to be enough evidence to give up the concept of the subphylum Mandibulata.

The crustaceans are **gill-breathing arthropods,** with **two pairs of antennae, two pairs of maxillae** on the head, and usually a pair of appendages on each body segment. Primitively the appendages are **biramous** (two-branched), and at least some of the appendages of present-day adult crustaceans are biramous.

MATERIALS

Living materials
  Crayfish or lobsters
  Cultures of developing *Artemia* in different developmental stages (instructions on pp. 457-458)
  Other crustaceans, such as *Eubranchipus, Daphnia, Cyclops,* ostracods, barnacles, and crabs, as available
Preserved materials
  Crayfish or lobsters (may be injected)
  Barnacles, crabs, and the like, as available
Bowls or small aquariums
Dissecting pans
Slides and coverslips
Compound microscopes

CRAYFISH OR LOBSTER

Phylum Arthropoda
  Subphylum Mandibulata
    Class Crustacea
      Subclass Malacostraca
        Order Decapoda
          Genus *Cambarus*

---

*See Hickman, C.P., Jr., L.S. Roberts, and F.M. Hickman. 1984. Integrated principles of zoology, ed. 7. St. Louis, The C.V. Mosby Co., pp. 380-404.

The following description will apply to either the crayfishes *(Cambarus, Procambarus, Astacus, Orconectes)* or the lobster *(Homarus,* others), for crayfishes and lobsters are very similar except in size.

Where Found

The crayfish is found in freshwater streams and ponds all over the world. *Astacus* is found mainly west of the Rocky Mountains, *Procambarus* in the southern states and *Cambarus* and *Orconectes* are widely distributed east of the Rocky Mountains. There are about 100 species of crayfish in the United States and about 300 worldwide. They are omnivorous, feeding on fish, tadpoles, worms, insects, and plants.

The lobster is a marine form. *Homarus americanus* is found along the eastern North American coast from Labrador to North Carolina. Both lobsters and crayfish are widely used for food in various parts of the world. The spiny, or rock, lobster, *Panulirus,* from the West Coast and southern Atlantic coast, has no pincers and differs in several other respects.

Behavior

☞ Place a live crayfish or lobster in a bowl or small aquarium partly filled with water. Provide some shells or stones for shelter.

How are the antennae and the smaller antennules used? Determine the sensitivity of the animal by touching various parts of the body. What is the response when you stroke the antennae? Notice the compound eyes on the ends of stalks. Are the eyestalks movable? Of what advantage would such eyes be to the animal? What evidence of segmentation do you see on the dorsal side of the animal?

Note the five pairs of legs. Are they all used in walking? The first pair are called chelipeds because they bear the large claws (chelae). How are the chelipeds used? When the animal is startled, what does it do? Is the escape movement forward or backward? How is it accomplished? Is the tail fan involved?

Lift up a crayfish and notice the swimmerets on the abdomen. Release it near the surface of the water. Can it swim? How? The females use the swimmerets to carry and aerate their eggs during the breeding season.

Drop a bit of meat or fish near an undisturbed crayfish in an aquarium. What appendages are used to pick up and handle food? Notice the activity of the small mouth appendages as the animal feeds.

External Features

☞ Place a preserved crayfish (or lobster) in a dissecting pan and add water to the pan.

The **exoskeleton** is a cuticle secreted by the epidermis and hardened with an organic substance called **chitin,** with the addition of mineral salts such as calcium carbonate. The cuticle must be shed or **molted** several times while the crayfish is growing up, each time being replaced by a new soft exoskeleton that soon hardens. The tagmata are the cephalothorax and abdomen.

Cephalothorax. A hard **carapace** covers the cephalothorax. It is marked off by grooves into four areas—the **head;** a median **cardiac region** lying over the heart; and, laterally, two branchial regions that cover the gills. Lift the lower edge of the carapace to disclose the **gill chamber** with its feathery **gills.**

On each side of a spiny pointed **rostrum** is a stalked movable **compound eye.** Examine the eye with a hand lens. The small surface **facets** of the eye are terminations of the visual cells of the eye. The short **antennules** and the long **antennae** bear tactile and chemical sense organs.

The five fused somites of the head bear five pairs of appendages—the antennules, the antennae, and three pairs of mouthparts that you will dissect later. The thorax, made up of eight somites, bears three pairs of **maxillipeds** and five pairs of long **walking legs.**

Note the thin, flexible cuticle at the leg joints.

How many segments do the walking legs have? Do they all have pincers? Lift up the carapace and move a leg to see the relationship of the legs to the gills.

Abdomen. The abdomen is made up of six somites. Each is covered with a hard dorsal **tergum,** thin lateral **pleurons,** and a ventral **sternum.** The abdomen ends with a flat process, the **telson,** bearing the **anus** on the ventral side.

The first five abdominal somites bear **swimmerets.** Each swimmeret is typically **biramous** (has two branches). In the male the first two pairs of swimmerets are modified into stiff, grooved structures used in copulation (Fig. 14-7). In the female the first pair is reduced in size. The last somite bears a pair of fan-shaped **uropods,** also biramous. Uropods and telson together make up the **tail fan.**

> As you identify the external structures, add the appropriate labels to Fig. 14-6.

Genital Openings. In the male the genital openings of the sperm ducts are located medially at the base of each of the fifth walking legs (Fig. 14-7). In the female the genital openings of the oviducts are located at the base of each third walking leg. Also in the female is a midventral **opening to the seminal receptacle.** During copulation the male turns the female over and sheds the sperm directly onto her abdomen.

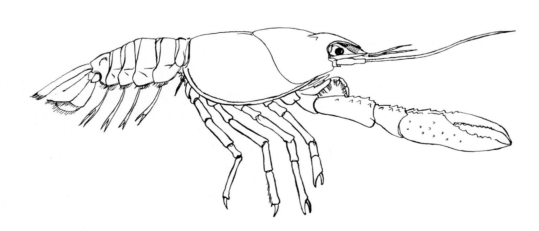

Fig. 14-6. External structure of the crayfish (to be labeled by student).

Serial Homology in Crayfish Appendages

The theory of **serial homology** is well illustrated in crustaceans. Arthropods have probably descended from an annelid-like ancestor whose metameres were all similar and whose appendages were all **biramous** (two branched) and unspecialized. During the evolution of the arthropods the metameres and appendages became specialized for special purposes, with differentiations in structure and function. But, since they are all derived from parts that were essentially the same, they are considered **homologous**. Homology refers to structures that are similar in both origin and development. Arthropod appendages are not only homologous to the unspecialized appendages of the annelids, but each appendage is homologous to the other appendages of a single individual. Where repetitive or metameric structures are fundamentally similar in origin and development, the term "serial homology" is used.

Dissection of Appendages

☞ From the animal's left side, remove and study each of the appendages described below. You will find sketches of them on p. 217 for labeling. However, do not remove any appendages until you have read the instructions for doing so.

To remove an appendage, grasp it at the base with forceps as near the body as possible and work it loose gradually by gently manipulating the forceps back and forth. When you remove an appendage, be sure to identify its **medial** and **lateral** sides. Although appendages are numbered consecutively beginning at the anterior end of the animal, it is easier to remove them by beginning at the **posterior** end and proceeding forward. Pin the appendages in order on a sheet of paper in the bottom of a dissecting pan. Keep covered with water.

Uropods. The broad uropods on the last segment are biramous, and together with the telson, they make up the strong tail fan that is used in the rapid backward swimming that is so important in escape movements.

Swimmerets (Pleopods). Swimmerets are typical of the **biramous plan.** This is considered to be the general type of primitive appendage from which all the other arthropod appendages have evolved.

The first two pairs in the male are modified as **copulatory organs.** They are large and grooved and used to direct sperm onto the female (Fig. 14-7). The first pair in the female is usually reduced in size. Of the others, the part attached to the body is called the **protopod** (Gr. *protos*, first, + *pous*, foot). The two branches are the medial branch, or **endopod,** and the lateral **exopod.** Both are attached to the protopod.* Swimmerets are used for creating currents over the gills and, in the female, for holding the eggs.

Walking Legs (Pereiopods). The first pair of pereiopods (per-rye′o-podz) are the **chelipeds,** with enlarged claws (**chelae**) used in defense. The other four pairs are used in walking and food handling. There is no exopod so they are called **uniramous** (single branched). The first four pairs are attached to gills; the second and third bear smaller chelae used for grasping.

Note the **genital pores** on the medial side of the protopods of the male's fifth walking legs and on the female's third walking legs (Fig. 14-7).

Maxillipeds. Each of the three paired maxillipeds has an endopod and a very slender exopod. Be sure to obtain both in removing the appendage. Remove the third maxilliped and its attached gill.

Remove the second and then the first maxilliped. They are similar to the third but smaller. The first

---

*Alternative terms are *protopodite*, *endopodite*, and *exopodite*.

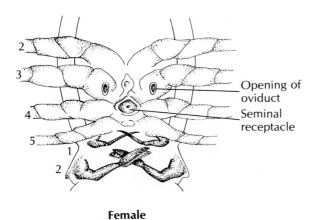

Female

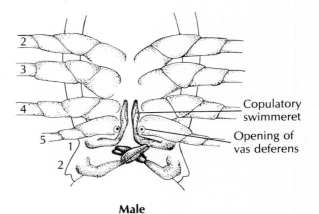

Male

Fig. 14-7. Ventral views of crayfish, showing sexual differences in genital openings and swimmerets. The first two swimmerets and the bases of the last four walking legs are shown.

maxilliped bears no gill. The maxillipeds are food handlers that break up food and move it to the mouth.

Maxillae. The maxillae are head appendages that direct food toward the mouth. On the second maxilla, the protopod is expanded into four little processes; the endopod is small and pointed; and the exopod forms a long blade, the **gill bailer,** that beats to draw currents of water forward over the gills.

The tiny first maxilla is a reduced foliaceous appendage.

Mandibles. The mandibles on either side of the mouth are heavy triangular structures, bearing teeth on their inner edges. These are the protopods. A little palp folded above each tooth margin represents the endopod. The exopod is absent. The mandibles work from side to side to direct food into the mouth and to hold it while the maxillipeds tear it up. Pry the mandibles apart and probe the mouth region. Remove the left mandible carefully. A strong mandibular muscle attached to the base makes the mandible difficult to remove. The muscle may be removed with the mandible.

Antennae. The endopod of each antenna is a very long, many-jointed filament. The exopod is a broad, sharp, movable projection near the base. On its broad protopod on the ventral side of the head is the **renal opening** from the excretory gland. Antennae and antennules are sensitive to touch, to vibrations, and to the chemistry of the water (taste).

Antennules. Each antennule has a three-jointed protopod as well as two long, many-jointed filaments. The antennules are also concerned with equilibrium.

Summary. Note that the crayfish appendages all fall into one of three types: biramous, uniramous, or foliaceous. Of the two branches (exopod or endopod), which is often lost in the specialization of appendages? Is the endopod present in all appendages? What disadvantage might the crayfish have suffered if all its appendages had retained a primitive biramous plan such as that of the swimmerets?

DRAWINGS AND REPORT

1. Label the drawing of the appendages on p. 217 with the correct name of each appendage. On each appendage, color the protopod red, the endopod blue, and the exopod (if present) yellow, and leave the gills uncolored. Or, if colors are not available, designate as *Pr, En,* and *Ex.*

2. On p. 218 make a table of the appendages, placing each into one of the type groups. List all the functions each appendage is called on to do.

Branchial (Respiratory) System

Remove part of the carapace on the animal's right side and note the feathery **gills** lying in the branchial chamber. You have already seen that some of the gills are attached to certain appendages. These outer gills are called the foot gills. How many appendages have gills attached? Move the appendages to determine this. Separate the gills carefully, laying aside the foot gills. Another row of gills underneath is attached to membranes that hold the appendages to the body. These gills are called the joint gills. Some genera, but not *Cambarus,* have a third row of gills (side gills) attached to the body wall.

☞ Remove a gill, place it in a watch glass, cover with water, and examine with a hand lens. Now cut the gill in two and look at one of the cut ends.

The **central axis** bears **gill filaments** that give it a feathery appearance. Notice that the central axis and also the little filaments contain canals. These represent blood vessels (afferent and efferent) that enter and leave the filaments. What happens to the blood as it passes through the filaments? Water enters the gill chamber by the free ventral edge of the carapace and is drawn forward over the gills by the action of the gill bailer of the second maxilla, facilitated by movements of the other appendages.

Internal Structure

☞ Remove the dorsal portion of the exoskeleton as follows: insert the point of the scissors under the posterior edge of the carapace about 1.5 cm to one side of the medial line. Cut forward to a point about 1 cm back of the eye. Do the same on the other side, thus loosening a dorsal strip about 2.5 cm wide. Carefully remove this center portion of the carapace, a little at a time, being careful not to remove the underlying **epidermis** and **muscles,** which cling to the carapace, especially in the head region. Loosen such tissue with your scalpel and push it carefully back into place.

☞ Remove the dorsal portion of the abdominal exoskeleton in the same way, uncovering each somite carefully so as not to destroy the long **extensor muscle** lying underneath.

The thin tissue covering the viscera is the **epidermis,** which secretes the exoskeleton. Notice the position of the pinkish portion of the epidermis, which lies in the same position as the cardiac region of the carapace. This covers the **pericardial sinus** containing the **heart.** The sinus may be filled with colored latex if the circulatory system has been injected.

☞ Remove the epidermis very carefully with your forceps to expose the viscera.

CRAYFISH APPENDAGES—EXAMPLE OF SERIAL HOMOLOGY

Label each appendage. On each indicate, where appropriate, the protopod, exopod, and endopod.

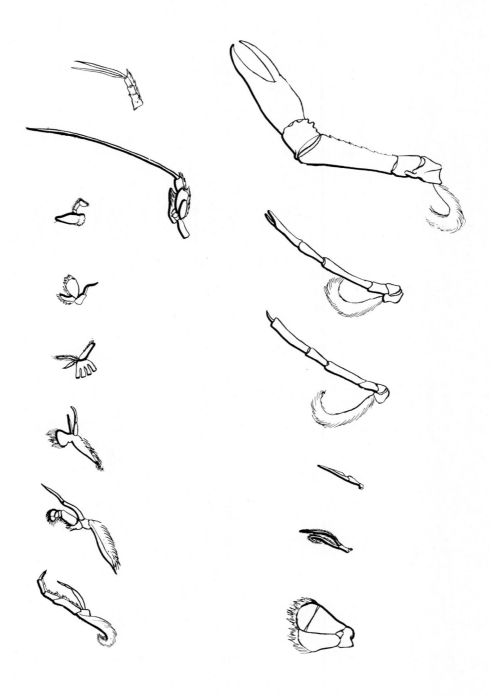

TABLE OF CRAYFISH APPENDAGES

No.	Appendage	Type	Function
1	Antennule	Biramous	Sensory—touch, taste, equilibrium
2			
3			
4			
5			
6			
7			
8			
9			
10			
11			
12			
13			
14			
15			
16			
17			
18			
19			

The large **stomach** lies in the head region, anterior to the heart (Fig. 14-8). On each side of the stomach and heart are large cream-colored lobes of the **hepatopancreas** (digestive gland). They extend the full length of the thorax. This gland is the largest organ in the body.

Muscular System. As you removed the carapace, you noticed the tendency of certain muscles in the

head region to cling to it. The⸍ ⸍ ⸍ which attach the stomach to the ca⸍ ⸍ ⸍ On each side of the stomach lies a little m⸍ ⸍ ⸍ the **mandibular muscles,** which move the ma⸍ ⸍ ⸍ One of these may have been removed when the m⸍ ⸍ ⸍ dible was removed. On each side of the thorax a narrow band of muscles runs longitudinally to the distal end of the abdomen. These are **extensor muscles** that

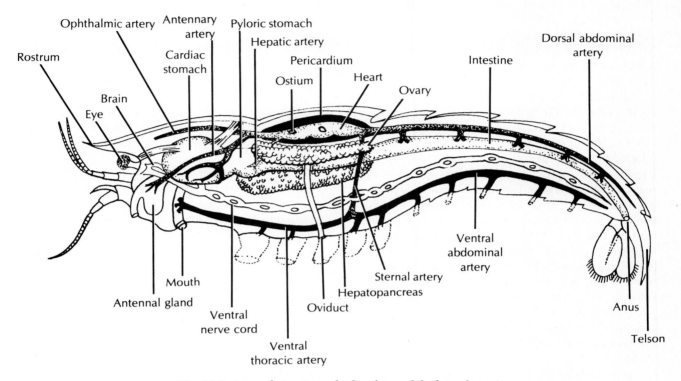

Fig. 14-8. Internal structure of a female crayfish (*lateral view*).

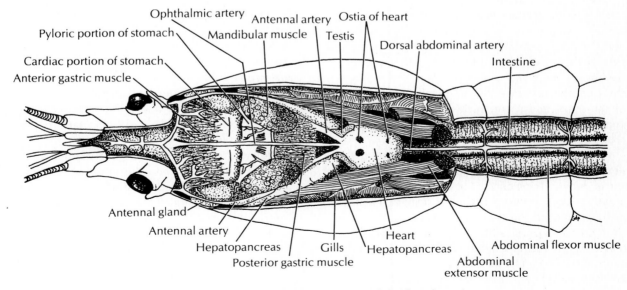

Fig. 14-9. Internal structure of a male crayfish (*dorsal view*).

straighten the abdomen. In the abdomen, beneath the extensors and nearly filling the abdomen, are the **flexor muscles,** which flex the abdomen (bend it ventrally). The flexors are large and important muscles used in the quick backward "escape mechanism" of the animal.

☞ Remove the exoskeleton from one cheliped and see if you can distinguish the extensor and flexor muscles in it. Tease out a tiny piece of muscle fiber in a drop of water on a slide and examine with the microscope.

Can you distinguish the striations on the muscle fibers? Does this skeletal muscle resemble the skeletal muscle of the vertebrates? Compare with Fig. 5-13, A, p. 68.

Circulatory System. The small angular **heart** is located just posterior to the stomach (Figs. 14-8 and 14-9). If the circulatory system has been injected, the heart may be filled and covered with a mass of colored injection fluid or latex filling the sinus. Remove the latex carefully bit by bit, being careful not to destroy the tiny blood vessels leaving the heart. The heart lies in a cavity called the **pericardial sinus,** which is enclosed in a membrane, the **pericardium.** The pericardium may have been removed with the carapace.

This is an "open" type of circulatory system. The hemolymph leaves the heart in arteries but returns to it by way of venous sinuses, or spaces, instead of veins. Hemolymph enters the heart through three pairs of slitlike openings, the **ostia,** which open to receive the hemolymph and then close when the heart contracts to force the hemolymph out through the arteries.

Your instructor may want to demonstrate the heartbeat of a living crayfish (instructions given on p. 223).

Five arteries leave the anterior end of the heart (Fig. 14-9): a median **ophthalmic artery** extends forward to supply the cardiac stomach, the esophagus, and the head; a pair of **antennal arteries,** one on each side of the ophthalmic, pass diagonally forward and downward over the digestive gland to supply the stomach, the antennae, the antennal glands, and parts of the head; and a pair of **hepatic arteries** from the ventral surface of the heart supply the hepatopancreas.

Leaving the posterior end of the heart are the **dorsal abdominal artery,** which extends the length of the abdomen, lying on the dorsal side of the intestine, and the **sternal artery,** which runs straight down (ventrally) to beneath the nerve cord where it divides into the **ventral thoracic** and **ventral abdominal arteries,** which supply the appendages and other ventral structures (Fig. 14-8). Do not attempt to find these ventral arteries now. They will be referred to later.

Reproductive System. The **gonads** in each sex lie just under the heart. Their size and prominence will depend on the season in which the animals were killed. To find them, lay aside the heart and abdominal extensor muscle bands. The gonads are very slender organs usually slightly different in color from the digestive glands, lying along the medial line between and slightly above the glands. The gonads are sometimes difficult to distinguish from the digestive glands.

In the female the gonads or **ovaries** are slender and pinkish, lying side by side, with the anterior ends slightly raised. In some seasons the ovaries may be swollen and greatly distended with eggs, appearing orange in color. A pair of **oviducts** leaves the ovaries and passes laterally over the digestive glands to the genital openings on the third walking legs (Fig. 14-8).

The male gonads, or **testes,** are white and delicate. The **sperm ducts** pass diagonally over the digestive glands and back to the openings in the fifth walking legs.

Digestive System. The **stomach** is a large thin-walled organ, lying just back of the rostrum. It is made up of two parts—anteriorly a large firm **cardiac chamber** and posteriorly a smaller soft **pyloric chamber.** These will be examined later.

Sometimes a mass of calcareous crystals (**gastroliths**) is attached to each side of the cardiac chamber near the time of molting. These limy masses are supposed to have been recovered from the old exoskeleton by the blood to be used in the making of the new exoskeleton.

The **intestine,** small and inconspicuous, leaves the pyloric chamber, bends down to pass under the heart, and then rises posteriorly to run along the abdominal length above the large flexor muscles. In the sixth abdominal somite it enlarges to form a blind **cecum** and ends at the **anus** on the telson.

The large **hepatopancreas,** also called **liver** or **digestive glands** (Fig. 14-9), furnishes digestive secretions that are poured through hepatic ducts into the pyloric chamber. The hepatopancreas also is the chief site of absorption and serves for storage of food reserves.

☞ To see the whole length of the digestive system, remove the left lobe of the hepatopancreas and gonad and push aside the left mandibular muscle.

You can now see the **esophagus,** connecting the stomach and mouth, and the intestine, arising from the stomach and running posteriorly to the abdomen where it lies just under the dorsal abdominal artery. You can also see the **sternal artery** descending ventrally.

☞ Now remove the stomach by severing it from the esophagus and intestine, turn it ventral side up, and open it longitudinally. Wash out the contents, if necessary.

The cardiac chamber contains a **gastric mill,** which consists of a set of three chitinous teeth, one medial and two lateral, that are used for grinding food. They are held by a framework of ossicles and bars in the stomach and operated by the gastric muscles.

In the cardiac stomach the food is ground up and partially digested by enzymes from the hepatopancreas before it can be filtered into the smaller pyloric stomach in liquid form. Large particles must be egested through the esophagus. Rows of setae and folds of the stomach lining strain the finest particles and pass them from the pyloric stomach into the hepatopancreas or into the intestine where digestion is completed.

Excretory System. In the head region anterior to the digestive glands and lying against the anterior body wall are a pair of **antennal glands** (also called **green glands,** though they will not appear green in the preserved material) (Figs. 14-8 and 14-9). They are round and cushion shaped. In the crayfish each gland contains an **end sac,** connected by an excretory tubule to a bladder. Fluid is filtered into the end sac by hydrostatic pressure in the hemocoel. As the filtrate passes through the excretory tubule, reabsorption of salts and water occurs, leaving the urine to be excreted. A duct from the bladder empties through a renal pore in the base of the antenna.

In the freshwater crayfish the urine is copious and hypotonic. In the marine lobsters, which live in an isotonic medium and would have no problem of salt preservation, the nephridial tubule is missing, and the urine is scanty and isotonic. The role of the antennal glands seems to be largely the regulation of the ionic and osmotic composition of the body fluids.

Excretion of nitrogenous wastes (mostly ammonia) occurs by diffusion in the gills and across thin areas of the cuticle.

Nervous System

☞ Carefully remove all of the viscera, leaving the esophagus and sternal artery in place. The brain is a pair of **supraesophageal ganglia** that lie against the anterior body wall between the antennal glands. Can you distinguish the three pairs of nerves running from the ganglia to the antennae, eyes, and antennules?

From the brain two connectives pass around the esophagus, one on each side, and unite at the **subesophageal ganglion** on the floor of the cephalothorax.

☞ Chip away the calcified plates that cover and conceal the double ventral nerve cord in the thorax and follow the cord posteriorly. By removing the big flexor muscles in the abdomen you can trace the cord for the length of the body. Note the ganglia, which appear as enlargements of the cord at intervals. Observe where the nerve cord divides to pass on each side of the sternal artery. Note the small lateral nerves arising from the cord.

In the annelids there is a ganglion in each segment, but in the arthropods there is some fusion of ganglia. The brain is formed by the fusion of three pairs of head ganglia, and the subesophageal ganglia is formed by the fusion of at least five pairs.

Sense Organs. The crayfish has many sense organs: **tactile hairs** over many parts of the body, **statocysts, antennae, antennules,** and **compound eyes.** The eyes have already been observed.

The tactile hairs are variously specialized for touch reception, detection of water currents, and orientation. A number of chemoreceptors are present on the antennae, antennules, and mouthparts.

A **statocyst** is located in the basal segment of each antennule. The pressure of sand grains against sensory hairs in the statocyst gives the crayfish a sense of equilibrium.

ORAL REPORT

Be prepared to demonstrate any phase of your dissection to your instructor. (1) Locate on the dissection any structure mentioned in the exercise, and (2) explain its function. (3) Explain how the appendages of the crayfish or lobster illustrate the principle of serial homology.

OTHER CRUSTACEANS
Subclass Branchiopoda

Fairy Shrimp, *Eubranchipus* **(Order Anostraca).** Fairy shrimp have 11 pairs of basically similar appendages used for locomotion, respiration, and egg carrying. They swim ventral side up. They have no carapace. Note the dark eyes borne on unsegmented stalks. The females carry their eggs in a ventral brood sac, which can usually be seen when the animal is moving (Fig. 14-10, A).

Water Flea, *Daphnia* **(Order Cladocera).** Water fleas are common in pond water. Mount a living *Daphnia* on a slide, using enough water to prevent crushing but not enough to allow free swimming. *Daphnia* is 1 to 3 mm long and, except for the head, is covered by a thin transparent carapace. The large biramous second antennae are the chief organs of locomotion. There are five pairs of small leaflike swimmerets on the thorax. These are used to filter microscopic algae from the water for food. The blood in the open system is red from the presence of hemoglobin. The paired eyes are fused. The brood pouch in the female is large and posterior to the abdomen. They can reproduce parthenogenetically (Fig. 14-10, B). (See p. 223 for demonstration of heartbeat.)

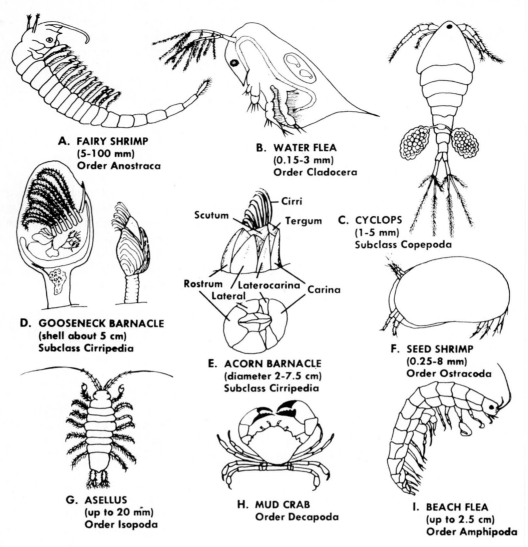

A. FAIRY SHRIMP
(5-100 mm)
Order Anostraca

B. WATER FLEA
(0.15-3 mm)
Order Cladocera

C. CYCLOPS
(1-5 mm)
Subclass Copepoda

D. GOOSENECK BARNACLE
(shell about 5 cm)
Subclass Cirripedia

Cirri

Scutum — Tergum

Rostrum / Laterocarina — Carina
Lateral

E. ACORN BARNACLE
(diameter 2-7.5 cm)
Subclass Cirripedia

F. SEED SHRIMP
(0.25-8 mm)
Order Ostracoda

G. ASELLUS
(up to 20 mm)
Order Isopoda

H. MUD CRAB
Order Decapoda

I. BEACH FLEA
(up to 2.5 cm)
Order Amphipoda

Fig. 14-10. Some representative crustaceans.

Subclass Ostracoda

Ostracods with their transparent bivalved carapace resemble tiny clams (1 to 2 mm). They have a median eye and seven pairs of appendages, including the large antennules and antennae (Fig. 14-10, F).

Subclass Copepoda

Cyclops and other copepods are found everywhere in fresh and brackish water. *Cyclops* has a median eye near the base of the rostrum and long antennae, modified in the male. The cephalothorax bears appendages; the abdomen has none. The sixth thoracic segment in the female carries large pendulous egg sacs. The last abdominal segment bears a pair of caudal projections covered with setae (Fig. 14-10, C).

Subclass Cirripedia

The barnacles (all marine) might be mistaken for molluscs because they are enclosed in a calcareous shell. *Balanus*, the acorn barnacle, has a six-piece shell that surrounds the animal like a parapet. The animal protrudes its six pairs of biramous appendages through the opening at the top to create water currents and sweep in plankton and particles of detritus (Fig. 14-10, E).

The gooseneck barnacle, *Lepas*, is attached to a long stalk. The body proper is enclosed in a bivalve carapace strengthened by calcareous plates. Six pairs of delicate filamentous and biramous appendages can be protruded from the carapace. Feathery with long setae, the appendages form an effective net to strain food particles from the water (Fig. 14-10, D).

Subclass Malacostraca

This is a large group. It includes the Isopoda (sow bugs, pill bugs, wood lice, and others with a dorso-ventrally flattened body); the Amphipoda (beach fleas, the freshwater *Gammarus*, and others that are laterally compressed); and the Decapoda (shrimps, crabs, crayfish, and lobsters). Examine a crab and compare its structure and appendages with those of the crayfish (Fig. 14-10, *G* to *I*).

CRUSTACEAN DEVELOPMENT, EXEMPLIFIED BY THE BRINE SHRIMP

The brine shrimp, *Artemia salina*, which is not really a true shrimp but a member of the more primitive order Anostraca, has a pattern of development typical of most marine crustaceans. If time permits, your instructor may have prepared a series of cultures, started 10 days before the laboratory period and every 2 days thereafter. The method is given on pp. 457-458. This should provide you with some five larval stages for study. The larvae hatch in 24 to 48 hours after the dry eggs are placed in natural or artificial seawater. The stages are classified on the basis of the number of body segments and the number of appendages present. Since growth cannot continue without molting (ecdysis), the larvae will go through a number of molts before reaching adult size. You may find some of the shed exoskeletons in the cultures.

The animals can be narcotized or killed by adding a few drops of ether or chloroform per 10 ml of culture water.

☞ To observe the larvae, mount a drop of the culture containing a few larvae on a slide and cover with a coverslip. Try to identify the following stages:

Nauplius. The newly hatched larva, called the nauplius, has a single median ocellus and three pairs of appendages (two pairs of antennae and one pair of mandibles), and the trunk is still unsegmented.

Metanauplius. The first and second maxillae have developed, and there is some thoracic segmentation.

Protozoea. There are now seven pairs of appendages, since the first and second pairs of maxillipeds have been added. The compound eyes are now developing.

Zoea. The third pair of maxillipeds have now appeared. The eyes are complete and there are several thoracic segments.

Mysis. Most or all of the 19 body segments and 11 pairs of appendages found in the adult are now present. In these transparent forms you should be able to see the digestive tract and, in live specimens, its peristaltic movements.

The brine shrimp attain adulthood at about 3 weeks and can be maintained on a diet of yeast, fed sparingly. The adults closely resemble their much larger freshwater cousins, the fairy shrimp, *Eubranchipus* (Fig. 14-10, *A*).

DRAWINGS (OPTIONAL)

Use separate paper for sketching crustaceans, other than the crayfish, or the developmental stages of crustacean larvae, as seen in the brine shrimp.

PROJECTS AND DEMONSTRATIONS

1. Heartbeat of Daphnia. Place a small drop of petrolatum in the center of a Syracuse watch glass. Fasten *Daphnia* by one of its valves to the petrolatum but be careful that water circulates between the valves. Observe the heartbeat. Start with water at 0° C. As the water slowly rises to room temperature, record the heartbeats at the different temperatures. Determine the number of heartbeats in 15 seconds at each 2° or 3° rise in temperature.

2. Heartbeat of crayfish. Anesthetize the crayfish in a suitable-sized dish by covering with club soda, freshly opened. Leave until the animal fails to respond to prodding; then remove the carapace and observe the beating heart. The crayfish should survive several hours under these conditions.

3. Female crayfish carrying eggs and young. Preserved specimens can be purchased at any time of the year from a biological supply house.

4. Demonstration of various kinds of barnacles.

5. Demonstrations of various kinds of crabs. Compare the appendages of a crab with those of the crayfish or lobster.

6. Blood cells in arthropods. Unlike vertebrate blood, arthropod blood contains no erythrocytes. The respiratory pigment, usually hemocyanin, is dissolved in the plasma rather than carried in cells. Arthropods possess a variety of leukocytes or amebocytes. Since the blood clots much more rapidly than does vertebrate blood, their observation is sometimes more difficult.

To observe in the natural state, place a drop of paraffin oil on a clean slide and focus on the drop. Snip off the tip of an antenna with petroleum-jellied scissors, place the tip in the oil drop, and observe and sketch the cells as they emerge.

For permanent blood smears see method on p. 203.

Phagocytosis in crayfish cells may be observed by injecting the animal with a suspension of carmine or India ink. Examination of the blood an hour or so later may reveal particles ingested by leukocytes.

7. Demonstration of antennal glands. The parts of the antennal gland can be identified by injecting into the hemocoel of a crayfish 0.25 ml of 0.5% aqueous solution of cyanol and an equal quantity of 0.5% solution of Congo red. After 18 to 24 hours remove one of the glands and examine under a dissecting microscope. The cells of the end sac (coelomosac) should be red and the channels of the labyrinth blue.

8. *Tropical land hermit crabs.* Hermit tree crabs, *Coenobita clypeatus,* spawn in the sea but live on land and in trees. They can be obtained from some of the biological supply houses (for example, Carolina Biological Supply Co., Burlington, N.C.) and can easily be kept in a sand-based terrarium, where they can be fed dry toast, lettuce, bananas, and the like.

9. *Demonstration of bioluminescence.* Dried *Cypridina,* a marine crustacean, can be obtained from biological supply houses. When crushed in a little water, they will produce a blue light that can be observed in a darkened room.

EXERCISE 14C
The terrestrial mandibulates*
THE MYRIAPODS AND THE INSECTS

The terrestrial mandibulates include the insects and the myriapods, a large group of mandibulate arthropods that breathe by means of **tracheae** and have only **one pair of antennae** and only **uniramous appendages.** There are few aquatic adults, and those are mostly freshwater forms. Some of the insect young are aquatic and possess gills, but their gills are not homologous to the gills of crustaceans.

MATERIALS
Living materials, if available
  Centipedes
  Millipedes
  Grasshoppers
  Vials of developing *Drosophila* (see Appendix B, p. 458, for preparation; should be started 2 weeks in advance)
Preserved materials
  Centipedes, such as *Scolopendra* or *Lithobius*
  Millipedes, such as *Spirobolus* or *Julus*
  Grasshoppers, such as *Romalea*
  Apis, workers
  Apis, queen, drones, larvae, pupae (for demonstration only)
Small terrariums or jars for living materials
Dissecting pans and instruments
Ether
Dissecting microscopes

---

*See Hickman, C.P., Jr., L.S. Roberts, and F.M. Hickman. 1984. Integrated principles of zoology, ed. 7. St. Louis, The C.V. Mosby Co., pp. 405-438.

THE MYRIAPODS

The term "myriapods" is a common name for several groups of mandibulate arthropods, the Chilopoda, Diplopoda, Pauropoda, and Symphyla. They have two tagmata—head and trunk. There is one pair of antennae, and the trunk bears paired appendages on all but the last segment (Fig. 14-11).

Class Chilopoda—Centipedes
Centipedes are active predators that live in moist places such as under logs, stones, and bark, where they feed on worms, larvae, and insects.

If living examples are available, watch their locomotion and note the agile use of the body and legs. Note the general shape of the body and the arrangement of segments and appendages. Is the body circular or flattened in cross section?

On a preserved specimen of *Lithobius* (Fig. 14-11, *A*) or *Scolopendra* examine the head. Some species have simple **ocelli**; others have large faceted **eyes** resembling the compound eyes of insects. Which does your specimen have? Pull aside the large poison claws (maxillipeds) to uncover the mouth area and examine under a dissecting microscope. Find the **antennae;** the **labrum,** anterior to the mouth; the **mandibles** and **first maxillae,** lateral to the mouth; and the **second maxillae,** bearing a long palp and a short labial portion just posterior to the mouth.

The first trunk appendages are the prehensile **maxillipeds,** each bearing a terminal **poison fang.** Do the rest of the trunk appendages bear legs? The last segment bears the **gonopores,** and the **anus** is located on a short telson. Find the **spiracles** near the bases of the legs.

Class Diplopoda—Millipedes

The millipedes, or "thousand foot worms," are found throughout the world, usually hiding in damp woods under bark, leaves, rocks, or logs. They are herbivorous, feeding on decaying wood or leaves. Their most distinguishing feature is the presence of diplosegments, which are double trunk segments probably derived from the fusion of two single segments and each bearing two pairs of legs. The name Diplopoda comes from Greek *diploos*, double, and *pous, podos*, foot. (See Fig. 14-11, *B*).

If living millepedes, such as *Spirobolus* or *Julus*, are available, note the shape of the body, the use of the antennae, the use of the legs, and the animal's ability to roll up. Notice the metachronal rhythm of the power and recovery strokes of the legs with the opposite sides of each segment exactly out of phase. It is similar to that seen in the polychaete *Nereis*. The movements of the legs pass in regular successive waves toward the anterior end. If a specimen has been chilled, you should be able to determine the number of legs working together in each wave and the number of segments between waves.

Studying a preserved specimen, locate on the head the **ocelli, antennae, labrum, mandibles, and labium (fused second maxillae).** Are there appendages on the first trunk segments? How many pairs on the next three segments? The **gonopores** open on the third trunk segment at the bases of the legs. How does this compare with the centipede? The dorsal overlapping of the exoskeletal plates provides full protection for the animal, even when rolled into a ball. Notice that the diplosegments each have two pairs of **spiracles.**

THE INSECTS
Class Insecta—the Grasshopper, the Honeybee, and Metamorphosis of the Fruit Fly

The insects are the most extensive group of animals in the world. Approximately 800,000 species of insects have been recorded, and probably as many more remain to be discovered and named.

Among their chief characteristics are **three pairs of walking legs, one pair of antennae,** body typically divided into **head, thorax,** and **abdomen,** and respiratory system of **tracheal tubes.** Most insects are also provided with one or two pairs of **wings.** They show many variations of adaptive structures according to their habits and life cycles. Their sense organs are often specialized and perhaps account for much of their success in the competition for ecological niches.

Most insects are less than 2.5 cm long, but they range from 1 mm to 20 cm, with the largest insects usually living in tropical areas.

Some insects are highly specialized; others have a generalized plan of body structure. We shall study and compare an example of each. The grasshopper is a fairly generalized, or primitive, type of insect. The honeybee, on the other hand, has become specialized for particular conditions. Not only is its morphology modified for special functions and adaptations, but it belongs to the social insects in which patterns of group organization involve different types of individuals and division of labor.

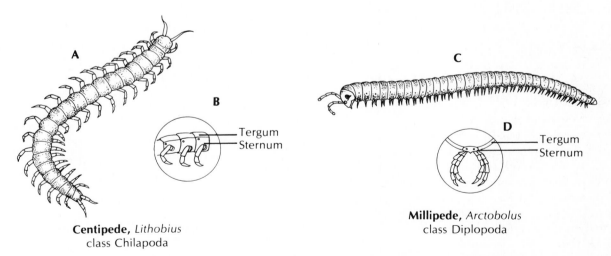

Centipede, *Lithobius*
class Chilapoda

Millipede, *Arctobolus*
class Diplopoda

Fig. 14-11. Examples of myriapods, showing typical arrangement of legs. **A,** The centipede, *Lithobius* (Class Chilopoda). **B,** Centipede arrangement of legs and spiracles, *lateral view*. **C,** The millipede, *Arctobolus*, (Class Diplopoda). **D,** Ventral portion of a middle segment of millipede, showing typical arrangement of legs.

ROMALEA, THE LUBBER GRASSHOPPER

Phylum Arthropoda
    Subphylum Mandibulata
        Class Insecta
            Order Orthoptera
                Genus *Romalea*

APIS, THE HONEYBEE

Phylum Arthropoda
    Subphylum Mandibulata
        Class Insecta
            Order Hymenoptera
                Genus *Apis*
                    Species *A. mellifera*

Behavior of the Grasshopper

☞ If living grasshoppers are available, observe them in a terrarium or place one in a glass jar with moist paper toweling in the bottom and a stick or other object to perch upon.

Is the color adaptive to its habit of living? Can it move its head?

Observe how the grasshopper moves and how it uses its legs. How does it crawl up a stick? How does it make use of its claws? What position does it assume when quiescent in the jar? How does it jump? How are its legs adapted for jumping? Does it use its wings?

Note the movements of the body while it is at rest. Are the movements related to breathing? How does the grasshopper breathe? How does it get air into its body?

What is the common food of grasshoppers? Observe how one eats a piece of lettuce leaf. Watch how it moves its mouthparts.

Take a specimen from the jar and induce it to regurgitate its greenish digestive juices and food on a glass plate. Is this a defensive adaptation?

Record your observations on p. 232.

External Structure

☞ Study and compare preserved specimens of the grasshopper and the honeybee. Both are models of compactness, so use the hand lens freely to observe the smaller structures.

Note on each insect the division of the body into **head, thorax,** and **abdomen.** Are these animals segmented throughout, or is segmentation more apparent in certain regions of the body?

The chitinous **exoskeleton** is secreted by the underlying epidermis. It is made up of hard plates, called **sclerites,** that are bounded by sutures of soft cuticle.

The Head of the Grasshopper. The head of the grasshopper is freely movable. Notice the **compound** eyes, the **antennae,** and three **ocelli,** one dorsal to the base of each antenna and one in the groove between them.

The head consists of a dorsal **epicranium,** the cheeks, or **genae,** the front of the face, or **frons,** the **clypeus** below the frons, and the movable bilobed upper lip, or **labrum** (Fig. 14-12). Lift up the labrum and observe the toothed **mandibles.** The mouth contains a membranous **hypopharynx** for tasting food. The bilobed lower lip or **labium** is the result of the fusion of the second maxillae. The labium bears on each side a three-jointed labial palp. Between the mandible and labium are the paired **maxillae,** each with a maxillary palp, a flat lobe, and a toothed jaw. Note how the mouthparts are adapted for biting and chewing.

☞ After studying the rest of the external features, if time permits, you may be asked by the instructor to use the forceps and teasing needles and carefully remove all the mouthparts and arrange them in their relative positions on a sheet of paper. Permanent demonstration mounts can be made very easily by method no. 1, p. 234.

The Head of the Honeybee. Now examine the head of the honeybee. Identify the **antennae,** the **compound eyes,** and the three **ocelli** near the top of the head. The mouthparts can be used for both chewing and sucking (Fig. 14-13, *C*). Observe the narrow upper lip or **labrum** with a row of bristles on its free margin. Below the labrum are the **mandibles,** which are brownish in color. From its mouth projects a sucking apparatus made up of the long slender hairy **tongue** (or **labium**), the paired **labial palps,** one on each side of the tongue, and the paired broad **maxillae,** one on each side of the labial palps.

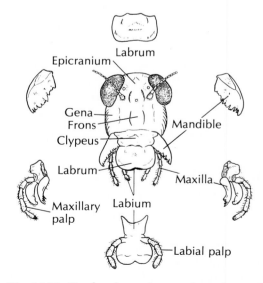

Fig. 14-12. Head and mouthparts of grasshopper.

In summary, the insect head bears four pairs of true appendages; the antennae, the mandibles, the maxillae, and the labium (fused second maxillae). There are at least six somites in the head region although some of them are apparent only in the insect embryo.

The Thorax. In both the grasshopper and the honeybee, the thorax is made up of three somites: the **prothorax,** the **mesothorax,** and the **metathorax,** but the division lines are less distinguishable on the bee. The honeybee's first abdominal somite is also a part

of the thorax, lying anterior to its narrow "waist." Each somite bears a pair of **legs,** and the mesothorax and metathorax each bear a pair of **wings. Spiracles** are located above the legs in the mesothorax and metathorax.

The Wings of the Grasshopper. Note the leathery **forewings** and the membranous **hindwings.** Which is the more useful for flight? What appears to be the chief function of the forewings? The small **veins** in the wings are really tracheal tubes.

The Wings of the Honeybee. How do the wings

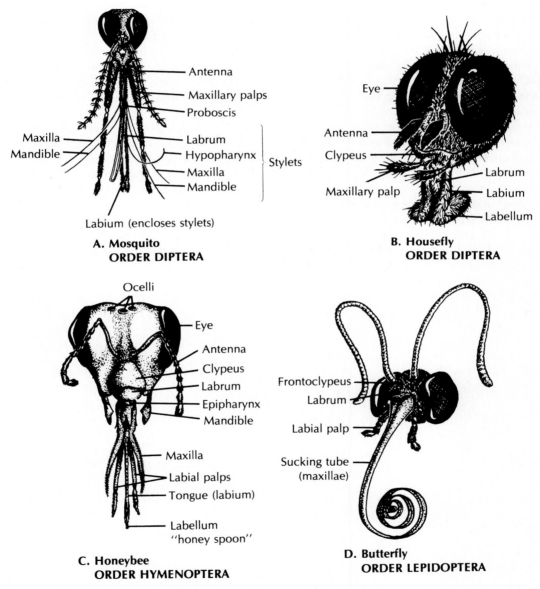

A. Mosquito
ORDER DIPTERA

B. Housefly
ORDER DIPTERA

C. Honeybee
ORDER HYMENOPTERA

D. Butterfly
ORDER LEPIDOPTERA

Fig. 14-13. Some representative insect mouthparts. **A,** Mosquito has labrum, maxillae, mandibles, and hypopharynx modified for piercing; the groved labrum and hypopharynx form a food channel, and the labium is a sheath that folds back when the stylets are used. **B,** Lapping or sponging mouthparts of fly; grooves in the labellum form food channels. **C,** Labium and maxillae form a tonguelike sucking structure in honeybees. **D,** Food channel for sucking lies within the long fused maxillae of butterflies; mandibles and hypopharynx are lacking.

of the honeybee differ from those of the grasshopper? How does this difference affect their function? Notice the small hooks on the front margin of the hindwing. During flight the hooks catch hold of a groove near the margin of the forewing. The wings of the bee may vibrate 400 times or more per second during flight.

The Legs of the Grasshopper. Examine the legs of the grasshopper and identify the basal **coxa**, small **trochanter**, large **femur**, slender spiny **tibia**, and five-jointed **tarsus** with two **claws** and a terminal pad, the **arolium**. Which pair of legs is most specialized, and for what function?

The Legs of the Honeybee. The legs of the honeybee have the same segments as those of the grasshopper but are highly adapted for specific functions. Use your hand lens freely, or remove the legs from one side of the bee and examine under a dissecting microscope.

On the **foreleg**, note the hairs. Are they branched? These hairs serve to collect pollen. The **pollen brush** (Fig. 14-14) consists of the long hairs on the proximal end of the tarsus. The pollen brushes on the forelegs and middle legs brush pollen off the body hairs and deposit it on the pollen brushes of the hindlegs (Fig. 14-14).

At the distal end of the tibia is a movable spine, the **velum**. The velum covers a **semicircular notch** that bears a row of stiff bristles, the **antenna comb**. The velum, notch, and comb together make up the **antenna cleaner**. The antenna is freed from pollen as it is drawn through this antenna cleaner.

The **middle leg** has a long sharp **spur** projecting from the end of the tibia that is used to remove pollen from the pollen basket of the leg behind. This leg also bears a pollen brush.

The **hindleg** is the largest of the legs and the most specialized. One of its striking adaptations is the **pollen basket**, a wide groove with bristles on the outer surface of the tibia. By keeping these bristles moist with mouth secretions the bee can use the basket for carrying pollen. On the inner surface of the metatarsus are **pollen combs**, which are composed of rows of stout spines. The large spines found along the distal end of the tibia and the proximal end of the metatarsus make up the **pollen packer**. The pecten removes the pollen from the pollen brush of the opposite leg, and then, when the leg is bent, the auricle packs it into the pollen basket. The bee carries her baskets full of pollen back to the hive and pushes it into a cell to be cared for by other workers.

The Abdomen of the Grasshopper. There are 11 somites in the abdomen of the grasshopper. Notice the large **tympanum**, the organ of hearing, located on the first abdominal segment. On which of the abdominal segments are the paired spiracles located? In both sexes somites 2 to 8 are similar and unmodified and somites 9 and 10 are partially fused.

The eleventh somite forms the genitalia (secondary sex organs). On each side behind the tenth segment is a projection, the **cercus** (pl. **cerci**). In the female the posterior end of the abdomen is pointed and consists of two pairs of plates with a smaller pair between, with the whole forming the **ovipositor**. Between the plates is the opening of the **oviduct**. The end of the

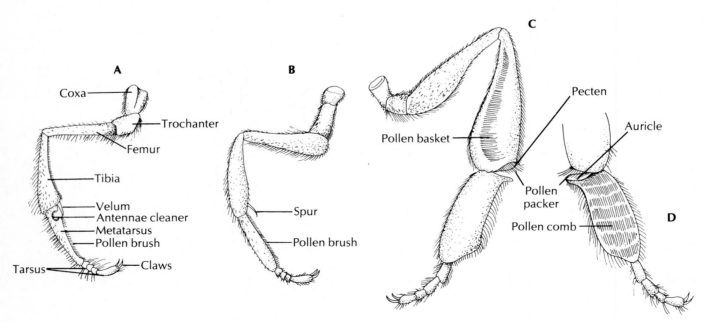

Fig. 14-14. Adaptive legs from the left side of the honeybee worker. **A,** Foreleg. **B,** Middle leg. **C,** Hindleg. **D,** Inner surface of hindleg.

abdomen in the male is rounded. What is the sex of your specimen?

The Abdomen of the Honeybee. The abdomen of the bee has 10 segments, the first of which is really part of the thorax, as mentioned before. The last three segments are modified and hidden within the seventh segment. Can you identify five pairs of spiracles on the abdomen?

The **sting** is a modified ovipositor, so it is found only in the queen bee and female workers and is absent in the drones. Anterior to the sting is a large **poison sac** connected with an alkaline gland and a pair of acid glands. The sting is made up of barbed stylets surrounding a **poison canal,** and the whole is enclosed in a sheath with a barbed tip. The barbs prevent removal of the sting, and secretions from the poison sac are pumped into the wound. Because the sting remains in the wound, the posterior end of the abdomen is pulled off and the bee dies. The queen bee has a less-barbed sting used only against rival queens, and it can be withdrawn and used many times.

DRAWING

Label the external view of the grasshopper on p. 231.

The Honeybee as a Social Animal. In the hive of the honeybee there are three **castes:** the **workers,** the **queens,** and the **drones.** The **workers** are sexually immature females and make up most of the society. They do most of the work of the hive, except lay the eggs. They collect the food, clean out the hive, make the honey and wax, care for the young, guard the hive, ventilate the hive, and so on.

The **queen** lays the eggs and probably guides the swarm when it leaves the hive.

The **drones** are males, and there are usually only a few hundred of them in a hive. Their main duty is to fertilize the eggs, although only one is required to fertilize the queen's eggs during the nuptial flight.

Internal Structure of the Grasshopper

The study of the internal structures of a grasshopper is not always satisfactory because the various organs and systems are somewhat vague and poorly defined.

☞ Remove the wings and cut off a strip from the dorsal body wall. Do not cut too deeply. Pin back the walls and cover with water.

Note the muscles attached to the inside of the body wall and those that move the appendages. A large yellowish **fat body,** probably for storage, may be found between the muscles and the viscera. The exposed cavity, which contains the viscera, is not a true coelom but a **hemocoel.**

Digestive and Circulatory Systems. The digestive system fills most of the internal cavity. It includes a short **esophagus,** a large **crop** in the thorax, an elongated **stomach,** which is partially concealed by six pouchlike **gastric ceca** that secrete digestive juices, and an **intestine, rectum,** and **anus** (Fig. 14-15, A).

The threadlike **malpighian tubules,** lying over the intestine, are excretory in function.

The **circulatory system** is poorly developed, and the **heart,** which is very difficult to see, is a long tube in the dorsal part of the abdomen. Much of the circulatory system is made up of **blood sinuses.**

Tracheal System. The tracheal system is found only in insects and a few related forms. To find the tracheae, look on the inside of the body wall near the location of the **spiracles.** Push the viscera away from the body wall and look for small tubes. The system is best seen in fresh specimens or in injected ones. Note that the tracheae branch to form a network throughout the body (Fig. 14-16). Why is this an efficient system for breathing? Examine a piece of a tracheal tube mounted in water on a slide. Prepared slides of the tracheae may be studied.

Reproductive System. The **gonads** of the male lie dorsal to the intestine in the region of the malpighian tubules. From each side of the fused **testes** a slender **vas deferens** passes ventrally to unite with the other vas deferens to form the common **ejaculatory tube,** which passes into the **penis** (Fig. 14-15, C). Accessory glands secrete fluid into the ejaculatory tube.

In the female the paired **ovaries** consist of egg tubules (**ovarioles**) (Fig. 14-15, B). From the ovaries **oviducts** run posteriorly and ventrally and unite to form the **vagina,** which is located under the intestine. Associated with the vagina and dorsal to it is a sac, the **spermatheca** (**seminal receptacle**), which receives spermatozoa from the male at copulation.

Adult grasshoppers deposit their eggs in the ground usually in late summer or fall. The young nymphs hatch in the spring. Although their heads are large and they lack wings, they look very much like the adults. They molt periodically to allow for growth, and after the fifth molt they attain full size with wings. What is this type of development called?

Nervous System. The nervous system lies under the digestive system and is similar to that of the crayfish.

METAMORPHOSIS OF DROSOPHILA*

Class Insecta
  Order Diptera
    Genus *Drosophila*

---

*Metamorphosis in the blowfly may be substituted here if desired. Culture instructions are given on p. 234.

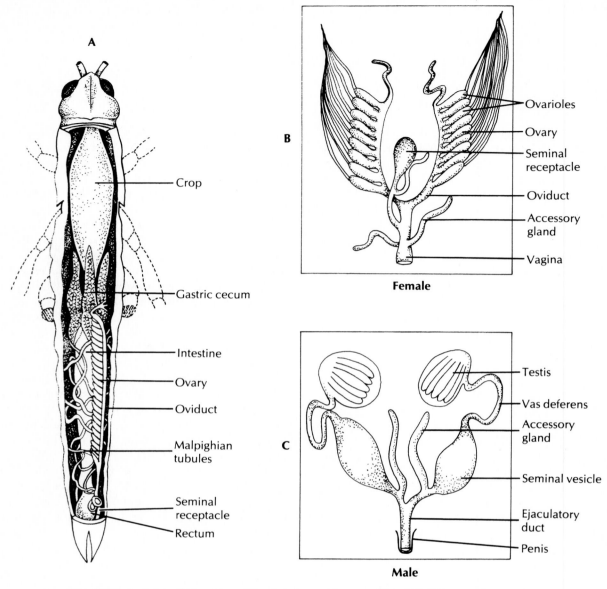

A

Crop

Gastric cecum

Intestine

Ovary

Oviduct

Malpighian
tubules

Seminal
receptacle

Rectum

B

Ovarioles

Ovary

Seminal
receptacle

Oviduct

Accessory
gland

Vagina

Female

C

Testis

Vas deferens

Accessory
gland

Seminal vesicle

Ejaculatory
duct

Penis

Male

Fig. 14-15. A, Internal structure of the female grasshopper. **B** and **C,** The reproductive system.

The early development of an insect occurs within the egg. Most insects change form during the growth stages after hatching. This is called **metamorphosis.** As they grow, they undergo a series of molts, and each stage between molts is called an **instar.** Insect young vary in the degree of development at hatching.

Epimorphosis. Some of the wingless insects, such as the silverfish (Thysanura) and springtails (Collembola), undergo **epimorphosis** rather than metamorphosis. The young are like the adults except in size and sexual maturity and are called juveniles. The stages are egg, juvenile, and adult.

Gradual Metamorphosis (Hemimetabola). In grad-

ual or incomplete, metamorphosis the immature forms are called **nymphs** if terrestrial, or **naiads** if aquatic. Their wings develop as external outgrowths that increase in size as the animal grows by successive molts. The aquatic naiads have tracheal gills or other modifications for aquatic life. Terrestrial forms include the grasshoppers, locusts, mantids and the like (Orthoptera), termites (Isoptera), and true bugs (Hemiptera). Those having aquatic naiads include mayflies (Ephemeroptera), stoneflies (Plecoptera), and dragonflies (Odonata). The stages are egg, nymph, naiad (several instars), and adult.

Complete Metamorphosis (Holometabola). Most

LUBBER GRASSHOPPER

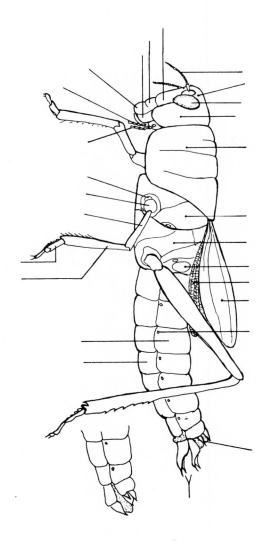

External view of female (male abdomen at bottom left)

DROSOPHILA METAMORPHOSIS

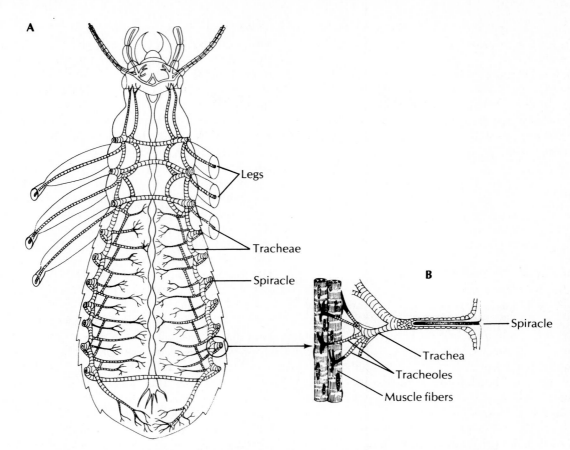

A

Legs

Tracheae

Spiracle

B

Spiracle

Trachea

Tracheoles

Muscle fibers

Fig. 14-16. A, Generalized arrangement of the tracheal system. **B,** Trachea sends branches (tracheoles) to the body tissues.

of the insects (about 88%) have complete metamorphosis. There is a series of larval instars that are different from the adult, followed by a pupal stage, and finally the adult. The wormlike larvae, called caterpillars, maggots, bagworms, grubs, and so on, usually have chewing mouthparts. The wings develop internally during the larval stages. The larva forms a cocoon or case about itself and becomes a pupa. At the final molt the adult emerges fully grown. Flies and fruit flies (Diptera), butterflies and moths (Lepidoptera), beetles (Coleoptera), and the like undergo complete metamorphosis. The stages are egg, larva, pupa, and adult.

Procedure

Beginning 7 to 10 days before your laboratory period, your instructor will have placed in each of several vials a pair of fruit flies and a microscope slide bearing a thick layer of banana-agar medium (see Appendix B, p. 458) and a drop of yeast solution. The vials will have been stoppered with cotton (Fig. 14-17). There should now be eggs, larvae, and pupae in the medium on the various slides.

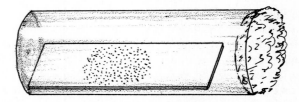

Fig. 14-17. Vial for culture of fruit fly metamorphosis. Warm culture medium was placed on one side of the microscope slide.

☞ Remove a slide from a vial, place it under a dissecting microscope, and study its contents.

The female deposits her **eggs** on food appropriate to feed her growing larval young. They hatch in about 24 hours into wormlike **larvae.** The larvae of flies are often called maggots. The larvae eat and grow for about 5 days and then become inactive **pupae,** each enclosed in a brownish case, within which the final metamorphosis occurs. About 4 days after the pupae are formed the **adults** emerge. Adults are full sized when they appear.

To study the adults, you may kill them by placing a few drops of ether on the cotton plug in the vial.

Can you distinguish sexual differences in the adults? (See p. 429.) Compare the structure of the fruit fly with that of the honeybee and the grasshopper. What difference do you note in the wings? Fruit flies belong to the order Diptera (Gr., *di-*, two, + *pteron*, wing). Is the order well named?

DRAWING

On p. 232 sketch eggs, larvae, and pupa of developing *Drosophila*.

PROJECTS AND DEMONSTRATIONS

1. Demonstration mounts. Demonstration mounts of insect parts such as legs, wings, antennae, and mouthparts can be made very simply. Clean a microslide with acetone, spray on a very thin coat of clear lacquer or varnish from a pressurized can from enough distance to prevent formation of bubbles, orient the parts on the slide while the lacquer is still wet, and label. A coverslip is not necessary. The mount is dry and usable within 30 minutes and can be stored flat in a dust-protected tray for future use.

Slides that show the various types of insect antennae can be purchased or may be prepared by this method.

2. Demonstration of adult and immature insects. This illustrates the types and stages of metamorphosis—epimorphosis and gradual and complete metamorphosis.

3. Culture of the blowfly for insect metamorphosis study. For demonstration of metamorphosis in an insect, blowflies have been found to be excellent material, for their life cycle is very short. These can be purchased from biological supply houses, or you can collect the eggs by exposing in a dish in a suitable location two pieces of fresh beef so that one strip of meat slightly overlaps the other. The meat should be kept moist, and some sugar should be added. Flies will usually lay their eggs in the crevices between the two pieces of meat within 3 to 6 days. Place the meat with the eggs in culture test tubes that contain a small amount of raw lean meat and stopper the tubes with cotton. The eggs will hatch within 24 hours. When the maggots are fully grown, remove the cotton stoppers and place the test tubes in a quart glass jar that contains sand or paper to a depth of 1 inch. Cover the jars with muslin. The maggots will leave the test tube and pupate in the sand or paper (4 or 5 days). Then place the pupae in small screen cages, in which they will emerge as adults.

4. Observation of circulation in the roach. Using wax or modeling clay (or Plasticene) and a microscope slide, construct a cell large enough to hold an adult roach. Hold the insect down with two strips of paper, one across the thorax and the other crossing beneath the wings. Stretch a wing out on a bit of white paper or foil and adjust the light for a combination of reflected and transmitted illumination. The flow of blood in the wing may then be observed under a dissecting microscope.

By pinning the wings out at the side and focusing the light up through the body you may see the beating of the dorsal tubular heart in the abdomen. What is the direction of blood flow?

5. Real-life study of cecropia or polyphemus moth life cycles. Both the large larvae and the cocoons of the cecropia and polyphemus moths can be collected in the fall. The larvae feed on the leaves of the dwarf willow.

To observe the spinning of the cocoons, collect the large larvae and place them in a rearing cage in the laboratory. The rearing cage is an aquarium tank with 1 inch of moist sand in the bottom and a jar of water or wet sand to contain fresh willow branches. Place cotton around the mouth of the jar to prevent the larvae from falling into it. Place the larvae on the willow leaves and cover the aquarium with a glass plate. Replace the willow leaves when they wilt or are eaten by the larvae.

Collected cocoons can be stored in a box of moss, which must always be kept slightly moist. If stored in a cool cellar or unheated room, the adults will emerge in the spring. If kept in a warm room, they will probably emerge in midwinter. Before the adults emerge, place the cocoons in a large cage or screened box containing branches to which the moths can cling while drying their wings.

6. Demonstration of insect eggs and egg cases. The eggs of walkingsticks and the egg masses (oothecae) of the praying mantids can be obtained from biological supply houses and kept in a refrigerator until a few weeks before their hatching is desired. Place in a terrarium or any glass container that can be covered with cheesecloth or fine-meshed wire screening.

Walkingsticks are herbivorous and will need fresh vegetation as soon as they hatch, so it is well to have preparations well in hand ahead of time. About 2 inches of sandy soil in the terrarium should be planted with grass or white clover seed and kept moist and in a warm sunny place until the plants are growing well before the eggs are put in. Avoid letting excessive moisture gather on the grass or glass while the nymphs are still small. Older walkingsticks will feed on fresh twigs of willow or hazelnut or leaves of juneberry, blueberry, or strawberry plants.

Mantids are carnivorous and have voracious appetites. If not provided with sufficient food (*Drosophila* or other insects), they will devour each other. The nymphs should not be kept in direct sun or drafts as they require enough natural humidity to protect them

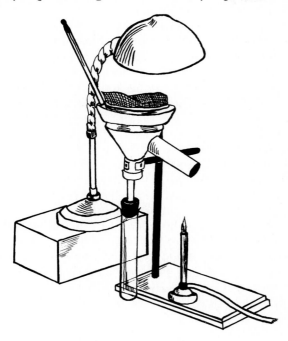

Fig. 14-18. Berlese funnel.

during their molts. If the skins they are molting are too dry, they may not be able to free themselves and so die during the process.

Grasshopper eggs are sometimes available and will hatch if kept in damp soil at room temperature. The nymphs can be fed fresh lettuce.

7. *Firefly luminescence*. Dried firefly lanterns together with ATP and all necessary materials for demonstrating the emission of light can be obtained from several biological supply houses. The ATP supplies the energy needed for the oxidation of the luciferin. Instructions are included.

8. *Berlese funnel*. Small terrestrial arthropods can be collected from soil, moss, and surface litter by means of a Berlese funnel. The apparatus is a double-walled copper funnel in which warm water is circulated by applying a small flame to a projecting arm (Fig. 14-18). Place wire gauze of small mesh at the bottom of the inner glass funnel. Attach the stem of the glass funnel with modeling clay to a large test tube or bottle. Place moss, soil, or surface litter in the glass funnel and support it by the gauze. Then lower a gooseneck lamp with a 15-watt bulb over the mass. The surface temperature of the material should never be above 60° C, nor the water in the double-walled funnel above 30° to 40° C. As the organisms are driven out by the heat, they drop into the collecting tube, where they may be examined from time to time.

EXERCISE 14D
Collection and classification of insects

MATERIALS

For collecting
   Insect nets (aerial, sweep, and water nets)
   Cheesecloth
   Collecting bottles
   Cellucotton or envelopes
For killing
   Small screw-top bottles
   Ethyl acetate or carbon tetrachloride
   Cotton
   Cardboard or blotting paper
For preserving
   Mounting boxes
   Cotton
   Transparent cover (glass or acetate)
   Insect pins
   Cork pinning boards or insect spreading boards
   Labels
   70% alcohol
   KAAD (see Appendix B, p. 462) (optional)

Where to Collect Insects

1. Using a sweep net, sweep over grass, alfalfa, weed patches.
2. Spread a cloth, newspaper, or inverted umbrella under a bush or shrub; then beat or shake the plant vigorously.
3. Overturn stones, logs, bark, leaf mold, and rubbish; look under dung in pastures.
4. Watch for butterflies to alight; then drop a net over them.
5. Look around outdoor lights at night.
6. At night, suspend a sheet from a limb or clothesline with the lower part of the sheet spread on the ground. Direct headlights or spotlight on the sheet. The insects will be attracted to the white light, hit against the sheet, and drop onto the cloth below.
7. Moths may be baited by daubing a mixture of crushed banana or peach and molasses or sugar on the bark of trees; then visit the trees at night with a flashlight to collect the moths.
8. Locate soil insects by placing humus and leaf matter in a Berlese funnel (Fig. 14-18).
9. Water insects may be seined with a water net. Aquatic insect larvae may be found in either quiet or running water, attached to plants and leaves or under stones, or sieved out of bottom mud.
10. In early spring the sap exuding from stumps or tree trunks attracts various insects.

Killing the Insects

If an insect is to be preserved it must be killed in such a way as not to injure it. Killing bottles of various sizes and shapes may be used. Wide-mouthed jars of thick glass with screw-top lids are best. Each jar should be conspicuously labeled POISON.

A temporary killing bottle may be made quickly by wetting a piece of cotton with a fumigant such as ethyl acetate or carbon tetrachloride, placing it in the bottom of a bottle, covering it with several discs of blotting paper, and screwing on the lid. *Do not inhale* the fumigant. Half pint or pint jars are satisfactory. Wrapping the bottom of the bottle with adhesive or masking tape will help prevent breakage. The fumigant must be replenished before each collecting trip. A few loose strips of paper toweling added to the bottle will help absorb moisture and protect small insects. Replace the paper as it becomes moist.

A more permanent killing jar may be made by placing several rubber bands or small pieces of inner tube and a layer of sawdust 0.6 to 1.5 cm thick in the bottom of the jar and then pouring in a layer of plaster of Paris of about the same depth. Before the plaster sets completely, poke several holes in it with a toothpick or wire to allow the fumigant to reach the sawdust. When it dries, add a 1 cm layer of carbon tetrachloride, and screw on the lid. When the fluid has been absorbed, some strips of paper towel can be added and the lid screwed on.

Hard-bodied insects may require an hour or so to ensure death. If you are collecting many kinds of insects, it is well to have several killing bottles so that large insects such as butterflies, dragonflies, and large beetles may be kept in separate jars, away from smaller, more fragile insects.

To transfer an insect from a net to the kill bottle, fold the net over, uncover the jar, work the jar into the net, and place its mouth over the insect as it clings to the cloth; then hold the lid over the jar for a few seconds until the insect is quiet before removing it from the net and placing it in the jar. Remove the insect from the killing jar when it is dead and place it between layers of paper toweling (or Cellucotton) or in paper envelopes to prevent damage.

Soft-bodied insects and especially larval forms should not be put into the killing bottle. They can be killed by being dropped into 70% to 80% ethyl alcohol, or by injecting their bodies with alcohol. Or better yet, they may be fixed first in KAAD for ½ to 4 hours before preserving in alcohol. If killed in alcohol, the alcohol should be changed after a few days, since it becomes diluted by the body fluids of the animals put into it.

Relaxing

Insects that have dried out are fragile and easily broken. If an insect cannot be mounted before it has become dry, it can be relaxed. A relaxing jar can be prepared by placing a layer of wet sand or wet paper toweling in the bottom of a glass jar, adding a few drops of formalin or carbolic acid to discourage mold, and covering the sand with a disc of blotting paper to protect the insect. The insect may be left in the relaxing jar for 1 to 3 days, as necessary.

Displaying the Insects

1. Insects may be mounted in shallow boxes on a layer of cotton and covered with glass plates or sheets of acetate. An attractive box of this kind can be made by cutting out the main portion of the lid of the box (so that a 1.5 to 2.5 cm rim or "frame" remains around each side) and then gluing a piece of glass or acetate to the inside of the lid behind the frame. The box itself is lined with cotton, and the insects are held in place on the cotton by the acetate or glass in the lid.

2. Pinned insects can be displayed in purchased Schmitt boxes, cigar boxes, or cardboard boxes deep enough to take an insect pin. Balsa wood or layers of soft corrugated cardboard should be glued to the bottom of the box.

Insects should be pinned with regular insect pins, which are longer and thinner than common pins. Sizes 2 and 3 are convenient sizes for general use.

Pin grasshoppers through the posterior part of the pronotum, a little to the right of the midline (Fig. 14-19). Butterflies and moths are pinned through the center of the thorax, whereas bees, wasps, and flies are pinned through the thorax but a little to the right of midline. Pin bugs (hemipterans) through the scutellum, a bit to right of midline, and beetles through the right elytron (forewing), halfway between the two ends of the body. To insert the pin, hold the insect between the thumb and forefinger of one hand and insert the pin with the other. Mount all specimens at

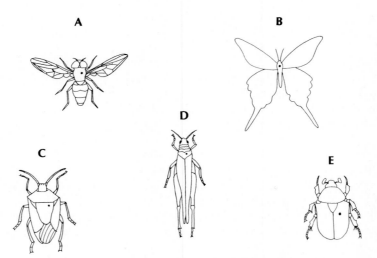

Fig. 14-19. How to pin insects. The dots indicate the location of the pins. **A,** Flies. **B,** Butterflies and moths. **C,** Bugs. **D,** Grasshoppers. **E,** Beetles.

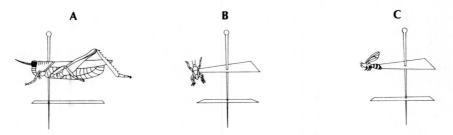

Fig. 14-20. Pinning and labeling insects. Larger insects (**A**) are pinned through the body. Minute ones may be glued to a paper point, either dorsal side up (**B**) or laterally (**C**). Dates, places, and scientific names may be printed on slips of paper beneath the insect.

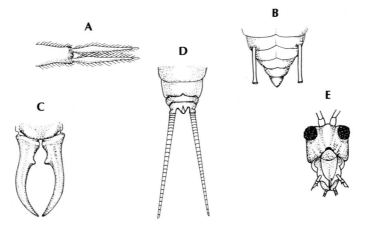

Fig. 14-21. Some diagnostic characters used in identifying insects. **A,** Ventral view of furcula (Collembola). **B,** Cornicles on abdomen (Homoptera). **C,** Forcepslike cerci (Dermaptera). **D,** Long cerci (Plecoptera). **E,** Conelike beak (Thysanoptera).

a uniform height—about 2.5 cm above the point of the pin.

A tiny label with the date, place, and name of the collector may be added to the pin under the insect, and another label giving the scientific name of the insect may be placed under that.

A very small insect may be mounted on the point of an elongated triangular piece of light cardboard about 8 to 10 mm long and 3 to 4 mm wide at the base (Fig. 14-20). After putting the triangular cardboard mount on the pin, hold the pin and touch the tip of the cardboard triangle to the glue; then touch the insect with the gluey tip. Use as little glue as possible to hold the insect.

Many winged insects, such as butterflies, moths, and mayflies, are usually mounted first on a spreading board, which has a groove into which the body is pinned. The wings are then spread out and held in the proper position to dry. Care must be taken not to damage the insect, or to rub off the scales of a lepidopteran. The wings are carefully moved into position by pins and held by strips of paper pinned to the board. In butterflies, moths, and mayflies the rear margins of the front wings should be straight across, at right angles to the body; the hindwings should be far enough forward so that there is no gap between the forewings and hindwings. The front wing should overlap the front edge of the hindwing a little. With damselflies, dragonflies, grasshoppers, and most other insects, the front margins of the hindwings should be straight across, with the front wing far enough forward that the forewings and hindwings do not touch.

Try to maneuver the wings by means of a pin held near the base of the wings—not through the wings, since that would make holes in the wings. Pin the wings securely with paper strips, using several strips if necessary. When completed, hold the body down with forceps and remove the pin from the thorax.

If the specimen is to be displayed on cotton under glass, the spreading board is not necessary. Pin the insect upside down (feet up) on a flat pinning surface, such as corrugated cardboard, and pin the wings with strips of paper as before.

Drying may take several days. If the abdomen, when touched gently with a pin, can be moved independently of the wings, the specimen is not dry enough; wait till the body is stiff before removing the pins and strips of paper.

Classifying Insects

There are a number of good insect guides on the market. The Key to Orders of the More Common Insects (pp. 239-247) will help you place your specimens into the proper order. Use this key according to the method given on p. 78. Fig. 14-21 shows some of the diagnostic features used in insect identification.

Key to orders of the more common insects*

Terms used in the Key:
Cercus (pl., cerci) One of a pair of jointed anal appendages.
Elytron (pl., elytra) Horny veinless front wing of Coleoptera and a few others.
Haltere Pair of small knobbed organs on metathorax, representing hindwings of Diptera.
Tarsus (pl., tarsi) The leg segment distal to the tibia, consisting of one or more segments or subdivisions.

In each couplet the number in parentheses refers back to the number of the couplet from which that couplet was reached, thus making it possible to work backward if a mistake is made.

1.　　 With functional wings . . . . 　2
　　　 Without wings, or with rudimentary or vestigial wings . . . . 　22
2. (1) Both wings entirely membranous . . . . 　3
　　　 Front wings horny, leathery, or parchmentlike, at least at base; hindwings,
　　　 if present, membranous . . . . 　18
3. (2) With one pair of wings . . . . 　4
　　　 With two pairs of wings . . . . 　6
4. (3) Body grasshopper-like, pronotum extending back over abdomen and point-
　　　 ed apically; hindlegs enlarged (pygmy grasshoppers) . . . . . . Orthoptera
　　　 Not as above . . . . 　5
5. (4) Abdomen with 2 or 3 threadlike "tails"; mouthparts vestigial; antennae
　　　 short, bristlelike; wings with numerous veins; halteres absent (mayflies)
　　　 . . . . Ephemeroptera
　　　 Abdomen without "tails", sucking mouthparts, halteres present (true flies)
　　　 . . . . Diptera

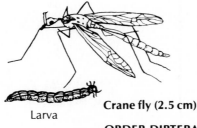

Crane fly (2.5 cm)
Larva
ORDER DIPTERA

6. (3) Wings largely or entirely covered with scales; mouthparts usually a coiled
　　　 proboscis; many-segmented antennae (butterflies and moths) . . . .
　　　 . . . . Lepidoptera

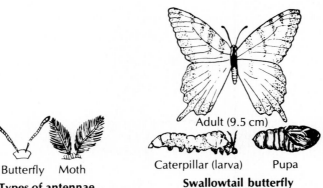

Adult (9.5 cm)

Butterfly　Moth
Types of antennae

Caterpillar (larva)　Pupa
Swallowtail butterfly

ORDER LEPIDOPTERA

*The following orders have been omitted from this key: Protura, Diplura, Embioptera, Zoraptera, and Strepsiptera.

239

Wings not covered with scales (body may be hairy); mouthparts not in form of coiled proboscis; antennae variable   7

7. (6) Wings long, narrow, and fringed with long hairs; tarsi with only one or two segments and swollen at tip; minute, usually less than 5 mm in length (thrips) . . . . Thysanoptera

Larva   **Female thrips (0.5-5.0 mm)**

ORDER THYSANOPTERA

Wings not as above; if wings are somewhat linear, then the tarsi have more than 2 segments   8

8. (7) Front wings triangular, and larger than hindwings; hindwings rounded; wings at rest held together above body; wings with many cross veins; antennae inconspicuous; abdomen with 2 or 3 long threadlike "tails"; soft-bodied delicate insects (mayflies) . . . . Ephemeroptera

Mayfly (2.75 cm)   Nymph

ORDER EMPHEMEROPTERA

Not exactly as above   9

9. (8) Tarsi with 5 segments   10

Tarsi with 4 or fewer segments   13

10. (9) Front wings noticeably hairy; antennae as long or longer than body; rather soft-bodied (caddis flies) . . . . Trichoptera

Caddis fly (15 mm)

Larva in case
ORDER TRICHOPTERA

Front wings not noticeably hairy; antennae shorter than body   11

11. (10) Rather hard-bodied and wasplike, the abdomen often constricted at base; front wings larger than hindwings and with more veins (sawflies, ichneumons, ants, wasps, and bees) . . . . Hymenoptera

Paper wasp (2 cm)

ORDER HYMENOPTERA

Soft-bodied, not wasplike; forewings and hindwings similar in size and venation 12

12. (11) Costal area (just behind the anterior border) of front wings usually with many cross veins; or if not, then the hindwings are shorter than front wings; head not elongated (fishflies, dobsonflies, lacewings, and ant lions) . . . . Neuroptera

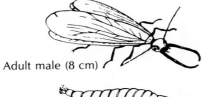

Adult male (8 cm)

Nymph

Dobson fly

ORDER NEUROPTERA

Costal area of front wing with only 2 or 3 cross veins; chewing mouthparts; head prolonged into a beaklike structure (scorpionflies) . . . . Mecoptera

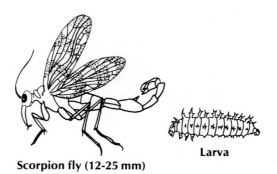

Scorpion fly (12-25 mm)

Larva

ORDER MECOPTERA

13. (9) Hindwings as long as front wings and of same shape or wider at base; many-veined wings; wings at rest held together above body or outstretched (never held flat over abdomen); tarsi with 3 segments; abdomen long and slender; 2 to 8 cm in length (dragonflies and damselflies) . . . . Odonata

Nymph

Dragonfly (5 cm)

ORDER ODONATA

        Not exactly as above 14

14. (13) Sucking mouthparts 15
        Chewing mouthparts 16

15. (14) Beak arising from front part of head; minute in size (gnat bugs, not likely to be found in general collections) . . . . Hemiptera
        Beak arising from hindpart of head (cicadas, hoppers, aphids, psyllids, and whiteflies) . . . . Homoptera

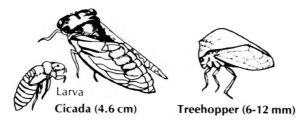

Larva
Cicada (4.6 cm)   **Treehopper (6-12 mm)**

ORDER HOMOPTERA

16. (14) Front wings and hindwings usually similar in size, shape, and venation; cerci minute or absent (termites) . . . . Isoptera

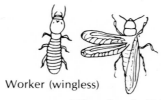

Worker (wingless)

Winged male (7.5 mm)

Termites

ORDER ISOPTERA

        Hindwings usually smaller than front wings; tarsi with 3 or fewer segments 17

17. (16) Hindwings with posterior lobe enlarged and folded fanwise at rest; cerci often fairly long (Fig. 14-21, D); mostly 10 mm or more in length; usually found near water (stoneflies) . . . . Plecoptera

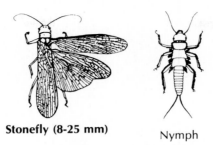

Stonefly (8-25 mm)

Nymph

ORDER PLECOPTERA

Hindwings without enlarged posterior lobe and not folded at rest; cerci absent; antennae usually long and hairlike and with 13 or more segments; usually 7 mm or less in length (book lice, bark lice) . . . . Psocoptera

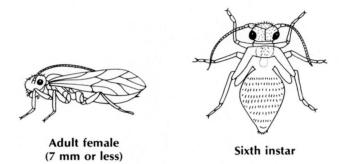

**Adult female
(7 mm or less)**

Sixth instar

ORDER PSOCOPTERA

18. (2) Sucking mouthparts; the beak elongate and usually segmented   19
    Chewing mouthparts   20

19. (18) Beak arising from front part of head; front wings usually thick and leathery at base and membranous at tip; wings at rest held flat over abdomen with tips of front wings overlapping (true bugs) . . . . Hemiptera

Chinch bug (5 mm)
(winged)

ORDER HEMIPTERA

Beak arising from hindpart of head; front wings uniform in texture throughout, the tips not, or only slightly, overlapping (hoppers) . . . . Homoptera

20. (18) Abdomen with forcepslike cerci (Fig. 14-21, *C*); elytra short, leaving most of abdomen exposed; three-segmented tarsi (earwigs) . . . . Dermaptera

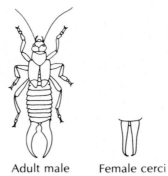

Adult male   Female cerci

Earwig (4-20 mm)

ORDER DERMAPTERA

Cerci not forcepslike, or if they are, then wings cover most of abdomen; tarsi variable 21

21. (20) Elytra veinless, usually meeting in a straight line down middle of back; hindwings narrow, usually longer than front wings when unfolded, and with few veins; antennae usually with 11 or fewer segments (beetles) . . . . Coleoptera

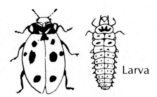

Larva

Ladybird beetle (7.5 mm)

ORDER COLEOPTERA

Front wings veined; wings at rest held either rooflike over abdomen or overlapped over abdomen; hindwings broad, usually shorter than front wings, and many-veined; antennae usually with more than 12 segments (grasshoppers, crickets, cockroaches, and mantids) . . . . Orthoptera

22. (1) Body with more or less distinct head and segmented legs 23
        Body without distinct head or legs; flattened, elongate-oval insects covered with hard exoskeleton or wax (scale insects) . . . . Homoptera

23. (22) Ectoparasites of birds or mammals and usually found on host; body some-what leathery and usually flattened 24
        Freeliving (not ectoparasitic), terrestrial, or aquatic 28

24. (23) Tarsi with 5 segments; antennae short and usually concealed in grooves in head; sucking mouthparts 25
        Tarsi with fewer than 5 segments; antennae and mouthparts variable . . . 26

25. (24) Body flattened laterally; legs relatively long and adapted for jumping (fleas) .... Siphonaptera

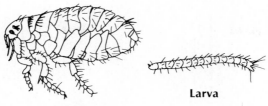

Larva

Adult flea
ORDER SIPHONAPTERA

Body flattened dorsoventrally; usually with short legs, not adapted for jumping (louse flies, bat flies, and bee lice, not likely to be found in general collections) .... Diptera

26. (24) Antennae distinctly longer than head: tarsi with 3 segments (bedbugs and bat bugs) .... Hemiptera

Bedbug (7.5 mm)
(wingless)
ORDER HEMIPTERA

Antennae not longer than head; tarsi with 1 segment (lice) 27

27. (26) Head narrower than prothorax; sucking mouthparts; parasites of mammals (sucking lice) .... Anoplura

Head as wide or wider than prothorax; chewing mouthparts; parasites of birds and mammals (chewing lice) .... Mallophaga

28. (23) Abdomen distinctly constricted at base; antennae usually elbowed; hard-bodies (ants and wingless wasps) .... Hymenoptera

Carpenter ant (1.3 cm)
(worker)
ORDER HYMENOPTERA

Not as above 29

29. (28) Abdomen of six or fewer segments and usually a forked furcula (spring) near posterior end (Fig. 14-21, *A*) (springtails) .... Collembola

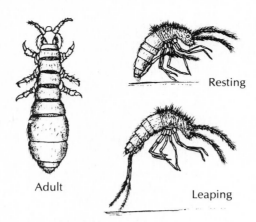

Resting

Adult

Leaping

Springtail (2.6 mm)

ORDER COLLEMBOLA

Abdomen with more than six segments and without forked furcula 30

30. (29) Abdomen with three "tails" (long cerci and median caudal appendage); abdomen with 11 segments; some with stylelike appendages; body nearly always covered with scales (silverfish and bristletails) .... Thysanura

Silverfish (15 mm)

ORDER THYSANURA

Abdomen not "three-tailed"; body not covered with scales 31

31. (30) Tarsi with five segments; antennae with more than 5 segments; (some cockroaches, walkingsticks) .... Orthoptera
Tarsi with 4 or fewer segments 32

32. (31) Cerci forcepslike; tarsi with three segments 33
Cerci absent or, if present, not forcepslike 34

33. (32) Antennae more than half as long as body; cerci short; found in western United States (some walkingsticks) .... Orthoptera
Antennae usually less than half as long as body; cerci long (Fig. 14-21, *D*) (earwigs) .... Dermaptera

34. (32) Sucking mouthparts with beak long and extended backward from head, or cone-shaped and directed ventrally 35
Chewing mouthparts 37

35. (34) Body long and slender, beak cone-shaped (Fig. 14-21, *E*); tarsi with 1 or 2 segments, often without claws; usually less than 5 mm in length (thrips) .... Thysanoptera
Body more or less oval; tarsi with two or three segments with well-developed claws 36

36. (35) Beak arising from front part of head; abdomen without cornicles; antennae with 4 or 5 segments (wingless bugs) .... Hemiptera
Beak arising from rear of head, abdomen usually with a pair of cornicles (Fig. 14-21, *B*); antennae usually with more than five segments (aphids and others)

.... Homoptera

Corn-root aphid (2 mm)

ORDER HOMOPTERA

.... Homoptera
37. (34) Grasshopper-like insects, with hindlegs adapted for jumping; over 15 mm in length (wingless grasshoppers) .... Orthoptera
Not as above; length less than 10 mm 38
38. (37) Tarsi with four segments; whitish, soft-bodied (termites) ...... Isoptera
Tarsi with two or three segments; antennae hairlike, with 13 or more segments; compound eyes and three ocelli usually present (psocids)
.... Psocoptera

CHAPTER 15

THE ECHINODERMS

PHYLUM ECHINODERMATA*
A Deuterostome Group

The echinoderms are an all-marine phylum that includes sea lilies, sea stars, brittle stars, sea urchins,
and sand dollars. They rank high in the evolutionary
blueprint, having a well-developed coelom and a **dermal endoskeleton** made up of calcareous ossicles and
spines. The endoskeleton is the basis of the phylum
name (Gr., *echinos*, hedgehog, sea urchin, + *derma*,
skin).

Although they have **pentaradiate symmetry** as
adults, this is secondarily acquired, for the larvae are
bilaterally symmetrical (Fig. 15-1). Echinoderms and
chordates have probably arisen from common ancestors.

*See Hickman, C.P., Jr., L.S. Roberts, and F.M. Hickman, 1984.
Integrated principles of zoology, ed. 7. St. Louis, The C.V. Mosby
Co., pp. 464-487.

Unique features of echinoderms are their **water-
vascular system, tube feet,** pincerlike **pedicellariae,**
and **skin gills (dermal branchiae).** Their body parts are
arranged radially in fives or multiples of five.

The nervous system and sense organs are primitive; there are no excretory organs; a definite head is
lacking; locomotion is slow (some forms are sessile);
and they are unsegmented.

The echinoderms have traditionally been divided
into two subphyla: the Eleutherozoa, which contain
most of the living species and are all free-moving forms
oriented with the oral side down; and the Pelmatozoa,
which contain most of the extinct classes and are mostly stemmed forms with the oral side up. However,
because many authorities now consider these groups

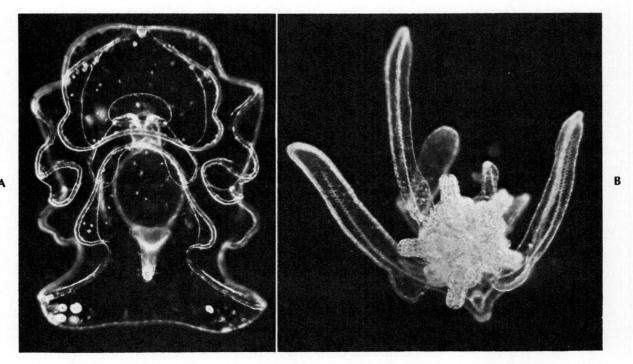

Fig. 15-1. Larval forms of some echinoderms. **A,** *Auricularia* larva of the sea cucumber. (×100.)
B, *Ophiopluteus* larva of a brittle star. (×120.) Note that this larva is in the process of
metamorphosis from the bilateral larva to the five-rayed juvenile form. (Photographs by D.P.
Wilson, Plymouth, England.)

to be polyphyletic, we have adopted the classification proposed by H.B. Fell and used in Moore.*

Subphylum Homalozoa (ho-mal′o-zo′a). Fossil carpoids, which were not radially symmetrical.

Subphylum Crinozoa (krin′o-zo′a). Radially symmetrical, with rounded or cup-shaped theca and brachials, or arms; attached by stem during all or part of life; oral surface up. Four extinct classes and one living class.

    Class Crinoidea (krin-oi′de-a)—**sea lilies and feather stars.** Aboral attachment stalk of dermal ossicles; mouth and anus on oral surface; five branching arms with pinnules; ciliated ambulacral groove on oral surface with tentacle-like tube feet for food collecting; spines, madreporite, and pedicellariae absent. Examples: *Antedon, Florometra*.

Subphylum Asterozoa (as′ter-o-zo′a). Radially symmetrical; star-shaped; unattached as adults.

    Class Stelleroidea (stel′ler-oi′de-a)—**sea stars, brittle stars, and basket stars.** Characteristics of subphylum; body of central disc with radially arranged rays or arms.

        Subclass Somasteroidea (som′ast-er-oi′de-a). Mostly extinct sea stars with primitive skeletal structure; one living species. *Platasterias latiradiata*.

---

*Moore, R.C. (ed.). 966, 1967. Treatise on invertebrate paleontology. Echinodermata; Parts U and S. Lawrence, Kan., Geological Society of America and University of Kansas Press.

        Subclass Asteroidea (as′ter-oi′de-a)—**sea stars.** Star shaped, with arms not sharply marked off from central disc; ambulacral grooves open, with tube feet on oral side; tube feet often with suckers; anus and madreporite aboral; pedicellariae present. Example: *Asterias*.

        Subclass Ophiuroidea (o′fe-u-roi′de-a)—**brittle stars and basket stars.** Star-shaped, with arms sharply marked off from central disc; ambulacral grooves closed, covered by ossicles; tube feet without suckers and not used for locomotion; pedicellariae absent. Examples: *Ophiura, Gorgonocephalus*.

Subphylum Echinozoa (ek′in-o-zo′a). Mostly unattached; globoid or discoid; without arms. Four extinct and two living classes.

    Class Echinoidea (ek′i-noi′de-a)—**sea urchins, sea biscuits, and sand dollars.** More or less globular or disc-shaped, with no arms; compact skeleton, or test, with closely fitting plates; movable spines; ambulacral grooves closed and covered by ossicles; tube feet with suckers; pedicellariae present. Examples; *Arbacia, Strongylocentrotus, Lytechinus, Mellita*.

    Class Holothuroidea (hol′o-thu-roi′de-a)—**sea cucumbers.** Cucumber-shaped; with no arms; spines absent; microscopic ossicles embedded in thick muscular wall; anus present; ambulacral grooves closed; tube feet with suckers; circumoral tentacles (modified tube feet); pedicellariae absent; madreporite plate internal. Examples: *Thyone, Stichopus, Cucumaria*.

EXERCISE 15A

Subclass Asteroidea—the sea stars

MATERIALS

Living sea stars (*Asterias*, *Pisaster*, and others) in seawater
Preserved (or anesthetized) sea stars for dissection
Dishes for live material
Fresh or frozen seafood for feeding sea stars
Carmine suspension
Pieces of dried sea star tests
Dissecting pans and tools
Dissecting microscopes or hand lenses

ASTERIAS

Phylum Echinodermata
  Subphylum Asterozoa
    Class Stelleroidea
      Subclass Asteroidea
        Order Forcipulatida
          Genus *Asterias*

Where Found

Sea stars are commonly found along rocky seacoasts, often in tidewater regions. They may also be found farther out and at considerable depths in the ocean.

Behavior

☞ Examine a living sea star in a dish of seawater. Using the following suggestions, observe its behavior.

Note the general body plan of the star with its **pentaradial** (Gr., *pente*, five, + L., *radius*, ray) symmetry, its five **arms**, or **rays**, its **oral-aboral flattening**, and the **mouth** on the under (oral) side. Lift up the dish and look at the oral surface. Notice the rows of **tube feet**. How are they used? How are the ends of the tube feet shaped? What is the sequence of action of a single tube foot as the animal is moving? The tube feet are filled with fluid and are muscular, providing the necessary components for a hydraulic skeleton. When a foot contracts, the water flows into a bulblike **ampulla** inside the arm. The tube feet and ampullae are parts of the **water-vascular system.**

Tilt the dish to one side (pour out a little water if necessary) and watch the animal's reaction. Does it move up or down the inclined plane? Now tilt the dish in the opposite direction. Does the animal change direction? Is it positively geotactic (moves toward the earth) or negatively geotactic?

Place a piece of fresh seafood (oyster, fish, shrimp) near one of the arms. Is there any reaction? Are the arms flexible? If the sea star makes no move toward the food, touch the tip of an arm with it, or slip it under the end of an arm. Hold the dish up and look underneath. How is the food grasped? How is it moved toward the mouth? What position does the animal assume when feeding?

Examine the aboral surface with a hand lens or dissecting microscope. The **epidermis** is ciliated. Place a drop of carmine suspension on the surface and see which direction the ciliary currents take. Notice the calcareous **spines** protruding through the skin. These are extensions of the skeletal ossicles. Do the spines move? Small fingerlike bulges in the epidermis are **skin gills** (also called **dermal branchiae, papulae**), concerned with gaseous exchange. Around the spines you will see small **pedicellariae.** These are calcareous two-jawed pincers that are modified spines and are concerned with capturing tiny prey and protecting the dermal branchiae from collecting sediment or small parasites. Irritate the skin a little with a small brush or a bit of paper toweling and try to observe the pincer action.

Look at the tip of each arm to see a small red **eyespot** and **sensory tentacle.**

Record your observations on 258.

External Structure

☞ Place a preserved sea star in a dissecting pan and cover with water.

The star-shaped body is composed of a **central disc** and five **rays** (arms). Are all the rays alike? What would account for some of the rays being shorter than others? Compare your specimen with those of your neighbors. Preserved specimens may seem rigid but live stars can bend their arms by means of muscles.

Aboral Surface. The central disc bears a small porous **madreporite plate** composed of calcium carbonate (Fig. 15-2). It allows seawater to seep into an intricate **water-vascular system,** which provides the means of locomotion. The rays on each side of the madreporite are called the **bivium**; the other three are the **trivium.** The **anus** opens in the center of the central disc, but it may be too small to see.

☞ Submerge one of the arms under water and examine the dorsal body wall under a dissecting microscope. Compare with a piece of dried-up test.

The body is covered with a thin ciliated **epidermis,** through which white calcareous **spines** extend from the endoskeleton beneath. These spines are not movable. Notice that around the base of each spine is a raised ring of skin bearing tiny calcareous pincerlike **pedicellariae.** Some are also found between the spines.

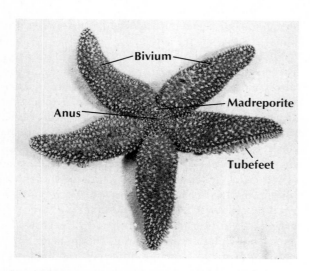

Fig. 15-2. External aboral view of the starfish *Asterias*. (Photograph by F.M. Hickman.)

Sometimes pedicellariae can be seen more easily on a dried-up piece of test. Some pedicellariae have straight jaws and others have curved jaws. They are moved by tiny muscles. What is their function? Between the spines are soft transparent fingerlike projections, the **skin gills** or **dermal branchiae**. These are hollow evaginations of the coelomic cavity through which respiration takes place.

A small pigmented **eyespot** is located at the tip of each ray.

Oral Surface. The **ambulacral** groove of each ray contains four rows of **tube feet (podia)**. The **ambulacral spines** bordering the groove are movable and can interlock when the groove is contracted to protect the tube feet. Note the size, shape, number, and arrangement of the tube feet.

☞ Scrape away some of the tube feet and note the arrangement of the pores through which they extend.

The tube feet are a part of the water-vascular system. They are hollow, and their tips form suction discs for attachment (Fig. 15-3). These are effective not only in locomotion but also in opening bivalves for food.

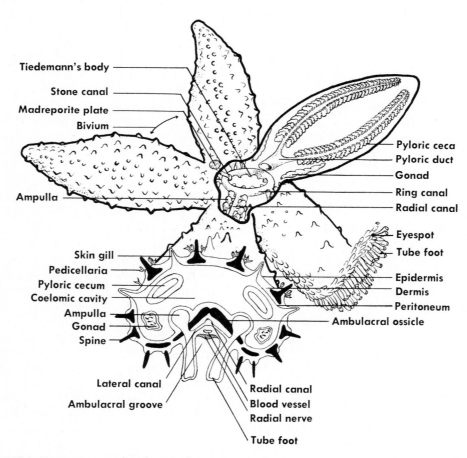

Fig. 15-3. Diagrammatic aboral view of a sea star, with stomach removed to show the water-vascular system and with one arm shown in transverse section.

The central **mouth** is surrounded by five pairs of movable spines. Push the spines outward, bend the arms back slightly, and note the thin **peristomial membrane**, which surrounds the mouth. Sometimes the mouth is filled with part of the everted **stomach.**

DRAWINGS

Sketch some pedicellariae on p. 257.

Endoskeleton

Echinoderms are the first of the invertebrates to have a mesodermal endoskeleton. It is formed of calcareous plates, or **ossicles**, bound together by connective tissue.*

 Cut off part of one of the rays of the bivium, remove the aboral wall from the cut-off piece, and study its inner surface. Compare with a piece of dried body wall, which is excellent for examining the arrangement of ossicles.

Note the skeletal network of irregular ossicles. Now look at the cut edge of the body wall and identify the outer layer of **epidermis**, the thicker layer of **dermis** or connective tissue in which the ossicles are embedded, and the thin inner layer of ciliated **peritoneum.** The dermis and peritoneum are mesodermal in origin. Are the spines part of the endoskeleton? Hold the piece up to the light. The thin places you see between the ossicles are where the dermal branchiae extend through the connective tissue.

Look at the oral surface of the cut-off ray. Note how the **tube feet** extend up between the **ambulacral ossicles**, emerging on the inside surface of the wall as bulblike **ampullae.** Compress some of the ampullae and note the effect on the tube feet. Press on the tube feet and see the effect on the ampullae. Both are muscular and can regulate the water pressure by contraction. Scrape away some of the ampullae and tube feet and examine the shape of the ossicles in the ambulacral groove. How do they differ from the ossicles elsewhere? Note the alternating arrangement of the **ambulacral pores** through which the tube feet extend. Whereas the ambulacral ossicles form a groove on the oral surface of each arm, they form an **ambulacral ridge** on the inner surface.†

On p. 257 sketch a series of ambulacral ossicles, showing the arrangement of the openings for the tube feet.

---

*A method for mounting skeletal ossicles on microslides is given on p. 256.
†A demonstration of the action of the tube feet of a live sea star is suggested on p. 255.

Internal Structure (Dissection)

☞ Place the specimen aboral side up in a dissecting pan and cover with water. Select the three rays of the trivium and snip off their distal ends. Insert a scissors point under the body wall at the cut end of one of the rays. Carefully cut along the dorsolateral margins of each ray to the central disc. Lift up the loosened wall and carefully free any clinging organs. Uncover the central disc, but cut around the madreporite plate, leaving it in place. Be careful not to injure the delicate tissue underneath. The tissue around the anal opening in the center of the disc will cling. Cut the very short intestine close to the aboral wall before lifting off the body wall.

The **coelomic cavity** inside the rays and disc contains **coelomic fluid**, which bathes the visceral organs.* What is the lining of the coelom called?

Digestive System. A pentagonal **pyloric stomach** (Fig. 15-4) lies in the central disc, and from it a **pyloric duct** extends into each arm where it divides to connect with a pair of large much-lobulated **pyloric ceca** (digestive glands) (Fig. 15-4). A very short **intestine** leads up from the center of the stomach to the anus in the center of the disc. Attached to the intestine are two **rectal ceca**, small branched sacs of uncertain function. Below the pyloric stomach is the larger five-lobed **cardiac stomach**, which fills most of the central disc (Fig. 15-4). Each lobe of the stomach is attached to the ambulacral ridge of one of the arms by a pair of **gastric ligaments**, which prevent too much eversion of the stomach.

When the sea star feeds on a bivalve, it folds itself around the animal, attaches its tube feet to the valves, and exerts enough pull to cause the shell to gape a little. Then by contraction of its body walls, causing pressure of the coelomic fluid, it everts its stomach and inserts it into the open clam shell. There it digests the soft parts of the clam by juices from the pyloric ceca. Partly digested material is drawn up into the stomach and pyloric ceca, where digestion is completed. There is little waste fecal matter. When the sea star is finished with feeding, the stomach withdraws into the coelom by contraction of stomach muscles and relaxation of the body wall, which allows coelomic fluid to flow back into the arms.

Many stars feed on small bivalves by engulfing the entire animal, digesting out its contents, then casting the shell out through the mouth.

Reproductive System. The sexes are separate in the sea star. Remove the pyloric ceca from one arm to find the paired **gonads** attached to the sides of the

---

*A method for demonstrating the phagocytic action of coelomic cells is given on p. 256.

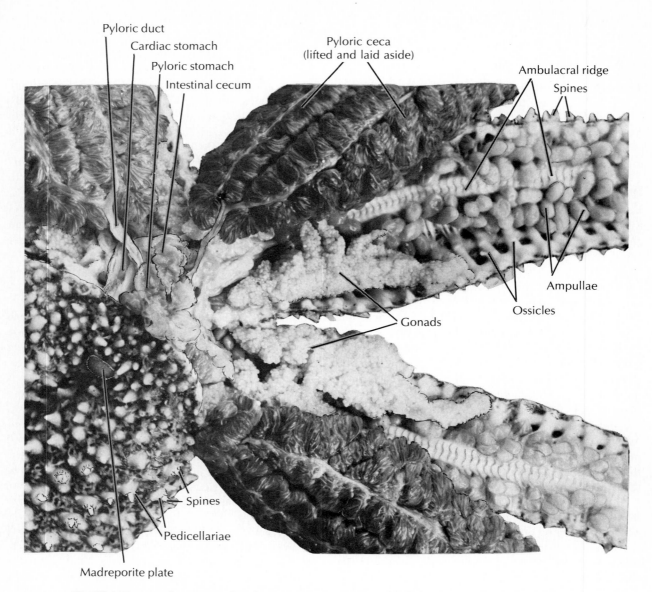

Pyloric duct

Cardiac stomach

Pyloric stomach

Intestinal cecum

Pyloric ceca
(lifted and laid aside)

Ambulacral ridge

Spines

Gonads

Ampullae

Ossicles

Spines

Pedicellariae

Madreporite plate

Fig. 15-4. Sea star dissection, aboral view. Aboral body wall has been removed from three of the rays, and pyloric ceca have been laid aside on two of the rays *(upper and lower right)* to expose the ampullae, ambulacral ridge, and gonads. (Photograph by F.M. Hickman.)

ray where the ray joins the disc (Fig. 15-4). During the breeding season the gonads are larger than at other times. Each gonad opens aborally to the exterior at the point of attachment by a very small **reproductive duct** and **genital pore**. The sex can rarely be determined by simple inspection of the gonads, although the female gonads may be a little coarser in texture and more orange than the male gonads.

☞ Make a wet mount of a mashed bit of gonad and examine with a microscope.

In the ovary eggs with large nuclei will be found; and in the testes many small sperm are to be found. In early summer large streams of eggs and sperm are shed into the water, where fertilization occurs exter-

nally. Review Exercise 4B (p. 53) for the development of the sea star.

Nervous System The nervous system of the sea star is somewhat primitive and actually consists of three interrelated systems.

The **ectoneural** system consists of a network of nerve cell bodies and their processes found just beneath the epidermis all over the body. Specifically, this part of the nervous system consists of a nerve ring around the mouth in the peristomial membrane; a radial nerve to each arm running along the ambulacral groove to the eyespot; and the nerve plexus, which is found underneath the epidermis.

☞ To find the nerve ring, remove the tube feet and movable spines around the mouth and expose the peristomial membrane.

The nerve ring is a whitish thickening on the outer margin of this membrane. To see one of the **radial nerves,** bend an arm aborally and look along the oral surface of the ambulacral groove for a whitish cord (Fig. 15-3). Trace the nerve from the ring to its termination in the arm.

There is also a similar system near the upper surface and a deeper-lying one. You will be unable to see these.

The **sense organs** are not well developed. Cells sensitive to touch are found all over the surface.

Each pigmented **eyespot** consists of a number of light-sensitive ocelli.

Water-Vascular System. The water-vascular system is found only in echinoderms, which use it for locomotion and, in the case of sea stars, for opening clam shells. If this system in your specimen has been injected with a colored injection mass, its features can be studied to greater advantage.

☞ Carefully remove the stomach from the central disc.

The **madreporite plate** on the aboral surface (Fig. 15-2) contains ciliated grooves and pores. From it a somewhat curved **stone canal** leads to a **ring canal** (Fig. 15-3), which is found around the outer edge of the peristomial membrane next to the skeletal region of the central disc. The ring canal may be difficult to find if not injected.

Five **radial** canals, one in each arm, radiate out from the ring canal, running along the apex of the ambulacral groove just below the ambulacral ossicles and above the radial nerve. The position of the radial canal is best seen in a cross section of one of the arms (Fig. 15-3). Short **lateral canals,** each with a valve, connect the radial canal with each of the tube feet. Now look on the inside of an arm and study the alternating arrangement of the ampullae. Note how each ampulla connects with a tube foot through a **pore** between the ambulacral plates.

Water can enter the madreporite plate, pass down the stone canal to the ring canal, from there to the radial canals, and finally through the lateral canals to the ampullae and the tube feet. During life, water brought through the madreporite plate fills this system. It is suggested that the madreporite plate allows rapid adjustment of hydrostatic pressure within the water-vascular system in response to changes in external hydrostatic pressure resulting from depth changes, as in tidal fluctuations.

Tube feet have longitudinal muscles; ampullae have circular muscles. When the tube feet are contracted, most of the water is held in the ampullae. When the ampullae contract, the water is forced into the elastic tube feet, which elongate because of the hydrostatic pressure within them. When the cuplike ends of the extended tube feet contact a hard surface, they attach with a suction force and then contract, pushing the water back into the ampullae and pulling the animal forward. Valves in the lateral canals prevent the backflow of water into the radial canals. Although a single foot is not very strong, hundreds of them working together can move the animal along slowly and can create a tremendous pull on the shell of a mussel. Suckers are of little use on a sandy surface, where the tube feet serve as tiny legs. Some species have no suckers but use the stiff podia like little legs to "walk." Sea stars can travel about 15 cm per minute.

DRAWING

On p. 257 complete the internal aboral view as you see it in your dissection. Label the ambulacral ridge, ampullae, gonads, pyloric and cardiac stomachs, pyloric duct, and pyloric cecum.

CROSS SECTION OF THE ARM OF A SEA STAR (MICROSLIDE)

Identify the **epidermis** covering the entire animal, the **dermis** containing the **ossicles,** the muscular tissue, the **peritoneum,** and the **coelomic cavity.** The spines have not yet erupted, for these are young sea stars. Observe the **dermal branchiae** projecting from the coelom through the body wall. The **pyloric ceca** hang from the aboral wall by **mesenteries.** Notice the ampullae in the coelom and their connection to the **tube feet** and **lateral canals.** The canals may not always be seen. Why? Locate the **ambulacral ossicles** with the pores through which tube feet extend. Find the **radial canal,** a small tube under the **ambulacral groove,** and the **radial nerve** beneath the radial canal (Fig. 15-3).

PROJECTS AND DEMONSTRATIONS

1. Various types of sea stars. Examine several types of sea stars.

2. Microslides of asteroid larval forms. Examine some larval of asteroids such as the bipinnaria, brachiolaria, and young sea star.

3. Microslides of sea star spermatozoa smears and sections of ovary showing mature and developing eggs.

4. Microslides of pedicellariae.

5. Sea stars with regenerating rays.

6. Demonstration of stepping action of tube feet. Fasten a living sea star, oral side up, on a piece of plate glass by means of rubber bands. Lay a small piece of celluloid or thick polyethylene film on top of the tube feet. Note the action of the tube feet and the direction in which the celluloid moves.

7. *Phagocytic action of coelomic cells*. Inject into the coelom of a live sea star 5 ml of carmine suspension in seawater. Set the animal aside for about 8 hours; then examine under a dissecting microscope for the appearance of circulating particles in the skin gills. Pinch off some of the gills and examine under a compound microscope. Some of the carmine particles will have been picked up by phagocytic cells of the coelom, and these may be seen migrating through the thin walls of the gills. Such cells appear to have an excretory function. Examine drops of the coelomic fluid for the presence of other coelomic cells.

8. *Echinoderm ossicles and pedicellariae*. Put left-over skeletal parts (or whole specimens, if desired) into 5% to 10% potassium hydroxide solution, or into full-strength commercial bleach. Warm over a Bunsen burner or boiling water to dissolve away the flesh. Let stand several hours; then decant carefully. Add fresh water, let stand again, and decant. The very tiny skeletal fragments settle slowly. Repeat until free of potassium hydroxide and debris. Cover with alcohol and again wash and decant, using two or three changes of alcohol. Now add 90% alcohol, shake, and put a drop on a clean slide. Quickly ignite the alcohol with a match. Add a drop or two of Canada balsam and cover.

SEA STAR RAY AND CENTRAL DISC

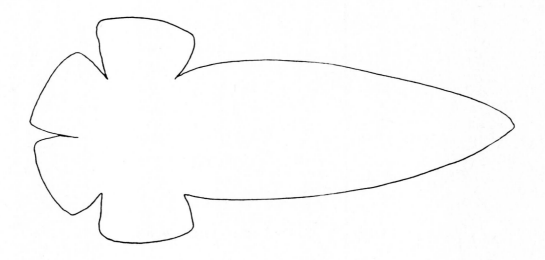

Internal structure, aboral view

Pedicellariae of sea star   Ambulacral ossicles of sea star

NOTES ON BRITTLE STAR BEHAVIOR

EXERCISE 15B
Subclass Ophiuroidea—the brittle stars

MATERIALS
Preserved brittle stars
Dissecting microscopes
Living brittle stars, if available
Fresh seafood for feeding, if living forms are used

BRITTLE STARS
Phylum Echinodermata
  Subphylum Asterozoa
    Class Stelleroidea
      Subclass Ophiuroidea
        Order Ophiurida
          Available genera

Where Found
Brittle stars are widely distributed in all oceans and at nearly all depths. Many are shallow-water forms, hiding in mud or seaweed or attached to other animals. Common forms along North American shores are *Ophiothrix*, *Ophioderma*, *Ophiopholis*, and *Ophioplocus*.

External Features
General Body Form. Note that the arms of brittle stars are sharply marked off from the central disc—a characteristic of ophiuroids (Fig. 15-5). The arms are more flexible than those of the sea stars. Does their appearance give a clue as to why? In both appearance and function the arms resemble vertebral columns. In fact, internally they consist of a series of calcareous vertebral **ossicles,** each joined to the next by two pairs of muscles. Externally the arm is encased in a series of aboral, lateral, and oral **plates.**

☞ Pull an arm off a preserved specimen and examine it in cross section. Locate the muscles and note the vertebral articulations.

Are there spines on the arms? How do the spines compare with those of asteroids? Do you find any pedicellariae or skin gills?

Oral Surface. On the oral side note the five triangular **jaws** around the mouth. Find the five **oral shields** (also called **buccal shields**), which are oval plates located on the interradial area between the rays (Fig. 15-6). One of these is slightly modified as a **madreporite plate,** and its tiny pores connect with a madreporic canal inside. Compare the location of the madreporic plate in asteroids and ophiuroids.

Distal to the oral shields and close to each arm is a pair of grooves, representing the openings of the **bursae.** (In *Ophioderma* a second pair is located distal to the first.) The bursae are 10 saclike cavities within the disc peculiar to the ophiuroids. Water is pumped in and out of them for respiratory purposes, and the gonads discharge their products into them. In some species they also serve as brood pouches, but in *Ophioderma* development is external.

The tube feet are often called tentacles in ophiuroids. They are small, do not have suckers, and project between the skeletal plates. They are largely sensory in function. Does the brittle star seem to have true ambulacral plates? Examine the rough spines used in gripping the substrate. How do they compare with those of asteroids?

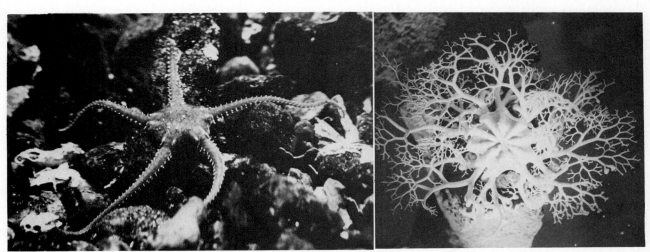

A   B

Fig. 15-5. A, The brittle star *Ophiura albida*. **B,** Basket star, *Gorgonocephalus*. Ophiuroids have sharply marked-off arms and are very agile. They are bottom feeders. (**A,** Photograph by B. Tallmark. **B,** Courtesy Vancouver Public Aquarium, Vancouver, B.C.)

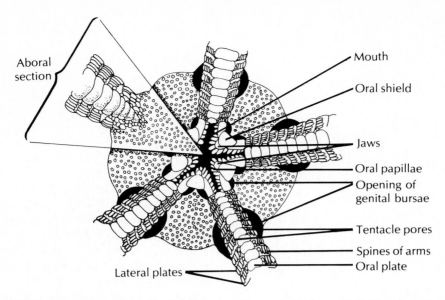

Fig. 15-6. Semidiagrammatic oral view of the brittle star *Ophioderma* with aboral section shown at upper left.

Behavior

Locomotion. How many arms are used in locomotion? Watch closely. Do they work in pairs? Does one arm always lead or follow? Is it always the same arm? Are the podia used in locomotion? How does the brittle star right itself when turned over?

Feeding. Ophiuroids generally combine detritus feeding with carnivorous feeding. The arms sweep over the substrate and small particles are passed along by the podia to the mouth, with possibly some ciliary help. Mucus is not used. Drop small bits of fish into an aquarium near (but not touching) some brittle stars to see their method of carnivorous feeding. Do the animals seem to sense the food before they touch it? How do they "capture" the food and move it to the mouth? The digestive system is restricted to a stomach in the central disc; there are no hepatic ceca, intestine, or anus.

Reactions to Other Stimuli. Avoid rough handling of the brittle star because this may cause it to "freeze" and become immobile or even to throw off one or more of its arms.

Note how the animal reacts to mechanical stimulation. Touch the tip of an arm. Does it retreat or advance toward the source of the stimulus? Touch a more proximal part of the arm. Is the reaction the same or different? Stimulate the base of an arm or the central disc. What happens?

Can you determine whether ophiuroids—at least the species you are studying—respond positively or negatively to light?

The water-vascular, hemal, and nervous systems are on a plan similar to that of the asteroids.

Record your observations on p. 258.

EXERCISE 15C
Class Echinoidea—the sea urchin

The sea urchins, although armless, are pentamerous. Typically they have a more or less globular test of interlocking plates; anus and genital pores located on the aboral side; five meridional ambulacral rows with tube feet; a typical water-vascular system; movable spines; and a complex five-toothed chewing mechanism (Aristotle's lantern). The class also includes the heart urchins, sea biscuits, and the more flattened sand dollars.

MATERIALS

Living sea urchins in seawater*
Preserved sea urchins
Dried or preserved sand dollars and/or sea biscuits
Dried tests of sea urchins (p. 264, number 6)
Glass plates
Large finger bowls
Carmine suspension
Dissecting pans and tools

---

*If living sand dollars are also available, see study suggestions on p. 264.

ARBACIA

Phylum Echinodermata
  Subphylum Echinozoa
    Class Echinoidea
      Genus *Arbacia*
        Species *A. punctulata*

Where Found

Sea urchins are benthic marine animals common along rocky coastlines, where they may be found in crevices, often in tide pools. Sometimes they may be found in sandy regions and coral reefs. They tend to congregate. *Arbacia*, the purple sea urchin, is found from Cape Cod to Florida and Cuba. Various species of *Strongylocentrotus* and *Lytechinus* (Fig. 15-7) are common along both U.S. coasts. Sea urchins are called "regular urchins," in contrast to the more bilaterally arranged "irregular urchins" such as sea biscuits, sand dollars, and heart urchins. (Fig. 15-8).

Fig. 15-7. The sea urchin *Lytechinus*. Note the slender, suckered tube feet. They often attach to bits of shell, seaweed, and so forth to themselves for camouflage. Stalked pedicellariae can be seen between the spines (Photograph by R.O. Hermes.)

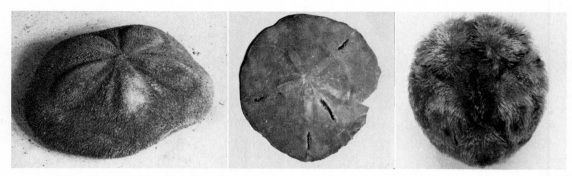

A   B   C

Fig. 15-8. Some "irregular" echinoids. **A,** Sea biscuit. **B,** Sand dollar. **C,** Heart urchin. They differ from "regular" urchins in having become secondarily bilateral, having very short spines, and moving chiefly by their spines instead of tube feet. (Photographs by F.M. Hickman.)

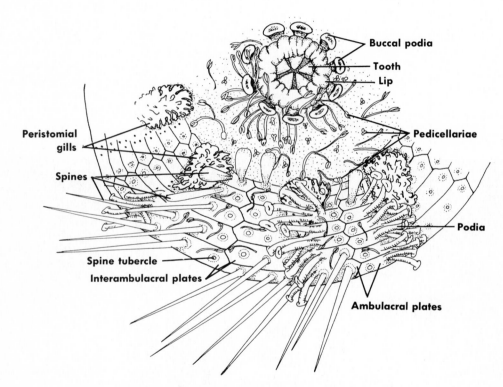

Fig. 15-9. External structure of the sea urchin, oral surface.

Behavior and External Structure

The external features of the sea urchin are best observed in a living animal. However, if living forms are not available, submerge a preserved specimen in a bowl of water and use the following account, directed toward live urchins, to identify the external structures.

☞ Place a living sea urchin on a glass plate, submerge in a bowl of seawater, and observe under a dissecting microscope.

Spines. Examine the long movable **spines** attached at a ball-and-socket joint by two sets of ring muscles.

Remove a spine and note that its socket fits over a rounded **tubercle** on the test (Fig. 15-9). An inner ring of **cog muscles** holds the spine erect. Hold the tip of a spine and try to move it. Do you feel the locking mechanism of the cog muscles? Of what advantage is such a locking mechanism? Now with a probe touch the epidermis near a spine. Does the spine move? In which direction? The outer ring of muscles is responsible for directional movement. Are all the spines the same length? Are they all pointed on the distal ends?

Tube Feet. Notice that the tube feet all originate

from rows of perforations in the five **ambulacral regions.** The podia can be extended beyond the ends of the spines. Do any of them possess suckers? Are any of them suckerless? Do they move? What happens when you touch one? When you jar the bowl?

Mouth and Peristome. Examine the oral side of the urchin and find the **mouth** with its five converging **teeth** and its collarlike **lip** (Fig. 15-9). The teeth are part of a complex chewing mechanism called the **Aristotle's lantern,** which is operated internally by several sets of muscles. The lip contains circular, or "purse-string," muscles.

The membranous **peristome** surrounding the mouth is perforated by five pairs of large oral tube feet called **buccal podia.** Do these podia have suckers? They are probably sensitive to chemical stimuli. The peristome also bears some small spines.

Pedicellariae. Notice on the peristome a number of three-jawed **pedicellariae** on the ends of long, slender stalks. Smaller but more active pedicellariae are located among the spines. Stimulate some of them by touching gently with the tip of a teasing needle. You may want to pinch off some of the pedicellariae to examine more closely on a slide, particularly if you are using a preserved specimen. Their function is to discourage intruders and to help keep the skin clean.

Locomotion. Note how the urchin uses its spines and tube feet in locomotion. Carefully, so as not to injure its tube feet, turn the urchin over (oral side up) and see whether it uses its spines or tube feet in righting itself. Notice which ambulacra turned first and mark that row by removing some of its spines or by marking some of the spines with thread; then turn the animal over again and see if the same ambulacra turn first. Repeat once more.

Tilt the glass plate to determine if the urchin moves up or down. Tilt in the opposite direction and see what happens. Is it positively or negatively geotactic?

Some sea urchins are adapted for burrowing into rock or other hard material by using both their spines and their chewing mechanisms. *Strongylocentrotus purpuratus,* common on the North American Pacific coast, excavates cup-shaped depressions in stone.

Most echinoids have tiny modified spines called **sphaeridia,** believed to be organs of equilibrium. In *Arbacia* these are minute glassy bodies located one in each ambulacrum close to the peristome. Try to find and remove the sphaeridia to see whether their removal affects the urchin's righting reaction or its geotactic responses.

Direct a beam of bright light toward an urchin. How does it react?

Epidermis. The test, podia, pedicellariae, and spines are covered with **ciliated epidermis,** although the epidermis may have become worn off from the exposed spines. Drop a little carmine suspension on

various parts of the sea urchin and find the direction of the ciliary currents. Of what advantage are such currents to the urchin?

Gills. At the outer edge of the peristome, between the ambulacra, find five pairs of branching peristomial **gills,** which open into the coelomic cavity.

Record your observations on sea urchin behavior on p. 266.

Test

☞ Examine a dried sea urchin test from which the spines have been removed and also a dried sand dollar and/or sea biscuit.

The test, or endoskeleton, is composed of calcareous plates (ossicles) that are symmetrically arranged and interlocked or fused so as to be immovable. Note the tubercles to which the spines were attached. Note the arrangement of the plates into ten meridional double columns—five double rows of **ambulacral plates** alternating with double rows of interambulacral plates. What is the function of the perforations in the ambulacral plates? In asteroids the tube feet were extended between the plates rather than through them.

Examine the test of a sand dollar or a sea biscuit, or both. Can you find perforations in them similar to those of the urchin? Do they have ambulacra, and is their arrangement pentamerous? The ambulacra in echinoids are homologous to the ambulacra of asteroids and ophiuroids.

On the aboral surface, note the area that is free from spines, the **periproct.** The **anus** is centrally located, surrounded by four (sometimes five) valvelike **anal plates** (Fig. 15-10). Around the anal plates are five **genital plates,** so-called because each bears a **genital pore.** Note that one of the genital plates is larger than the others and has many minute pores. This is the **madreporite plate,** which has the same function in sea urchins as it does in sea stars, for the echinoids have a **water-vascular system** in common with all echinoderms.

The test grows both by the growth of plates and by the production of new plates in the ambulacral area near the periproct.

WRITTEN REPORT

Make a table of the comparative characteristics of the external anatomy of the sea star, brittle star, and sea urchin on p. 265. Include such items as shape, integument, ambulacra, tube feet, skin gills, pedicellariae, skeleton, and so on. After you have observed the sea cucumber, add that to the table.

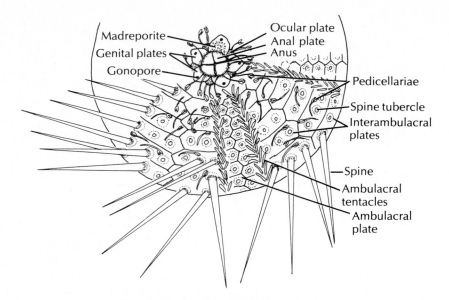

Madreporite
Genital plates
Gonopore
Ocular plate
Anal plate
Anus
Pedicellariae
Spine tubercle
Interambulacral plates
Spine
Ambulacral tentacles
Ambulacral plate

Fig. 15-10. Aboral surface of the sea urchin.

PROJECTS AND DEMONSTRATIONS

1. Various echinoids. Examine such forms as sand dollars, sea biscuits, heart urchins, and various genera of sea urchins, such as the club urchin, rock urchin, or long-spined urchin.

2. Demonstration of Aristotle's lantern, both preserved and dried.

3. Microslides of sea urchin pedicellariae.

4. Microslides of pluteus larvae.

5. Demonstration of the internal organs of the sea urchin in situ. To view the internal organs of the sea urchin, treat the aboral half of a preserved specimen by soaking it in 2% nitric acid for 24 to 48 hours. With the specimen completely submerged in water, use fine scissors to cut a circle outside the periproct; then make five meridional cuts through the ambulacral areas. The untreated portion of the test holds the animal together. Keep submerged for study.

6. Preparation of dried tests. Dried tests may be prepared by placing the animals in fresh commercial bleach (sodium hydrochlorite). Use a toothbrush to complete the removal of organic matter.

7. A study of living sand dollars. If live sand dollars, such as the east coast *Mellita*, the Caribbean *Leodia*, or the west coast *Dendraster*, are available, they should be kept in sandy-bottomed containers. Place one in a bowl of seawater and examine with a hand lens. Examine the spines on the aboral surface. Are they movable? Note the petal-shaped ambulacra; they are called petaloids. The podia are adapted for gas exchange rather than locomotion. Look at the flattened oral surface. Are the spines movable? How do they differ from those of *Arbacia?* Note the central mouth and the five-toothed chewing apparatus. Return the animal to the sandy bottom and observe how it burrows under the sand, using its oral spines to move the sand. Sand dollars feed on minute organic particles from the sand. They are passed back by tiny club-shaped aboral spines, caught in mucus, and carried by ciliary currents to food grooves on the oral side that lead to the mouth.

Phylum _____   Name _____

Class _____   Date _____ Section _____

Genus _____

COMPARATIVE TABLE OF ECHINODERM CHARACTERISTICS

Characteristic	Sea star	Brittle star	Sea urchin	Sea cucumber

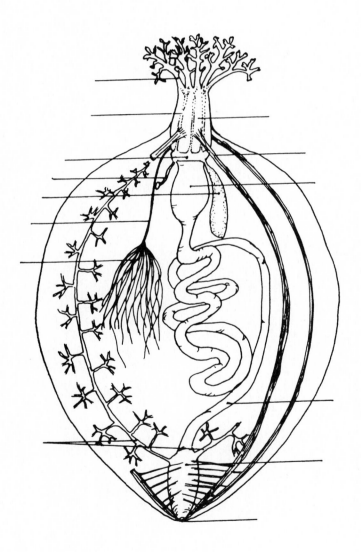

Internal structure of the sea cucumber (to be labeled by the student).

EXERCISE 15D
Class Holothuroidea—the sea cucumber

MATERIALS

Living sea cucumbers in aquarium or in bowls of seawater
Preserved or relaxed sea cucumbers for dissection*
Prepared slides of holothurian ossicles

SEA CUCUMBER

Phylum Echinodermata
  Subphylum Echinozoa
    Class Holothuroidea
      Order Dendrochirotida
        Genus *Cucumaria* (or *Thyone*)
          or
      Order Aspidochirotida
        Genus *Parastichopus*

Where Found

The sea cucumbers are benthic and generally sluggish animals common in intertidal muddy or sandy bottoms, usually where there is decaying marine vegetation. The two common genera of the east coast of the United States are *Thyone* and *Cucumaria* (Fig. 15-11); *Parastichopus* is a familiar genus on the West Coast.

---

*Sea cucumbers can be relaxed for dissection by injecting the living animals with 5 to 10 ml of 10% magnesium chloride an hour before use.

Fig. 15-11. The sea cucumber *Cucumaria* with tentacles extended for feeding. This is a suspension feeder. Minute plankton organisms adhere to the tentacles, which when loaded bend over and wipe off the food into the pharynx. (Photograph by T. Lundalv.)

Behavior and External Structure

☞ If possible, study living specimens in an aquarium or in a bowl of seawater containing a generous layer of sand, and then examine a preserved specimen. Sea cucumbers are slow to react. They should be left undisturbed for some time before the laboratory period if you are to see them relaxed and feeding.

Note that the holothurian, unlike the other echinoderms, is orally-aborally elongated and has a cylindric body with the **mouth** encircled by **tentacles** at one end and the **anus** at the other.

The more detailed description and the behavior of the animal will depend somewhat on the species you happen to be observing. Notice the tentacles. Are they branched and extensible (as in *Cucumaria* and *Thyone*), or short and shield-shaped (as in *Parastichopus*), or of some other type? The tentacles, which are modified tube feet, are hollow and a part of the **water-vascular system;** they are connected internally with the radial canals. The type of tentacle structure is related to feeding habits. *Cucumaria* and *Thyone* are plankton feeders that stretch their mucus-covered tentacles into the water or over the substrate till they are covered with tiny food organisms, and then they thrust the tentacles into the mouth, one by one, to lick off the food. Can you observe these actions? *Parastichopus* and some others simply shovel the mud and sand into the mouth, digest out organic particles, and void the remainder.

Have you noticed any rhythmic opening and closing of the anus? This is a respiratory movement coordinated with the pumping action of the cloaca, which pumps water into and out of the **respiratory trees** (internal respiratory organs).

Does the animal try to burrow into the sand? Does it cover itself completely, or leave the ends exposed? *Thyone* may take 2 to 4 hours to bury its middle by alternate circular and longitudinal muscle contractions. It is likely to be more active in the late afternoon and night than in the morning. If you are watching *Parastichopus* move, what does its muscular action remind you of? Are there waves of contraction?

Does the sea cucumber react to mechanical stimulus? Try touching a tentacle. Does one or more than one tentacle react? Touch several tentacles. What happens when you stroke the body or gently pick the animal up? Did it expel water when you picked it up? Having observed its movement and reactions, can you see the advantages of the hydraulic skeleton? What other phyla used the hydraulic skeleton to advantage?

Note the **podia.** Are they scattered all over the

body or arranged in ambulacral rows? This pattern differs among different species. Are the podia all alike, that is, are ventral podia any different from dorsal podia? Are any of them suckered? If the pentamerous arrangement of ambulacra is in evidence in your specimen, how many rows make up the ventral **sole?** How many are on the dorsal surface? If you place the animal on a solid surface, do the podia attach themselves? Can the animal right itself if turned over? Are the podia involved in the righting action? Are muscles involved?

Does the animal show any geotactic reaction if placed on a vertical or sloping surface? Some burrowing forms are positively geotactic and move downward; other species are negatively geotactic and climb upward. *Thyone* gives no geotactic response.

Do you find any pedicellariae or skin gills? Do you feel the presence of a test under the epidermis? That is because the skeleton of the holothurian is usually limited to tiny ossicles embedded in the tough, leathery body wall.

Internal Structure

 When the sea cucumber is held by both ends, the upper concave surface is the dorsal side, and the lower convex surface is the ventral side. Locate five longitudinal ridges caused by muscle bands in the radii. There are two dorsal radii and three ventral radii. With *scissors*, open the body longitudinally along the *ventral* side just below the right ventral radius. Pin down the walls and cover with water.

Digestive System. Note the large **coelomic cavity.** Just behind the mouth is the **pharynx** supported by a ring of calcareous plates. It is followed by a short muscular **stomach** and a long convoluted **intestine,** held in place by mesenteries and expanding somewhat at the end to form a **cloaca,** which empties at the **anus.**

Respiratory System. Two branched **respiratory trees** are attached to the cloaca; they serve as both respiratory and excretory organs. They are aerated by a rhythmic pumping of the cloaca. Several inspirations 1 minute or more apart are followed by a vigorous expiration that expels all the water.

Note the muscles between the cloaca and the body wall. Long **retractile muscles** (how many?) run from the pharynx to join the longitudinal **muscle bands** in the body wall.

Water-Vascular System. A **ring canal** surrounds the pharynx. One or more rounded or elongated sacs called **polian vesicles** hang from the ring canal into the coelom and open into the ring canal by a narrow neck. Polian vesicles are believed to function as expansion chambers in maintaining pressure within the water-vascular system. One or more **stone canals** also open into the ring canal from the body cavity. In adult sea cucumbers the water-vascular system has usually lost contact with the seawater outside. Coelomic fluid rather than seawater enters and leaves the system.

Five **radial canals** extend from the ring canal forward along the walls of the pharynx to give off branches to the tentacles, which are actually modified tube feet. From there the radial canals run back along the inner surface of the ambulacra, where each gives off **lateral canals** to the **podia** and **ampullae.** Valves in the lateral canals prevent backflow. Note the ampullae along the ambulacra in the inner body wall.

Reproductive System. The **gonad** consists of numerous tubules united into one or two tufts on the side of the dorsal mesentery. These become quite large at sexual maturity. A **gonoduct** passes anteriorly in the mesentery to the **genital pore.** The sexes are separate, and fertilization is external.

Endoskeleton. The **endoskeleton** consists largely of tiny calcareous ossicles scattered about in the dermis. These can be seen on a prepared slide. A method for making such slides is given below.

DRAWING

Label the diagram of the internal structure of the sea cucumber on p. 266.

WRITTEN REPORT

Complete the summary, using the table on p. 265, of the likenesses and differences between the external anatomy of the sea star, brittle star, sea urchin, and sea cucumber, mentioning such things as shape, ambulacra, tube feet, skin gills, skeleton, integument, and so on.

PROJECTS AND DEMONSTRATIONS

1. Various types of sea cucumbers. Some of the transparent burrowing sea cucumbers are particularly interesting.

2. Microslides of sea cucumber ossicles.

3. Microslides of holothurian larval forms. Examine such forms as the auricularia and doliolaria larvae.

4. Preparation of holothurian ossicles. Place the body wall and tentacles in a 20% potassium hydroxide solution and let stand until all the flesh has dissolved. Or hasten the process by boiling the solution. Then let the solution stand for several hours until a sediment of whitish particles settles to the bottom. Pour off the supernatant fluid carefully and add clean water. Allow to settle again and change the water. Let settle again and then pipette some of the sediment to a slide and examine. The preparation may be preserved by two changes of 90% alcohol, absolute alcohol, and finally xylol. Pipette into a drop of Canadian balsam on a slide and add a coverslip. The same method can be used for the ossicles and pedicellariae of other echinoderms.

5. *Evisceration in the sea cucumber*. Some sea cucumbers are able, as a defense mechanism, to eviscerate most of their internal organs and then to re-generate them. Either *Thyone* or *Parastichopus* can be induced to eviscerate by submerging in 0.1% NH_4OH in seawater for a minute or so. If spontaneous contractions do not occur at once, hold the animal up for a few seconds by the posterior end. Some regeneration should be evident within 2 weeks, but it may take months for complete regeneration.

EXERCISE 15E
Class Crinoidea—the feather stars and sea lilies

MATERIALS
Antedon—preserved or plastic mounted

FEATHER STAR
Phylum Echinodermata
  Subphylum Crinozoa
    Class Crinoidea
      Order Articulata
        Genus *Antedon*

Where Found
Crinoids are found in all seas except the Baltic and Black seas. Most are subtidal, although some are littoral, and some are found as deep as 5000 m. The free-living feather stars (Fig. 15-12) prefer rocky bottoms and are most abundant in shallow tropical lagoons. The sea lilies (about one-eighth of the species) are stalked and sessile and prefer muddy sea bottoms and deep waters.

General Structure
The feather star passes through a stalked stage but is free moving as an adult (Fig. 15-12, *A*). It is made up of the following three general regions: (1) 10 long, jointed **arms**, bearing jointed **pinnules**, (2) the **calyx**, containing the digestive and other organs, and (3) a

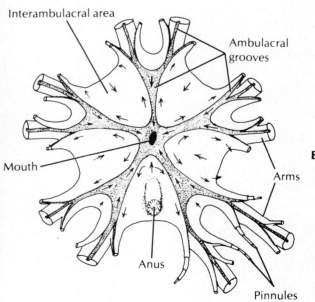

Fig. 15-12. A, The feather star *Florometra*, a crinoid. Feather stars hold on to objects with the adhesive ends of pinnules and swim by raising and lowering alternate sets of arms. **B,** Oral view of the calyx of *Antedon*, showing direction of ciliary food movements. (**A** courtesy Vancouver Public Aquarium, Vancouver, B.C.)

circlet of short-jointed **cirri** on which the animal can rest or move about. The calyx and arms together are called the **crown.** The leathery skin covering the calyx is called the **tegmen.**

Ambulacral grooves radiate out from the central mouth to the anus and pinnules, so that each arm and pinnule has a ciliated food groove. Tube feet are located along the food grooves in the pinnules.

Crinoids are suspension feeders, using a mucus-ciliary method of capture and food movement. Small organisms are caught in mucous nets and moved along the ambulacral grooves to the mouth (Fig. 15-12, *B*).

The pinnules can secrete a narcotizing toxin for quieting the prey.

Feather stars can crawl by holding onto objects with the adhesive ends of their pinnules and pulling themselves along by bending their arms. They can also swim by raising and lowering alternate sets of arms.

DEMONSTRATIONS
1. *Various fossil crinoids.*
2. *Preserved sea lilies.*

CHAPTER 16

PHYLUM CHORDATA

THE PROTOCHORDATES
Subphylum Urochordata
Subphylum Cephalochordata

THE CHORDATES*

The chordates show a remarkable diversity of form and function, ranging from the simple protochordates to humans. Although all organ systems are well developed in this group, the nervous system has been chiefly responsible for giving the phylum its eminence among animals. Most members of this phylum belong to the vertebrates, but chordates actually include a few invertebrate groups. All animals that belong to the phylum Chordata must have at some time in their life cycle the following characteristics:

1. Notochord. The notochord (Fig. 16-1) is a slender rod of cartilage-like connective tissue lying near the dorsal side and extending most of the length of the animal. It is regarded as an early endoskelton and has the functions of such. In higher vertebrates it is found only in the embryo.

2. Pharyngeal Gill Slits. The pharyngeal gill slits (Fig. 16-3) are a series of paired slits in the pharynx, serving as passageways for water to the gills. In higher vertebrates they appear only in the embryonic stages.

3. Dorsal Tubular Nerve Cord. The dorsal tubular nerve cord, with its modification, the brain, forms the central nervous system. It lies dorsal to the alimentary tract and is hollow, in contrast to the invertebrate nerve cord, which is ventral and solid.

4. Postanal Tail. The postanal tail projects beyond

---

*See Hickman, C.P., Jr., L.S. Roberts, and F.M. Hickman. 1984. Integrated principles of zoology, ed. 7. St. Louis, The C.V. Mosby Co., pp. 496-516.

the anus at some stage and serves as a means of propulsion in water. It may or may not persist in the adult. Along with the body muscles and stiffened notochord, it provides motility for a free-swimming existence.

These features vary in chordates. Some lower forms have all of these structures throughout life. In many higher forms the gill slits never break through from the pharynx but merely form pouches that have no function; the notochord is replaced by the vertebral column, and only the dorsal nerve cord actually persists in the adult as a diagnostic chordate character. The **protochordates** are primitive forms, but they demonstrate the chief chordate characteristics in simple form.

The phylum is divided as follows:

PHYLUM CHORDATA

Group Protochordata (pro'to-kor-da'ta). (**Acrania**). Chordates with no cranium or vertebral column.

    Subphylum Urochordata (u'ro-kor-da'ta) (**Tunicata—tunicates.** Only larval forms have all chordate characteristics; adults sessile, without notochord and dorsal nerve cord; body enclosed in tunic. Example: *Molgula*, a sea squirt.

    Subphylum Cephalochordata (sef'a-lo-kor-da'ta)—**lancelet.** Notochord and nerve cord persist throughout life; lance-shaped. Example: *Branchiostoma* (amphioxus).

Group Craniata. Animals with a cranium and vertebral column, that is, the vertebrates.

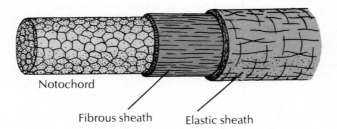

Notochord

Fibrous sheath   Elastic sheath

Fig. 16-1. Structure of a notochord and its surrounding sheaths. The cells of the notochord proper are thick walled, pressed together closely, and filled with semifluid. The stiffness is caused mainly by the turgidity of fluid-filled cells and the surrounding tissue sheaths.

Subphylum Vertebrata (ver'te-bra'ta). Enlarged brain enclosed in cranium; nerve cord surrounded by bony or cartilaginous vertebrae; notochord in all embryonic stages and persists in adults of some fishes; typical structures include two pairs of appendages and body plan of head, trunk, and postanal tail.

Superclass Agnatha (ag'na-tha) (**Cyclostomata**). No jaws or ventral fins; notochord persistent.

Class Cephalaspidomorphi (sef-a-lass'pe-do-morph'e) (**Petromyzontes**)—lampreys.

Class Myxini (mik-sy'ny)—hagfishes.

Superclass Gnathostomata (na'tho-sto'ma-ta). Jaws present; usually paired limbs; notochord persistent or replaced by vertebral centra.

Class Chondrichthyes (kon-drik'thee-eez)—sharks, skates, rays, chimaeras.

Class Osteichthyes (os'te-ik'thee-eez). The bony fishes.

Class Amphibia (am-fib'e-a)—amphibians. The frogs, toads, and salamanders.

Class Reptilia (rep-til'-a)—reptiles. The snakes, lizards, turtles, crocodiles, and others.

Class Aves (ay'veez)—birds.

Class Mammalia (ma-may'lee-a)—mammals.

EXERCISE 16A
Subphylum Urochordata*—*Molgula,* an ascidian

MATERIALS
Living (or preserved) *Molgula,* or other small, translucent ascidian, such as *Perophora*
Whole mounts of ascidian larvae
Carmine suspension in seawater
Finger bowls
Compound and dissecting microscopes

MOLGULA
Phylum Chordata
  Subphylum Urochordata
    Class Ascidiacea (sea squirts)
      Order Pleurogona
        Genus *Molgula*

The urochordates are commonly called tunicates because of their leathery covering or tunic. They are divided into three classes: Ascidiacea, the sea squirts; Thaliacea, the salpians; and Larvacea, the appendicularians. The largest group is the ascidians, which are also the most generalized. They are called sea squirts because of their habit, when handled, of squirting water from the atrial siphon. Any of the small translucent ascidians may be used for this exercise. Adult tunicates are all sessile, whereas the larvae undergo a brief free-swimming existence.

Where Found
Tunicates are found in all seas and at all depths. *Molgula manhattensis* is a common Atlantic coast ascidian found in shallow water on wharf pilings, on anchored and submerged objects, and on eelgrass. It grows to 20 to 25 mm in its largest dimension.

*See Hickman, C.P., Jr., L.S. Roberts, and F.M. Hickman. 1984. Integrated principles of zoology, ed. 7. St. Louis, The C.V. Mosby Co., pp. 509-511.

External Features and Behavior
Sea squirts are fairly hardy in the marine aquarium.

☞ Examine, in a fingerbowl of seawater, a living solitary tunicate such as *Molgula* or *Ciona,* or a portion of a colony of *Perophora* or other ascidians as available.

Observe the use of the two openings, or **siphons.** When fully submerged and undisturbed the siphons are open and respiratory water, kept moving by ciliary action, enters the more terminal siphon (called the **incurrent,** or **oral siphon**) at the mouth, circulates through a large pharynx, and leaves through the **excurrent siphon** on one side (the dorsal side) (Figs. 16-2 and 16-3).

☞ Release a little carmine suspension near the animal to verify this.

The outer covering of the tunicate is called the **test,** or **tunic** (Fig. 16-3). It is nonliving, is secreted by the **mantle,** which lies underneath it, and is chemically similar to cellulose—a rare substance in animals. The mantle contains the muscle fibers by which the body can contract. If the test is translucent enough and the light is properly adjusted, you may be able to see some of the internal structure.

Neuromuscular System. The tunicate nervous system is reduced and not well understood. There is a cerebral ganglion (closely associated with a subneural gland of uncertain function) located between the siphons. The test probably has no sensory nerves, but pressure on the test may be transmitted to nerves in the mantle. Both direct and crossed reflexes have been observed in some ascidians and can be tested in a living *Molgula* or other tunicate by touching selected areas with the tip of a glass, rod, or dissecting needle.

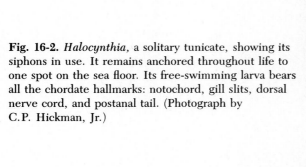

Fig. 16-2. *Halocynthia,* a solitary tunicate, showing its siphons in use. It remains anchored throughout life to one spot on the sea floor. Its free-swimming larva bears all the chordate hallmarks: notochord, gill slits, dorsal nerve cord, and postanal tail. (Photograph by C.P. Hickman, Jr.)

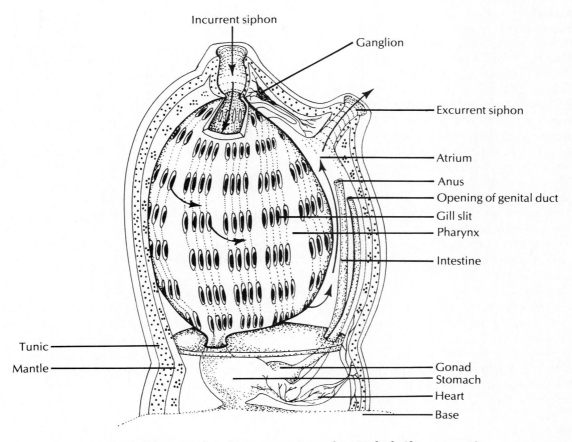

Fig. 16-3. Structure of a solitary sea squirt, such as *Molgula* (diagrammatic).

Direct reflexes result from the mechanical stimulation of the **outer** surface of the siphons or test.

☞ *Gently* stimulate various areas of the tunic and note the response. Touch gently the outer surface of one of the siphons, note the response, and then stimulate the other siphon.

What areas of the body are most sensitive? **Crossed reflexes** result from mechanical stimulation of the **inner** surface of the siphons.

☞ Gently touch the inner surface of the oral siphon. What happens? Try a stronger stimulus of the same siphon. Do you get the same response? How do these responses differ from those in which the outer surface was stimulated? Repeat with the atrial siphon.

Of what protective value would these reflexes be to the animal?

Internal Structure

☞ Internal structure can be observed on either a living or preserved specimen. If the test is too opaque for such study, use fine scissors to slit the test longitudinally in several places, using great care not to cut the mantle beneath it. Remove the test in strips; then return the animal to the bowl of seawater and study with a dissecting microscope.

Respiratory System. The **branchial sac,** or **pharynx,** is the largest internal structure, and the space between it and the mantle is the **atrium** (Fig. 16-3). The pharyngeal wall is perforated with many pharyngeal slits through which water passes into the atrium to be discharged through the excurrent siphon. The vascular wall of the pharynx serves as a gill for gaseous exchange.

Circulatory System. In a living specimen, with test removed and a light properly adjusted, you should be able to see the beating of the **heart** located near the posterior end on the right side. The tubular heart empties into two vessels, one at each end. Its peristaltic waves are of unusual interest, because they send the blood in one direction for awhile and then reverse direction and pump the blood in the opposite direction. Apparently there are two pacemakers that initiate contractions, one at each end of the heart, and they alternate in dominance over each other. The tunicate vascular system is an open type of system. Blood cells are numerous and colorful. There are no respiratory pigments.

The Digestive System. At the junction of the **mouth** and the **pharynx** there is a circlet of **tentacles,** which forms a grid that screens the incurrent water. By dropping a grain of sand into the incurrent siphon of a living tunicate, one may be able to observe the ejection reflex.

Inside the pharynx along the midventral wall is the **endostyle,** which is a ciliated groove that secretes a great deal of mucus. Cilia on the walls of the pharynx distribute the mucus. Food particles become tangled in the mucus and are propelled dorsally and then posteriorly to the esophagus and stomach.

☞ If living ascidians are available, examine one that has been submerged for some time in a suspension of carmine particles in seawater. Open the pharynx by cutting through the oral siphon and downward, a little to one side of the midventral line. Then cut around the base and lay the animal open in a pan of water. You should be able to see the concentration of carmine particles in the middorsal area. Cut out a small piece of the pharyngeal wall (free from the test and mantle) and mount on a slide to observe the gill slits and the beating cilia.

It will be difficult to differentiate the **esophagus, stomach,** and **intestine.** The **anus** empties into the atrium near the exhalent siphon.

Excretion. A ductless structure near the intestine is assumed to be a type of nephridium and to be excretory in function.

Reproduction. Most tunicates are hermaphroditic and bear a single ovary and testis, each with a gonoduct that opens into the atrium. *Molgula* and some others have paired gonads.

Some tunicates are colonial. In some the zooids are separate and attached by a stolon, as in *Perophora*. In some the zooids are regularly arranged and partly united at the base by a common tunic; all the oral siphons are at one side and the atrial siphons at the other. Another type has a system of zooids that share a common atrial chamber. Colonies are formed asexually by budding.

Solitary tunicates generally shed their eggs from the atrial siphon, and development occurs in the sea. Colonial species usually brood their eggs in the atrium, and the microscopic larvae leave by the atrial siphon. For a very brief period the larvae are nonfeeding and live a planktonic free-swimming existence; then they settle down, attach to the substrate, and metamorphose.

Ascidian Larvae (Study of Stained Slides)

Ascidian larvae are free swimming and are often called tadpole larvae because of their shape (Fig. 16-4). They do not look like the sessile adult sea squirts but are actually more characteristic of the chordates than are the adults. They possess not only gill slits but also a notochord and a dorsal tubular nerve cord, structures that have been lost in the adult. You may be able to identify some of the following structures on a stained slide.

Adhesive papillae in the anterior end of the larva

are used to attach to some object during metamorphosis, which occurs within a short time after hatching.

The **notochord** can be identified in the long **tail.** A **nerve cord** dorsal to the notochord enlarges anteriorly into a neural vesicle. Can you identify a pigmented photoreceptive **eyespot?** A smaller pigmented area anterior to the eye is an **otocyst,** which is an organ of equilibrium. At metamorphosis these portions of the nervous system degenerate, and a ganglion serves as the nerve center.

Look for the anterior **oral (incurrent) aperture** and the more posterior **atriopore (excurrent aperture).** Perhaps you can identify the **pharynx, gill slits, stomach, intestine, atrium,** and **endostyle.**

PROJECTS AND DEMONSTRATIONS

1. Variety of preserved tunicates, both single and colonial.

2. Microslides of various stages of metamorphosis of the ascidian tadpole.

3. Early development of the tunicate. Slit open the test of living *Molgula* or *Ciona* and extend the animals by cutting the superficial muscles. With a Pasteur pipette remove eggs from the oviduct (eggs are peach-colored when mature) to a watch glass of seawater. Put through several changes of seawater. From another individual remove sperm from the sperm duct and make a suspension of the sperm in seawater. Add a drop or two of sperm suspension to the eggs to impart a milky appearance, and let stand for 15 minutes. Now wash away the sperm. The tadpole larvae develop in about 24 hours.*

4. Obtaining living Amaroucium larvae. Brooding colonial tunicates usually release their larvae at dawn. In the laboratory healthy specimens of *Amaroucium* or *Botryllus* may be kept overnight in a dark room or dark container and exposed to the light 15 to 20 minutes before needed, at which time swarms of larvae should appear. Locate with a dissecting microscope.

Or, squeeze a portion of a colony over a container of seawater in order to force out the eggs and larvae. Locate live larvae with a dissecting microscope and transfer to clean seawater.

5. Preserved specimens of the hemichordate Saccoglossus. The Hemichordata were formerly included with the chordates but are now considered as a separate phylum. See p. 490 of *Integrated Principles of Zoology,* ed. 7, for a description.

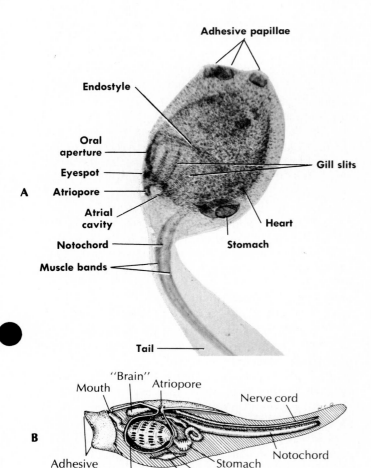

Fig. 16-4. A, Microphotograph of a tadpole larva of the ascidian *Distaplia.* **B,** Structure of an ascidian tadpole larva (diagrammatic). The ascidian larva is believed to resemble closely the ancestor of all vertebrates; it contains all four chordate characteristics. (**A,** Photograph by F.M. Hickman.)

*For more details see Costello, D.P., M.E. Davidson, A. Eggers, M.H. Fox, and C. Henley. 1957. Methods for obtaining and handling marine eggs and embryos. Woods Hole, Mass., The Marine Biological Laboratory.

EXERCISE 16B
Subphylum Cephalochordata—amphioxus*

MATERIALS
Preserved mature amphioxus
Slides
  Stained and cleared whole mounts of immature amphioxus
  Stained cross sections of amphioxus
Watch glasses
Microscopes

AMPHIOXUS
Phylum Chordata
  Subphylum Cephalochordata
    Genus *Branchiostoma* (= *Amphioxus*)
      Species *B. lanceolatus*

The little lancelet, *Branchiostoma*, commonly called amphioxus, illustrates the basic chordate structure in simple form and is considered similar to the primitive ancestor of the vertebrates. Besides the basic characteristics—**notochord, pharyngeal gill slits, dorsal hollow nerve cord,** and **postanal tail**—it also possesses the beginning of a **ventral heart,** and **metameric arrangement** of muscles and nerves. There are only two genera of Cephalochordates—*Asymmetron* and *Branchiostoma*.

Where Found
Branchiostoma is common along the southern California coasts and the southern Atlantic coasts of the United States, as well as the coasts of China and the Mediterranean Sea. It burrows, tail first, in the sand of shallow water, leaving the anterior end exposed.

---

*See Hickman, C.P., Jr., L.S. Roberts, and F.M. Hickman. 1984. Integrated principles of zoology, ed. 7. St. Louis, The C.V. Mosby Co., pp. 511-513.

External Structure
☞ Place a preserved mature specimen in a watch glass and cover with water. Do not dissect or mutilate the specimen.

How long is it? Why is it called the lancelet? Does it have a distinct head? Observe the **dorsal fin,** which broadens in the tail region (**caudal fin**) and continues around the end of the tail to become the **ventral fin.**

The anterior tip is the **rostrum** (Fig. 16-5). With the hand lens, find the opening of the **oral hood,** which is fringed by a number of slender oral tentacles, also called buccal cirri, that strain out large particles of sand and are sensory in function.

On the flattened ventral surface are two **metapleural folds** of skin extending like sled runners to the ventral fin. Anterior to the ventral fin, find the **atriopore.** It is the opening of the **atrium,** a large cavity surrounding the pharynx. The **anus** opens slightly to the left of the posterior end of the ventral fin.

In mature specimens little blocklike **gonads** (testes or ovaries) lie in the atrium anterior to the atriopore and just above the metapleural folds on each side. They can be seen through the thin body wall.

Study of the Whole Mount
☞ Examine with low power a stained and cleaned whole mount of an immature specimen.

The chevronlike **myotomes** along the sides of the animal are segmentally arranged muscles. Identify the various parts of the **fin** and note its skeletal support, the transparent **fin rays.** You may have to reduce the light to see the fin rays.

Beneath the rostrum is a large chamber called the **buccal cavity,** which is bounded laterally by fleshy curtainlike folds and is open ventrally. The rostrum

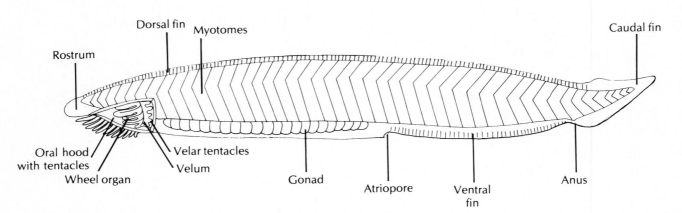

Fig. 16-5. External structure of amphioxus, with a section of the oral hood cut away to show wheel organ, velum, and velar tentacles.

and lateral folds together make up the **oral hood** (Fig. 16-5). The roof of the oral hood bears the notochord, which may have a supporting function in spreading the hood open. Each of the oral tentacles is stiffened by a skeletal rod of fibrous connective tissue. Behind the buccal cavity is an almost perpendicular membrane, the **velum,** pierced ventrally by a small opening, the true **mouth,** which is always open and leads into the **pharynx.** On the walls of the buccal cavity, projecting forward from the base of the velum, are several fingerlike ciliated patches that comprise the **wheel organ.** The rotating effect of its cilia helps maintain a current of water flowing into the mouth. Around the mouth, projecting posteriorly from the velum, are about a dozen delicate **velar tentacles,** also ciliated. Both the oral tentacles and the velar tentacles have chemoreceptor cells for monitoring the incurrent water.

The large **pharynx** narrows into a straight **intestine** extending to the **anus** (Fig. 16-6). The sidewalls of the pharynx are composed of a series of parallel oblique **gill bars,** between which are the **gill slits.** Just back of the pharynx is a diverticulum of the intestine called the **hepatic cecum** or liver, which extends forward on one side of the pharynx. Surrounding the pharynx is the **atrium,** a large cavity that extends back to the **atriopore.** Water entering the mouth filters through the gill slits into the atrium and then out the atriopore.

Cephalochordates, in common with sponges and clams, are filter feeders. They use a mucus-ciliary method of feeding on minute organisms. As the animal rests with its head out of the sand, the ciliated, tentacles, wheel organ, and gills draw in a steady current of food-laden water, from which the cirri and velar tentacles strain out large or unwanted particles. On the floor of the pharynx is an **endostyle** (Fig. 16-7) consisting of alternating rows of ciliated cells and mu-cus-secreting cells, and in the roof of the pharynx is a ciliated **hyperbranchial groove** (= epipharyngeal groove). Particles of food are entangled in the stream of mucus secreted by the endostyle and carried upward by cilia on the inner surface of the gill bars, then backward toward the **intestine** by cilia in the hyperbranchial groove. Digestion occurs in the intestine.

Oxygen–carbon dioxide exchange occurs in the epithelium covering the gill bars. The **notochord** just dorsal to the digestive system is transversely striated and is best seen in the head and tail regions. It provides skeletal support and a point of attachment for the muscles. Above and parallel to the notochord is the **dorsal nerve cord** (Fig. 16-6). How far does it extend? The row of black spots in the nerve cord are pigmented **photoreceptor cells.** Chemoreceptors are scattered over the body but are particularly abundant on the oral and velar tentacles. Touch receptors are located over the entire body.

DRAWING

Using the outline on p. 281, draw and label all the structures you can identify in the anterior two-thirds of an immature amphioxus as seen on the stained whole mount.

Circulatory System

Amphioxus does not have a heart; peristaltic contractions of the **ventral aorta** keep the colorless blood in motion, sending it forward and then upward through **afferent branchial arteries** (Fig. 16-6) to capillaries in the gill bars for gas exchange. Blood is carried from the gills by **efferent branchial arteries** up to a pair of **dorsal aortas.** These join back of the gills to form a **median dorsal aorta,** which gives off **segmented arteries** to the capillaries of the myotomes and to the capillaries in the wall of the intestines. From the intestinal wall,

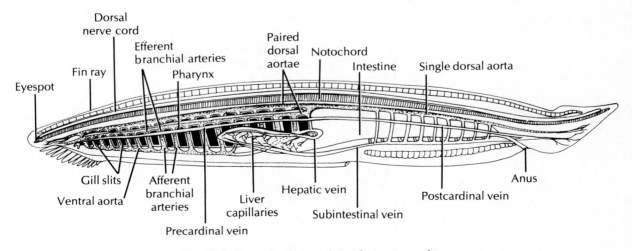

Fig. 16-6. General scheme of circulation in amphioxus.

blood now rich in digested food nutrients is picked up by the **subintestinal vein** and is carried forward to the **hepatic portal vein,** which enters the hepatic cecum. In the capillaries of the liver, nutrients are removed and stored in liver tissue or processed and returned to the blood as needed. Blood returns to the ventral aorta by way of the **hepatic vein.** Blood with waste products from the muscular walls returns by way of left and right **precardinal** and **postcardinal veins,** which empty into left and right **ducts of Cuvier** and then into the ventral aorta.

ORAL REPORT

Compare circulation in amphioxus with that in the earthworm and crayfish. Be able to trace a drop of blood to various parts of the body and back to the ventral aorta, and explain what the blood gains and loses in each of the capillary beds through which it passes on its journey. Locate as many of the main vessels as you can on any transverse sections you may have.

Cross Section—Stained Slide

Look at a cross section through the pharynx with the unaided eye and understand how the section is cut with reference to the whole animal. Note the **dorsal fin,** its supporting **fin ray,** and the ventral **metapleural** folds (Fig. 16-7). With the microscope, examine the **epidermis,** a single layer of columnar epithelial cells, and the **dermis,** a gelatinous connective tissue layer. The large **myotomes** or muscles are paired, but members of a pair are not opposite each other. The myotomes are separated by connective tissue **myosepta.**

The **nerve cord,** enclosed within the **neural canal,** has in its center a small **central canal,** which is prolonged dorsally into a slit. In some sections dorsal **sensory nerves** or ventral **motor nerves** may be seen. These are given off alternately from the cord to the myotomes. The large oval **notochord** with the vacuolated cells is surrounded by the **notochordal sheath.**

The cavity of the **pharynx** is bounded by a ring of triangular **gill bars** separated by **gill slits** that open into the surrounding **atrium.** From your study of the whole mount why do you think the cross section shows the gill bars as a succession of cut surfaces? The somewhat rigid gill bars contain blood vessels and are covered by ciliated **respiratory epithelium,** where the gaseous exchange is made. On the dorsal side of the pharynx, find the ciliated **hyperbranchial groove** and on the ventral side, the **endostyle.** The latter secretes mucus in which food particles are caught. What is the function of the cilia in the grooves?

The **gonads** (ovaries or testes) lie on each side of the atrial cavity. The reduced **coelom** consists of spaces, usually paired, on each side of the notochord

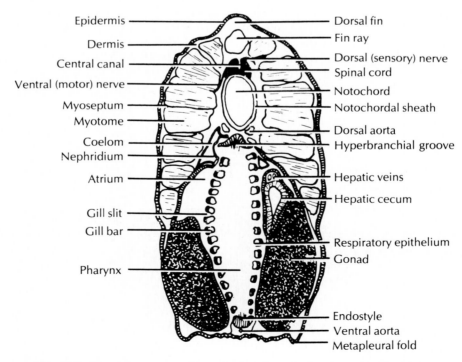

Fig. 16-7. Transverse section of amphioxus through the pharynx and gonads.

Epidermis — Dorsal fin
Dermis — Fin ray
Central canal — Dorsal (sensory) nerve
— Spinal cord
Ventral (motor) nerve — Notochord
Myoseptum — Notochordal sheath
Myotome — Dorsal aorta
Coelom — Hyperbranchial groove
Nephridium
Atrium — Hepatic veins
— Hepatic cecum
Gill slit
Gill bar — Respiratory epithelium
— Gonad
Pharynx
— Endostyle
— Ventral aorta
— Metapleural fold

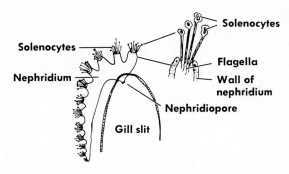

Fig. 16-8. The solenocyte type of nephridial structure found in amphioxus. *Upper right,* Enlarged section through one of the nephridia.

and the epipharyngeal groove, as well as on each side of the gonads and below the endostyle. In favorable specimens little **nephridia** tubules (of the solenocyte type) may be found in the dorsal coelomic cavities (Figs. 16-7 and 16-8). What is their function?

Ventral to the notochord are the paired **dorsal aortas.** The **ventral aorta** is below the endostyle.

DRAWINGS

Use p. 281 and the balance of p. 282 to make drawings of such sections of amphioxus as your instructor requests.

IMMATURE AMPHIOXUS FROM WHOLE MOUNT

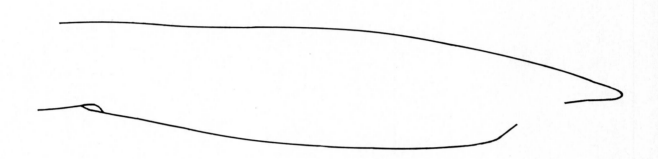

TRANSVERSE SECTIONS OF AMPHIOXUS

CHAPTER 17

THE CARTILAGINOUS FISHES—LAMPREYS AND SHARKS

EXERCISE 17A
Class Cephalaspidomorphi (= Petromyzontes)*
The Lamprey (ammocoete larva and adult)

MATERIALS
Preserved material
   Ammocoetes (lamprey) larvae
   Adult lamprey specimens
   Longitudinal and transverse sections of lampreys†
Watch glasses
Slides
   Stained whole mounts of ammocoetes
   Cross sections of ammocoetes
Microscopes, compound and dissecting

LAMPREY
Phylum Chordata
  Subphylum Vertebrata
    Superclass Agnatha
      Class Cephalaspidomorphi (= Petromyzontes)
        Order Petromyzontiformes
          Genus *Petromyzon*

The most primitive of the living vertebrates are the Agnatha, represented by the lampreys, hagfishes, and slime eels. Although not considered direct ancestors of modern vertebrates, lampreys retain some primitive features probably found in vertebrate ancestors. Members of the Agnatha are jawless and have round sucking mouths.

The genus *Petromyzon* includes the sea lampreys, which grow to be 1 m long and can live in both fresh and marine water. Sea lampreys breed in freshwater

---

*See Hickman, C.P., Jr., L.S. Roberts, and F.M. Hickman, 1984. Integrated principles of zoology, ed. 7, St. Louis, The C.V. Mosby Co., pp. 523-526.
†For preparation of such sections see Projects and Demonstrations, p. 288.

streams. The young larvae, known as **ammocoetes** (sing. ammocoete), live in the sand for 3 to 5 years and then metamorphose rapidly and become dangerous parasites of fishes. They rasp holes in the flesh of the fishes and suck the blood. The adults grow rapidly for a year, spawn in the spring, and soon die. These marine lampreys have invaded the Great Lakes, where they remain a threat to the commercial fishing trade.

The freshwater lampreys, known as brook lampreys, belong to the genera *Entosphenus* and *Ichthyomyzon*, of which there are about 19 species. They have larval habits similar to the marine form, but the adults in about half the species are not parasitic. The nonparasitic forms do not eat as adults but live only a month or so after emerging from the sand and spawning.

The lamprey larva, commonly called ammocoete, have many resemblances to the adult amphioxus *Branchiostoma*, as to both form and methods of living, but it lacks the atrium and the metapleural folds.

AMMOCOETE LARVA
☞ Examine a preserved ammocoete larva as well as a stained whole mount of a small specimen.

Study of the Preserved Larva
☞ Cover the preserved larva with water in a watch glass. Use a hand lens or dissecting microscope.

How does the preserved specimen compare with that of the mature amphioxus? Note the **myotomes** (segmental muscles) appearing faintly on the surface. Do they have the same arrangement as those of the amphioxus? Note the **oral hood** with **oral papillae** attached to the roof and sides of the hood. They are used, as in amphioxus, for filter feeding. The **lateral groove** on each side contains seven small **gill slits.** The anus or **cloacal opening** is just anterior to the **caudal fin.** Are the caudal and **dorsal fins** continuous? Note the **chromatophores** scattered over the body.

Study of the Stained Whole Mount

☞ With low power, examine a stained whole mount of an ammocoete larva. As you locate and identify the structures described here, you may add the appropriate labels to the figure on p. 289.

On the whole mount, find the darkly stained dorsal **nerve cord,** enlarged anteriorly to form the **brain.** Immediately below it is the lighter **notochord.**

The oral hood encloses a **buccal cavity** to the back and sides of which are attached oral papillae. Back of the buccal cavity is the **velum,** a large pair of flaps that create water currents. The large **pharynx** has seven pairs of **internal gill slits,** which open into **gill pouches.** The gill pouches open to the outside by small **external gill slits.** Focus upward onto the outer surface of the animal to see the row of small external gill slits. Between the internal gill slits the pharynx walls are strengthened by cartilaginous rods, the **gill bars.** Note the **gill lamellae** on the pharynx walls. They are rich in capillaries, in which the blood gives up its carbon dioxide and takes up its oxygen from the water.

The ammocoete, like the amphioxus, is a filter feeder. Water is kept moving through the pharynx by muscular action both of the velum and of the whole branchial basket. This contrasts with the amphioxus, in which water is moved by ciliary action.

The **endostyle (subpharyngeal gland)** in the floor of the pharynx is a closed tube the length of four gill slits. It secretes mucus into the pharynx. Food particles brought in by water currents are trapped in the mucus and carried by ciliary action to the **esophagus.**

During the metamorphosis of the larva a portion of the endostyle becomes a part of the thyroid gland of the adult.

The narrow esophagus widens to become the **intestine,** the posterior end of which, called the **cloaca,** receives also the kidney ducts. The **anus** empties a short distance in front of the postanal tail.

The **liver** lies under the posterior end of the esophagus, and embedded in it is the **gallbladder,** appearing as a clear, round vesicle. The two-chambered **heart** lies under the forepart of the esophagus.

Over the heart and around the esophagus is the **pronephric kidney,** consisting of a number of small tubules that empty into the cloaca by pronephric ducts. Later a mesonephric kidney will develop above the intestine, using the same ducts, and the pronephros will degenerate.

The tubular dorsal **nerve cord** enlarges anteriorly into a three-lobed **brain,** visible in most slides. The **forebrain** contains the **olfactory lobe.** In front of the forebrain is the **nasohypophyseal canal,** opening dorsally to the outside by a median **nostril.** The darkly pigmented **eyes** connect with each side of the **midbrain.** At this stage the eyes are covered with skin and muscle and have little sensitivity to light; however, the ammocoetes larva, which usually remains head down in the mud, does have photoreceptors in the tail. Find the **hindbrain** and, with careful focusing, try to see one of the clear oval **ear vesicles** that flank each side of it. Which lobe of the brain is chiefly concerned with the sense of smell? Of sight? Of hearing and equilibrium? How is the nervous system of this cyclostome advanced over that of the cephalochordate amphioxus?

Circulation in Ammocoetes

This system is similar basically to that of amphioxus, but a **two-chambered heart** is the pumping organ, and the dorsal aorta is the chief distributing vessel (Fig. 17-1). The **ventral aorta** carries blood forward, bifur-

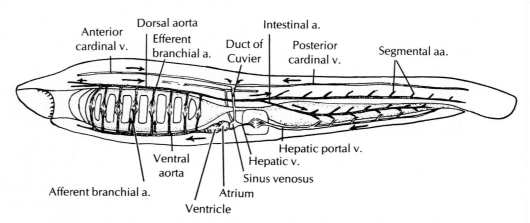

Fig. 17-1. General scheme of circulation in ammocoetes larva of the lamprey.

cating at about the fourth gill pouch and giving off eight pairs of **afferent branchial arteries** to the capillary beds of the gills, from which arise eight pairs of **efferent branchial arteries** that empty into the single median **dorsal aorta.** The dorsal aorta sends a pair of arteries forward to the brain and head. The rest of the dorsal aorta carries blood posteriorly, giving off **segmental arteries** to the capillaries of the body walls and one large **intestinal artery** to the capillary bed in the intestinal wall. The **hepatic portal vein** carries blood from the intestine to the liver capillaries; the **hepatic vein** carries blood from the liver capillaries to a thin-walled sac, the **sinus venosus,** at the back of the heart. Blood from the body wall and myotomes is picked up by cardinal veins. On each side an **anterior cardinal vein,** lateral to the notochord and a **posterior cardinal vein,** lateral to the dorsal aorta, unite to form

a **duct of Cuvier,** which, with its mate from the other side, enters the sinus venosus. The sinus venosus is connected to the thin-walled **atrium** of the heart by valves that prevent backflow. Valves also guard the opening from the atrium to the thicker more muscular **ventricle,** which pumps the blood through valves into the ventral aorta.

Transverse Sections of Ammocoetes

☞ Your slide may contain four typical sections—one each through the brain, the pharynx, the intestine, and the postanal tail. Or it may contain 15 to 20 sections through the body, arranged in sequence. As you study each section, refer to the whole mount again to interpret relationships. Use the low power of the microscope.

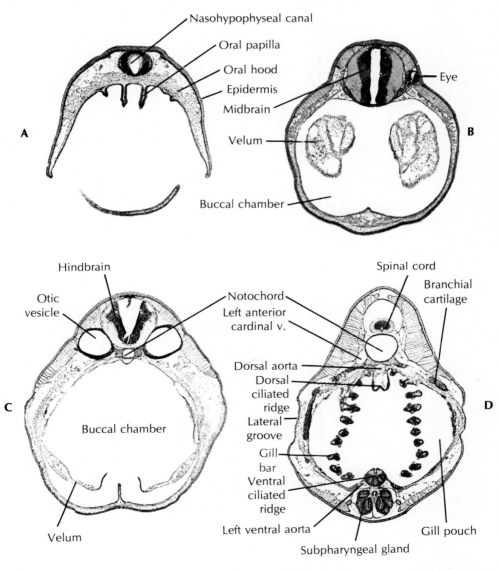

Fig. 17-2. Transverse sections of ammocoetes larva through **A,** oral hood; **B,** buccal chamber; **C,** posterior part of the buccal chamber; **D,** pharynx.

285

Sections Anterior to the Pharynx. Sections through the **forebrain** may include the **oral papilla** and **oral hood.** You may find sections anterior to the brain showing the **nasohypophyseal canal** (Fig. 17-2, *A*). Sections through the **midbrain** may include the **eyes**, the **buccal chamber**, and portions of the **velar flaps** (Fig. 17-2, *B*). Sections through the **hind-brain** may include the **ear** (**otic**) **vesicles** and the buccal chamber (Fig. 17-2, *C*) or the forepart of the pharynx. Compare the size of the brain with that of the spinal cord in more posterior sections. Do you find a **notochord** lying just below the brain in any of the sections?

Sections Posterior to the Pharynx. Choose a section through the trunk posterior to the pharynx and identify the following (Fig. 17-3): **epidermis; my-**otomes (lateral masses of muscles); **nerve cord** surrounded by the **neural canal** and containing a cavity, the **neurocoel; notochord** with large vacuolated cells; **dorsal aorta** (probably contains blood cells); **posterior cardinal veins,** one on each side of the aorta (blood cells are usually present). You may find the cardinals joining to form the duct of Cuvier and the **coelomic cavity,** lined with peritoneum and containing the visceral organs.

The visceral contents of the coelomic cavity will vary according to the location of the section.

1. Just behind the pharynx you will find the **esophagus,** of columnar epithelium; the paired **pronephric kidneys,** appearing as sections of small tubules; and the chambers of the **heart** (Fig. 17-3, *A*).

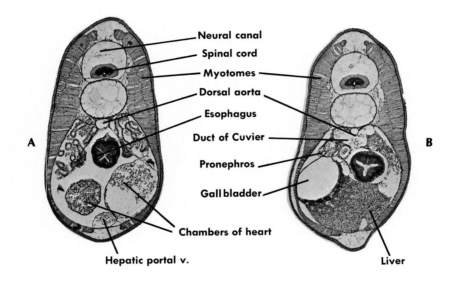

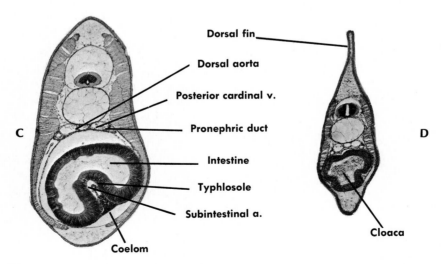

Fig. 17-3. Transverse sections of ammocoetes larva through **A,** esophagus and heart; **B,** esophagus and liver; **C,** intestine; **D,** cloaca.

2. Sections cut posterior to the heart will show the dark **liver** and possibly the hollow **gallbladder** ventral or lateral to the esophagus, with sections of pronephric kidneys or their ducts located under the posterior cardinals (Fig. 17-3, *B*).

3. Sections through the intestine will reveal it as a large tube of columnar epithelium with a conspicuous infolding, the **typhlosole**, carrying the **subintestinal artery** (Fig. 17-3, *C*). Above the intestine you may find the **mesonephric kidneys,** with their tubules, and a small **gonad** between the kidneys. The **cloaca** is located farther back (Fig. 17-3, *D*).

4. In sections posterior to the anus (postanal tail), identify the caudal fins. What other structures can you identify?

Sections through the Pharynx. The body wall, nerve cord, and notochord will be similar to the preceding sections. The central part of the section is taken up by the large pharynx, on whose walls are the **gills,** with their platelike **gill lamellae** extending into the pharynx or into the **gill pouch,** depending on how the section has been cut. Each lamella has lateral ridges. Some sections, such as that in Fig. 17-2, *D*, may show only the gill bars without the feathery lamellae. The gills are liberally supplied with blood vessels. Outside the gill chambers you will find sections through cartilage rods that give support to the branchial basket.

In the middorsal and midventral regions of the pharynx are **ciliated ridges** bearing grooves. The cilia are concerned with the movement toward the esophagus of mucous strands in which food particles are caught. Below the pharynx in certain sections is the bilobed **subpharyngeal gland** whose function is probably the secretion of mucus. Later, certain portions of this gland are incorporated into the adult thyroid gland. Between the ventral ciliated ridge and the subpharyngeal gland is the single or paired **ventral aorta.** Note the **gill pouches** lateral to the pharynx, the lateral groove, and in some sections the **external gill slits** to the outside. Locate the **anterior cardinal veins** on each side of the notochord and the **dorsal aorta** beneath it.

DRAWINGS

On p. 289 label the drawing. On p. 290 draw such transverse sections as are required by your instructor. Label fully.

Which of the **basic types of tissues** have you been able to identify on these slides? What is the advantage to the animal of an **endoskeleton** rather than an exoskeleton as an attachment for muscles? For what type of locomotion are these animals best fitted? How does

the nerve cord differ in location from that of the flatworms? The annelids and arthropods? Compare the circulation with that of the earthworm.

How many of the features of the primitive amphioxus adult do you find repeated in the ammocoete, the larval form of a vertebrate? Do you think this might be interpreted as an example of the **biogenetic law?*** It will be well to keep in mind the structure of these primitive chordate forms and be able to compare them with those of the fish, amphibian, and mammal forms that we will study later.

ADULT LAMPREY

 Using whole specimens and prepared longitudinal and transverse sections of the sea lamprey, compare the adult lamprey with the larva. Do not injure the specimens, which will be used by other laboratories.

External Structure

Identify the dorsal and caudal fins; the round mouth with its horny "teeth" and rasping tongue used in puncturing the body of its prey; the eyes, functional in the adult; the median nostril; the seven external gill slits; the lateral line on each side, made up of sensory pits, which probably record water vibrations or currents; the urogenital sinus, with its projecting urogenital papilla at the ventral juncture of the trunk and tail; the anal opening in front of the urogenital sinus; and the segmented nature of the musculature (myotomes).

Internal Structure

Identify the notochord and portions of cartilaginous skeleton; the brain, spinal cord, and olfactory canal; and the myotomes and large muscles of the tongue. Follow the digestive tract of oral hood, mouth, tongue, esophagus, and intestine with typhlosole. Note that the esophagus is now connected with the roof of the pharynx, leaving the pharynx (respiratory tube) with its spherical gill pouches as a blind sac now devoted only to respiratory function. Note the arrangements of the gill lamellae in the gill pouches. Find the liver; the median gonad, from which eggs or sperm are freed into the coelom to enter the urogenital sinus; and the kidneys, each with a ureter to the urogenital sinus. Locate the two-chambered heart and as many of the blood vessels as possible. The circulatory system is similar to that of the larva.

---

*See Hickman, C.P., Jr., L.S. Roberts, and F.M. Hickman, 1984. Integrated principles of zoology, ed. 7. St. Louis, The C.V. Mosby Co., p. 12, Principle 7.

PROJECTS AND DEMONSTRATIONS

1. Life-history set of a lamprey. The life history set of a lamprey is available from biological supply houses.

2. Examples of adult brook lampreys.

3. Preparation of transverse sections of large laboratory animals. Freeze the preserved and injected animal; then cut it into 5 cm transverse sections with a sharp hacksaw. Clean the sections carefully and secure the viscera in place with insect pins. Place each section in a container of slightly larger diameter than the section and cover with a melted 2.5% solution of agar, being careful while the solution is still warm to place the organs properly and to expel trapped air bubbles. When they have cooled, cut away excess agar from the surface. These preparations can be stored in a formaldehyde solution for long periods. Longitudinal sections can be made by the same method.

INTERNAL STRUCTURE OF THE LAMPREY LARVA

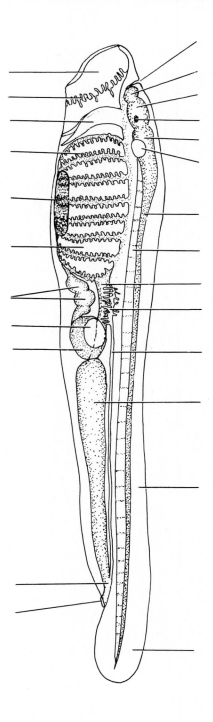

AMMOCOETES LARVA OF THE LAMPREY

(section through pharynx)

EXERCISE 17B
Class Chondrichthyes—the dogfish shark

MATERIALS
Preserved dogfish sharks
   or
Longitudinal and transverse sections of the shark*

SQUALUS
Subphylum Vertebrata
  Superclass Gnathostomata
    Class Chondrichthyes
      Subclass Elasmobranchii
        Order Squaliformes
          Genus *Squalus*

The cartilaginous or elasmobranch fishes are characterized by their cartilaginous skeleton and exposed gill slits. They include the sharks, rays, skates, and chimaeras. The sharks are very generalized vertebrates. Their basic structural plan is also the basic embryonic plan of higher vertebrates.

Where Found
The dogfish sharks are marine and are common along both the Atlantic and Pacific coasts. They grow to about 1 m in length.

External Structure
The body is divided into the **head** (to the first gill slit), **trunk** (to the cloacal opening), and **tail.** The **fins** include a pair of **pectoral fins** (anterior), which control changes in directions during swimming; a pair of **pelvic fins,**

---

*Prepared sections may be made by the method given on p. 288.

which serve as stabilizers and which in the male are modified to form claspers used in copulation; two median **dorsal fins,** which also serve as stabilizers; and a **caudal fin,** which is **heterocercal** (asymmetric dorsoventrally).

Identify the **mouth** with its rows of **teeth** (modified placoid scales), which are adapted for cutting and shearing; two ventral **nostrils,** which lead to olfactory sacs and which are equipped with folds of skin that allow continual in-and-out movement of water; and the lateral **eyes,** which lack movable eyelids but have folds of skin that cover the outer margin of the eyeballs. The part of the head anterior to the eyes is called the **snout.** A pair of dorsal **spiracles** posterior to the eyes are modified gill slits that open into the pharynx. They can be closed by folds of skin during part of the respiratory cycle to prevent the escape of water. Five pairs of **gill slits** are the external openings of the gill chambers. Insert a probe into one of the slits and notice the angle of the gill chamber. The **pharynx** is the region in back of the mouth into which the gill slits and spiracles open. A **lateral line,** appearing as a white line on each side of the trunk, represents a row of minute, mucus-filled sensory pores used to detect differences in the velocity of surrounding water currents, and thus to detect the presence of other animals, even in the dark. Note the **cloacal opening** between the pelvic fins.

> Complete and label the outline of the external view in Fig. 17-4.

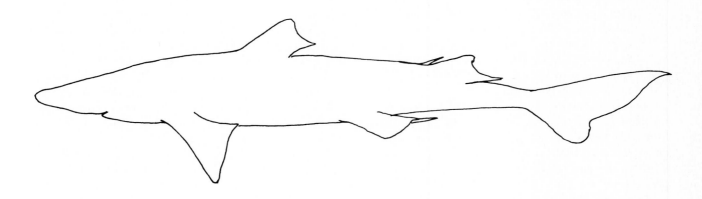

Fig. 17-4. External view of the dogfish shark—to be completed and labeled.

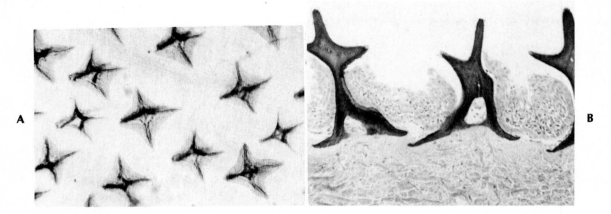

Fig. 17-5. A, Surface view of shark skin showing dermal denticles, or placoid scales. (×50.) **B,** Placoid scales of shark, *side view,* showing spines upright and dental bases with dental tubercles. (×120.) (Photographs by F.M. Hickman.)

The skin consists of an outer layer of epidermis covering a much thicker layer of dermis densely packed with fibrous connective tissue. The leathery skin is covered with **placoid scales (denticles)** (Fig. 17-5). Each scale has a wide base in the dermis and a spine that projects from the epidermis pointing posteriorly. The teeth are modified placoid scales. The dark dorsal and light ventral coloration is protective when viewed from either above or below.

Internal Structure

If a dissection is not to be made, examine longitudinal and transverse sections of the shark. Note the cartilaginous skull and vertebral column. With the help of Fig. 17-6, identify as many of the internal structures as possible.

Dissection of the Shark

☞ Open the coelomic cavity by extending a midventral incision posteriorly from the pectoral girdle through the pelvic girdle and then around one side of the cloacal opening to a point just posterior to it. On each side, make a short transverse cut just posterior to the pectoral fins and another one just anterior to the pelvic fins. Rinse out the body cavity.

The abdominal cavity is lined with smooth **peritoneum** (parietal peritoneum). This peritoneal epithelium also forms part of (1) the double-membraned dorsal mesentery that supports the digestive tract and (2) the outer covering of the visceral organs (visceral peritoneum).

Digestive system. Identify the large **liver** (Fig. 17-6). It has two large lobes and a small median lobe bearing the elongated **gallbladder.** Dorsal to the liver is the large **esophagus,** which leads from the pharynx to the J-shaped **stomach.** Between the stomach and the first part of the intestine (**duodenum**) is a muscular constriction called the **pyloric valve.** Ducts from the liver and pancreas empty into the duodenum. The **pancreas** lies close to the ventral side of the duodenum, with a slender dorsal portion extending posteriorly to the large **spleen** (not a part of the digestive system). The **valvular intestine** (ileum) is short and wide and contains a **spiral valve.** The short narrow **rectum** has extending from its dorsal wall a **rectal gland,** which regulates ion balance. The cloaca receives the rectum and urogenital ducts.

☞ Make a longitudinal incision in the ventral wall of the esophagus and stomach. Remove and save the contents of the stomach, if any. Extend the longitudinal incision along the wall of the ileum (taking care not to destroy the blood vessels) to expose the spiral valve. Rinse out the exposed digestive tract.

Examine the contents taken from the stomach and compare with those of others at your table. What do you infer about the shark's eating habits? Examine the inner surface of the digestive tract. The walls of the esophagus bear large papillae, whereas the walls of the stomach are thrown into longitudinal folds called **rugae.** Note the structure of the pyloric valve. Observe the cone-shaped folds of the **spiral valve** and see if you can determine how materials pass through. The spiral valve provides extra surface area for absorption of nutrients. (If the intestine has been everted into the cloacal region, carefully pull it back into the pleuroperitoneal cavity.)

Urogenital System. Although the excretory and reproductive systems have quite different functions, they are closely associated structurally, and so are studied together.

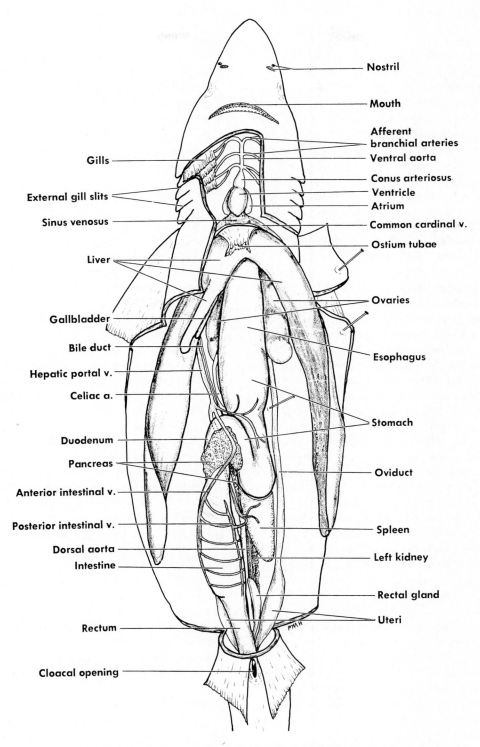

Nostril

Mouth

Afferent branchial arteries

Ventral aorta

Gills

Conus arteriosus

External gill slits

Ventricle

Atrium

Sinus venosus

Common cardinal v.

Ostium tubae

Liver

Ovaries

Gallbladder

Bile duct

Esophagus

Hepatic portal v.

Celiac a.

Stomach

Duodenum

Pancreas

Oviduct

Anterior intestinal v.

Posterior intestinal v.

Spleen

Dorsal aorta

Intestine

Left kidney

Rectal gland

Uteri

Rectum

Cloacal opening

Fig. 17-6. Internal structure of the female dogfish shark, *Squalus*.

Male. The soft, elongated **testes** lie along the dorsal body wall, one on each side of the esophagus. They are held in place by a mesentery called the **mesorchium.** A number of very fine tubules (vasa efferentia) in the mesentery run from each testis to a much-convoluted **wolffian duct** (also called sperm duct or archinephric duct). The **kidneys** (also called opisthonephroi) are long and narrow and lie behind the peritoneum on each side of the dorsal aorta. They extend from the pectoral girdle to the **cloaca** (p. 297). The wolffian ducts, which serve as both urinary ducts and sperm ducts, take a twisting course along the length of the kidneys to the cloaca and collect wastes from the kidneys by many fine tubules. The wolffian ducts widen posteriorly into **seminal vesicles,** which dilate terminally into **sperm sacs** before entering the cloaca. The cloaca is a common vestibule into which both the rectum and the urogenital ducts empty. In the center of the cloaca, dorsal and posterior to the rectum, is a projection called the **renal papilla,** which is larger in the male than in the female. The seminal vesicles empty into the renal papilla, which empties into the cloaca. Slit open the cloaca to see the renal papilla. A groove along the inner edge of each **clasper** is used in conducting spermatozoa to the female at copulation.

Female. A pair of **ovaries** lies against the dorsal body wall, one on each side of the esophagus. Enlarged ova may form several rounded projections on the surface of the ovaries. A pair of **oviducts** (Müllerian ducts) run along the dorsal abdominal mesentery. The anterior ends join to form a common opening into the abdominal cavity, called the **ostium tubae** (p. 298). This lies anteroventral to the liver but may be difficult to find except in large females. Ova, rupturing from the ovaries, fall into the abdominal cavity and are drawn through the ostium into the oviducts, where fertilization may occur. An expanded area of each oviduct dorsal to the ovary is a **shell gland (oviducal gland),** which in *Squalus* secretes a thin membrane around several eggs at a time. The posterior end of each oviduct enlarges into a **uterus,** the caudal end of which opens into the **cloaca.** In immature dogfish the shell gland and uterus may not be apparent.

The slender **kidneys** (opisthonephroi) extend the length of the dorsal abdominal wall, dorsal to the peritoneum. A very slender **wolffian duct (archinephric duct)** embedded on the ventral surface of each kidney empties into the cloaca through a **renal papilla.** Slit open the cloaca and identify the renal papilla, the entrance of the rectum, and, on the dorsal side, the openings from the uteri.

The eggs, which develop in the ovaries, are about 2.5 cm in diameter at maturity. At ovulation they burst through the ovarian walls into the coelom close to the funnel-shaped ostium. Fertilization occurs in the oviducts. When the eggs reach the shell gland, they receive a thin horny shell, which later dissolves. One to six or seven eggs, depending on the species, may develop in each uterus. Vascularized villi on the wall of the uterus come in contact with the yolk sac of the embryo in a placenta-like manner. Because the dogfish shark embryo receives very little nourishment from this source, it is called **ovoviviparous;** that is, it gives birth to living young without dependence on placental nourishment. Other sharks include some that are dependent on the mother for nourishment through the placental connection (viviparous), and some primitive sharks that lay shelled eggs containing a large amount of yolk (oviparous). Gestation periods vary from 16 to 24 months, and the young at birth range from 12 to 30 cm in length.

DRAWINGS

On pp. 297 and 298 are diagrams of the male and female urogenital systems drawn in on one side only. On the other side of each diagram, draw in and label the urogenital organs as they appear in the specimens you are studying. Your specimens may be immature and the reproductive organs not fully developed.

Circulatory System. The basic plan of circulation in the shark is similar to that in the ammocoetes larva.

☞ Spread the walls of the **pericardial cavity** and lift up the **heart** to see the thin-walled triangular **sinus venosus.**

Blood passes from the sinus venosus to the **atrium,** which surrounds the dorsal side of the muscular **ventricle;** it then flows into the ventricle, which pumps it forward into the **conus arteriosus** (Fig. 17-6). **Valves** prevent backflow between the compartments.

Venous system

☞ Slit open the sinus venosus transversely, extending the cut somewhat to the left; then wash out its contents.

Look for openings into the sinus venosus of one of each of the following paired veins: (1) **common cardinal veins (ducts of Cuvier)** (Fig. 17-6), which extend laterally and into which empty the large **anterior cardinal sinuses, posterior cardinal sinuses,** and **subclavian veins;** (2) the **inferior jugular veins** from the floor of the mouth and gill cavities; and (3) the **hepatic sinuses,** which empty near the middle of the posterior wall of the sinus venosus and bring blood from the liver.

The **hepatic portal vein** (Fig. 17-6) gathers blood chiefly from the digestive system through a system of **gastric, pancreatic,** and **intestinal veins.** The hepatic portal vein enters the right lobe of the liver and divides into several small **portal veins** (trace some of these subdivisions); it then divides into a system of capillaries, from which some of the carbohydrates brought from the intestine may be stored in liver cells as glycogen or animal starch until needed. Blood from capillaries flows into the **hepatic sinuses** and from there into the sinus venosus.

The **renal portal veins** arise from the **caudal vein** in the tail and carry blood to the capillaries of the kidneys. Many small renal veins carry blood from the kidneys to the posterior cardinal sinuses and from there to the sinus venosus.

Arterial system. The arterial system includes (1) the afferent and efferent branchial arterial system and (2) the dorsal aorta and its branches.

From the conus arteriosus, trace the **ventral aorta** forward, removing most of the muscular tissue to the lower jaw. The ventral aorta gives off three paired branches, which give rise to the **five pairs of afferent branchial arteries** (Fig. 17-6). In injected specimens you can follow these arteries into the interbranchial septa, where each gives off tiny arteries to the gill lamellae.

☞ The **efferent branchial arteries** are more difficult to dissect. With scissors, cut through the left corner of the shark's mouth and backward through the centers of the left gill slits, continuing as far as the transverse cut you made earlier at the base of the pectoral fins. Now cut transversely across the floor of the pharynx straight through the sinus venosus, and turn the lower jaw to one side to expose the gill slits and the roof of the pharynx. Locate the spiracle internally. It represents the degenerated first gill slit. Dissect the mucous membrane lining from the roof of the mouth and pharynx to expose the four pairs of **efferent branchial arteries**, which carry oxygen-rich blood from the gill filaments and unite to form the **dorsal aorta**. By cutting the cartilages under which they pass, trace these arteries back to the gills.

The dorsal aorta extends posteriorly along the length of the body ventral to the vertebral column. It gives rise to the **subclavian arteries,** which connect to the pectoral region, a **celiac artery** (Fig. 17-6), which gives off branches to the intestinal tract and gonads, numerous **parietal arteries** to the body walls, **mesenteric arteries** to the intestine and the rectum, **renal arteries** to the kidneys, and **iliac arteries** to the pelvic fins. The aorta continues to the tip of the tail as the **caudal artery.**

Respiratory System. In the sharks, water ingested through both the mouth and the spiracles is forced laterally through the five pairs of gills and leaves through the five pairs of external gill slits (some elasmobranchs have a different number of gills).

☞ On the shark's right (intact) side, separate the gill units by cutting dorsally and ventrally from the corners of each gill slit. Now you can ex-

amine the structure of the intact gills on this side and observe the gills in cross section on the other side.

The area between the **external gill slits** and **internal gill slits** comprises the **gill chambers (gill pouches, branchial chambers).** The incomplete rings of heavy cartilage supporting the gills and protecting the afferent and efferent branchial arteries are called **gill arches.** Short spikelike projections extending medially from the gill arches are the **gill rakers,** which filter the respiratory water and direct food toward the esophagus. Cartilaginous **gill rays** fan out laterally from the gill arches to support the gill tissues.

☞ Remove an intact half of a gill arch along with its gill tissue. Examine with a hand lens. Float a small piece of gill in water and examine with a dissecting microscope.

Primary lamellae (gill filaments) are the small plate-like sheets of epithelial folds arranged in rows along the lateral face of each gill. Under the microscope the primary lamellae are seen to be made up of rows of tiny plates, called **secondary lamellae,** which are the actual sites of gas exchange. Blood capillaries in the secondary lamellae are arranged to carry blood inward, or in the opposite direction of the seawater, which is flowing outward. This counter-current flow encourages gas exchange between the blood and water. The gill lamellae are arranged in half-gills, or **demibranchs,** on each side of the branchial arch. The two demibranchs together form the gill unit, or **holobranch.** The spiracles are believed to be remnants of the gill openings found in more primative chordates. They are usually larger in the slow-moving, bottom-dwelling sharks than in the fast-swimming sharks in which, because of their motion, there is a more massive flow of water through the mouth.

ORAL REPORT

Be able to identify both the external and internal features of the shark and give the functions of each organ or structure.

Be able to trace the flow of blood from the heart to any part of the body (such as the pectoral region, the kidneys, the tail, and so on) and back to the heart.

DEMONSTRATIONS

1. *Dogfish uterus with developing pups.* Or dogfish embryos with attached yolk may be used.
2. *Various sharks, rays, and skates.*
3. *Preparation of the skull or skeleton of a shark.*
4. *Corrosion preparation of the arterial system.*
5. *Shark teeth.*
6. *Microscopic slides of shark skin.*

MALE REPRODUCTIVE SYSTEM OF SHARK

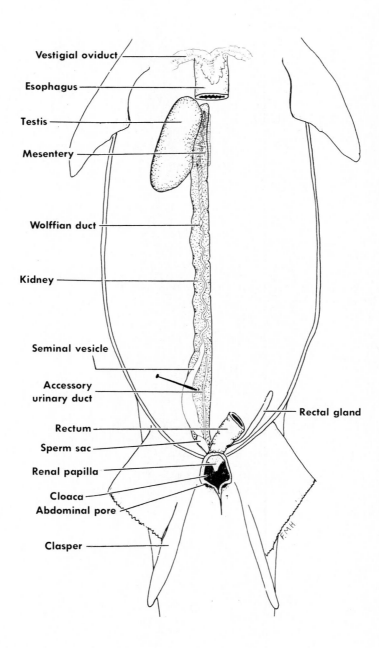

Vestigial oviduct

Esophagus

Testis

Mesentery

Wolffian duct

Kidney

Seminal vesicle

Accessory urinary duct

Rectum

Sperm sac

Renal papilla

Cloaca

Abdominal pore

Clasper

Rectal gland

FEMALE REPRODUCTIVE SYSTEM OF SHARK

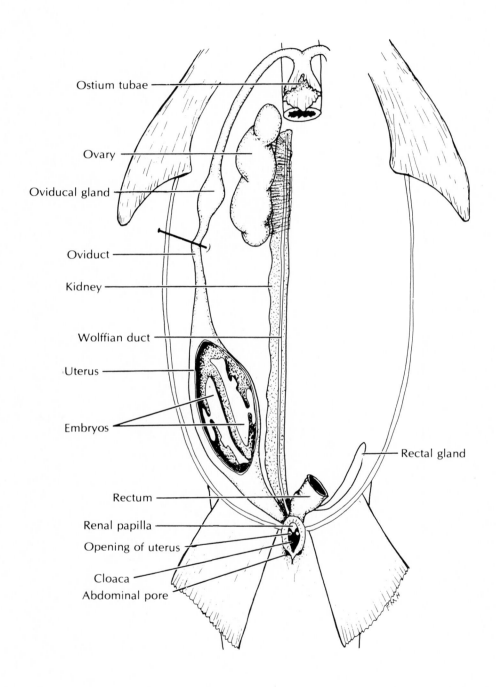

Ostium tubae

Ovary

Oviducal gland

Oviduct

Kidney

Wolffian duct

Uterus

Embryos

Rectal gland

Rectum

Renal papilla

Opening of uterus

Cloaca

Abdominal pore

CHAPTER 18

CLASS OSTEICHTHYES — THE BONY FISHES

EXERCISE 18
The yellow perch*

MATERIALS
Preserved, injected perch†
Living fishes, any kind, in aquarium
Stained slides of fish blood
Mounted fish skeletons and skeletons of other vertebrates
Longitudinal and cross sections of preserved perch

PERCA, THE YELLOW PERCH
Subphylum Vertebrata
   Class Osteichthyes (Teleostomi)
     Subclass Actinopterygii
       Superorder Teleostei
         Order Perciformes
           Genus *Perca*
             Species *P. flavescens*

Where Found
The yellow perch is a common freshwater form widely distributed through the lakes of the American Midwest and parts of Canada. A closely related species is found in Europe and Asia.

Characteristics
Class Osteichthyes (bony fishes) represents the largest group of vertebrates both in number of species (at least 20,000) and in number of individuals. By adaptive radiation they have developed an amazing variety of forms and structures. Their chief characteristics usually are **dermal scales, operculum over the gill chamber** of each side, **bony skeleton, terminal mouth, swim bladder, homocercal tail,** and both **median** and **paired fins.** Bony fishes flourish in freshwater and saltwater environments and are usually present wherever fishes can exist.

---

*See Hickman, C.P., Jr., L.S. Roberts, and F.M. Hickman. 1984. Integrated principles of zoology, ed. 7. St. Louis, The C.V. Mosby Co., pp. 532-533, 535-553.
†Although this study is based on the perch, it can easily be adapted to bass, trout, sunfish, crappie, or others, either fresh or preserved.

External Structure
☞ Obtain a preserved fish and, after you have studied its external anatomy, compare it with the living fishes in the aquarium. See what their structure can tell you of their living habits.

The body of the perch is fusiform, or torpedo shaped. Is it compressed in any of its planes? Identify the **head** extending to the posterior edge of the operculum, the **trunk** extending to the anus, and the **tail.** Identify the **pectoral, pelvic, anal, dorsal,** and **caudal fins.** How many of each are there? Which are paired? Note the **fin rays,** which support the thin membrane of each fin. Are some of the rays soft and some spiny? Where are the spiny rays located? The caudal fin is **homocercal,** meaning that the upper and lower halves are alike. Fins such as those of the dogfish shark, which are asymmetric, are called heterocercal.

The terminal **mouth** is adapted for overtaking prey while swimming. Fishes with superior mouths are usually surface feeders, whereas those with inferior mouths are usually bottom feeders. Do the **eyes** have lids? Could the perch have binocular vision? Why? On each side in front of the eye a pair of **nostrils** open into an olfactory sac. Water enters the sac through the anterior aperture, which is provided with a flaplike valve, and leaves through the posterior aperture. The **ears** are located behind the eyes, but they are not visible externally. The **lateral line,** along the side of the body, is a row of small pores or tubules connecting with a long tubular canal bearing sensory organs. These are sensitive to pressure and temperature changes and so are responsive to water currents. Many microscopic sense organs are found in the skin.

Lift an **operculum** and study its structure. Along the ventral margin of the operculum, find a membrane supported by bony rays. This membrane fits snugly against the body to close the branchial cavity during certain respiratory movements. With your probe, examine the **gills** beneath the operculum.

Find the **anus** near the base of the anal fin and the small slitlike **urogenital opening** just posterior to the anus.

Note the arrangement of the **scales.**

☞ Remove a scale from the lateral line region, mount in water on a slide, and examine with low power.

The anterior or embedded side of the perch scale has radiating grooves. The posterior, or free, edge has very fine teeth. These are **ctenoid scales** (from Gr., *kteis, ktenos*, comb). Note the fine concentric lines of growth. The scales are covered with a very thin epidermis that secretes mucus over the scales. This reduces friction in swimming and makes capture more difficult. Ctenoid scales are usually found on fishes with spiny rays in the fins, whereas the soft-rayed fishes usually have cycloid scales, which lack the marginal teeth.

Skeletal System

The bony skeleton of the perch consists of the **axial skeleton** (which includes the bones of the skull, the vertebral column, the ribs, and the medial fins) and the **appendicular skeleton** (which includes the pectoral girdle and fins and the pelvic girdle and fins). Examine the mounted perch skeletons on display and compare them with the skeletons of other fishes and of amphibians, birds, and mammals. Do you see any basic similarity?

Muscular System

Though the muscles of the perch are less complex than those of land vertebrates, they make up a much larger mass in relation to body size. Tetrapod locomotion results largely from direct action of muscles on bones of the limbs, but fish locomotion results from the indirect action of the segmental muscles—**myotomes**—on the vertebral column, a method by which a large muscle mass produces a relatively small amount of action. This type of movement is efficient in a water medium, but it would be less effective on land. The myotomes consist of blocks of longitudinal muscle fibers placed on each side of a central axis, the vertebral column. Their contraction, therefore, bends the body, and the action passes in waves down the body, alternating on each side.

☞ After cutting off the sharp dorsal and ventral spines, skin one side of the body and note the shape of the myotomes.

They resemble **W**'s that are turned on their sides and stacked together. A horizontal septum of connective tissue divides the muscles into dorsal **epaxial muscles** and ventral **hypaxial muscles** (Fig. 18-1). Posteriorly both epaxial and hypaxial muscles are active in locomotion, but anteriorly the hypaxial muscles serve more for support of body viscera than for locomotion. Try to separate some of the myotomes. Observe the direction of the muscle fibers. Do they run zigzag as the myotomes seem to? Or are they all directed horizontally—or vertically?

☞ Now watch the swimming motions of fishes in the aquarium and try to visualize the use of the body muscles in locomotion. What part do the fins play in locomotion?

Dissection of individual muscles in the fish is difficult and will not be attempted here. Muscles operating the jaws, opercula, and fins are often named according to their function and, as in other vertebrates, include adductors, abductors, dilators, levators and so on (Fig. 18-2).

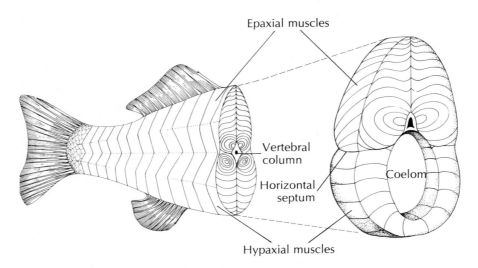

Fig. 18-1. Diagram of the skeletal musculature of a teleost fish.

Mouth Cavity, Pharynx, and Respiratory System

Before starting the dissection, it is well, if you have not already done so, to cut off the sharp dorsal and ventral fins to protect the hands.

☞ Cut away the operculum from the left side, exposing the gill-bearing bars or arches. How many arches are there?

Cut one gill arch, place in water and examine with the hand lens or binocular microscope.

If injected, the branchial arteries will be colored. Note the double row of **gill filaments** borne on the posterior or aboral side of the arch. It is in these filaments, containing capillaries from the branchial arteries, that exchange of gases takes place. **Gill rakers** on the oral surface of each gill bar strain out food organisms and offer some protection to the gill filaments from food passing through the pharynx.

☞ Cut through the angle of the left jaw, continuing the cut through the middle of the left gill arches to expose the mouth cavity and pharynx.

Open the mouth wide and note the **gill slits** in the pharynx. In the mouth, locate the fine **teeth**. Would they be effective in chewing? In holding the prey? Just inside the teeth, across the front of both the upper and lower jaw, find the **oral valves**. These are transverse membranes that prevent the outflow of water during respiration. An inflexible **tongue** is supported by the hyoid bone. Explore the spacious pharynx, noting the size and arrangement of the gill bars and gill slits.

The mechanics of water movement involve a continuous pumping action in which, by muscular action, the gill arches are pushed out laterally, and the opercula are pressed against the body; this enlarges the branchial cavity, at the same time closing its exit, so that water flows into the mouth and pharynx. Then, by the closing of the flaplike oral valves, the opening of the gill covers, and the compressing of the gill chamber, water is forced out over the gills (Fig. 18-2).

☞ Now go watch the fishes in the aquarium, observing their respiratory movements until you understand their meaning and functions.

Abdominal Cavity

☞ Starting near the anus and being careful not to injure the internal organs, cut anteriorly on the midventral line to a region anterior to the pelvic fins. Now on the animal's left body wall, make a transverse cut, extending dorsally from the anal region; make another cut dorsally between the pectoral and pelvic fins; then remove the left body wall by cutting between these two incisions. On the right side, make similar transverse cuts so that the right wall can be laid back, but do not remove it.

You have now exposed the abdominal cavity. This, together with the **pericardial cavity**, which contains the heart, makes up the **coelomic cavity**. Note the shiny lining of **peritoneum**.

Probably the first organ you will see will be the **intestine** encased in yellow fat. Carefully remove enough of the fat to trace the digestive tract anteriorly. Find the **stomach**, lying dorsal and somewhat to the left of the intestine. Anterior to the stomach is the **liver**, dark red in life but bleached to a cream color by preservative. The **spleen** is a dark, slender organ lying between the stomach and the intestine. The **gonads** are in the dorsoposterior part of the cavity. The **swim bladder** lies dorsal to these organs and to

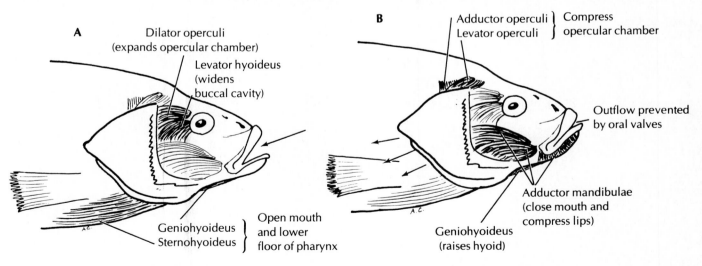

Fig. 18-2. Diagram of muscles used in respiratory movements. **A,** The mouth opens, the operculum to the gill slits closes, the pharynx expands, and water is drawn in. **B,** The mouth closes, the operculum opens, the pharynx constricts, the water is expelled through the operculum.

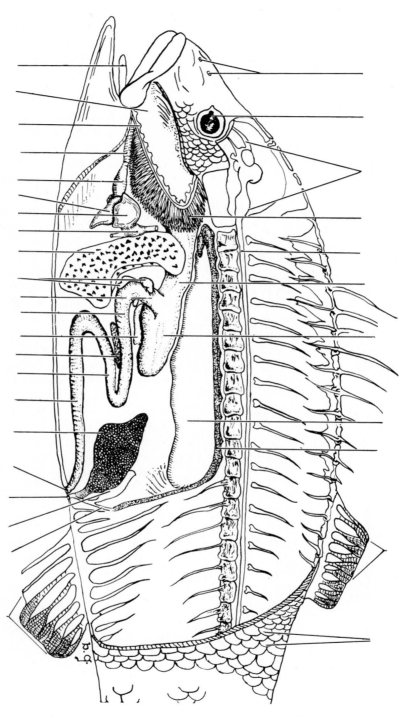

Fig. 18-3. Internal structure of the perch (to be labeled by the student).

the peritoneal cavity. It is long and thin walled. Do not injure its walls or any of the blood vessels lying among the viscera. The **kidneys** are located dorsal to the swim bladder and will be seen later.

☞ As you continue the dissection and identify anatomical structures, you may fill in the appropriate labels on Fig. 18-3.

Digestive System. Run your probe through the mouth and into the opening of the **esophagus** at the end of the pharynx. Now lift up the liver and trace the esophagus from the pharynx to the large **cardiac portion** of the stomach. This ends posteriorly as a blind pouch. The short **pyloric portion** of the stomach, opening off the side of the cardiac pouch, empties by way of a **pyloric valve** into the **duodenum,** the S-shaped proximal part of the intestine. Three intestinal diverticula, the **pyloric ceca,** open off the proximal end of the duodenum near the pyloric valve. Follow the intestine to the **anus.** Note the supply of blood vessels in the **mesentery.** The **pancreas,** a rather indistinct organ, lies in the fold of the duodenum. The **liver** is large and lobed, with the **gallbladder** located under the right lobe. The liver is drained by tubules into the gallbladder, which in turn opens by several ducts into the duodenum posterior to the pyloric ceca.

Cut open the stomach and place its contents in a watch glass of water. Compare with your neighbor's findings. Examine the stomach lining. How is its surface increased? Cut open and examine the pyloric valve. Open a piece of the intestine, wash, and examine the lining with a dissecting microscope. What type of muscle would you expect to find in the intestine?

Swim Bladder (Air Bladder). The swim bladder is a long, shiny, thin-walled sac that fills most of the body cavity dorsal to the visceral organs. In some fishes (not the perch) it connects with the alimentary canal. Cut a slit in it and observe its internal structure. In its anterior ventral wall, look for the **red body** (gas gland), a network of capillaries (rete mirabile) that secrete gases into the bladder. Another capillary bed, the **oval,** lies in the dorsal wall. Gases may be reabsorbed from the swim bladder in this area. The swim bladder is a hydrostatic organ that adjusts the specific gravity of the fish to varying depths of water.

Reproductive System. The sexes are separate, but it is difficult to distinguish them externally.

The **ovary** is single and lies back of the stomach, just below the swim bladder and dorsal to the intestine. The size of the ovary varies seasonally, being largest during the winter months prior to spawning. A prolongation of the ovary posteriorly serves as a sort of **oviduct** for carrying eggs to the **urogenital pore** just posterior to the anus (Fig. 18-4, *A*).

In the male two elongated **testes** are attached to the swim bladder by mesenteries. They become greatly enlarged prior to spawning and are usually smallest during the summer months. A **sperm duct** (**vas deferens**) runs along a longitudinal fold in each testis adjacent to the spermatic artery. The two ducts join in the posterior midline and extend to the **genital pore** just posterior to the anus (Fig. 18-4, *B*).

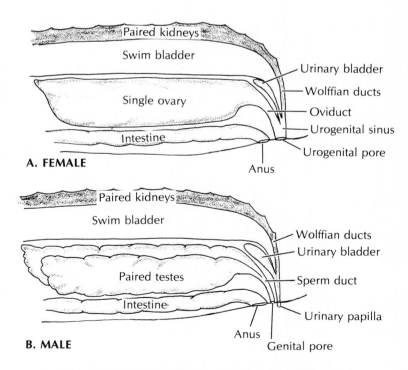

Fig. 18-4. Diagram of the urogenital system of the perch.

Excretory System. The **kidneys** (**mesonephroi**) are paired masses that lie against the dorsal body wall and extend the whole length of the abdomen above the swim bladder. They are often fused posteriorly, but the anterior parts are usually separated by the dorsal aorta. The anterior ends consist largely of blood sinuses and have lost their renal function. In the posterior end they follow the body wall ventrally. Here the posterior ends of the **wolffian ducts** (**mesonephric ducts**) may be seen extending the short distance from the kidneys to the small **urinary bladder,** which lies posteriorly between the gonad and the swim bladder. In the female the urinary bladder joins the oviduct to form a **urogenital sinus,** emptying through the **urogenital pore.** In the male the bladder empties separately through a **urinary pore,** around which there may be a small external projection of the bladder called **the urinary papilla.** The male urinary and genital pores lie close together posterior to the anus.

Circulatory System

☞ Extend the midventral incision to the jaw to expose the heart. Enlarge the opening by removing a triangular piece of body wall on each side of the cut.

Heart. The **pericardial cavity** is separated from the abdominal cavity by a **transverse septum.** The sep-tum is not homologous to the diaphragm of mammals.

The **heart** has two chambers—a thin-walled **atrium** and a muscular **ventricle** (Fig. 18-5). Blood collected from the venous system enters the **sinus venosus,** a thin-walled sac adjoining the atrium posteriorly. Blood flows from this into the atrium and from there to the ventricle, which lies ventral to it and is the most prominent part of the heart. The ventricle pumps into a short swollen **bulbus arteriosus,** which is really the first part of the ventral aorta.

Arterial System. From the bulbus arteriosus the blood flows into the short **ventral aorta.** Remove the operculum and trace the aorta forward. It gives off four pairs of **afferent branchial arteries** to the branchial arches. From the capillaries in the gills the oxygenated blood is collected by **efferent branchial arteries** and is emptied into the two roots of the **dorsal aorta** above. These roots join immediately to form the dorsal aorta. Remove the swim bladder membrane carefully and look along the middorsal wall for the dorsal aorta. Dissect carefully above the branchial arches to see where the roots of the dorsal aorta arise from the efferent branchial arteries. Extending forward into the head from the first (most anterior) efferent artery on each side is a **carotid artery,** which branches before reaching the eye.

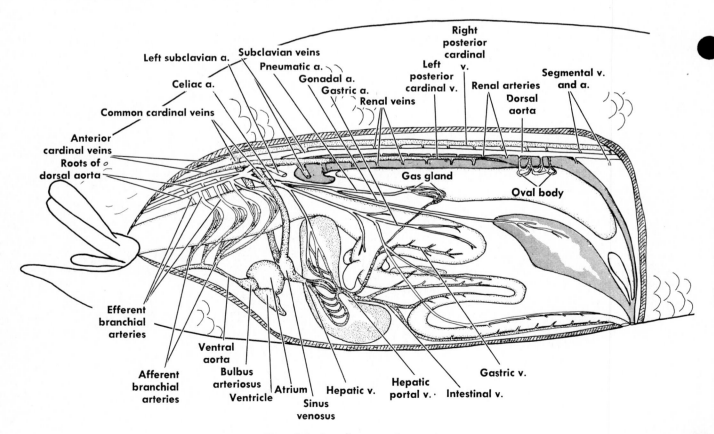

Fig. 18-5. Circulation in the perch.

The **celiac artery** (single) arises from the dorsal aorta just back of the efferent branchial arteries. Branches from the celiac supply the stomach (**gastric**), intestine (**intestinal**), liver (**hepatic**), gonads (**spermatic or ovarian**), spleen (**splenic**), and swim bladder (**pneumatic**).

The **subclavian arteries** (left and right) arise posterior to the beginning of the celiac artery and supply the pectoral fins.

Spinal and **renal arteries** are given off regularly along the aorta to the dorsal wall musculature and to the kidneys. The dorsal aorta continues posteriorly into the tail as the **caudal artery**.

Venous System. If your specimen is doubly or triply injected, you will be able to dissect the venous circulation. Otherwise it will be extremely difficult.

All blood returning to the heart both from the **cardinal venous system** and from the **hepatic portal system** passes first through the sinus venosus.

The paired **anterior cardinal veins** drain blood from the head into the **common cardinal veins** (**ducts of Cuvier**). The paired **posterior cardinal veins** lie on each side of the dorsal aorta between the kidneys and empty also into the common cardinals. The posterior cardinals collect from a number of **renal** and **segmental veins**. One of the posterior cardinal veins begins in the tail as the **caudal vein**. Small **subclavian veins** also empty into the common cardinals. The common cardinals are large sinuses passing ventrally on each side of the esophagus and just in front of the liver to empty into the sinus venosus.

☞ Examine some fish blood on a slide; note oval **erythrocytes** and various types of **leukocytes**. Do the red cells have nuclei?

Resume of the Circulatory Plan. To review the general plan of the circulatory system of the perch, we find blood driven by a muscular two-chambered heart through a system of arteries into an intricate network of capillaries that are microscopic and are enclosed in walls of simple squamous epithelium. These capillary beds are found in all tissues of the body, where they deliver oxygen and nutrients to the cells and collect from the cells carbon dioxide and other metabolic waste products. From the capillary beds the blood is carried off by veins, which are thin walled and nonelastic, to the heart. Arteries always spring from the heart or branch off from larger arteries and end in capillaries. Veins begin in capillaries and end in larger veins or in the heart. There are two exceptions to this pattern: (1) gill capillaries are interposed in the arterial system between afferent and efferent arteries; and (2) the liver interrupts the venous system between the hepatic portal vein and the hepatic vein. Thus the hepatic portal vein, which begins in capillaries (intestine and others) and also ends in capillaries (liver), comprises a portal system. All blood in the circuit around the body passes through at least two capillary beds, one in the gills and one in the body tissue it serves. That which goes to the digestive tract and related organs passes through three capillary beds—in the gills, in the organ tissue, and in the liver—before returning to the heart.

ORAL REPORT

Be able to identify the organs and the external structures of the perch and give their functions.

Be able to trace the flow of blood to and from any part of the body (for example, the head, pectoral fins, dorsal muscles, intestine, and others), noting the position of any capillary beds within each circuit.

Make a comparison between the various body systems of the perch and those of the ammocoetes larva. Compare with those of the dogfish shark.

DEMONSTRATIONS

1. Various teleost fishes.

2. Aquarium with live fishes. An aquarium is the best way to observe swimming and respiratory movements of live fishes.

3. Chromatophores in fish scales. With fine forceps, remove a few scales from a living fish, for example, goldfish, minnow, or perch (it does not need to be anesthetized). Mount in a drop of water on a slide under a coverslip and observe under low power. Note the different types of chromatophores and the amount of pigment dispersal.

If desired, the effect of drugs or hormones on the chromatophores can be studied by adding a drop of the drug or hormone to the water on the slide. Any of the following can be used: epinephrine (1 mg/ml), acetylcholine (100 mg/ml), melatonin (0.5 mg/ml), or pituitary extract (1 g beef pituitary powder [whole gland] shaken up in 10 ml of frog Ringer's solution and filtered after letting stand 30 minutes).

CHAPTER 19

CLASS AMPHIBIA*

THE FROG

The amphibians are a transition group between the aquatic and the strictly land animals. They still have the soft moist epidermis of the aquatic forms and therefore cannot stray too far from water or moist surroundings. Amphibian eggs lack the tough protective shell characteristic of terrestrial forms and so must still be laid in water. Many of the amphibians have developed lungs for air breathing but still have aquatic larvae with external gills. The moist skin is also a respiratory organ. The change to lungs has brought a change in circulation that includes a pulmonary circuit as well as a systemic circuit. Amphibians have a three-chambered heart. They are usually four limbed for walking or jumping but often have webbed feet for swimming.

Frogs, although more specialized and less typical of amphibians than salamanders and newts, are more commonly used in the general laboratory because of their ready availability. The large bullfrog *Rana catesbeiana* and the small leopard or grass frog *Rana pipiens* are favorites for dissection.

MATERIALS
General
  Live frogs
  Preserved frogs
  Dissecting pans
  Dissecting tools
  Pins
  Mounted frog skeletons
  Microscopes

---

*See Hickman, C.P., Jr., L.S. Roberts, and F.M. Hickman. 1984. Integrated principles of zoology, ed. 7, St. Louis, The C.V. Mosby Co., pp. 556-577.

Exercise 19A
  Jars or bowls
  Paper toweling
  Fruit flies (for feeding frogs)
Exercise 19B
  Frog and other vertebrate skeletons
  Individual vertebrae
  Frog skulls
Exercise 19D
  Prepared slides
    Frog kidney
    Frog testis
    Frog ovary
    Sperm smears
Exercise 19E
  Frog Ringer's solution
  Frog holders
  Ice
  Pins
  Paper towels
  Masking tape

THE FROG
Phylum Chordata
  Subphylum Vertebrata
    Class Amphibia
      Order Salientia (= Anura)
        Family Ranidae
          Genus *Rana*

Where Found
Ranid frogs (family Ranidae) are almost worldwide in distribution. Their favorite habitats are swamps, low meadows, brooks, and ponds, where they feed on flies and other insects. Their young, the tadpoles, develop in water and are herbivorous.

Behavior and adaptations

☞ Place a live frog on a piece of wet paper toweling in a jar large enough not to cramp it. Do not let its skin become dry. Do not excite it unnecessarily. Observe its adaptations. Make notes of your observations. Have a mounted frog skeleton at your table for comparison.

Is the skin smooth or rough? Moist or dry? Why is the frog difficult to hold? Examine the frog's feet. How are they adapted for jumping? For swimming? For landing on a slippery rock or log? Note the sitting position and compare with the mounted skeleton.

Compare the color of the dorsal and ventral sides. What is the adaptive advantage in this difference?

How is the position of the eyes advantageous to the animal? How is the nictitating membrane used? How do the eyes close? Examine the skeleton and see how this is possible.

Feeding Reactions. Place some live fruit flies in the jar with the frog. If the frog is hungry and is not excited, it may feed. Describe how it seizes the prey.

Breathing. Observe the movements of the throat and nostrils. Air is drawn into the mouth; then the nostrils close and throat muscles contract to force the air into the lungs. The nostrils (nares) can be opened or closed at will. Record the number of movements per minute of the throat and then of the nostrils. Do their rates coincide? Do the sides of the body move in breathing? Excite the animal by prodding; then as soon as it becomes quiescent, count the rate of breathing again.

Righting Reaction. Place the frog on its back, release it, and note how it rights itself. Repeat this experiment but hold the frog down gently with your hand until it ceases to struggle. When you release it, the frog may remain in this so-called hypnotic state for some time.

Locomotion. Observe a frog in an aquarium. What is the floating position? How does it use its limbs in swimming? How does it dive from a floating position? Note how it jumps.

WRITTEN REPORT

Record your observations on pp. 309-310.

External Structure

☞ Study a preserved frog and compare it with a live frog and a mounted skeleton.

Note the **head** and **trunk**. Note the **sacral hump** produced by the protrusion of the pelvic girdle. Find this hump on the mounted skeleton. The **cloacal opening** is at the posterior end of the body.

On the **forelimbs**, identify the arm, forearm, wrist, hand, and digits. On the **hindlimbs**, find the thigh, shank, ankle, foot, and digits. How does the number of digits compare with your own? Can you find the rudimentary thumb (prepollux) and the rudimentary sixth toe (prehallux)? During the breeding season the inner (thumb) digit of the male is enlarged into a nuptial pad for clasping the female. Observe the **webbed** toes of the hindfoot. How is the long ankle advantageous to the frog?

The **eyes** are protected by **eyelids** and by a transparent **nictitating membrane.** Look in the corner of your neighbor's eye for a vestige of this membrane, the semilunar fold.

The **tympanic membrane** (eardrum) is a circular region of tightly drawn skin just back of the eye. The frog has no external ear—only the middle and internal ears.

Label the external parts of the frog in the drawing on p. 309.

PROJECTS AND DEMONSTRATIONS

1. Demonstration of common species of frogs, toads, and salamanders. Learn the names and distinguishing characteristics of the species common to your territory.

2. Preserving the color patterns of frog skin. Kill a frog with ether fumes. Slit the ventral side of the skin from jaw to anus, extending the cut down the legs and digits. Loosen and remove the skin. Float the skin in water and remove any clinging tissue. Float again in clean water, pigment side uppermost. Slip a piece of wet cardboard under the skin and bring it up so that the skin is spread out on it. Straighten the skin and smooth it out to remove air bubbles. Place the cardboard on a blotter or newspaper until most of the moisture has escaped and the skin has become firmly attached. Then dry thoroughly between blotters under pressure or in botanic driers. When the skin is dry, clinging fibers of the dry blotting paper may be removed with a moist cloth. Such skins may be mounted under glass and kept for long periods in excellent color.

EXTERNAL ANATOMY OF LEOPARD FROG

OBSERVATIONS ON BEHAVIOR OF FROG

Feeding reaction _____

_____

_____

_____

Breathing _____

_____

_____

_____

Reaction to touch _____

_____

_____

_____

Righting reaction _____

_____

_____

_____

Locomotion _____

_____

_____

_____

_____

Other observations _____

_____

_____

_____

EXERCISE 19B
The skeleton*

The skeletal system serves as a **supporting framework** for the body, as a **protection** for delicate structures, and as a **place of attachment for muscles.** It also furnishes **levers** for muscular action and serves as a **factory for blood corpuscles** and a **storehouse for minerals.**

☞ Prepared skeletons are brittle and delicate. Handle them with care and do not deface the bones in any way. Use a probe, not a pencil, in pointing.

The **axial skeleton** includes the skull, the vertebral column, and the sternum. The **appendicular skeleton** includes the pectoral and pelvic girdles and the forelimbs and hindlimbs.

At what points are the girdles articulated with the axial skeleton? What parts of the body are best protected by skeletal parts? What causes the hump in the frog's back? Is the pectoral girdle free to move around the backbone to some extent? Of what advantage is this to the frog? The pelvic girdle is firmly attached to the vertebral column. Is this an advantage in jumping?

For a discussion of the structure and growth of bones and their articulations, see Exercise 20A, pp. 339-340.

*See Hickman, C.P., Jr., L.S. Roberts, and F.M. Hickman. 1984. Integrated principles of zoology, ed. 7, St. Louis, The C.V. Mosby Co., pp. 569-570, 673-679.

Axial Skeleton

Skull. The skull (Fig. 19-1) includes the **cranium,** or brain case, and the **visceral skeleton,** made up of the bones and cartilages of the jaws, the hyoid apparatus, and the columellae (little bones of the ears). The **orbital fossae** and the **nasal fossae** are the dorsal openings where the eyes and external nares (nostrils) are located.

Locate on the dorsal side of the skull the **nasal** bones; the single **sphenethmoid;** the long **frontoparietals,** which cover much of the brain; the **prootics,** which enclose the inner ears; the **exoccipitals,** which surround the hindpart of the brain; and the **foramen magnum,** the opening for the spinal cord.

The upper jaw is formed by the **premaxillae,** the **maxillae,** and the **quadratojugals.** Which bones bear teeth? The **squamosal** supports the cartilaginous **auditory capsule.** The three-pronged **pterygoid** articulates with the maxillary, the prootic, and the quadratojugal.

On the ventral surface of the skull, find the wing-shaped **vomers** (the **vomerine teeth** are projections of these bones); the slender **palatines;** and the dagger-shaped **parasphenoid,** which forms the floor of the brain case. Cartilages form the sides of the braincase.

The lower jaw (**the mandible**) consists of small **mentomeckelians,** long **dentary** bones, and the **angulosplenials.**

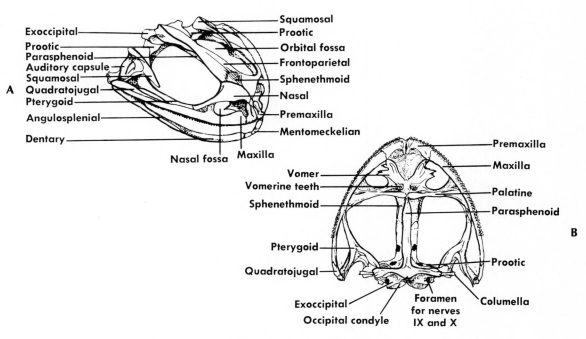

Fig. 19-1. The skull of a frog. **A,** Dorsolateral view. **B,** Ventral view.

311

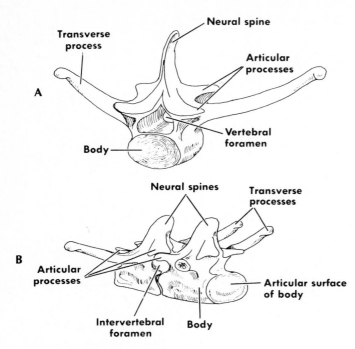

Fig. 19-2. A, Frog vertebra, anterior view. **B,** Two frog vertebrae, posterolateral view, showing articulations.

The **hyoid** apparatus lies in the floor of the mouth, but in the prepared frog skeleton it may be mounted separately. It is cartilaginous and supports the tongue and larynx.

In some preparations the **columella**, a small bone used in the transmission of sound, is found in the auditory capsule.

Vertebral Column. The backbone of the frog consists of nine vertebrae and a urostyle, which probably represents the fusion of several vertebrae. The first vertebra, **atlas**, articulates with the skull. The ninth or **sacral vertebra** has transverse processes for articulation with the ilia of the pelvic girdle. With the help of Fig. 19-2 identify the parts of a vertebra. Frogs are usually described as having no ribs, but their embryology indicates that the ends of the transverse processes originate separately and so are actually short ribs fused to the transverse processes.

Sternum. The sternum provides ventral protection for the heart and lungs and a center for muscular attachment. Its five parts, beginning at the anterior end, are the **episternum, omosternum, epicoracoid** (a thin cartilage between the clavicles and coracoids), **mesosternum,** and **xiphisternum** (Fig. 19-3).

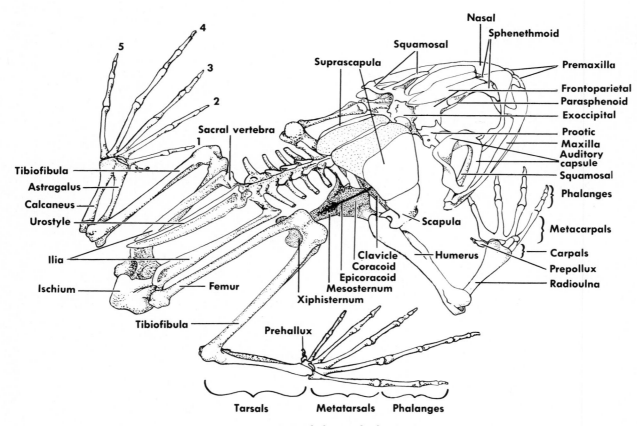

Fig. 19-3. Skeleton of a frog.

Appendicular Skeleton

Pectoral Girdle and Forelimbs. The pectoral girdle supports the forearms and articulates with the sternum ventrally. Each half of the girdle includes a **suprascapula** and **scapula** (shoulder blade) and a **clavicle** (collar bone) lying anterior to the **coracoid.** Consult Fig. 19-3 for the bones of the forelimb.

Pelvic Girdle and Hindlimbs. The pelvic girdle supports the hindlimbs. Each half is made up of the long **ilium,** the anterior **pubis,** and the posterior **ischium.** Consult Fig. 19-3 for the bones of the hindlimb.

ORAL REPORT

Be familiar with the parts of the skeleton and the purpose served by each part.

EXERCISE 19C
The skeletal muscles*

For a general discussion of skeletal muscles and for the terminology used in their study, see pp. 340-341.

DIRECTIONS FOR STUDY OF FROG MUSCLES

☞ To skin the frog, slit the skin midventrally from anal region to chin, keeping the scissors point up to prevent injuring the underlying muscles. Make a transverse cut completely around the body just above the hindlegs and another one anterior to the forelegs. A middorsal cut the length of the back will divide the skin into portions that can be peeled off easily. Loosen the skin with a blunt instrument and carefully pull off over a leg wrong side out. Be careful not to tear thin muscle attached to the skin. The skin can be pulled over the head and eyes in the same way.

The large spaces between skin and muscle where the skin is not attached are **subcutaneous lymph sacs.**

☞ In separating the muscles from each other, first observe the direction of the muscle fibers and the extent of the muscle; then use your fingers, a blunt probe, or the *handle* of a scalpel to loosen the tissues. *Never use scissors, scalpel blade, or needle for dissecting muscles.* They

*See Hickman, C.P., Jr., L.S. Roberts, and F.M. Hickman. 1984. Integrated principles of zoology, ed. 7. St. Louis, The C.V. Mosby Co., pp. 570, 680-687.

PROJECTS AND DEMONSTRATIONS

1. Skeletons of other vertebrates. Compare fish, reptile, amphibian, bird, and mammal skeletons with the human skeleton or bones.

2. Staining skeletons. Skeletons of embryos and small animals in situ may be stained by the following method. First, fix small specimens (skinned small frogs are excellent) in 95% alcohol for 2 to 4 days. Then place them in 2% potassium hydroxide solution until the bones are visible through the tissue. Check frequently to make sure that the specimens are not macerated. Now, transfer them to the following solution for 24 hours: 1 part of alizarin to 10,000 parts of 2% potassium hydroxide. Allow the stain to act until the desired intensity is obtained. It may take longer than 24 hours. Finally, clear the specimens in increasing concentrations of glycerin (10% to 50%). Excessively stained bones may be destained in 1% sulfuric acid made up in 95% alcohol.

3. Other suggestions. See p. 340 for other suggestions.

tear the muscles. Never cut a muscle unless instructed to do so. If necessary to cut superficial muscles to find deep muscles, cut squarely across the belly (middle fleshy portion) of the muscle, leaving the origin and insertion in place.

Ventral Trunk Muscles

Pectoralis. Large fan-shaped muscles (Fig. 19-4). **Origin**—sternum, also fascia of the abdominal wall. **Insertion**—on the humerus. **Action**—to flex, adduct, and rotate the arm.

Rectus Abdominis. Longitudinal muscles along midventral line, separated by the **linea alba** on median line. **Origin**—pubic border. **Insertion**—on the sternum. **Action**—support the abdominal viscera; hold the sternum in place.

External Oblique. Sheet muscles. **Origin**—dorsal fascia of vertebrae, also the ilium. **Insertion**—on the linea alba. **Action**—help constrict abdomen and support the viscera.

Transversus. Sheet muscles beneath the external oblique and the rectus abdominis. **Origin**—ilium and vertebrae. **Insertion**—linea alba. **Action**—help support the abdominal viscera.

Ventral Muscles of the Thigh

Sartorius. Long, strap-shaped muscle (Fig. 19-4). **Origin**—pubis. **Insertion**—tibiofibula. **Action**—flexes the shank and adducts the thigh.

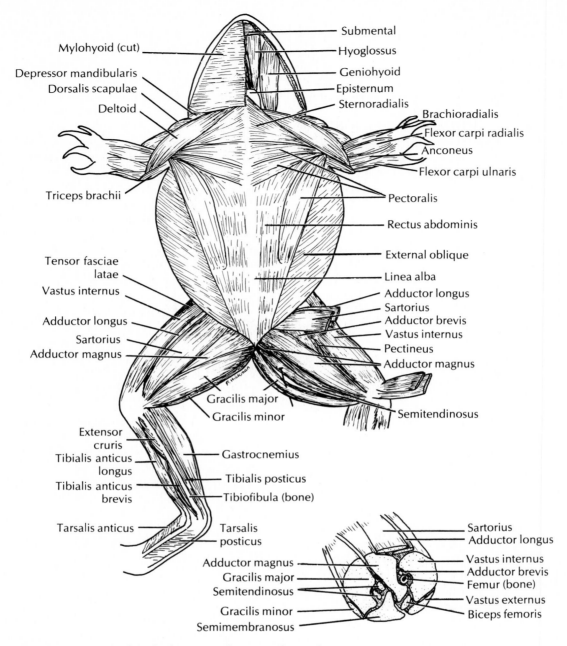

Labels on main figure (ventral view):

Mylohyoid (cut) — Submental
Depressor mandibularis — Hyoglossus
Dorsalis scapulae — Geniohyoid
Deltoid — Episternum
— Sternoradialis
— Brachioradialis
— Flexor carpi radialis
— Anconeus
Triceps brachii — Flexor carpi ulnaris
— Pectoralis
— Rectus abdominis
— External oblique
Tensor fasciae latae — Linea alba
Vastus internus — Adductor longus
Adductor longus — Sartorius
Sartorius — Adductor brevis
Adductor magnus — Vastus internus
— Pectineus
— Adductor magnus
Gracilis major
Gracilis minor — Semitendinosus
Extensor cruris
Tibialis anticus longus — Gastrocnemius
Tibialis anticus brevis — Tibialis posticus
— Tibiofibula (bone)
Tarsalis anticus — Tarsalis posticus

Transverse section labels:

Adductor magnus — Sartorius
Gracilis major — Adductor longus
Semitendinosus — Vastus internus
Gracilis minor — Adductor brevis
Semimembranosus — Femur (bone)
— Vastus externus
— Biceps femoris

Fig. 19-4. Muscles of a frog, ventral view. *Below right,* transverse section of the left thigh.

Adductor Magnus. Seen as a triangle near the groin when the sartorius is in place. **Origin**—pubic and ischial symphysis. **Insertion**—distal end of femur. **Action**—adducts and flexes the thigh.

Gracilis Major (Rectus Internus Major). Partly covers adductor magnus. **Origin**—ischium. **Insertion**—tibiofibula. **Action**—adducts the thigh and flexes or extends the shank, according to its position.

Gracilis Minor (Rectus Internus Minor). Strap-shaped muscle. **Origin, insertion, and action**—same as for the gracilis major.

Adductor Longus. Thin muscle under sartorius.

Origin—ilium. **Insertion**—femur. **Action**—adducts the thigh.

Semitendinosus. Not a surface muscle; lies under and between the gracilis major and the adductor magnus. **Origin**—two heads on the ischium. **Insertion**—tibiofibula. **Action**—adducts the thigh and flexes the leg.

Dorsal Muscles of the Thigh

Triceps Femoris. Covers lateral surface of thigh; has three divisions: (1) **vastus internus**, the ventral portion, (2) **vastus externus**, the dorsal portion (Fig.

314

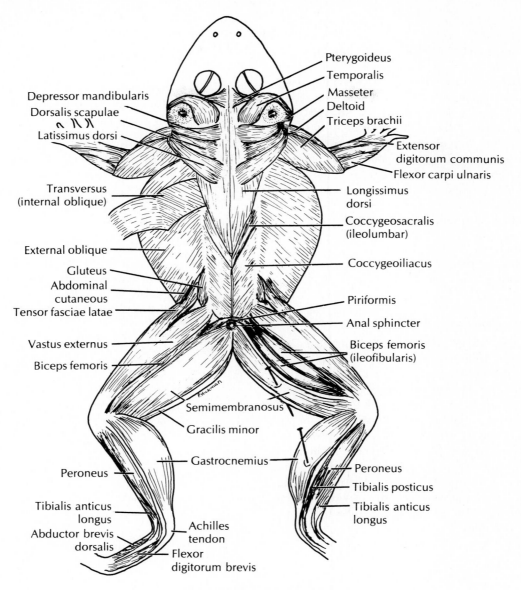

Fig. 19-5. Muscles of a frog, dorsal view.

Labels in figure:
Depressor mandibularis
Dorsalis scapulae
Latissimus dorsi
Pterygoideus
Temporalis
Masseter
Deltoid
Triceps brachii
Extensor digitorum communis
Flexor carpi ulnaris
Transversus (internal oblique)
Longissimus dorsi
Coccygeosacralis (ileolumbar)
External oblique
Coccygeoiliacus
Gluteus
Abdominal cutaneous
Tensor fasciae latae
Piriformis
Anal sphincter
Vastus externus
Biceps femoris (ileofibularis)
Biceps femoris
Semimembranosus
Gracilis minor
Gastrocnemius
Peroneus
Tibialis posticus
Tibialis anticus longus
Peroneus
Tibialis anticus longus
Abductor brevis dorsalis
Achilles tendon
Flexor digitorum brevis

19-4), and (3) **tensor fascia lata (rectus femoris)**, the lateral portion that inserts on the other two. **Origin**—one head on the acetabulum, two on the ilium. **Insertion**—tibiofibula. **Action**—abducts the thigh and extends the shank.

Biceps Femoris (Iliofibularis). Slender muscle; the sciatic nerve lies underneath. **Origin**—ilium. **Insertion**—tibiofibula. **Action**—flexes the shank (Fig. 19-5).

Semimembranosus. Large dorsal muscle. **Origin**—ischium. **Insertion**—tibiofibula. **Action**—adducts the thigh and flexes or extends the shank, according to its position.

Gluteus. Small muscle between the ilium and the tensor fascia lata. **Origin**—ilium. **Insertion**—femur. **Action**—rotates the thigh forward.

Piriformis. Small muscle between biceps and semi-membranosus. **Origin**—tip of urostyle. **Insertion**—femur. **Action**—draws the urostyle ventrally and laterally and the femur dorsally.

Muscles of the Shank

Gastrocnemius. Forms the calf of the leg (Fig. 19-5). **Origin**—two heads: distal end of femur and tendon from triceps femoris. **Insertion**—by tendon of Achilles to sole of foot. **Action**—flexes the shank and extends the ankle and foot.

Peroneus. On dorsal side of shank. **Origin**—femur. **Insertion**—distal end of the tibiofibula; head of the calcaneus. **Action**—extends the shank; when foot is extended, extends it further; when flexed, flexes it further.

Tibialis Anticus Longus. On the tibiofibula. **Ori-**

315

gin—femur; divides into two bellies. **Insertion**—by two tendons on the ankle bones (astragalus and calcaneus). **Action**—extends the shank and flexes the ankle.

Extensor Cruris. On the upper two-thirds of the tibiofibula. **Origin**—femur. **Insertion**—ventral surface of the tibiofibula. **Action**—extends the shank.

Tibialis Anticus Brevis. Small muscle on tibiofibula. **Origin**—middle of tibiofibula. **Insertion**—ankle (astragalus). **Action**—flexes the ankle.

Tibialis Posticus. Between gastrocnemius and tibiofibula. **Origin**—side of tibiofibula. **Insertion**—ankle (astragalus). **Action**—extends the foot when flexed; flexes the foot when fully extended.

Other Muscles

With the help of Figs. 19-4 and 19-5 you can identify many of the muscles of the back, shoulder, head, and arm.

How the Muscles Act

A muscle has only one function—to contract. For effective action, muscles must be arranged in antagonistic pairs. The gastrocnemius and the tibialis anticus longus represent such an antagonistic pair. Loosen the body of each of these muscles, pull on the gastrocnemius, and see what happens. Now pull on the tibialis anticus longus. Which of these muscles would be used in jumping or diving? Which in sitting? See if you can locate other antagonistic pairs.

For most movements groups of muscles rather than single muscles are required. By varying the combination of these groups, many complicated movements are possible.

ORAL REPORT

Be able to demonstrate a careful dissection of the muscles and to name the muscles and their actions.

WRITTEN REPORT

On p. 318 tell (1) what principal muscles of the hindleg are involved when a frog leaps and (2) when the frog resumes a sitting position, what principal muscles contract.

DRAWINGS

On p. 317 draw and label the following muscles, locating the origin and insertion of each muscle in the proper place on each diagram:

On **A,** draw the gracilis muscles; add and label an antagonistic muscle (ventral view).

On **B,** draw the gastrocnemius muscle; add and label an antagonistic muscle.

On **C,** draw the gluteus, biceps femoris, and peroneus muscles.

On **D,** draw in muscles that would be used to straighten the leg as in diving—at least one each that would adduct the thigh, extend the shank, and extend the foot.

RIGHT HINDLEG OF FROG

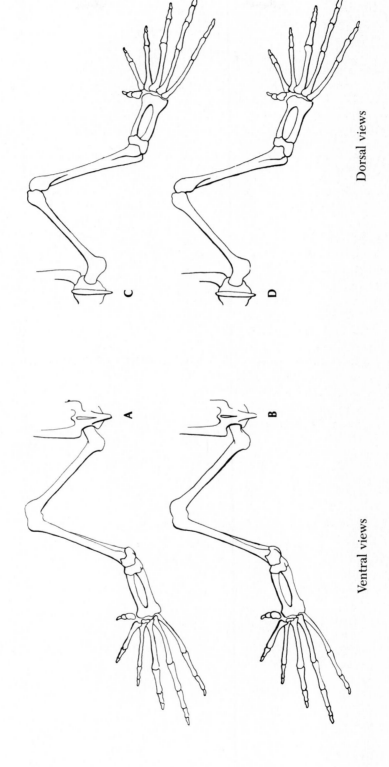

Dorsal views

Ventral views

PRINCIPAL LEG MUSCLES INVOLVED IN LEAPING

PRINCIPAL LEG MUSCLES INVOLVED IN SITTING

EXERCISE 19D
The digestive, respiratory, and urogenital systems

MOUTHPARTS

☞ Pry open the mouth, cutting the angle of the jaw if necessary, and wash in running water.

The posterior portion of the mouth cavity is the **pharynx**, which connects with the **esophagus**. Feel the **maxillary teeth** along the upper jaw and the **vomerine teeth** in the roof of the mouth (Fig. 19-6). Are these better adapted for biting and chewing or for holding the prey to prevent its escape? Find the **internal nares** (sing. **naris**) in the roof of the mouth and note how they connect with the external nares. Note how the ridge on the lower jaw fits into a groove in the upper jaw to make the mouth closure airtight. This is important in the frog's respiratory movements.

Eustachian tubes, which connect with and equalize the air pressure in the middle ear, open near the angle of the jaws. In male frogs openings on the floor of the mouth slightly anterior to the eustachian tubes lead to **vocal sacs,** which, when inflated, serve as resonators to intensify the mating call. Examine the **tongue** and note where it is attached. Which end of the tongue is flipped out to catch insects? Feel the **sensory papillae** on the tongue surface. Behind the tongue is a slight elevation in the floor of the mouth, containing the **glottis,** a slitlike opening into the **larynx.**

Dissection of the Frog

☞ Make an incision through the abdominal wall from the junction of the hindlegs to the lower jaw, cutting through the bones of the pectoral girdle as you go. Make transverse cuts anterior to the hindlegs and posterior to the forelegs and pin back the flaps of muscular tissue.

Note the three layers of the body wall: **skin, muscles** (with enclosed skeleton in some places), and **peritoneum,** which lines the large **coelom.**

In a mature female the ovaries with their dark masses of eggs may fill much of the coelomic cavity. In this case, remove the left ovary and its white convoluted oviduct.

Note the **heart** enclosed in its **pericardial sac** and surrounded by lobes of the **liver.** Lift up the heart to find the **lungs.**

Digestive System*

You have seen the **mouth** and **pharynx.** Lift the heart, liver, and lungs to see where the **esophagus empties into the stomach** (Fig. 19-7). A **pyloric valve** controls movements of food into the **small intestine.** Note the blood vessels in the mesentery, which holds the stomach and small intestine in place. Why must the digestive tract be so well supplied with blood?

The **liver,** the largest gland in the body, secretes bile, which is carried by a small duct to the **gallbladder** for storage. Find the gallbladder between the right and median lobes of the liver. The **pancreas** is thin and inconspicuous, lying in the mesentery between the stomach and duodenum.

The **large intestine** narrows down in the pelvic region to form the **cloaca,** which also receives urine from the kidneys and products from the reproductive or-

---

*See Hickman, C.P., Jr., L.S. Roberts, and F.M. Hickman. 1984. Integrated principles of zoology, ed. 7. St. Louis, The C.V. Mosby Co. pp. 572, 744-751.

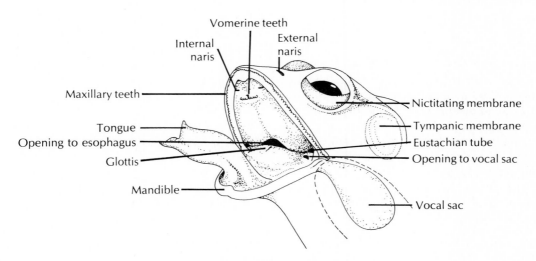

Fig. 19-6. Mouthparts of a frog.

Labels: Vomerine teeth; Internal naris; External naris; Maxillary teeth; Nictitating membrane; Tympanic membrane; Tongue; Eustachian tube; Opening to esophagus; Glottis; Opening to vocal sac; Mandible; Vocal sac

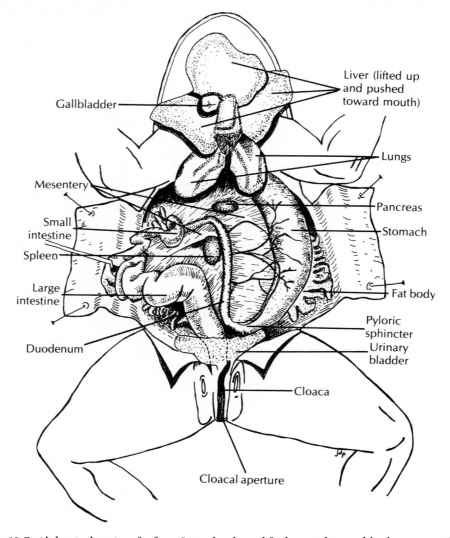

Gallbladder

Liver (lifted up and pushed toward mouth)

Lungs

Mesentery

Pancreas

Small intestine

Stomach

Spleen

Large intestine

Fat body

Duodenum

Pyloric sphincter

Urinary bladder

Cloaca

Cloacal aperture

Fig. 19-7. Abdominal cavity of a frog. Liver has been lifted up and turned back to expose lungs.

gans. It empties through the **cloacal opening (anus)**. Not all vertebrate animals have cloacas; humans do not.

Respiratory system*

Amphibians breathe through their skin (**cutaneous respiration**) as well as their lungs. Some respiration also occurs through the lining of the mouth.

The frog has no diaphragm. Thus it draws air into its mouth cavity through nares by closing the **glottis** and depressing the floor of its mouth. Then by closing the nares and raising the floor of its mouth cavity it forces air from the mouth through the glottis into the lungs. Air is expelled from the lungs by the contraction of the muscles of the body wall and the elastic recoil of the stretched lung.

Probe through the **glottis** into the **larynx**. Find the short **bronchus**, connecting each **lung** to the larynx. Slit open a lung and observe its internal structure. Note the little pockets or **alveoli** (sing. **alveolus**) in the lining. What is the purpose of this arrangement?

Did you find any parasites in the lungs of your specimen? Can you identify them?

Urogenital System*

Functionally, the urogenital system is two systems, the **urinary**, or **excretory**, **system** and the **reproductive system.** However, since some structures function in both systems, they are usually considered together.

Be careful not to injure the blood vessels as you study this system.

---

*See Hickman, C.P., Jr., L.S. Roberts, and F.M. Hickman. 1984. Integrated principles of zoology, ed. 7, St. Louis, The C.V. Mosby Co., pp. 570-571, 710-719.

*See Hickman, C.P., Jr., L.S. Roberts, and F.M. Hickman. 1984. Integrated principles of zoology, ed. 7. St. Louis, The C.V. Mosby Co., pp. 574-576, 729-737, 842, 856.

Excretory System. The **kidneys** separate the urine from the blood. They rid the body of metabolic wastes (aided by the lungs and skin), and they maintain a proper water balance in the body and a general constancy of content in the blood.

The kidneys lie close to the dorsal body wall separated from the coelom by a thin peritoneum. The **urinary bladder,** when collapsed, appears as a soft mass of thin tissue just ventral to the large intestine. It is bilobed and empties into the cloaca. The **ureters** connect the kidneys with the cloaca.

☞ To expose the cloaca, cut through the ischiopubic symphysis with a scalpel and push the pelvic bones aside. If you wish, you may open the cloaca just left of the midventral line, separate the cut edges, and find the bladder opening on the ventral wall of the cloaca and the openings of the ureters on the dorsal wall.

The **adrenal glands,** a light stripe on the ventral surface of each kidney, are endocrine glands, not urogenital organs. **Fat bodies** attached to the kidneys, but lying in the coelom, are for fat storage. They may be large in the fall and small or absent in the spring. Why?

Male Reproductive System. A small pale **testis** lies on the ventral side of each kidney. Sperm pass from the testis into some of the kidney tubules and then are carried by the ureter to the cloaca and outside. Thus the male ureters serve also as genital ducts. In leopard frogs a small **vestigial oviduct** runs parallel to the ureter.

Female Reproductive System. The **ovaries** are attached by mesenteries to the dorsal wall of the coelom. In winter and early spring the ovaries are distended with eggs. If the specimen was killed in summer or early fall, the ovaries will be small, pale, and fan shaped. Convoluted **oviducts** widen anteriorly (dorsal to the lungs) into funnel-like **ostia** and posteriorly into **uteri,** which empty into the cloaca. Eggs are released from the ovary into the coelom, carried in coelomic fluid to the ostia, and then down the oviducts by ciliary action to the outside. At amplexus (the copulatory embrace) the male clasps the female and fertilizes the eggs externally as they are laid in the water.

DRAWINGS

Complete and label the drawings of the urogenital systems on p. 323. Draw your own specimen first; then exchange with someone who has one of the other sex and draw that. Some of the structures have been included on one side as they might appear in some specimens. Draw the other side as it appears in your specimen.

Histology of the Urogenital Organs

1. Kidney. Examine a stained slide of a cross section of frog (or other vertebrate) kidney. Use both the low and the high power of the microscope.

The kidney is made up of renal units, called **nephrons,** embedded in connective tissue. Each is made up of a **renal (malpighian) corpuscle** and a **uriniferous tubule.** The tubules drain into **collecting tubules.** First, note the renal corpuscles that appear on your slide as conspicuous islands of tissue surrounded by clear spaces. Each renal corpuscle is made up of a network of capillaries called a **glomerulus,** which is surrounded by a capsule of simple squamous epithelium called **Bowman's capsule.** The uriniferous and collecting tubules, which are coiled and twisted in the kidney, will be cut at all angles in a cross section, thus showing up as circles, ellipses, and longitudinal sections of cuboidal and columnar epithelium. Read your textbook to understand the structure and physiology of the kidney.*

2. Testis. Examine the stained cross section of frog testis. Note the sections through the **seminiferous tubules.** Most of the cells within the tubules are developing sex cells. These differentiate progressively from periphery to lumen in the tubule, with the **spermatogonia** lying nearest the periphery, the **spermatocytes** a little nearer the lumen, and the **spermatids** close to the lumen where they are transformed into functional spermatozoa, become detached, and are carried into the ducts. Do you see any nearly mature sperm in or near the lumen?

3. Spermatozoa. Examine some stained smears of frog testis, showing spermatozoa.

4. Ovary. Look at a prepared slide of a frog ovary. In an ovary containing mature eggs or those nearing maturity you will notice a few round objects of varying size with a scanty surrounding mass of tissue. These are the **eggs** in the thin walls of the ovary. In a less mature ovary each ovum or egg lies embedded in a

---

*See Hickman, C.P., Jr., L.S. Roberts, and F.M. Hickman. 1984. Integrated principles of zoology, ed. 7. St. Louis, The C.V. Mosby Co., pp. 729-737.

small chamber, the **follicle.** Connective tissue and blood vessels may also be seen.

DRAWINGS (OPTIONAL)

Make drawings of the histological sections of the kidneys, the testis, and the ovary, labeling each fully. Use p. 324 for these drawings.

PROJECTS AND DEMONSTRATIONS

1. Cross sections of the frog body. Cross sections should be made after the frog has been thoroughly frozen at a low temperature. Sections are easily cut with a hacksaw and completed according to directions given on p. 288. Revealing relations of organs may be seen from sections made at the level of (a) a region a short distance anterior to the hind legs and (b) a region just posterior to the forelegs.

2. Study of cross sections of the intestine. Compare slides of the cross section of the human intestine with that of the frog (See pp. 72-73 and Fig. 5-18.)

3. Survey of feeding habits of the frog. Have each student empty into a watch glass of water the contents of the stomach of a freshly collected frog and identify the kinds of food that the frog has eaten. Make a compilation of the results for the class.

4. Peristalsis in the frog. Pith the brain of a frog (Appendix B, p. 463) that has been fed an hour or two previously. Open the abdominal cavity and flood with warm 0.6% saline solution. Can you observe any peristaltic movement in the digestive tract? What type of muscle is involved here? In what direction(s) do the muscle fibers run?

Tie a thread tightly around the pyloric end of the stomach. Open the stomach near the cardiac end and with a pipette introduce physiological salt solution into the lumen. Close the opening with a second ligature; then cut out the stomach. Suspend the stomach by the pyloric end in Ringer's solution. Observe the wavelike peristaltic movement passing over the stomach. Record the rate per minute.

5. Ciliary movement in the mouth cavity of the frog. Pith the brain and spinal cord of a frog. (The same frog used for peristalsis can be used here.) Remove all the viscera except the digestive tract. Cut through the symphysis of the lower jaw with scissors and extend the cut through the esophagus to the stomach. Pin back the flaps of the lower jaw and pin out the esophagus to form a surface level with the roof of the mouth. Keep the mucous membrane moist. Place a tiny bit of wet cork on the mucous membrane and note what happens. By extending a thread between two pins placed on each side of the pharynx and another across the esophagus, mark a suitable start and finish line. (1) Time the passage of the cork between these marks. (2) Tilt the board and time the passage of the cork up the incline. (3) Bathe the ciliated area with quantities of warm 0.6% saline solution and time again. (4) Bathe with iced saline and time. (5) If ether vapor tubes are available (see Appendix B, p. 457, for preparation), try blowing ether vapor on the ciliated surface and note the results.

UROGENITAL SYSTEM OF FROG

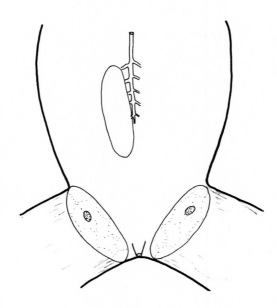

Male

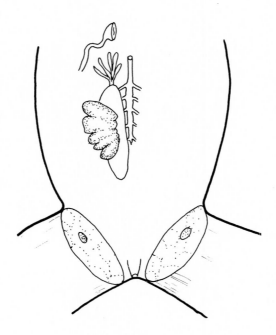

Female

EXERCISE 19E
The circulatory system*

In injected frogs the arteries are usually filled with red or yellow latex and the veins with blue latex. If only the arteries are injected, the veins may be filled with dark clotted blood.

☞ Dissect carefully and do not cut or injure the blood vessels. With a probe you may loosen the connective tissue that holds them in place. Cut away a midsection of the pectoral girdle and pin back the arms so that the heart is fully exposed. Carefully remove the **pericardium.**

---

*See Hickman, C.P., Jr., L.S. Roberts, and F.M. Hickman. 1984. Integrated principles of zoology, ed. 7. St. Louis, The C.V. Mosby Co., pp. 570-572.

Identify the thick-walled conical **ventricle;** the thin-walled **left and right atria;** the **conus arteriosus,** arising from the ventricle and dividing to form the **truncus arteriosus** on each side; and on the dorsal side of the heart the thin-walled **sinus venosus,** formed by the convergence of three large veins—two **precaval veins** and one **postcaval vein.**

Venous System

The **precava (anterior vena cava)** (Fig. 19-8) is formed by the union of (1) the **external jugular** from the tongue and floor of the mouth, (2) the **innominate** made up from the **subscapular vein** from the shoulder and the **internal jugular** from the brain, and (3) the **subclavian,**

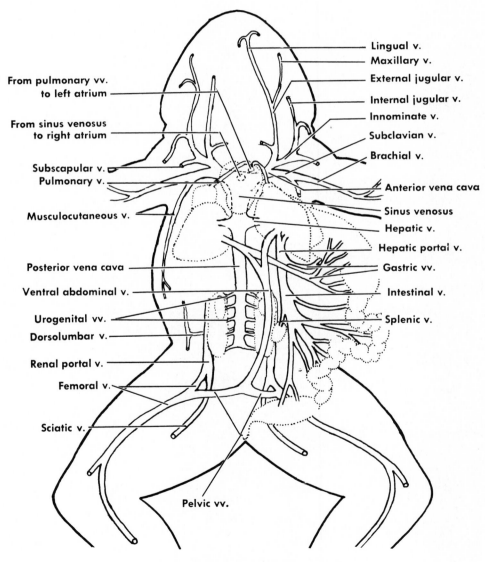

Fig. 19-8. Venous system of a frog.

which receives blood from the arm's **brachial vein** and the dorsal body wall's **musculocutaneous vein.**

The **postcava (posterior vena cava)** extends from the sinus venosus through the liver to the region between the kidneys. It receives **hepatic veins** from the liver, **renal veins** from each kidney, and the **ovarian** or **spermatic veins** from the gonads.

Pulmonary veins bring blood from the lungs to the left atrium. How does the blood in these veins differ from that in any other vein?

Portal Systems. Ordinarily veins carry blood directly from a capillary bed to the heart. This plan is interrupted in amphibians by capillary beds in two portal systems—the hepatic portal and the renal portal systems.

Hepatic portal system. In the hepatic portal system blood is carried to the capillaries of the liver by two veins: (1) The **ventral abdominal vein** in the ventral body wall collects from the pelvic veins, which are branches of the femoral veins; it empties into the liver. (2) the **hepatic portal vein** receives the splenic, pancreatic, intestinal, and gastric veins. From the capillary bed in the liver the blood is picked up by the **hepatic veins,** carried to the postcava, and so to the sinus venosus.

Renal portal system. Much of the blood from the hindlegs passes through the renal portal system. The **renal portal vein,** found along the outer margin of each kidney, is formed by the union of the **sciatic** and **femoral veins.** Blood from the kidneys is picked up by the **renal veins** and carried to the **postcava.** The renal portal system is absent in mammals.

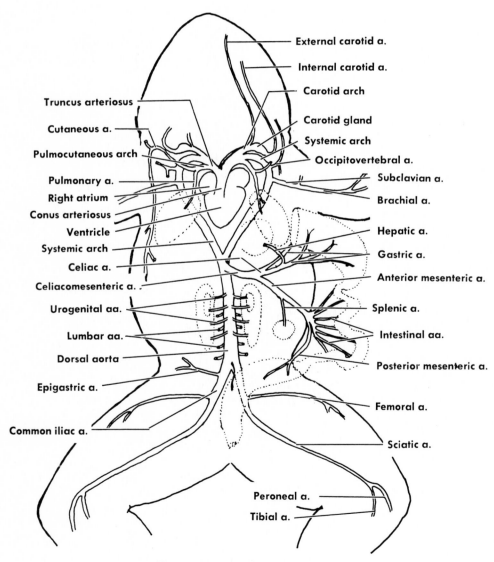

Fig. 19-9. Arterial system of a frog.

Arterial System

The **carotid, systemic,** and **pulmocutaneous arches** (known collectively as the **aortic arches**) arise from the **truncus arteriosus** (Fig. 19-9).

Carotid Arch. The **common carotid artery** divides into (1) the **internal carotid** to the roof of the mouth, eye, brain, and spinal cord, and (2) the **external carotid** (**lingual**) to the floor of the mouth, tongue, and thyroid gland.

Pulmocutaneous Arch. The third arch divides into the short **pulmonary artery** to the lungs and the longer **cutaneous artery** to the skin.

Systemic Arch. Each systemic arch gives off the following:

1. Small **esophageal arteries.**

2. The **occipitovertebral artery** (divides to nourish the skull and vertebral column).

3. The **subclavian artery** to the shoulder.

The two **systemic arteries** join to form the **dorsal aorta** (lift the kidneys to see it). The dorsal aorta gives off the following:

1. The large **celiacomesenteric artery,** which gives rise to the **celiac artery** (to the stomach, pancreas, and liver) and the **anterior mesenteric artery** (to the spleen and intestines).

2. **Urogenital arteries,** to the kidneys, fat bodies, and gonads.

3. **Lumbar arteries,** to the muscles of the back.

4. The **common iliac arteries** formed by the division of the dorsal aorta. Each iliac gives off (a) the **epigastric** to the abdominal wall, rectum, and bladder and (b) the **femoral** to the thigh, and (c) continues as the **sciatic** to the leg.

Heart

In the three-chambered amphibian heart there is a separate pulmonary system, but some mixing of venous and arterial blood can occur.

☞ Make a frontal section of the heart (Fig. 19-10), dividing it into dorsal and ventral valves.

Find the opening from the sinus venosus into the **right atrium** and the opening from the pulmonary veins into the **left atrium.** Why is the ventricle more muscular than the atria? The **conus arteriosus,** which receives blood from the **ventricle,** divides to form a left and right truncus arteriosus. Valves to prevent backflow of blood guard the entrances to the atria and the conus. What are these valves?

Lymphatic System

Blood does not come into direct contact with the tissues. Some of its constituents are filtered through the walls of the smaller blood vessels. This, plus what the cells give off, constitutes **lymph.** The cells select from the lymph the constituents they need and give off to the lymph their own excretions. The lymphatic system of the frog consists mainly of large irregular spaces so arranged that they carry the lymph back to the blood vessels. You noted some of these subcutaneous lymph spaces when you skinned the frog.

WRITTEN REPORT

Fill in the report on blood circulation on pp. 329-330.

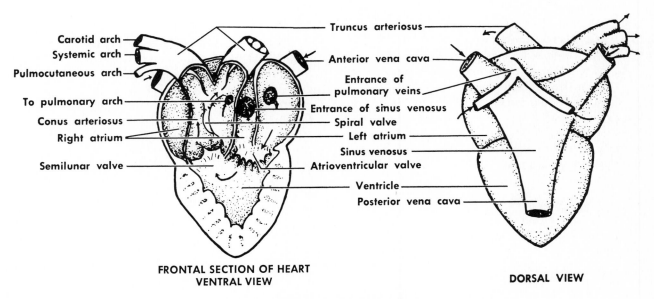

Carotid arch
Systemic arch
Pulmocutaneous arch
To pulmonary arch
Conus arteriosus
Right atrium
Semilunar valve

Truncus arteriosus
Anterior vena cava
Entrance of pulmonary veins
Entrance of sinus venosus
Spiral valve
Left atrium
Sinus venosus
Atrioventricular valve
Ventricle
Posterior vena cava

FRONTAL SECTION OF HEART
VENTRAL VIEW

DORSAL VIEW

Fig. 19-10. Structure of the frog heart.

Observing the Heartbeat

☞ Open a pithed frog as you did the preserved frog, but avoid cutting the abdominal vein. Cut carefully through the pectoral girdle, keeping the scissors well up to avoid injuring the heart. Pin back the forelimbs and keep the heart well moistened with frog Ringer's solution.

Identify the ventricle, left and right atria, truncus arteriosus, and aortic arches. Watch the heartbeat and note the series of alternating contractions—first the two atria, then the ventricle, and finally the arterial trunk. Raise the ventricle carefully to view the contraction of the sinus venosus immediately before the contraction of the atria.

The frog is an **ectothermic** animal, that is, its temperature is governed by the environmental temperature, rather than by internal means (endothermic). You can examine the effect of temperature changes on the heart rate by a simple experiment.

☞ Count and record the number of beats per minute at room temperature.

Flood the abdominal cavity with ice-cold frog saline. Count the beat again immediately, then wait a little while till the maximum effect of the cold is achieved, and then count and record again.

Replace the cold water with frog saline warmed to about 40° C. Count the beat immediately and record. Allow the warmth to take effect; then count and record again. Replace the water with saline at room temperature and record again. Record your results on p. 330.

What effect does temperature change have on the heart rate? What advantage would this reaction to cold have for the frog? Why should the solution not be warmed to more than 40° C? What is the effect of fever on human heart rate?

PROJECTS AND DEMONSTRATIONS

1. Record of a frog's heartbeat. Pith a frog, and make a median ventral incision through the body wall in the region between the forelegs. Loosen the heart from the surrounding blood vessels and tissues. Fasten the heart in place to a heart lever so that a record can be made on a smoked kymograph drum or chart recorder. Keep the heart moist with Ringer's solution. Determine the rate of the beat. By adding cold or warm Ringer's solution to the heart you can make a recording of the effect of temperature on the heartbeat.

2. Microcirculation in the frog's tongue. See Exercise 22B.

3. Blood typing. See pp. 385-386.

4. Capillaries in latex-injected material. If a latex-injected frog is available, remove a piece of skin and examine it under the low-power lens of the microscope. The capillary system should show up very well.

5. Effect of drugs on circulation. Pith a frog and place it on a frog board with the web of the foot exposed to show capillary circulation. Place on the stage of a microscope. Inject at various intervals certain depressant and stimulating drugs such as digitalis, caffeine, epinephrine, aconite, and others. Report the effects of these drugs on the circulation.

6. Effects of stimulants or depressants on frog heart rate. Heart muscle is autonomous; it has the power to contract without connection to body nerves and without external stimulation. The heartbeat originates within specialized cardiac muscle (the pacemaker) located in the right atrium. To demonstrate the autonomous nature of the heartbeat, remove the heart from an ectothermic (cold-blooded) animal such as a frog, turtle, or snake, taking care not to injure it. Place it in a watch glass of frog Ringer's solution. It should beat for some time if kept submerged in Ringer's.

To test the effects of stimulants or depressants, allow the heart a few minutes to adjust and then count the number of beats per minute. You can then add a drop of any of several substances, such as alcohol, ether, epinephrine, atropine, nicotine, or caffeine, to the watch glass. Check the heartbeat rate to determine any change. Be sure to rinse well in fresh Ringer's before trying a different substance.

7. Other suggestions. For further suggestions, see pp. 369-370. See also Exercise 22A, p. 385.

Name _____

Date _____ Section _____

1. Trace the shortest route a corpuscle could take on each of the following trips, underscoring each place where it would go through a **capillary bed.**

a. Ventricle to lung and return _____

_____

_____

b. Ventricle to brain and return _____

_____

_____

_____

c. Systemic arch to intestine to right atrium _____

_____

_____

_____

d. Hindleg to left atrium by way of renal portal _____

_____

_____

_____

_____

2. What are the chief **gains** and **losses** that take place in the blood in the following organs?

a. Lung _____

b. Kidney _____

_____

c. Intestinal wall _____

d. Liver _____

_____

_____

e. Muscles _____

_____

_____

3. Describe the shortest route from the **left atrium** to the **arm** and back to the **left atrium.**

_____

_____

_____

_____

_____

4. Effect of temperature on heart rate.

Rate of contraction

at room temperature _____ /min (first count)

of cooled heart _____ /min (first count)

of cooled heart _____/min (later count)

of warmed heart _____ /min (first count)

of warmed heart _____/min (later count)

at room temperature _____/min (later count)

EXERCISE 19F
The nervous and endocrine systems

NERVOUS SYSTEM*

The nervous and endocrine systems are the coordinating systems of the body, integrating the activities of the various organ systems.

The nervous system is composed of (1) the **central (cerebrospinal)** nervous system, consisting of the brain and spinal cord, which are housed in the skull and spinal column, and which are concerned with integrative activity, and (2) the **peripheral** nervous system, made up of the paired cranial and spinal nerves and the **autonomic** nervous system. These make up a sys-

---

*See Hickman, C.P., Jr., L.S. Roberts, and F.M. Hickman. 1984. Integrated principles of zoology, ed. 7. St. Louis, The C.V. Mosby Co., pp. 573-574, 769-790.

tem for the conduction of sensory and motor information throughout the body. The autonomic system consists of a pair of autonomic nerve cords together with their ganglia and nerve fibers, which innervate the viscera.

Spinal Nerves

☞ Remove any remaining viscera carefully. Do not molest the dorsal aorta and systemic arches.

The 10 pairs of spinal nerves (Fig. 19-11) emerge through small openings between the vertebrae and appear as white threads in the dorsal body wall. The first three nerves on each side form a **brachial plexus** to the arm, neck, and shoulder. The next three are in the body wall. The seventh to tenth form the **sciatic**

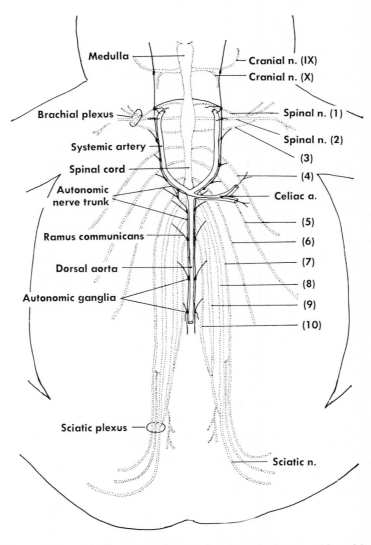

Fig. 19-11. Nervous system of a frog, ventral view, showing spinal nerves *(dotted lines)* and autonomic trunks *(solid lines)*.

plexus to the leg. A plexus is a network of nerves that interchange fibers.

The sciatic plexus on each side gives off to the hindleg a **femoral nerve** and a large **sciatic nerve,** the largest nerve in the body.

☞ To see the sciatic plexus and sciatic nerve, cut through the pelvic girdle, if you have not already done so, spread the legs apart, and trace the nerves into the leg.

Autonomic Nervous System

The parts of the autonomic system are small. Immersing the frog in water and using the hand lens or dissecting microscope may help you find them.

The **autonomic nerve trunks** follow the systemic arches and dorsal aorta (Fig. 19-11). Lift up the aorta with forceps to see on each side a small whitish cord, probably most evident where the systemic arches join to form the aorta. Separate one of the trunks carefully from the blood vessel and connective tissue and lift it up on a series of pins resting on the backbone. Note the **autonomic ganglia** and the **rami communicantes,** which connect the ganglia with the spinal nerves. Note that a branch of the autonomic nerve trunk also follows the celiacomesenteric artery. Study the function of this system in the text.

Central Nervous System
(Brain and Spinal Cord)

☞ Cut through the skull just back of the nares. Beginning here, use the tip of the scalpel or forceps to chip away small pieces from the top of the cranium, being careful not to injure the delicate brain tissue beneath. Expose the entire brain from the olfactory nerves to the vertebral column.

Now, beginning with the first vertebra, snip through each side of each vertebra between the neural spine and the articular processes (Fig.

19-2, *A*) and remove the dorsal piece to expose the spinal cord. Continue until the whole cord is exposed.

The central system is enclosed in two membranes called the **meninges.** The tough **dura mater** usually clings to the cranial wall and neural canal; the thinner **pia mater** adheres to the brain and cord.

☞ Remove the dura mater from brain and cord and identify the following parts.

Dorsal View of the Brain and Spinal Cord

1. The **forebrain** consists of the cerebral hemispheres and the diencephalon (Fig. 19-12, *A*). The **cerebral hemispheres** constrict anteriorly to form **olfactory lobes,** from which the **olfactory nerves** (cranial nerves I) extend to the nares. The **diencephalon** is a depressed region behind the cerebrum. The **epiphysis,** or **pineal gland,** ascending dorsally to the brow spot, is a rudimentary third eye. The optic nerves (II) arise on the ventral side.

2. The **midbrain** bears on the dorsal side two prominent **optic lobes.** Cranial nerves III and IV arise in the midbrain.

3. The **hindbrain** consists of the **cerebellum** and the **medulla.** The last six cranial nerves arise from the medulla.

4. The **spinal cord** is a continuation of the medulla and ends in the urostyle.

Ventral View of the Brain and Spinal Cord

☞ With the brain under water, cut the olfactory nerves and carefully lift the anterior end of the brain, gently working it loose. Continue loosening and lifting to remove the brain and spinal cord in one piece; then place them in a dish of water.

On the ventral side (Fig. 19-12, *B*), locate the **optic chiasma,** where the **optic nerves** meet and cross. Posterior to the optic chiasma is a slight extension of the diencephalon, to which is attached posteriorly the **hypophysis,** or **anterior lobe of the pituitary gland.** This

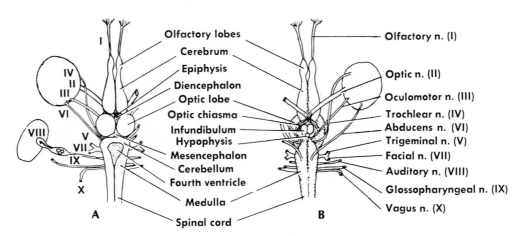

Fig. 19-12. Brain of a frog. **A,** Dorsal view. **B,** Ventral view.

is sometimes broken off in dissection, remaining in the floor of the cranium. There are 10 pairs of cranial nerves.

Consult the textbook for the functions of the various parts of the brain.

Sensory System (Sense Organs)

☞ Cut open the nasal chamber, under the external nares.

Note the convoluted lining of **olfactory epithelium.**

☞ Remove one of the eyes by cutting around the lid and through the optic nerve and muscles of the eye.

The eyeball is covered with a tough **sclerotic coat,** the transparent front part of which is the **cornea.** Under the cornea is a contractile **iris diaphragm,** with a center opening, the **pupil,** through which light enters the eye. Cut the eyeball open by removing the dorsal side. The **lens,** a firm and nearly spherical body, focuses light rays on the pigmented and sensory **retina,** which lines the area back of the lens.

The frog lacks an external ear. The **tympanic membrane,** or **eardrum,** is the external boundary of the middle ear.

☞ Cut around the periphery of the tympanic membrane and lift it up to locate the tiny ear bone, or **columella,** which transmits vibrations from the tympanic membrane to the inner ear, where the auditory nerve picks them up.

ORAL REPORT

Be prepared to demonstrate from your dissection all the component parts of the frog's nervous system so far studied. From your textbook, find out their various functions.

ENDOCRINE SYSTEM

Both the nervous and the endocrine systems carry messages from one part of the body to another. In contrast to the electrochemical messages carried by the nerves from receptors to effectors, the endocrine messages are chemical hormones carried by the blood from glandular or neurosecretory cells to target organs in the body. Endocrine glands are ductless and secrete their products directly into the bloodstream.

The frog has about the same endocrine glands as higher vertebrates. The pituitary gland (hypophysis) on the ventral side of the brain has been called the "master gland" because its secretions control the activities of several other endocrine glands. The ovary and testis have endocrine as well as reproductive functions. The thyroid gland in the throat, the adrenal glands on the ventral side of the kidneys, the islets of Langerhans in the pancreas, and the numerous neurosecretory cells all have varied functions too complex for a discussion here. Consult your textbook.

PROJECTS AND DEMONSTRATIONS

1. Maceration preparation of frog nervous system. Skin a frog and immerse it in a bath of 30% nitric acid solution, with the exception of the head, which is kept out of the bath by a glass hook. Leave in this solution for about 24 hours and then submerge the head, leaving the whole frog in the bath for about 8 hours. Check frequently to ascertain the degree of maceration. Then tease as much of the muscle as possible from the nervous system. Transfer the specimen to an empty dish and allow water from a faucet to fall gently on the tissue from a short distance. This will carry away the remaining part of the muscle from the nervous system. The critical part of this preparation is the length of time in the immersion bath. Too long a time will dissociate the nervous system along with the other tissues. Preserve the completed preparation in a flat dish containing glycerin.

2. Cross section of spinal cord (prepared slide). Note the surrounding meninges (**dura** and **pia mater**), the **ventral** and **dorsal fissures,** the **dorsal septum** from the dorsal fissure to near the center of the cord, the inner **gray zone** of cell bodies, the outer **white zone** of nerve fibers, and the **central canal.**

3. Cross section of frog eye (slide). Find and identify the **cornea,** the **anterior** and **posterior cavities,** the **crystalline lens,** and the **retina.**

4. Other suggestions. For other suggestions see p. 374.

CHAPTER 20

CLASS MAMMALIA—
THE FETAL PIG

MATERIALS

Fetal pigs, embalmed and injected, in plastic bags
Cat or dog skeletons
Dissecting pans
Kidneys, pig or sheep, fresh or preserved
Hearts, pig or sheep, fresh or preserved
Brains, pig, sheep, or dog, preserved
Prepared slides, if desired
    Intestine, mammalian, cross sections
    Stomach and/or esophagus, cross sections
    Kidney tissue, mammalian
    Seminiferous tubules, cross sections

FETAL PIG

Class Mammalia
  Subclass Theria
    Infraclass Eutheria
      Order Artiodactyla
        Genus *Sus*
          Species *S. domesticus*

The Mammalia* are those animals whose young are nourished by milk from the breasts of the mother. They have a muscular diaphragm, found in no other class, and a four-chambered heart. Most of them are covered with hair. Their nervous system is especially well developed. Their eggs develop in a uterus, with placental attachment for nourishment (except the monotremes, which lay eggs, and the marsupials, in which the placental attachment is only weakly and briefly developed).

The order Artiodactyla includes the even-toed hoofed mammals such as deer, sheep, cattle, and camels. These usually have two toes, but some, such as the hippopotamus and the pig, have four toes.

Fetal pigs are an especially desirable laboratory example of the mammal. They are easy to obtain, relatively inexpensive, and easily stored in individual plastic bags. Fetal pigs are obtained from the uteri of sows slaughtered for market. Because they are unborn, their bones are still largely cartilaginous, which

---

*See Hickman, C.P., Jr., L.S. Roberts, and F.M. Hickman. 1984. Integrated principles of zoology, ed. 7. St. Louis, The C.V. Mosby Co., pp. 635-665.

makes the specimens pliable and easy to handle. They have an umbilical cord, by which they were attached to the placenta in the uterus.

The embryo depends on maternal blood to bring it nutrients and oxygen and to carry off the waste products of metabolism since its own organs cannot serve these functions until birth. This exchange of materials between fetal blood and maternal blood takes place within the placenta of the mother's uterus. The difference between the fetus and the adult is largely physiological, but there are also a few morphological differences, especially in the circulatory system, which you will observe in your specimen.

The period of gestation in the pig is 16 to 17 weeks as compared with 20 days in the rat, 8 weeks in the cat, 9 months in the human being, 11 months in the horse, and 22 months in the elephant. Pig litters average 7 to 12 but may be as many as 18. The pigs are about 30 cm long at birth and weigh from 4.5 to 6.5 kg (about 2 to 3 pounds). The age of a fetus may be estimated from the length of its body:

> At 3 weeks the fetus is about 1.3 cm long.
> At 7 weeks the fetus is about 3.8 cm long.
> At 14 weeks the fetus is about 23 cm long.
> At full term the fetus is about 30 cm long.

External Structure

☞ Before proceeding with the regular exercises, take a look at the external structure of your pig.

On the head, locate the **mouth** with fleshy lips and the **nostrils** at the tip of the snout. The snout has a tough rim for rooting and bears **vibrissae**, stiff sensory hairs (whiskers). Each eye has two lids and a small membrane in the medial corner, representing the nictitating membrane. The **ears** have fleshy pinnae around the external auditory opening.

On the trunk, locate the **thorax** supported by ribs, sternum, and shoulder girdle with forelimbs attached; the **abdomen** supported by a vertebral column and muscular walls; the **sacral** region, comprising the pelvic girdle with hindlimbs attached; the **umbilical cord**; five to seven pairs of **mammae** or nipples on the abdomen; and the **anus** at the base of the tail.

Determine the **sex** of your specimen. In the **males** the **urogenital opening** is just posterior to the umbilical cord; the **scrotal sacs** form two swellings at the posterior end of the body (the **penis** can sometimes be felt under the skin as a long thin cord passing from the urogenital opening back between the hindlegs). In the **female** the urogenital opening is just ventral to the anus and has a fleshy tubercle projecting from it.

Examine the cut end of the umbilical cord. Note the ends of four tubes in the cord. These represent an umbilical vein, two umbilical arteries, and an allantoic duct, all of which during fetal life are concerned with transporting food, oxygen, and waste products to or from the placenta of the mother's uterus.

Notice that the entire body of the fetal pig is covered with a thin cuticle called the **periderm**.

EXERCISE 20A
The skeleton*

MATERIALS
Skeletons, dog or cat, and human
Sections of bones (if possible, a fresh joint)
Other vertebrate skeletons

The skeleton of the adult pig, although different in size and proportion from the skeletons of other mammals such as the dog, cat, or human, is, nonetheless, quite similar. The bones are homologous, and the origins and insertions of muscles as described for one can usually be traced on the others.

Because of its immature condition, the skeleton of the fetal pig is unsuitable for classroom study. However, if you are planning to dissect the muscles of the fetal pig, it is essential to be familiar with the bones. Skeletons of the cat or dog can be used quite satisfactorily for this purpose. Figs. 20-1 and 20-2 of the cat skeleton will help you in your identification.

☞ As you study the mounted skeleton of the dog or cat, compare the parts with those of the fetal pig skeleton (Fig. 20-3). Then try to locate these parts and visualize their relationships within the flesh of the preserved pig. As you do notice the similarities with the human skeleton (Fig. 20-4).

THE SKELETON
As in the frog, the pig skeleton consists of the **axial skeleton** (skull, vertebral column, ribs, and sternum) and the **appendicular skeleton** (pectoral and pelvic girdles and their appendages).

Axial Skeleton
Skull. The skull can be divided into a **facial region**, containing the bones of the eyes, nose, and jaws, and a **cranial region**, which houses the brain and ears. The

occipital condyles of the skull articulate with the ring-shaped first cervical vertebra (called the atlas). The foramen magnum is the opening at the posterior end of the braincase for the emergence of the spinal cord. Many of the important bones of the skull can be identified with the help of Fig. 20-1.

Vertebral Column. Note the five types of vertebrae—**cervical**, in the neck; **thoracic**, bearing the ribs; **lumbar**, without ribs but with large transverse processes; **sacral**, fused together to form a point of attachment for the pelvic girdle; and **caudal**, the tail vertebrae. In humans three to five vestigial caudal vertebrae are fused to form the coccyx of the tail bone. All of the vertebrae are built on the same general plan with recognizable individual differences (Figs. 19-2, 20-2, and 20-3).

Ribs. Observe the structure of a rib and its articulation with a vertebra. Each rib articulates with both the body of the vertebra and a transverse process. The **shaft** of the rib ends in a **costal cartilage,** which attaches to the sternum or to another costal cartilage and so indirectly to the sternum. The cat has one pair of free or floating ribs. The pig has 14 or 15 pairs of ribs, of which seven pairs attach directly to the sternum and seven or eight attach indirectly. How many does the human have?

Sternum. The sternum is composed of a number of ossified segments, the first of which is called the **manubrium,** and the last of which is called the **xiphisternum.** Those in between are called the **body** of the sternum.

Appendicular Skeleton
Pectoral Girdle and its Appendages. The pectoral girdle comprises a pair of triangular **scapulae**, each with a lateral **spine** and a **glenoid fossa** at the ventral point for attachment with the head of the humerus. The **forelimb** (Figs. 20-2 to 20-4) includes (1) the **humerus;** (2) the two forearm bones—a shorter more medial **radius** and a longer more lateral **ulna,** with an **olecranon process** at the proximal or elbow end; (3)

---

*See Hickman, C.P., Jr., L.S. Roberts, and F.M. Hickman. 1984. Integrated principles of zoology, ed. 7. St. Louis, The C.V. Mosby Co., pp. 673-679.

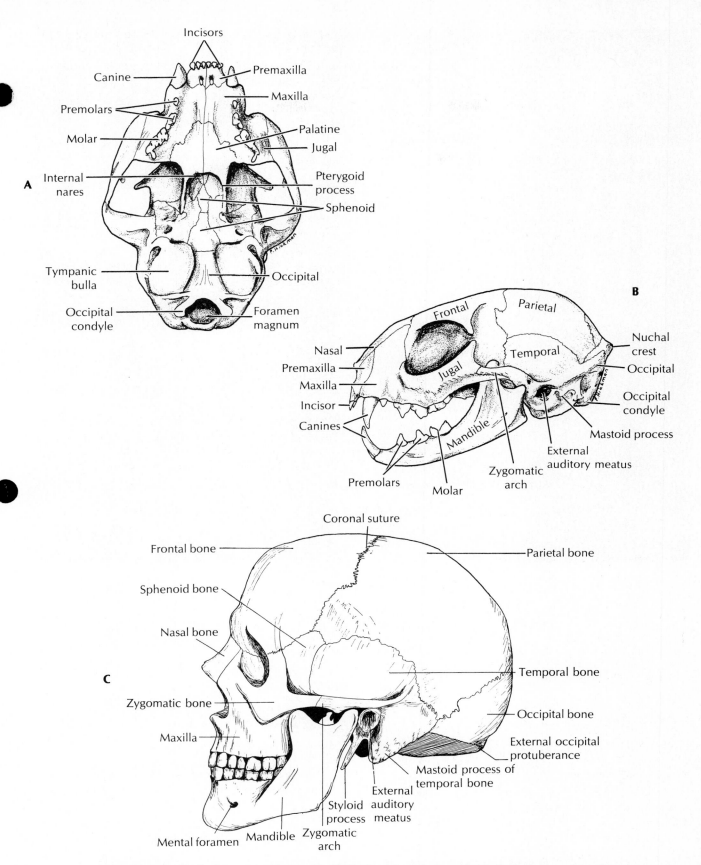

Fig. 20-1. A, Skull of cat, ventral view. **B,** Skull of cat, lateral view. **C,** Skull of a human, lateral view.

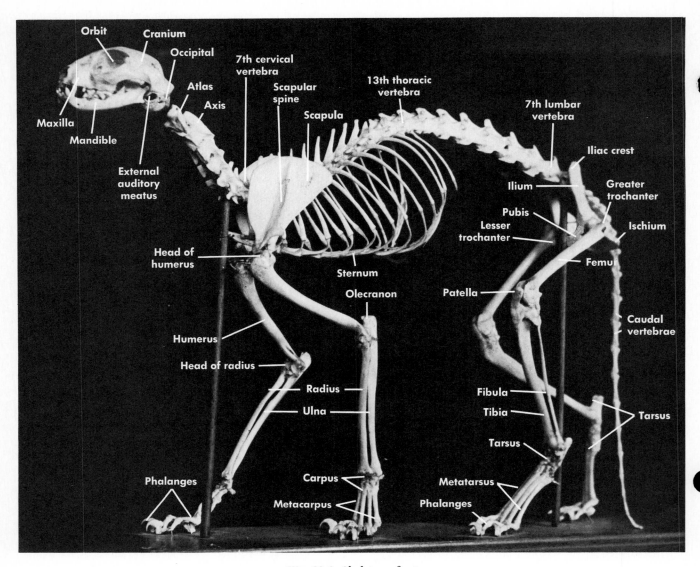

Fig. 20-2. Skeleton of cat.

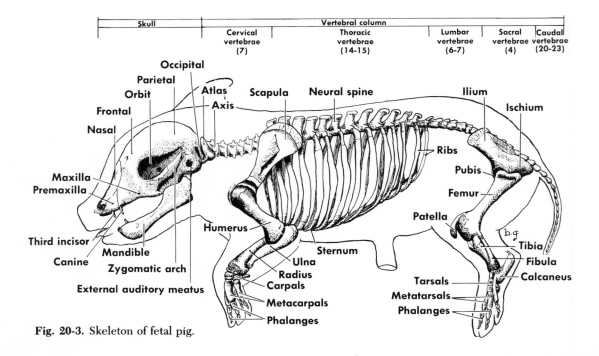

Fig. 20-3. Skeleton of fetal pig.

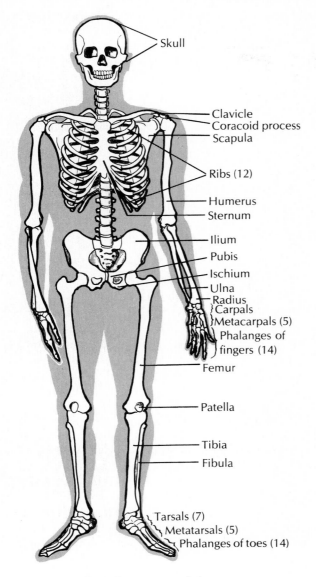

Fig. 20-4. Human skeleton.

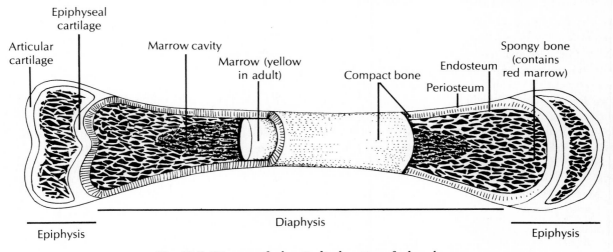

Fig. 20-5. Diagram of a longitudinal section of a long bone.

the **carpus**, consisting of two rows of small bones; (4) the **metacarpals** (five in the cat and human and four in the pig); and (5) the **digits** or toes, made up of **phalanges**.

Pelvic Girdle and its Appendages. The pelvic girdle in adult mammals consists of a pair of **innominate bones**, each formed by the fusion of the **ilium, ischium**, and **pubis**. A lateral cavity, the **acetabulum**, accepts the head of the femur. The pubic bones and the ischial bones of opposite sides unite at **symphyses**, and the ilia articulate with the sacrum so that the innominates and the sacrum together form a complete ring or **pelvic canal**. Each pelvic appendage includes (1) the **femur**, or thigh bone; (2) the larger **tibia** and more slender **fibula** of the shank; (3) the ankle, or **tarsus**, comprising seven bones, and the fibular tarsal (**calcaneus**) forming the projecting heel bone; (4) the **metatarsals** (five in the cat, of which the first is very small, and four in the pig); and (5) the **digits**, or toes (four in both the cat and the pig), composed of **phalanges**.

Structure of a Long Bone

Longitudinal and transverse sections through a long bone (Fig. 20-5) show it to consist of a shell of **compact bone** within which is a type of **spongy bone (cancellous bone)**, the spaces of which are filled with **marrow**. The **shaft (diaphysis)** is usually hollowed to form a **marrow cavity**. In the young animal there is only red marrow, a blood-forming substance, but this is gradually replaced in the adult with yellow marrow, which is much like adipose tissue. The extremities (**epiphyses**) of the bone usually bear a layer of **articular cartilage**. The rest of the bone is covered with a membrane, the **periosteum**. Arteries, veins, nerves, and lymphatics pass through the compact bone to supply the marrow.

Growth of a Bone

The primitive skeleton of the embryo consists of cartilage and fibrous tissue, in which the bones develop

339

by a process of ossification (pp. 66 and 67). Bones that develop in fibrous tissue, namely, some of the bones of the cranium and face, are called membrane bones. Most of the bones of the body develop from cartilage and are designated as cartilage bones.

In a typical long bone there are usually three primary centers of ossification—one for the diaphysis, or shaft, and one for each epiphysis, or extremity (Fig. 20-5). As long as this cartilage persists and grows, new bone may form, and the length may increase.

Articulations

An articulation or joint is the union of two or more bones or cartilages by another tissue, usually fibrous tissue or cartilage or a combination of the two.

Three types of joints are recognized:

Synarthrosis. A synarthrosis is an immovable joint. Interlocking margins of the bones are united by fibrous tissue. Example: sutures of the skull.

Diarthrosis. A diarthrosis is a movable joint (Fig. 20-6). Ends of articulating bones are covered with cartilage and enclosed in a joint capsule of fibrous tissue. The capsule contains a joint cavity lined with a vascular synovial membrane that secretes a lubricating fluid. Example: most of the joints—knee, elbow, and others.

Amphiarthrosis. An amphiarthrosis is a slightly movable joint. The bones are joined by a flattened disc of fibrocartilage. The bones of the joint are bound together by ligaments. These are strong bands or membranes composed mostly of white fibrous tissue and are pliable but not elastic (except for the nuchal ligament at the back of the neck). Examples: the pubic symphysis; joints between the vertebrae.

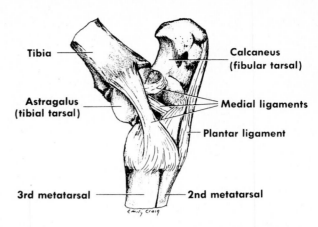

Fig. 20-6. Hock (tarsal) joint, showing articular ligaments.

PROJECTS AND DEMONSTRATIONS

1. Other vertebrate skeletons.

2. Fresh joint. Cut a fresh joint in longitudinal section.

3. Long bone. Cut a long bone in longitudinal section.

4. Piece of skull bone or face bone. Examine a piece of skull or face bone to see an example of membrane bone.

5. Tinting for comparison purposes. It is very convenient to have all the homologous bones of a series of skeletons tinted a certain color. For example, coloring the frontal lobes of each skeleton—cat, rabbit, turtle, frog, and others—blue and the parietals red, and so on, can make comparisons easy.

6. Other suggestions. Other suggestions are given on p. 313.

EXERCISE 20B
The muscular system*

Because the muscles of the fetal pig are softer and the separations of the muscles less evident than in a lean cat or frog, the pig has been less used for muscle dissection in beginning classes. However, if full-term pigs (30 cm or more in length) are used, and careful attention is given to the dissection, even beginning students can demonstrate a great many of the muscles along with their origins and insertions. Uninjected pigs are quite satisfactory for this work; in fact, in some ways they are easier to work with than injected pigs.

Many introductory courses lack sufficient laboratory time in which to dissect all the muscles mentioned here. In this case the instructor may be selective, or some of the dissection may be done on an optimal or extra-credit basis.

SKELETAL MUSCLES—GENERAL DISCUSSION

Muscle tissue is highly specialized and has greater powers of contractility than does any other tissue. In fact, muscles have only one function—they contract when stimulated. Some muscles can contract to half their relaxed length. Skeletal muscle, which is voluntary, is composed of bundles of striated fibers and makes up the bulk of the muscular system.

Fascia. Muscle fibers are bound together by a fibrous connective tissue (**fascia**), which also surrounds

---

*See Hickman, C.P., Jr., L.S. Roberts, and F.M. Hickman. 1984. Integrated principles of zoology, ed. 7. St. Louis, The C.V. Mosby Co., pp. 680-687.

the neck and face region this cutaneous layer is called the **platysma;** in the trunk region it is called the **cutaneous maximus.** These muscles are used in twitching the skin to shake off insects, dirt, or other irritants. This thin layer of muscle is *not shown in the illustrations.* After identifying this layer, *remove it carefully* to identify the superficial muscles underneath. Remember that these muscles are very thin; remove them carefully so as not to destroy the other muscles you are to identify.

☞ To locate the borders of muscles, scrape off the overlying connective tissue and fascia and look for the direction of the muscle fibers. A muscle edge may be seen where there is a change in direction of fibers. Try to slip the flat handle of the scalpel between the layers of muscle at this point. *Do not cut the muscles or tear them with a dissecting needle.* Try to loosen each muscle and find out where it is attached but *do not cut a muscle unless instructed to do so.* When you are told to cut a muscle to locate deeper muscles, cut through the belly of the muscle but leave the ends attached for identification.

Fig. 20-7 shows the more superficial muscles after removal of the cutaneous and platysma layer.

If, after skinning the animal, you find the muscles are still too soft to separate, exposure to air will help to harden them in a few hours. Dipping the pig in formalin or sponging a little formalin over its surface and keeping it overnight in a plastic bag should make it easier to handle.

The following muscles are grouped loosely into two groups—muscles of the forequarter and muscles of the hindquarter. The first group will contain some of the muscles of the face and neck as well as those of the shoulder, chest, and forelimb. The second group will include muscles of the back, abdomen, and hindlimb. This is not an exhaustive list, but it will include the chief superficial muscles and many of the muscles of the second layer.

MUSCLES OF THE FOREQUARTER
Muscles of the Face, Neck, Chest, and Shoulder

Sternohyoid. This pair marks the ventral midline of the neck and covers the larynx (Figs. 20-7, 20-8,

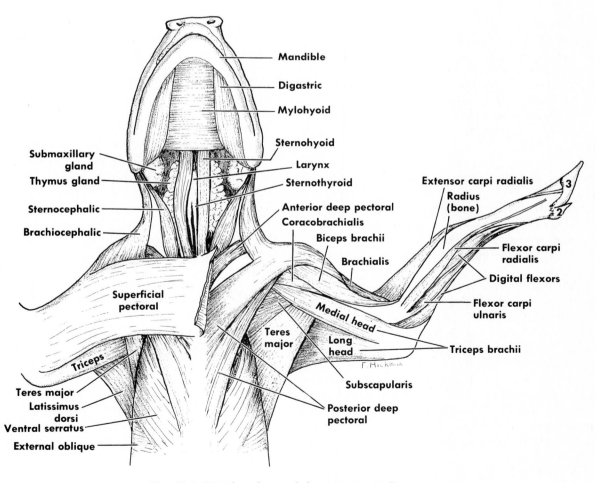

Fig. 20-8. Muscles of ventral thoracic region of pig.

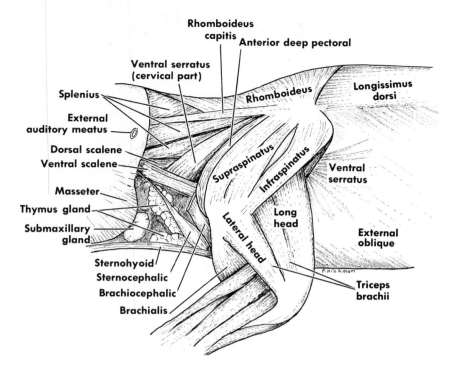

Fig. 20-9. Muscles of neck and shoulder, lateral view.

20-9 and 20-15); separate them to see the larynx, the sternothyroids, and, on each side, the thymus gland. **Origin**—anterior end of the sternum. **Insertion**—hyoid bone. **Action**—retracts and depresses the hyoid and the base of the tongue as in swallowing.

Sternothyroid. Immediately internal to the sternohyoids (Fig. 20-8); each has a lateral and a medial branch; separate these muscles to find the thyroid gland lying on the ventral side of the trachea. **Origin**—sternum. **Insertion**—larynx. **Action**—Retracts the larynx.

Digastric. Lies along the medial surface of each mandible (Fig. 20-8). **Origin**—by a tendon from the mastoid process of the skull. **Insertion**—medial surface of the mandible. **Action**—depresses the mandible.

Mylohyoid. Thin transverse sheet stretching between the mandibles (Fig. 20-8). **Origin**—medial surface of the mandibles. **Insertion**—hyoid bone. **Action**—raises the floor of the mouth, tongue, and hyoid bone.

Geniohyoid. Spindle-shaped muscle that lies with its partner in the midline between the mylohyoid and the tongue; the sublingual glands are lateral to these muscles (Fig. 20-15). **Origin**—inner surface of the mandible near the symphysis. **Insertion**—hyoid bone. **Action**—draws the hyoid bone and the tongue forward.

Masseter. Large muscle of the cheek; the parotid and submaxillary glands lie posterior to this muscle (Figs. 20-7 and 20-14). **Origin**—zygomatic arch. **Insertion**—lateral surface of the mandible. **Action**—elevates the jaw and closes the mouth.

Sternocephalic (Sternomastoid). Passes diagonally across the throat posterior to the submaxillary gland and beneath the parotid gland (Figs. 20-7 to 20-9). **Origin**—anterior end of the sternum. **Insertion**—mastoid process of the skull. **Action**—turns the head; the two muscles together depress the head.

Brachiocephalic (Figs. 20-7 to 20-9). **Origin**—one part (cleidooccipital) on the nuchal crest and the other (cleidomastoid) on the mastoid process of the skull. **Insertion**—proximal end of the humerus and fascia of the shoulder. **Action**—singly, inclines the head; when head is fixed, draws the limb forward; together, extend the head.

Superficial Pectoral (Fig. 20-8). Covers most of the chest region. **Origin**—sternum. **Insertion**—by a broad aponeurosis on the medial surface of the humerus. **Action**—adducts the humerus.

Posterior Deep Pectoral. Lies beneath the superficial pectoral (Fig. 20-8). **Origin**—posterior half of the sternum and cartilages of the fourth to ninth ribs. **Insertion**—proximal end of the humerus. **Action**—retracts and adducts the forelimb.

Anterior Deep Pectoral. Follows the anterior border of the scapula (Figs. 20-7 to 20-9). **Origin**—anterior part of the sternum. **Insertion**—scapular fascia and aponeurosis that covers the dorsal end of the supraspinatus. **Action**—adducts and retracts the limb.

Trapezius. Most superficial muscle of the back after removal of the cutaneous maximus; thin and triangular (Fig. 20-7). **Origin**—nuchal crest of the skull and neural spines of the first 10 thoracic vertebrae. **Insertion**—spine of the scapula. **Action**—elevates the shoulder.

Latissimus Dorsi (Figs. 20-7 and 20-8). **Origin**—some of the thoracic and lumbar vertebrae and the four ribs preceding the last rib. **Insertion**—medial surface of the humerus. **Action**—draws the humerus upward and backward and flexes the shoulder.

Deltoid (Spinodeltoid). Broad flat muscle on the surface of the scapula (Fig. 20-7). **Origin**—scapular aponeurosis. **Insertion**—by an aponeurosis on the proximal end of the humerus. **Action**—flexes the shoulder and abducts the arm.

Rhomboideus. Triangular muscle just beneath the trapezius (Fig. 20-9). **Origin**—second cervical to the ninth or tenth thoracic vertebrae. **Insertion**—medial surface of the dorsal border of the scapula. **Action**—draws the scapula mediodorsally or rotates it.

Rhomboideus Capitis. Straplike muscle on the dorsal side of the neck under the trapezius (Fig. 20-9). **Origin**—occipital bone. **Insertion**—dorsal border of the scapula. **Action**—draws the scapula forward and rotates the shoulder.

Splenius. Fleshy muscle of three or four muscle bands on the back of the neck (Figs. 20-7 and 20-9). **Origin**—first four or five thoracic neural spines. **Insertion**—occipital and temporal bones and first few cervical vertebrae. **Action**—singly, inclines the head and neck to one side; together, elevate the head and neck.

Ventral Serratus. Extensive fan-shaped chest muscle with two parts (Figs. 20-8 and 20-9). **Origin**—cervical part on the transverse processes of the last four or five cervical vertebrae; thoracic part on the lateral surfaces of the last eight or nine ribs. **Insertion**—medial surface of the scapula. **Action**—singly, cervical part draws the shoulder forward and the thoracic part backward; together, shift the weight to the limb of the contracting side; both sides together form an elastic support that suspends the trunk between the scapulas; they raise the thorax.

Ventral Scalene (Fig. 20-9). **Origin**—transverse processes of the third to the seventh cervical vertebrae; roots of the brachial nerve plexus pass through it, dividing it into bundles. **Insertion**—first rib. **Action**—flexes or inclines the neck.

Dorsal Scalene (Fig. 20-9). **Origin**—transverse processes of the third to the sixth cervical vertebrae. **Insertion**—first rib. **Action**—flexes or inclines the neck.

Supraspinatus. Fleshy muscle on the anterior surface of the scapula; bisect the deltoid to see it (Fig. 20-9). **Origin**—anterior and dorsal portion of the scapula and the scapular spine. **Insertion**—proximal end of the humerus. **Action**—extends the humerus.

Infraspinatus. Beneath the deltoid and posterior to the supraspinatus (Fig. 20-9). **Origin**—lateral surface and spine of the scapula. **Insertion**—lateral surface of the proximal end of the humerus. **Action**—abducts and rotates the arm.

Muscles of the Foreleg
Triceps Brachii (Figs. 20-7 to 20-9)

Long Head. Triangular shaped; at the posterior edge of the scapula. **Origin**—posterior border of the scapula.

Lateral Head. Between the long head and the brachialis. **Origin**—lateral side of the proximal end of the humerus.

Medial Head. On the medial surface of the humerus. **Origin**—medial surface of the proximal end of the humerus, covering the insertion of the teres major (Fig. 20-8).

Insertion (all three)—median and lateral surfaces of the olecranon process of the ulna. **Action** (all three)—extend the forearm.

Brachialis. Most anterior muscle of the humerus; ventral to the lateral head of the triceps (Figs. 20-7 to 20-9). **Origin**—proximal third of the humerus. **Insertion**—medial surface of the distal end of the radius and ulna. **Action**—flexes the elbow.

Biceps Brachii. Most anterior muscle on the medial surface of the humerus (Fig. 20-8). **Origin**—ventral surface of the scapula near the glenoid fossa. **Insertion**—proximal ends of the radius and ulna. **Action**—flexes the forearm.

Coracobrachialis. Between the biceps and the medial head of the triceps (Fig. 20-8). **Origin**—ventral side of the scapula near the glenoid fossa. **Insertion**—medial surface of the middle third of the humerus. **Action**—adducts the arm and flexes the shoulder joint.

Teres Major. Most posterior muscle on the medial surface of the scapula (Fig. 20-8). **Origin**—posterior border and angle of the scapula near the insertion of the thoracic portion of the ventral serratus. **Insertion**—in common with the latissimus dorsi on the medial surface of the shaft of the humerus. **Action**—rotates and flexes the humerus.

Extensor Carpi Radialis. Most anterior muscle on the forearm (Figs. 20-7 and 20-8). **Origin**—distal end of the humerus on the lateral side and the deep fascia of the arm. **Insertion**—distal end of the radius, with a tendon on the third metacarpal bone. **Action**—extends the carpal joint and flexes the elbow joint.

Flexor Carpi Radialis. On the medial side of the forearm, lying next to the radius (Fig. 20-8). **Origin**—distal end of the humerus on the medial side. **Insertion**—third metacarpal bone. **Action**—flexes the carpal joint and extends the elbow.

Digital Flexors of the Forelimb (Fig. 20-8). Include the superficial digital flexor, the deep digital flexor, and the flexors of the second and fifth digits. **Origin**—distal end of the humerus and proximal ends of the radius and ulna. **Insertion**—second to fifth digits. **Action**—flex the digits and carpal joint and extend the elbow.

Flexor Carpi Ulnaris (Fig. 20-8). **Origin**—distal end of the humerus and the olecranon process of the ulna. **Insertion**—on the lateral side of the carpus. **Action**—flexes the carpal joint and extends the elbow.

Ulnaris Lateralis. On the posterolateral side of the forearm (Fig. 20-7). **Origin**—distal end of the humerus on the lateral side. **Insertion**—by a tendon on the fifth metacarpal bone. **Action**—flexes the carpal joint and extends the forearm.

Digital Extensors of the Forelimb (Fig. 20-7). Include the common digital extensor, the lateral digital extensor, and the extensor of the second digit. **Origin**—proximal end of the humerus and distal end of the radius and ulna. **Insertion**—second to fifth digits. **Action**—flex the carpal joint and extend the digits.

Extensor Carpi Obliquus. Close to the bone and slightly distal and lateral to the extensor carpi radialis (Fig. 20-7). **Origin**—distal two-thirds of the radius and ulna on the lateral side. **Insertion**—second metacarpal bone. **Action**—extends the carpal joint.

MUSCLES OF THE HINDQUARTER
Muscles of the Abdomen, Back, and Hip

External Oblique. Superficial abdominal sheet muscle immediately under the cutaneous (Figs. 20-7, 20-9, and 20-10). **Origin**—lateral surface of the last nine or ten ribs and the lumbodorsal fascia. **Insertion**—linea alba, ilium, and femoral fascia. **Action**—singly, flexes the trunk laterally; together, compress the abdomen as in defecation, parturition, and expiration and arch the back.

Internal Oblique. Located internal to the external oblique; its fibers run in the opposite direction (Fig. 20-7).

Transverse Abdominal. Internal to the other two, with fibers running transversely; the thinnest of the three (Fig. 20-7). **Origin, insertion, and action** (of both)—similar to that of the external oblique.

Rectus Abdominis. Long muscle lying on each side of the midventral line (Fig. 20-10). **Origin**—pubic symphysis. **Insertion**—sternum. **Action**—constricts the abdomen.

Longissimus Dorsi. Covered by the lumbodorsal fascia; very long muscle extending from the sacrum to the neck (Figs. 20-9 and 20-11). **Origin**—sacrum, ilium, and neural processes of the lumbar and thoracic vertebrae. **Insertion**—transverse processes of most of the vertebrae and lateral surfaces of the ribs, except

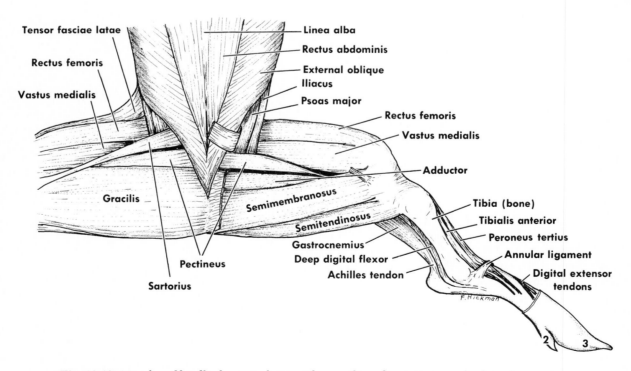

Fig. 20-10. Muscles of hindlimb, ventral view. The gracilis and sartorius muscles have been cut on the left leg.

the first. **Action**—singly, flexes the spine laterally; together, extend the back and neck; the rib attachments may aid in expiration.

Tensor Fasciae Latae. Most anterior superficial thigh muscle (Figs. 20-7 and 20-10). **Origin**—crest of the ilium. **Insertion**—fascia over the knee, the patella, and the crest of the tibia. **Action**—flexes the hip joint and extends the knee joint.

Biceps Femoris. Posterior superficial thigh muscle; covers the sciatic, tibial, and peroneal nerves and branches of the femoral vessels (Figs. 20-7 and 20-11). **Origin**—lateral part of the ischium and sacrum. **Insertion**—by a wide aponeurosis to the patella and the fascia of the thigh and leg. **Action**—complex; acting on all the joints of the limb except the digits; generally abducts and extends the limb, though may also flex the knee joint.

Gluteus Medius. Between and partially covered by the tensor fasciae latae and the biceps femoris (Figs. 20-7 and 20-11). **Origin**—fascia of the longissimus dorsi, the ilium, and the sacroiliac and sacrosciatic ligaments. **Insertion**—proximal end of the femur. **Action**—abducts the thigh.

Muscles of the Hindleg

Quadriceps Femoris. Large muscle group covering the anterior and lateral sides of the femur; composed of four muscles:

Rectus Femoris. Thick muscle on the anterior side of the femur (Figs. 20-10 and 20-11).

Vastus Lateralis. Lateral to the rectus femoris and partly covering it (Fig. 20-11).

Vastus Medialis. On the medial surface of the rectus femoris (Fig. 20-10).

Vastus Intermedialis. Deep muscle lying beneath the rectus femoris.

Origin—ilium (rectus femoris) and proximal end of the femur (other three). **Insertion** (all four)—patella and its ligament and the proximal end of the tibia. **Action** (all four)—extend the shank.

Sartorius. Thin muscle band anterior to the gracilis; easily destroyed if not identified; covers the femoral vessels (Fig. 20-10). **Origin**—iliac fascia and tendon of the psoas minor; external iliac vessels lie between the two heads. **Insertion**—patellar ligament and proximal end of the tibia. **Action**—adducts hindlimb and flexes hip joint.

Gracilis. Wide muscle covering most of the medial surface of the thigh (Fig. 20-10). **Origin**—pubic symphysis and ventral surface of the pubis; muscles united at their origin. **Insertion**—patellar ligament and proximal end of the tibia. **Action**—adducts the hindlimb.

Iliacus (Figs. 20-10 and 20-11). **Origin**—ventral surface of the ilium and wing of the sacrum. **Insertion**—proximal end of the femur together with the psoas major. **Action**—flexes the hip and rotates the thigh outward.

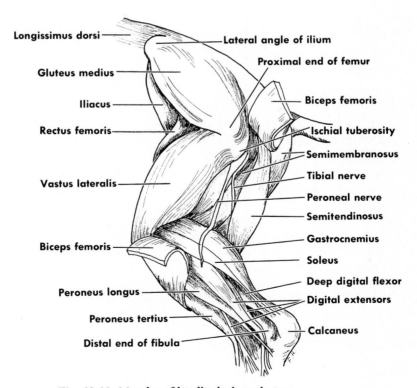

Fig. 20-11. Muscles of hindlimb, lateral view.

Psoas Major (Fig. 20-10). **Origin**—ventral sides of the transverse processes of the lumbar vertebrae and the last two ribs. **Insertion**—with the iliacus on the proximal end of the femur. **Action**—flexes the hip and rotates the thigh outward.

Pectineus (Fig. 20-10). **Origin**—anterior border of the pubis. **Insertion**—medial side of the shaft of the femur. **Action**—adducts the hindlimb and flexes the hip.

Adductor. Covered by the gracilis (Fig. 20-10). **Origin**—ventral surface of the pubis and ischium and the tendon of origin of the gracilis. **Insertion**—proximal end of the femur. **Action**—adducts the hindlimb and extends and rotates the femur inward.

Semimembranosus. Very large two-part medial thigh muscle (Figs. 20-7, 20-10, and 20-11). **Origin**—ischium. **Insertion**—distal end of the femur and proximal end of the tibia, both on the medial side. **Action**—extends the hip joint and adducts the hindlimb.

Semitendinosus. Thick muscle on the posterior side of the femur (Figs. 20-7, 20-10, and 20-11). **Origin**—first and second caudal vertebrae and the ischium. **Insertion**—proximal end of the tibia and the calcaneus. **Action**—extends the hip and the tarsal joint and flexes the knee joint.

Gastrocnemius. Large muscle on the back of the shank (Figs. 20-10 and 20-11). **Origin**—distal end of the femur. **Insertion**—by the tendon of Achilles onto the calcaneus. **Action**—flexes the digits and extends the ankle.

Soleus. Thick muscle lying beneath the gastrocnemius and having the same origin, insertion, and action (Figs. 20-7 and 20-11).

Deep Digital Flexor (Figs. 20-7 and 20-11). **Origin**—proximal ends of the tibia and fibula. **Insertion**—by a tendon that passes over the fibular tarsal (calcaneus) bone to the digits. **Action**—flexes the digits and extends the ankle.

Digital Extensors of the Hindlimb (Figs. 20-7 and 20-11). Includes the long digital extensor, lateral digital extensor, extensor hallucis longus, and extensor digitalis brevis. **Origin**—distal end of the femur and proximal ends of the fibula and tibia on the lateral side. **Insertion**—second to fifth digits. **Action**—extends the digits and flexes the ankle.

Peroneus Longus. Lateral side of the shank (Figs. 20-7 and 20-11). **Origin**—proximal end of the tibia on the lateral side. **Insertion**—first tarsal bone. **Action**—flexes the ankle.

Peroneus Tertius. On the anterior side of the shank (Figs. 20-7, 20-10, and 20-11). **Origin**—distal end of the femur on the lateral side. **Insertion**—first and second tarsal and third metatarsal bones. **Action**—flexes the ankle.

Tibialis Anterior. Anterior to the tibia (Fig. 20-10). **Origin**—proximal end of the tibia on the lateral side. **Insertion**—second tarsal and metatarsal bones. **Action**—flexes the ankle.

Be able to identify and to give the origin, insertion, and action of as many of the foregoing muscles as your instructor has assigned.

EXERCISE 20C
Preliminary dissection and the digestive system

DIRECTIONS FOR DISSECTION

Place the pig, ventral side up, in the dissecting pan. Tie a cord around one forelimb, loop the cord under the pan, and fasten it to the other forelimb. Do the same to the hindlegs.

☞ With a scalpel, make a midventral incision through the skin and muscles but not into the body cavity, beginning at the small papilla (with vibrissae) on the throat and continuing posteriorly to within 1 cm of the umbilical cord (incision 1, Fig. 20-12). Cut around each side of the cord (2). If your specimen is a female, continue on down the midline from the cord to the anal region (3). If it is a male, make two incisions, one on each side of the midline, to avoid cutting the penis, which lies underneath (3a). Now in either sex, deepen the incisions you

have made in the abdominal region through the muscle layer to reach the body cavity, taking care not to injure the underlying organs. Now, with scissors, make two lateral cuts on each side, one just anterior to the hindlegs (4), the other posterior to the ribs (5), and turn back the flaps of the body wall (6). Flush out the abdominal cavity with running water.

All visceral organs are invested in mesentery and held in place with connective tissue. Loosen this tissue carefully to separate organs, tubes, and vessels, being careful not to cut or tear them. *Do not remove any organs unless you are specifically directed to do so.* Be careful in all your preliminary dissection not to destroy blood vessels or nerves but to keep them intact for later dissection of the circulatory and nervous systems.

It is well to remember that instructions referring to the "right side" refer to the animal's right side, which will be on your left as the animal lies ventral side up on the dissection pan.

ABDOMINAL CAVITY

Notice that the umbilical cord is attached anteriorly by a tube, the **umbilical vein.**

☞ Tie a string around the umbilical vein in two places and sever the vein between the two strings.

The strings will identify this vein later. Lay the umbilical cord between the hindlegs and identify the following parts.

The **body wall** consists of several layers: (1) tough external **skin,** (2) two layers of **oblique muscle** and an inner layer of **transverse muscle** (try to separate the layers and determine the direction of the fibers), and (3) inner lining of thin transparent **peritoneum.**

The **diaphragm** is a muscular dome-shaped partition separating the peritoneal cavity (abdominal cavity) from the thoracic cavity. *Do not remove the diaphragm.*

The peritoneum is the smooth shiny membrane that lines the abdominal cavity and supports and covers the organs within it. That which lines the body walls is called the **parietal peritoneum.** It is reflected off the dorsal region of the body wall in a double layer to form the **mesenteries,** which suspend the internal organs, and then continues on around the organs as a cover, where it is called the **visceral peritoneum.**

The **liver** is a large reddish gland with four main lobes lying just posterior to the diaphragm. The greenish saclike **gallbladder** may be seen under one of the central lobes.

The **stomach** is nearly covered by the left lobe of the liver. The **small intestine** is loosely coiled and held by mesenteries. Note the blood vessels in the mes-

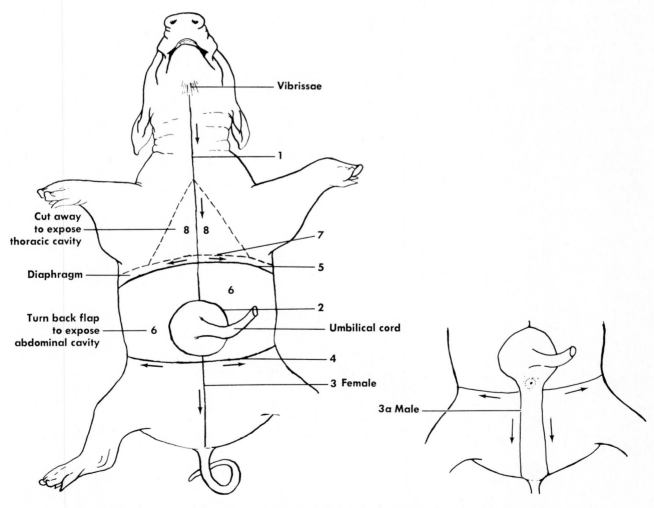

Fig. 20-12. Cutting diagram. The numbers indicate the order in which each incision is to be made; *1* to *6* expose the abdominal cavity; *7* and *8* expose the thoracic cavity.

entery. The **large intestine** is compactly coiled on the left side posterior to the stomach.

The **spleen** is a long reddish organ attached by a mesentery to the greater curvature of the stomach. The spleen is specialized for filtering the blood. It removes worn-out blood cells and salvages the iron and hemoglobin for reuse; it also produces antibodies and lymphocytes.

The **umbilical arteries** are two large arteries extending from the dorsal wall of the coelom to and through the umbilical cord.

The **allantoic bladder,** the fetal urinary bladder, is a large sac lying between the umbilical arteries. It connects with the allantoic duct in the umbilical cord.

The **kidneys** are two large bean-shaped organs attached to the dorsal wall dorsal to the intestines. They are outside the peritoneal cavity in the **cisterna magna** and are separated from the other abdominal organs by the peritoneum.

THORACIC CAVITY AND NECK REGION

☞ With scissors, begin just anterior to the diaphragm and cut along the midventral line through the sternum, to a point midway between the forelegs. Keep the lower blade of the scissors up to prevent injuring the heart underneath. Now make a lateral cut on each side just anterior to the diaphragm. (See incision 7, Fig. 20-12.) This exposes the thoracic cavity but leaves the diaphragm in place.

The **mediastinal septum,** which separates the right and left lung cavities, is a thin, transparent tissue attached to the sternal region of the thoracic wall (Fig. 20-13).

☞ Separate it carefully from the body wall. Now lift up one side of the thoracic wall and look for the small **internal thoracic artery** and **vein** (also called sternal or mammary) embedded in the musculature of the body wall. Carefully separate these vessels on each side and lay them down over the heart and lungs for future use. Now you may cut away some of the ventral thoracic wall (incision 8) to allow a better view of the thoracic cavity containing the left and right **lungs** and the **heart.**

The **pleura** is the name given to the peritoneum that lines each half of the thoracic cavity and covers the lung (Fig. 20-13). The peritoneum lining the thoracic cavity is the **parietal pleura;** the part applied to the lungs is the **visceral pleura.** The small space between is the **pleural cavity,** in which lubricating **pleural fluid** prevents friction. The portions of the parietal pleurae on the medial side next to the heart are called the **mediastinal pleurae.** The **mediastinum** is the region between the mediastinal pleurae. It contains the pericardium and heart and the roots of the big arteries and veins as well as the trachea, esophagus, part of the thymus, and other parts.

The double-walled **pericardium** enclosing the heart is made up of an outer **parietal pericardium** and a **visceral pericardium** applied to the heart, with pericardial fluid in the space between.

The **thymus gland** is a soft irregular mass of glandular tissue, part of which lies in the mediastinum overlying the heart. Loosen a portion of the thymus from the heart to distinguish it from lung tissue. It is an extensive gland in the fetus and young animal but regresses after puberty. It continues up through the

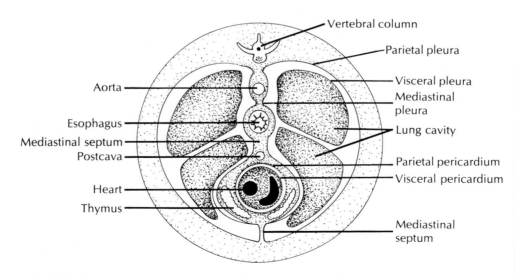

Labels: Vertebral column — Parietal pleura — Visceral pleura — Mediastinal pleura — Lung cavity — Parietal pericardium — Visceral pericardium — Mediastinal septum — Aorta — Esophagus — Mediastinal septum — Postcava — Heart — Thymus

Fig. 20-13. Diagrammatic transverse section through the thorax in the region of the ventricles to show the relations of the pleural and pericardial membranes, mediastinum, and so on.

small opening at the first ribs and extends well up into the neck region. Part the cut skin in the neck and probe between and on each side of the sternohyoid muscles for the elongated anterior lobes of the thymus gland (Figs. 20-9 and 20-15). The thymus produces protective antibodies.

The **thyroid gland** is a small dark oval gland lying on the trachea beneath the sternothyroid muscles (Fig. 20-8). Its hormones promote growth and development and regulate the metabolic rate.

DIGESTIVE SYSTEM

The digestive system consists of the alimentary canal, from mouth to anus, and the glands that help it in its function of converting food into a form assimilable by the body tissues and convertible to energy.

Salivary Glands

The paired salivary glands pour into the mouth secretions containing the enzyme ptyalin, which aids in the digestion of the carbohydrates. The glands may be identified in large fetuses.

☞ On the right side of the face, neck, and chin, carefully remove the skin if you have not already done so. A muscle layer will tend to adhere, but push this layer back into place gently so as not to destroy the glands beneath. Now carefully remove the thin muscles back of the angle of the jaw and beneath the ear to uncover the **parotid gland.** Do not destroy any large blood vessels.

The triangular parotid gland is broad, thin, and rather diffused, extending from almost the midline of the throat to the base of the ear (Figs. 20-7 and 20-14). Do not confuse the salivary glands, which are choppy and lobed in appearance, with the lymph nodes, which are more smooth and shiny. The **parotid duct** comes from the deep surface of the gland and follows the ventral border of the masseter (cheek) muscle along the external maxillary vein to the corner of the mouth (Fig. 20-14).

The **submaxillary (mandibular) gland** lies under the parotid gland and just posterior to the angle of the jaw. It is darker, compact, and oval. Its duct comes from the anterior surface of the gland and passes anteriorly, medial to the mandible, and through the sublingual gland to empty into the floor of the mouth. This duct is very difficult to trace.

☞ To find the **sublingual glands,** remove the mylohyoid muscle and the slender pair of geniohyoid muscles immediately beneath it.

On each side a whitish elongated sublingual gland is located between the digastric muscle, which lies inside the mandible, and the genioglossus, which is one of the muscles of the base of the tongue (Fig. 20-15). A sublingual artery and vein will be seen along the ventral side of each gland. The sublingual glands empty by way of several short ducts to the floor of the mouth.

Mouth Cavity and Pharynx

☞ On the right side, cut through the flesh at the corner of the mouth. Sever the two mandibles by cutting through the mandibular symphysis. Now draw the tongue down between the mandibles to expose the mouth for examination.

The **tongue** is attached ventrally to the floor of the

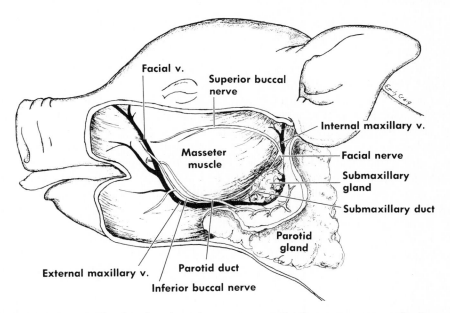

Fig. 20-14. Dissection of head and neck to show some superficial veins, nerves, and salivary glands.

351

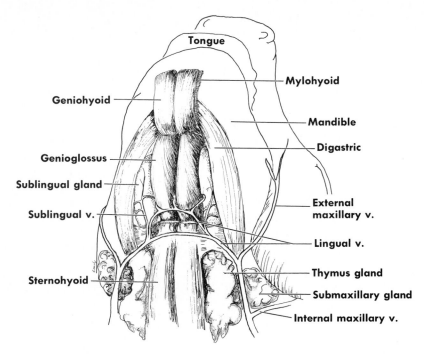

Fig. 20-15. Dissection to show sublingual glands and associated veins and muscles.

mouth by the **frenulum linguae,** a vertical fold of mucous membrane under the tongue, and posteriorly to the **hyoid cartilage.** The **papillae** on the surface of the tongue provide friction, and many of them contain taste buds. The **filiform papillae** are very small hairlike projections that stud the entire surface of the tongue, giving it a velvety pile. **Fungiform papillae** are larger and rounded and are most numerous along the sides of the tongue. Two or three large circular **vallate papillae** are found near the base of the tongue. Long soft **root papillae** are directed backward on the root of the tongue back of the vallate papillae.

The **teeth** may not be erupted yet, although the canines and third pair of incisors may be seen in older fetuses. The young pig will have on each side of each jaw three incisors, one canine, and four premolars. Remove the flesh along the right jaws and carefully cut away enough of the jawbone to expose the buds of the embryonic teeth. The third incisors and the canines are the first to erupt; the second incisors are the last.

The roof of the mouth is made up of a narrow bony **hard palate,** with its ventral mucous membrane covering ridged into transverse folds, and posterior to it a **soft palate,** which has no bone above it.

The **pharynx** is the posterior continuation of the mouth cavity and opens into the esophagus, the tube that leads to the stomach.

Open the mouth wide, drawing down the tongue, to locate at the posterior end of the soft palate the

opening into the **nasopharynx,** which is the space above the soft palate. It connects with the nasal passages from the nostrils.

The **pharynx** is the posterior continuation of the mouth cavity. At the back of the pharynx are the openings of the **esophagus,** the tube that leads to the stomach, and the **larynx,** which leads to the trachea and lungs. Locate the flaplike **epiglottis,** which folds up over the **glottis** (the open end of the larynx) to close it when food is being swallowed. Note that in the mouth the air passages are *dorsal* to the food passage. In the throat, however, the air is carried through the larynx and trachea, which are *ventral* to the food passage (esophagus). These passageways cross in the pharyngeal cavity (Fig. 20-16). While air is being breathed, the epiglottis fits up against the opening into the nasopharynx to allow air into the larynx but to prevent the entrance of saliva or food from the mouth. During swallowing, the larynx is pushed forward, causing the epiglottis to fold over the glottis, thus opening the food passage while closing off the air passage.

Digestive Tract

Read the textbook* and know the functions of each part of the system, the location and action of the diges-

---

*See Hickman, C.P., Jr., L.S. Roberts, and F.M. Hickman. 1984. Integrated principles of zoology, ed. 7, St. Louis, The C.V. Mosby Co., pp. 744-751.

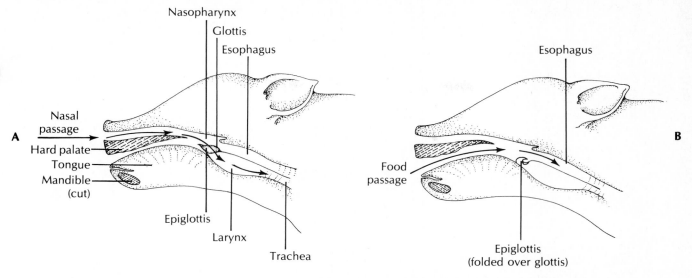

Fig. 20-16. Relationship of respiratory passage to the mouth and esophagus, in breathing and swallowing. **A,** During breathing the glottis is open to receive air from nostrils and is protected from food and saliva by the epiglottis. **B,** For swallowing the larynx is pushed anteriorly, causing epiglottis to fold over glottis, thus closing air passage to lungs. Feel your Adam's apple (larynx) as it moves up when you swallow.

tive glands, and the physiology of digestion and absorption.

The **esophagus** is a soft muscular tube that leads from the pharynx to the stomach. Locate it in the neck region posterior to the larynx, where it is attached to the dorsal side of the trachea by connective tissue. Find the esophagus in the thoracic cavity posterior to the lungs, and in the abdominal cavity find where it emerges through the diaphragm at the cardiac end of the stomach. The muscles at the anterior end of the tube are striated (voluntary), gradually changing to smooth muscle. How does this affect swallowing?

The **stomach** is a large muscular organ that breaks up food and thoroughly mixes it with gastric juice. Identify its **cardiac end** near the heart, its **pyloric end** that joins the duodenum, its **greater curvature** where the spleen is attached, and its **lesser curvature**. The **fundus** is the anterior blind pouch. The contents of the fetal digestive tract, made green by pigments in the bile salts, are called **meconium** and contain epithelium sloughed from the mucosa lining, sebaceous secretions, and amniotic fluid swallowed by the fetus. Open the stomach longitudinally, rinse out, and find (1) the **rugae**, or folds, in its walls, (2) the opening from the esophagus, and (3) the **pyloric valve**, which regulates the passage of food into the duodenum.

The **small intestine** includes the **duodenum**, or first portion, which lies next to the pancreas and receives the common bile duct and pancreatic duct, and the **jejunum** and **ileum**, indistinguishable in the fetal pig, which make up the remainder of the small intestine.

☞ Remove a piece of the intestine, open it, and examine it *under water* with a hand lens or dissecting microscope. Observe the minute fingerlike **villi**, which greatly increase the absorptive surface of the intestine.

Most of the digestion and absorption take place in the small intestine.

The **large intestine** is concerned largely with storage of undigested products awaiting defecation. It includes the long tightly coiled **colon** and the straight posterior **rectum**, which extends through the pelvic girdle to the **anus**.

Find the **cecum**, which is a blind pouch of the colon at its junction with the ileum. In humans and the anthropoid apes the cecum has a narrow diverticulum called the **vermiform appendix**. Open the cecum (on its convex side opposite the ileum) and note how the entrance of the ileum forms a ring-shaped **ileocecal valve**. The posterior end of the rectum will be exposed in a later dissection.

Digestive Glands. The **liver**, a large brownish gland posterior to the diaphragm, has four main lobes, left and right lateral lobes and left and right central lobes. The liver produces bile, which emulsifies fats. Bile is stored in the **gallbladder**, a small greenish oval sac embedded in the dorsal surface of the right central lobe of the liver. The liver is connected to the upper border of the stomach by a tough transparent membrane, the **gastrohepatic ligament**, in which are embedded blood vessels (in the left side) and ducts (in the right side). Carefully loosen the gallbladder

353

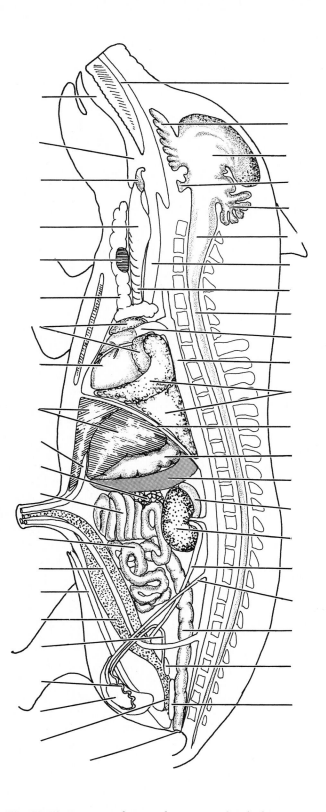

Fig. 20-17. Diagram of internal anatomy of male fetal pig.

354

and note its tiny **cystic duct.** This unites with **hepatic ducts** from the liver to form the **common bile duct,** which carries bile to the duodenum. Probe the gastrohepatic ligament and adjoining liver tissue carefully to find these ducts. Do not injure the blood vessels lying beside them.

The **pancreas** is a mass of soft glandular tissue in the mesentery between the duodenum and the end of the stomach. Push the small intestine, except the duodenum, to the animal's left to explore the gland. Its pancreatic juice is carried by a **pancreatic duct** to the pyloric end of the duodenum.

Other digestive juices are secreted by the **mucosa lining** of the **stomach** and of the **small intestine.**

Label as many of the structures on the diagram in Fig. 20-17 as you have identified so far. Add the others as you continue with the study of the systems.

Be able to locate and give the functions of the parts of the digestive system.

Histologic Study of the Intestine. Examine a cross section of human or other mammalian small intestine and compare it with the histological structure of amphibian intestine. Examine mammalian liver and pancreas slides.

PROJECTS AND DEMONSTRATIONS

1. Peristaltic movement. About 30 to 45 minutes after a rat has finished feeding, anesthetize it, open the abdominal cavity, and submerge the contents in warm physiologic saline solution. Note the peristaltic movement in the intestinal tract.

2. Stomach of a ruminant. The ruminants are cud-chewing animals (cattle, deer, camels, and the like). They have a four-compartmented stomach. Food swallowed after brief chewing passes immediately to the **rumen** for preliminary digestion, mostly bacterial. The rumen is the largest compartment and is marked off by grooves into several sacs. From here food passes to the **reticulum,** the smallest compartment. The lining of the reticulum is honeycombed by ridges and pits. From here small masses (cuds) are regurgitated, chewed thoroughly, and reswallowed to the third compartment, the **omasum,** where salivary enzymatic action continues. Finally in the **abomasum** gastric digestion occurs. The lining of the omasum and abomasum is folded into longitudinal ridges or rugae as in the stomachs of other vertebrates.

3. Crop and stomach of a bird (pigeon or chicken). Grain-eating birds have a **crop,** which is a membranous saclike diverticulum of the esophagus for initial food storage. The **stomach** is divided into a **proventriculus** (glandular stomach), which secretes a digestive enzyme, and a thick muscular **gizzard,** which grinds and macerates the food into a partially digested mash before passing it on to the small intestine.

4. Demonstration slides. Observe some demonstration slides of cross sections of mammalian esophagus, stomach, intestine, salivary glands, pancreas, and liver.

EXERCISE 20D
The urogenital system

MATERIALS
Fetal pigs
Pregnant pig (or dog or cat) uteri for dissection or demonstration
Preserved sheep kidneys

URINARY SYSTEM
☞ Read the directions carefully and dissect cautiously. Do not tear or remove any organs, blood vessels, or ducts. Instead, separate them carefully from the surrounding tissues.

The urinary system consists of a pair of kidneys, a pair of ureters, a urinary bladder, and a urethra (shared by the reproductive system in the male).

The fetal **urinary bladder** is the **allantoic bladder,** a long sac located between the umbilical arteries. It narrows ventrally to form the **allantoic duct,** which continues through the umbilical cord and is the fetal excretory canal. The bladder narrows dorsally to empty into the **urethra,** the adult excretory canal. The urethra will be dissected out later. After birth the allantoic end of the bladder closes to form the urinary bladder.

The **kidneys** are dark and bean shaped. They lie outside the peritoneum on the lumbar region of the dorsal body wall. Uncover the left kidney carefully. A depression on the median side of each kidney is called the **hilus.** Through it pass the renal blood vessels and the **ureter** or excretory duct. Follow the left ureter posteriorly to its entrance into the bladder. Be careful of small ducts and vessels that cross the ureter. (Note the small **adrenal gland,** an endocrine gland lying close to the medial side of the anterior end of the kidney and embedded in fat and peritoneum.)

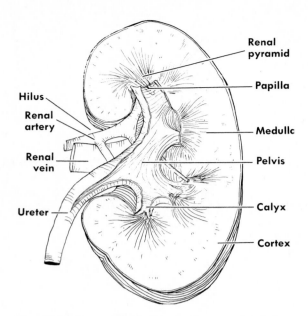

Fig. 20-18. Longitudinal section of mammalian kidney.

☞ Slit open the left kidney longitudinally, cutting in from the outer border. Remove the ventral half of the kidney and lay it in a watch glass of water.

Study with a hand lens or binocular scope. Identify the **cortex** or outer layer, containing the microscopic renal corpuscles; the **medulla** or deeper layer, containing the radially arranged blood vessels and collecting tubules; and the **renal pyramids,** which contain groups of collecting tubules coming together to empty through **papillae** into the **pelvis.** The pelvis is a thin-walled chamber that connects with the ureter. Divisions of the pelvis into which the papillae empty are called **calices** (sing. **calyx**) (Fig. 20-18).

☞ For a description of the anatomy and physiology of excretion, read your textbook.* Examine the demonstration specimens of sheep kidneys on display. You will be expected to understand the structure and functioning of the kidney and its functional unit, the nephron.

Male Reproductive System

The location of the testes in the fetus depends on the stage of development of the fetus. The testis originates in the abdominal cavity near the kidney. During the development of the fetus a prolongation of the peritoneum grows down into each half of the scrotum. This is called the **processus vaginalis.** Later the testis "decends" into the scrotum through the **inguinal canal,** where it lies within the sac or sheath formed by the

---

*See Hickman, C.P., Jr., L.S. Roberts, and F.M. Hickman. 1984. Integrated principles of zoology, ed. 7, St. Louis, The C.V. Mosby Co., pp. 729-737.

processus vaginalis. The inguinal canal is the tubular passage connecting the abdominal cavity with the scrotal sac (Fig. 20-19). The descent of the testes is usually completed shortly before birth.

Lay the umbilical cord and allantoic bladder down between the legs and locate the **urethra** at the dorsal end of the bladder. The urethra bends dorsally and posteriorly to disappear into the pelvic region. Find the sperm ducts (**vasa deferentia,** plural; **vas deferens,** singular), two white tubes that emerge from the openings of the **inguinal canals,** cross over the umbilical arteries and ureters, and come together medially to enter the urethra. Also emerging from each inguinal ring are the spermatic artery, vein, and nerve. Together with the vas deferens these make up the **spermatic cord,** which leads to the testis somewhere in the inguinal canal or scrotal sac.

Now lay the bladder up over the abdominal cavity and locate the thin, hard, cordlike **penis** under the strip of skin left posterior to the urogenital opening. The penis lies in a sheath in the ventral abdominal wall, ending at the **urogenital opening.**

☞ Carefully separate the penis from surrounding tissue; then cut away the skin and muscle that covered and surrounded it.

Now probe carefully into the **left scrotal sac** to locate the processus vaginalis, a transparent whitish sac that contains or eventually would contain the testis. Pick away the flesh carefully to free this sac. By pulling gently on the vaginalis while watching the spermatic cord in the abdominal cavity, you can locate the **inguinal canal,** through which the spermatic cord passes. Examining the processus vaginalis, you can feel the testis within it, if it has descended (Fig. 20-19).

Remove the membranous vaginalis from one of the testes. The **testis** is a small hard oval body containing hundreds of microscopic **seminiferous tubules** in which the sperm are formed. These tubules unite into a plexus and then pass out of the testis and are collected into a single much coiled **vas epididymis.** You will see the epididymis as a white mass coiled around the testis (Fig. 20-19). The **vas deferens,** a continuation of the epididymis, passes through the inguinal canal along with the spermatic artery and vein, then loops over the ureter, and enters the urethra, as you have already seen. A fibrous cord attaches the testis and the epididymis to the posterior end of the processus vaginalis. It is called the **gubernaculum.** A narrow band of muscle (the cremaster) runs along the lateral and posterior part of the processus vaginalis parallel with the vas deferens.

☞ Separate the tissues on each side of the penis in the pelvic region; then, being careful not to injure the penis or cut too deeply, use a scalpel to cut through the cartilage of the pelvic girdle.

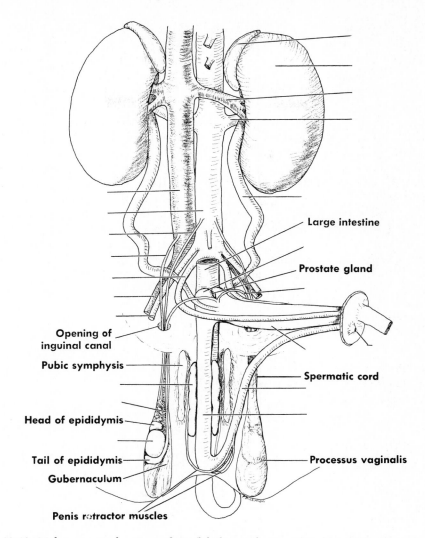

Large intestine

Prostate gland

Opening of
inguinal canal

Pubic symphysis

Spermatic cord

Head of epididymis

Tail of epididymis

Processus vaginalis

Gubernaculum

Penis retractor muscles

Fig. 20-19. Male urogenital system of pig (labeling to be completed by student).

Spread the legs apart to expose the **urethra** and its connection with the penis. The urethral canal extends throughout the length of the penis and serves as a common duct for both sperm and urine.

Now, beginning at its juncture with the bladder, follow the urethra posteriorly and locate the following **male glands:**

Seminal Vesicles. The seminal vesicles are a pair of small glands on the dorsal side of the urethra.

Prostate Gland. The prostate gland is poorly developed in the fetus; it lies between and often partly covered by the seminal vesicles, but it may be difficult to find(Fig. 20-19).

Cowper's Glands (Bulbourethral Glands). Cowper's glands are a pair of narrow glands about 1 cm long on each side of the urethra near its junction with the penis. Are these glands exocrine or endocrine? What are their functions?

Be able to follow the path of urine and of sperm to the outside. Compare the male urogenital system of the pig with that of the frog and with that of a man.

Add labels to the drawings in Figs. 20-17 and 20-19.

☞ Now exchange your specimen for a female and study the female reproductive system.

Female Reproductive System

The female reproductive organs are located in the lower lumbar region of the abdominal cavity.

Uterus. The uterus is Y shaped. The **horns of the uterus** are suspended by a wide mesentery called the **broad ligament.** The horns unite medially to form the **body of the uterus,** which leads to the vagina.

Fallopian Tubes. Each fallopian tube is a small convoluted duct that is a continuation of the distal end of one of the uterine horns. It coils around the **ovary** and terminates with a wide ciliated funnel called the **infundibulum.** The opening in the distal end of the tube, called the **abdominal ostium,** receives the ova as they leave the ovary at ovulation.

Ovaries. The ovaries are small pale organs at the free end of the fallopian tubes. Each is suspended by a mesentery called the **mesovarium.**

Urethra. Lay the allantoic bladder anteriorly over the abdominal viscera.

☞ Cut through the muscle in the midventral line between the legs and then through the cartilage of the pelvis. Spread the legs apart to expose the urethra. Separate the urethra carefully from surrounding tissues. (Note where the ureters join the dorsal end of the bladder.)

Vagina. Lay the bladder and urethra to one side. Follow the body of the uterus posteriorly to a slight constriction called the **cervix.** From here the tube widens and is called the vagina. The vagina and urethra soon join to form a **urogenital sinus,** which is a short comon passageway for the two systems.

Vulva. The vulva is the external opening of the urogenital sinus, ventral to the anus. The ventral side of the vulva extends out to form a pointed **genital papilla.** A small rounded **clitoris** may be seen extending from the ventral floor of the urogenital sinus. This is not always evident (*inset*, Fig. 20-20).

In adults, at copulation, the male penis places the spermatozoa, contained in a seminal fluid, into the vagina. The sperm must pass through the uterus to the fallopian tubes in order to fertilize the egg. After fertilization the zygote passes down to the horn of the uterus to develop. Note how the horns are adapted

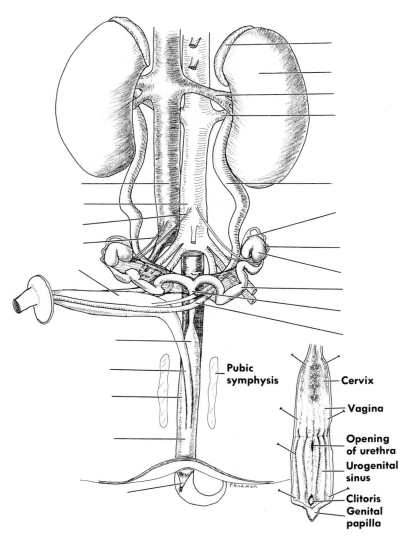

Pubic symphysis

Cervix

Vagina

Opening of urethra

Urogenital sinus

Clitoris

Genital papilla

P. Hickman

Fig. 20-20. Female urogenital system of pig, ventral view. *Right inset*, Vagina and urogenital sinus, opened from dorsal side, and pinned back.

for carrying a litter. How does this compare with the human uterus? How is the developing fetus nourished? How are waste products from the fetus disposed of?

Be able to trace the path of the unfertilized egg from the ovary to the outside. Compare with the frog.

Be able to trace the path of urine in the female. Are any parts shared by both urinary and reproductive systems?

Add labels to the drawing in Fig. 20-20.

If you have not studied the male system, trade your specimen for a male and make a thorough study of the male system.

Dissection of the Pregnant Pig Uterus

Compare the shape of a pregnant pig uterus with the fetal uterus. Locate the cervix, the muscular body of the uterus, the large horns of the uterus, the small fallopian tubes, and the ovaries (Fig. 20-21). How many fetuses can you detect in the uterus?

Cut open a portion of the uterine wall along the convex side. The **mucosa lining** of the uterine wall is arranged in folds and bears many tiny villi on its surface. Each fetus is enclosed in a long **fetal sac (chorioallantoic sac)**. The outer layer of this sac, the **chorion,** also bears many folds and villi, which increase the surface area. The uterine mucosa (the maternal part) and the chorion (the fetal part) together make up the **placenta** (Fig. 20-22). Both mucosa and chorion are highly vascular, but the blood vessels of the moth-

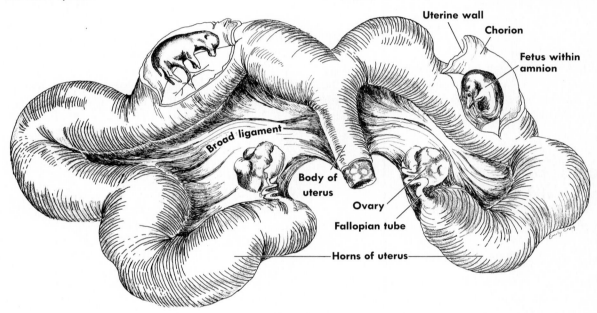

Fig. 20-21. Pregnant pig uterus.

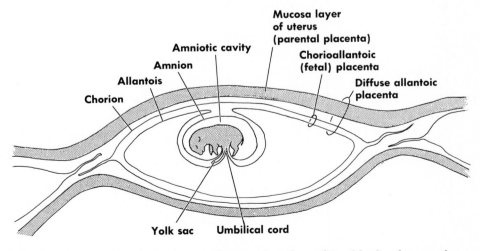

Fig. 20-22. Diagram of pig fetus in utero, showing the relationship of fetal and parental membranes in a diffuse placenta.

er and those of the fetus are separated by the placental membranes, which allow some materials to be exchanged but prevent crossing over of blood cells or blood proteins.

Carefully open one of the fetal sacs and note that the fetus is still enclosed in a thin transparent membrane, the **amnion,** along with the amniotic fluid (Fig. 20-21). Open the amnion to see how the umbilical cord connects the fetus with the blood vessels in the chorioallantoic sac.

The placenta of the pig is known as a **chorioallantoic** or **diffuse placenta** (Fig. 20-22). In the uterus of the dog and some other carnivores the placenta is called a zonary placenta because the chorionic villi are located in a girdlelike zone or band around the middle of each pup rather than over the whole surface of the chorion as in the pig. The human placenta is disc-shaped.

Histology of the Urogenital Organs

Descriptions of transverse sections of the frog kidney, testis, and ovary were given on pp. 321-322. Your instructor may want you to compare these with slides of mammalian tissue.

EXERCISE 20E
The circulatory system

MATERIALS
Fetal pigs
Pig or sheep hearts, fresh or preserved

The circulatory system of the pig is quite similar to that of other mammals, including humans. In contrast to the single-circuit system of the fish, with its two-chambered heart, and the incomplete double circuit of the three-chambered amphibian heart, the mammal has an effective four-chambered heart, which allows the blood two complete circuits: a **systemic** circuit through the body, followed by a **pulmonary** circuit to the lungs for oxygenation. The right side of the heart handles the oxygen-poor blood returning from the body tissues; the left side handles the oxygen-rich blood returning from the lungs.

The study of fetal pig circulation has the added advantage of illustrating not only typical mammalian circulation but typical **fetal circulation** as well. The changes in circulation necessary for the transition from a nonbreathing, noneating fetus to an independent individual with fully functioning organs are simple but very important and interesting.

Uninjected vessels will contain only dried blood (or they may be empty) and therefore will be fragile, flattened, and either brown or colorless. If injected,

Be able to identify and give the functions of the reproductive organs of both the male and female pig.

WRITTEN REPORT

On separate paper write a comparison of the mammalian and amphibian reproductive systems as illustrated by your study of the pig and frog. How is each adapted to its own type of reproduction?

PROJECTS AND DEMONSTRATIONS

1. Demonstration of the human fetus. In a preserved fetus of 3 to 5 months, observe (a) the placenta—disc-shaped embryonic attachment between the fetus and the mother's uterus (notice the irregular villi that serve as the means of attachment to the uterine wall); (b) the umbilical cord, twisted and convoluted, and (c) the navel, where the cord is connected until severed at birth.

2. Demonstration of pregnant dog uterus.

the arteries will have been filled with latex or a starchy injection mass through one of the arteries in the cut umbilical cord and will be firm and pink or yellow. The veins will have been injected through one of the jugular veins in the neck and will be blue. The lymphatic system will not be studied.

You should uncover and separate the vessels with a blunt probe and trace them as far into the body as possible, but be careful not to break or remove them. Nerves often follow an artery and vein and will appear as tough, shiny, white cords. Do not remove them. Try to dissect only on one side, leaving the other side for dissection of nerves later. As you identify a vessel, separate it and carefully remove investing muscle and connective tissue, taking care not to break the vessel or destroy other vessels that you have not yet identified. Do not remove any body organs unless specifically directed to do so.

As it is often difficult to trace the arterial system without damaging the venous system, which lies above it, you will study the veins first. They vary considerably in different individuals. Make notes or sketches of any variations that you find in your specimen. With careful dissection, both veins and arteries can be left intact.

Many students like to increase the usefulness of the accompanying diagrams by coloring the arteries red and the veins blue.

Heart

☞ Note carefully the shape and slope of the diaphragm and how it bounds the thoracic cavity posteriorly. Then cut the diaphragm away from the body wall to make entrance into the thoracic cavity more convenient.

Open the pericardial sac and examine the heart. It has two small thin-walled atria and two larger muscular-walled ventricles.

Right Atrium (Anterior and Ventral). Lift the heart (Fig. 20-23) and see the precaval and postcaval veins that empty into the right atrium. The **postcava** from the abdominal region emerges through the diaphragm; the **precava** comes through the space between the first ribs.

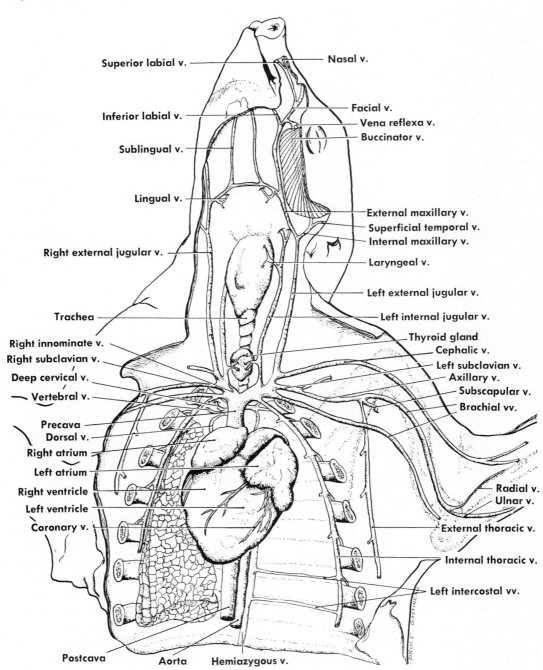

Superior labial v. — Nasal v.

Inferior labial v. — Facial v.
Vena reflexa v.
Buccinator v.

Sublingual v.

Lingual v.

External maxillary v.
Superficial temporal v.
Internal maxillary v.

Right external jugular v. — Laryngeal v.

Left external jugular v.

Trachea — Left internal jugular v.

Right innominate v. — Thyroid gland
Right subclavian v. — Cephalic v.
Deep cervical v. — Left subclavian v.
Vertebral v. — Axillary v.
Subscapular v.
Brachial vv.

Precava
Dorsal v.
Right atrium
Left atrium
Right ventricle
Left ventricle
Radial v.
Coronary v. — Ulnar v.

External thoracic v.

Internal thoracic v.

Left intercostal vv.

Postcava   Aorta   Hemiazygous v.

Fig. 20-23. Veins of head, shoulders, and forelimbs of pig. Internal thoracic veins, shown laterally, actually lie ventral to the heart.

Left Atrium (Anterior and Dorsal). The left atrium receives from **pulmonary veins** on the dorsal side.

Right Ventricle. The right ventricle is large and thick walled. The **pulmonary artery** leaves the right ventricle and passes over the anterior end of the heart to the left, where it divides back of the heart to go to the lungs.

Left Ventricle. The left ventricle is the apex of the heart (posterior). It gives off the large **aorta,** which rises anteriorly just behind (dorsal to) the pulmonary artery. The **coronary sulcus** is the groove between the auricles and ventricles. It contains the **coronary artery and vein,** which supply the tissues of the heart itself.

General Plan of Circulation

In mammalian circulation (after birth) the systemic and pulmonary systems of circulation are separate. Blood from all parts of the body except the lungs returns by way of the large **precava** (anterior vena cava) and **postcava** (posterior vena cava) to the **right atrium.** From there it goes to the **right ventricle** to be pumped out through the **pulmonary arteries** to the lungs to be oxygenated. The blood returns through the **pulmonary veins** to the **left atrium** and then to the **left ventricle,** which sends it through the **aorta** to branches that carry the blood finally to the capillaries of all parts of the body. The venous system returns it again to the right side of the heart to begin another circuit. The mammal has a **hepatic portal system** but no renal portal system as in the amphibian.

Before birth, when the lungs are not yet functioning, the fetus depends on the placenta for nutrients and oxygen, which are brought to it through the **umbilical vein.** Therefore the general plan of circulation is modified in the fetus so that most of the pulmonary circulation is short circuited directly into the systemic bloodstream to be carried to the placenta instead of the lungs. Only enough blood goes to the lungs to nourish the lung tissue until birth. These modifications will be mentioned as they arise in later descriptions.

Veins of the Head, Shoulders, and Forelimbs

☞ Carefully remove the remainder of the sternum and the first rib without damaging the veins beneath.

You will find considerable variation in the veins of individual specimens.

The precaval division of the venous circuit carries blood from the head, neck, thorax, and forelimbs to the heart. It includes the following:

Precava Vein (Anterior Vena Cava). The precava vein enters the right atrium. It is formed by the union of two short trunks, the **right** and **left innominate veins.** Follow the innominates through the opening at the

level of the first rib into the neck. Each innominate vein receives from the following:

1. Internal jugular vein. Lying along the trachea. It drains the brain, larynx, and thyroid. It lies next to the common carotid artery and the slender vagus nerve (Fig. 20-23).

2. External jugular vein. Lateral to the internal jugular (more superficial). It is formed by the union of the **internal maxillary** (dorsal) and the **external maxillary** (ventral) back of the angle of the jaw. It receives branches from the thyroid, trachea, and so forth (Figs. 20-14, 20-15, and 20-23).

3. Cephalic vein. May join the base of the external jugular (or sometimes the subclavian). It is a large superficial vein from the arm and shoulder. It lies just beneath the skin.

4. Subclavian vein. From the shoulder and forelimb. It is a deeper vein that follows the subclavian artery. (There may be from one to three subclavian or brachial veins. If there are more than one, there may be considerable anastomosing among them.) The subclavian is joined by the **subscapular** and **suprascapular** veins from the shoulder. The subclavian, as it continues into the arm, is called the **axillary** in the armpit and the **brachial** in the upper arm; it divides into **radial** and **ulnar** veins in the lower arm. Large nerves of the brachial nerve plexus overlie these veins. You may clip some of these in one arm, if necessary, but be sure to save them in the other arm.

Other Veins. Now go back to the chest cavity and find the following:

1. Internal thoracic vein (also called sternal, internal mammary). Joins the precava near the heart. This is the vein that, with the accompanying artery, you detached from the ventral muscle wall of the chest cavity. You will also find its continuation in the abdominal muscle wall.

2. Costocervical trunk. Enters the precava dorsolaterally close to the heart. It receives the **deep cervical** (from the neck region), the **vertebral,** and the **dorsal** veins. The dorsal receives from some of the intercostals. These veins vary considerably and may enter the precava separately or together. Look under the anterior end of the lungs for these veins, which lie close to the vertebral column.

Hemiazygous Vein. Lift the heart and left lung and push them to the animal's right. The unpaired hemiazygous vein lies in the chest cavity along the left side of the aorta and receives from the left and right intercostal veins. Follow the hemiazygous anteriorly to the point at which it crosses the aorta and goes under the heart. Lift the heart up to see the vein cross the pulmonary veins and enter the right atrium along with the postcava.

Arteries of the Head, Shoulders, and Forelimbs

Arteries carry blood away from the heart, and, with the exception of the pulmonaries, all branch from the main artery, the aorta.

The **aorta** begins at the left ventricle (ascending aorta), curves dorsally (aortic arch) behind the left lung, and extends posteriorly along the middorsal line (descending or dorsal aorta). The first two branches are the brachiocephalic trunk and, just to the left of it, the left subclavian artery (Fig. 20-24).

Brachiocephalic Trunk. The brachiocephalic is a large single artery that extends anteriorly a short distance and branches into the carotid trunk and the right subclavian artery.

1. Carotid trunk (or bicarotid artery). May extend anteriorly for as much as ½ inch or may divide at once into the **left** and **right common carotid** arteries, which form a Y over the trachea and extend up each side of it. The common carotid artery follows the internal jugular vein and the vagus nerve toward the

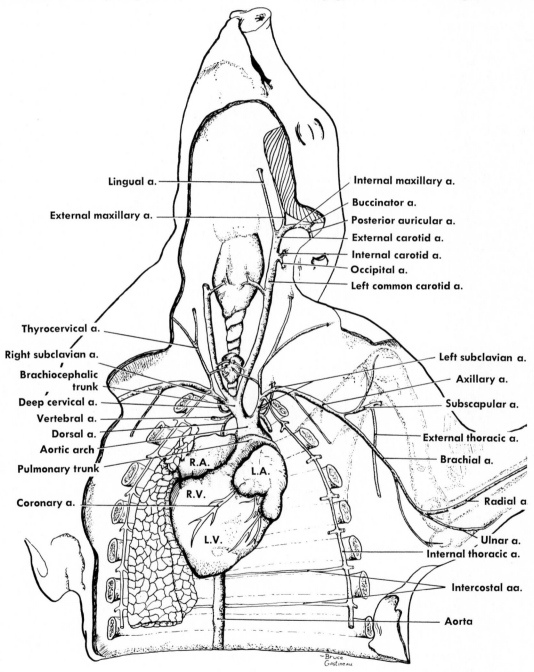

Fig. 20-24. Arteries of head, shoulders, and forelimbs of pig. Internal thoracic arteries, shown laterally, actually lie ventral to the heart.

head, giving off branches to the esophagus, thyroid, and larynx. Near the head each common carotid gives off an **internal carotid,** a deep artery passing dorsally to the skull and brain (it gives off at once an **occipital** artery) and an **external carotid,** a continuation of the common carotid that gives off deep branches to the tongue (**lingual**) and to the cheek and face (**internal** and **external maxillaries**).

2. *Right subclavian artery.* Arises from the brachiocephalic and continues into the right arm as the **axillary** in the armpit, the **brachial** in the upper arm, and the **radial** and **ulnar** in the lower arm. The sub-

scapular to the deep muscles of the shoulder branches off the axillary part of the artery.

Left Subclavian Artery. The left subclavian arises directly from the aorta. Otherwise it is similar to the right subclavian artery.

Each of the subclavian arteries gives off the following branches:

Internal thoracic artery (mammary or sternal). The one you detached earlier. It supplies the ventral muscular wall of the thorax and abdomen.

Thyrocervical artery (internal cervical). Arises from the subclavian opposite to the internal thoracic.

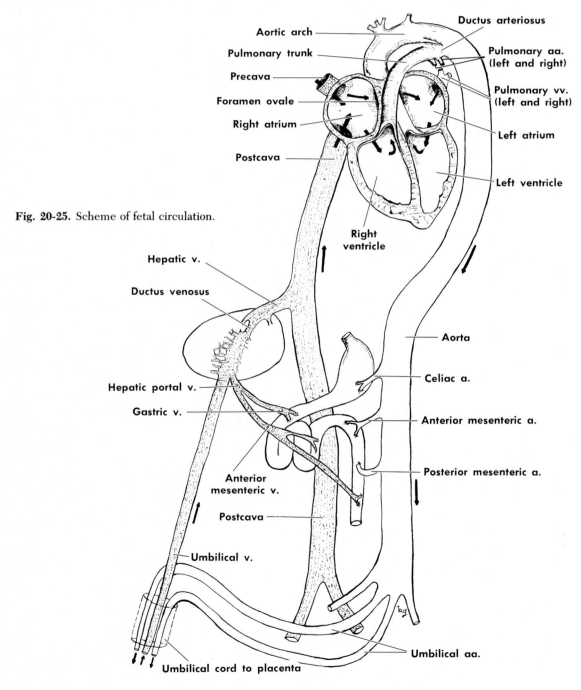

Fig. 20-25. Scheme of fetal circulation.

It supplies the thyroid and parotid glands and some of the pectoral muscles. Trace some of these branches.

Deep cervical, vertebral, and dorsal arteries (costocervical trunk). Supply the neck vertebrae and the deep muscles of the back and neck. They may arise together or separately from the subclavians. Look under the anterior end of each lung for them.

Pulmonary Circulation

Pulmonary Trunk. The pulmonary arises from the right ventricle, ventral to the aorta. Follow it as it branches into **right** and **left pulmonary arteries,** which carry oxygen-poor blood to the lungs. Scrape away some lung tissue to find branches of these vessels.

Pulmonary Veins. Pulmonary veins empty oxygen-rich blood into the left atrium. Probe gently under the left atrium to expose the veins. You should be able to find the large trunk entering the heart just under the hemiazygous vein and a pair of vessels servicing each lobe of the lungs.

Fetal Shortcuts. In the fetal heart there are two shortcuts that prevent most of the blood from making the complete circuit of the lungs, since at this time only enough blood is needed there to nourish the lung tissue itself.

1. Foramen Ovale. Fetal opening in the wall between the right and left atria. Through the foramen ovale part of the blood from the right atrium can pass directly to the left atrium where it can go to the left ventricle, to the aorta, and back into systemic circulation without going to the lungs at all (Fig. 20-25). The remainder of the blood can travel the regular pulmonary channel.

2. Ductus arteriosus. Short connection between the pulmonary trunk and the aorta. Trace this connection, which begins where the smaller pulmonaries branch off toward the lungs. Part of the blood from the right ventricle goes to the lungs, and part is sent directly through the ductus arteriosus to the aorta for systemic use (Fig. 20-25).

After birth the foramen ovale closes and leaves a permanent scar called the **fossa ovalis.** The ductus

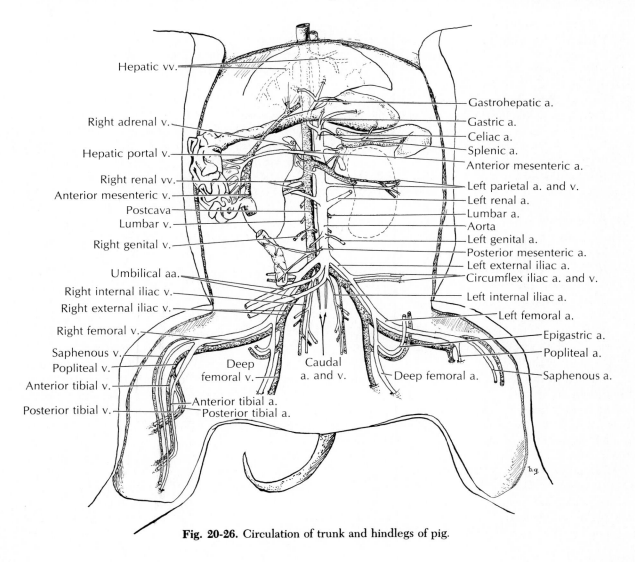

Fig. 20-26. Circulation of trunk and hindlegs of pig.

arteriosus closes and becomes a ligament, the **ligamentum arteriosum**, between the pulmonary artery and the aortic arch. Until birth, oxygenated blood comes from the placenta by way of the **umbilical vein** and is carried to the heart by the postcava.

Postcaval Venous Circulation

The **postcava** carries blood to the heart from the hindlimbs and trunk. Follow it from the right atrium through the intermediate lobe of the lung and through the diaphragm to the liver (Fig. 20-26). It receives from the following:

Phrenic Veins. The phrenics are two or three small veins from the diaphragm.

Hepatic Veins. The hepatics are three or four veins from the liver. Scrape away some of the liver tissue near the diaphragm to see them.

Umbilical Vein. The umbilical is the vein from the umbilical cord that you tied and cut. It carries oxygen and nutrients from the placenta to the growing fetus. The **ductus venosus,** another fetal shortcut, connects the umbilical vein with the postcava, passing through the liver tissue. During fetal life the anterior part of the postcaval vein carries a mixture of oxygen-poor blood from body tissues and oxygen-rich blood from the placenta. After birth both the umbilical vein and the ductus venosus degenerate.

☞ You will be able to see the arrangement of the vessels in the liver tissue by using the probe to scrape or "comb out" the liver tissue to separate it from the vessels. Wash out the loose tissue.

Some of the vessels you see will probably be hepatic ducts.

Other Postcaval Veins

Now push the intestines to the animal's left side, clean away most of the dorsal portion of the liver, and note where the **postcava** emerges and continues posteriorly. Clean away the peritoneum that covers the postcava, and find the following veins that enter it on each side.

Renal Veins. One or two renals extend from each kidney. Do both right and left veins enter at the same level?

Adrenal Vein. The adrenal may enter the renal vein or empty directly into the postcava.

Parietal (Phrenicoabdominal) Vein. The parietal extends from the body wall (lift up the kidney to see it) and may enter the renal vein or empty directly into the postcava.

Spermatic Vein. The spermatic arises as part of the spermatic cord from the testis in the male.

Ovarian Vein. The ovarian lies suspended in the mesovarium (female) along with the ovarian artery.

Lumbar Veins. Six or seven small lumbar veins

from the dorsal muscles of the lower back enter the postcava and common iliac.

Common Iliac Vein. Each common iliac from the leg comes together with its mate to form the postcava. It receives from an internal and an external iliac vein.

Internal Iliac Vein (Hypogastric). Medial branch extending dorsally and posteriorly into the deep tissue and draining blood from the rectum, bladder, and gluteal muscles.

External Iliac Vein. Lateral branch and the largest vein entering the common iliac. It drains the foot and leg and is formed by the union of the femoral and deep femoral veins. It receives the circumflex iliac.

Circumflex iliac vein. Drains the muscles of the abdomen and the upper thigh.

Deep femoral vein. Extends dorsally and posteriorly into deep muscles.

Femoral vein. Large vein, a direct continuation of the external iliac, extends ventrally toward the knee. It collects from the **saphenous vein,** a superficial vein that drains the medial side of the foot and leg. The femoral extends under the knee region, where it is called the **popliteal.** The popliteal vein is formed by the union of the **anterior** and **posterior tibial veins,** which are deep veins in the lower leg.

Caudal Vein. The caudal is a single small median vein from the tail.

Hepatic Portal System

The hepatic portal system is a series of veins that drain the digestive system and spleen. The blood is collected in the portal vein, which carries it to the liver capillaries. As the blood laden with nutrients from the intestine passes through the liver, the liver may store excess sugars in the form of glycogen, or it may give up sugar to the blood if sugar is needed. Some of the amino acids formed by protein digestion may also be modified here. Blood from the liver capillaries is collected by the **hepatic veins,** short veins that empty into the postcava.

The hepatic portal system in your specimen may not be injected. Unless these vessels are injected or are filled with dry blood, they may be difficult to locate. Try to examine an injected specimen.

Lift up the liver, stomach, and duodenum and draw the intestine posteriorly to expose the pancreas. Loosen the pancreatic tissue from the **hepatic portal vein** that runs through it. Now lay the duodenum over to your right and see how the portal vein enters the lobes of the liver near the common bile duct (in the gastrohepatic ligament). Lay the small intestine over to your left and fan out the mesenteries of the small intestine to see how the veins from the intestines are collected into the **anterior mesenteric vein** (Fig. 20-27). This joins with the small **posterior mesenteric vein** from the large intestine and the **gastrosplenic** from the

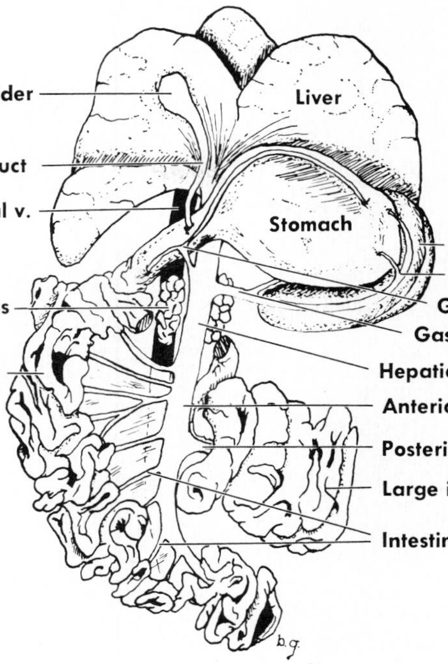

Fig. 20-27. Hepatic portal system of fetal pig. The small intestine has been pushed to one side, and the large intestine, where it crosses over the hepatic portal vein, has been cut.

Labels on figure: Gallbladder, Cystic duct, Postcaval v., Pancreas, Small intestine, Liver, Stomach, Spleen, Splenic v., Gastric v., Gastrosplenic v., Hepatic portal v., Anterior mesenteric v., Posterior mesenteric v., Large intestine, Intestinal vv.

stomach and spleen to form the **hepatic portal vein.** Remove some of the pancreas to find the gastrosplenic. Nearer the liver the portal also receives from smaller **gastric** and **splenic** veins.

Arteries of the Trunk and Hindlegs

Descending Aorta. The descending aorta follows the vertebral column posteriorly, first lying dorsal and then ventral to the postcava (Fig. 20-26). It gives off the following.

Intercostal Arteries. Fourteen or fifteen small vessels in the thoracic cavity run into the muscles between the ribs. These are called intercostal arteries.

Celiac Artery. A large single artery, the celiac, runs to the digestive tract. To find it, clip away the diaphragm and remove the tissue around the aorta at the anterior end of the abdominal cavity. Be careful of the celiac ganglion and sympathetic nerves. The very short celiac divides into three arteries:

Gastric artery. To the stomach.

Splenic artery. To the spleen, stomach, and pancreas.

Gastrohepatic artery. To the stomach, liver, pancreas, and duodenum.

Anterior Mesenteric Artery. The anterior mesenteric is an unpaired artery not far from the celiac; it sends branches to the pancreas, small intestine, and large intestine. It must be dissected out from surrounding tissue.

Parietal (Phrenicoabdominal) Arteries. The parietals run to the diaphragm and the body wall. Lift up a kidney to see one of them.

Renal Arteries. The renals extend to the kidneys, one to each kidney.

Adrenal Arteries. Each adrenal may branch from a renal artery or come directly from the aorta.

Spermatic or Ovarian Arteries. The paired spermatics or ovarians run to the gonads.

Lumbar Arteries. Six pairs of lumbars run from the dorsal side of the aorta to the muscles of the back; one pair comes from the caudal artery.

Posterior Mesenteric Artery. The posterior mesenteric is a small unpaired artery that divides to send branches to the colon and rectum.

Iliac Arteries. The posterior end of the aorta divides to form two large lateral **external iliac** arteries to the legs, two medial **internal iliac** arteries to the sacral region, and a **caudal** artery to the tail.

Internal Iliac Artery. Each internal iliac gives off at once the large **umbilical artery** and then continues dorsally and posteriorly into the sacral region beside the internal iliac vein, giving off branches to the bladder, rectum and gluteal muscles. After birth the umbilical arteries become small arteries to the urinary bladder.

External Iliac Artery. Each external iliac continues into the leg, giving off the following:

Circumflex iliac artery. Extends laterally to the abdominal muscles.

Deep femoral artery. Extends dorsally and posteriorly into the deep muscles of the thigh.

Femoral artery. Continuation of the external iliac toward the knee. It gives off the **saphenous artery** to

367

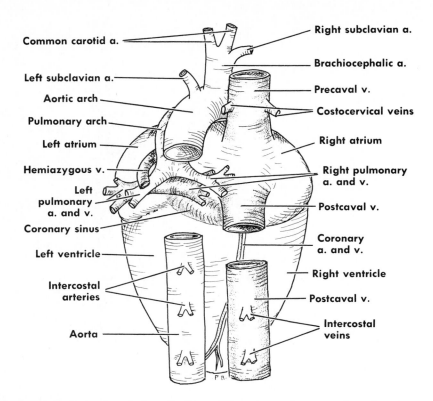

Fig. 20-28. Heart of pig, dorsal view. Portions of the aorta and postcava have been removed to show the underlying vessels.

the superficial muscles of the medial side of the leg and then continues into the deeper muscles to become the **popliteal** under the knee. The popliteal divides into the **anterior tibial** and **posterior tibial arteries in the lower leg.**

Caudal Artery (Middle Sacral). The caudal is a small continuation of the aorta into the tail.

Structure of the Heart

External View

☞ Use fresh or preserved pig or sheep hearts and compare with the heart of the fetal pig.

Locate the right and left **atria** and the right and left **ventricles.** Find the **coronary sulcus** and the **coronary artery** and **vein,** which supply the muscles of the heart. Now identify on the ventral side the large thick-walled **pulmonary trunk** leaving the right ventricle and the large **aorta** leaving the left ventricle. On the dorsal side, find the large thin-walled **precaval** and **postcaval veins** entering the right atrium and the **pulmonary veins** entering the left atrium (Fig. 20-28).

Frontal Section

☞ Now make a frontal section, dividing the heart into dorsal and ventral halves. Start at the apex and direct the cut between the origins of the pulmonary and aortic trunks. Leave the two halves of the heart attached at the top. Wash out (or, if injected, carefully pull out the latex filling).

The cavities of the heart are lined with a shiny membrane, the **endocardium.** Identify the chambers and the valves at the entrance to each chamber that prevent a backflow of blood.

Right Atrium. The right atrium is thin walled, with openings from the precava, postcava, and hemiazygous veins. Find the entrance of these veins (Fig. 20-29). Now, in the dorsal half of the fetal pig heart, probe for an opening between the two atria. This is the **foramen ovale,** one of the fetal shortcuts.

Right Ventricle. The right ventricle is thick walled. An atrioventricular valve, called the **tricuspid valve,** prevents backflow of blood to the atrium. The valve is composed of three **cusps** or flaps of tissue that extend from the floor of the atrium and are connected by fibrous cords, the **chordae tendineae** (ten-din′ee-ee), to **papillary muscles** projecting from the walls of the ventricle.

In the ventral half of the heart the opening into the pulmonary artery is guarded by **semilunar valves.**

☞ Cut the pulmonary artery close to the heart and remove the latex. Slit the vessel for a short distance into the ventricle and look into it to see the three cusps or pockets of tissue. Determine how they would work to allow passage into the vessel but prevent return of blood into the ventricle.

Left Atrium. The left atrium is thin walled. Find

368

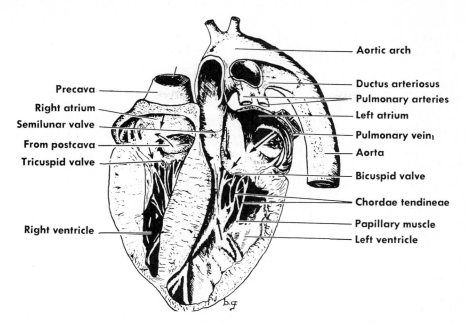

Fig. 20-29. Frontal section of fetal heart, ventral view. In this view the pulmonary trunk and part of the ascending aorta have been removed with the ventral half of the heart.

the entrance of the pulmonary veins that return oxygen-rich blood to the heart.

Left Ventricle. The left ventricle is the most muscular chamber of the heart, for it must send the freshly oxygenated blood at high pressure to all the tissues of the body. An atrioventricular valve, the **bicuspid valve** (also called the mitral valve), guards the entrance from the left atrium. Find the cusps of this valve. Push the valve open and closed to see how it works.

Find the three-cusped **semilunar valve,** which prevents backflow from the aorta (Fig. 20-29).

Notice the difference in the thickness of the walls of arteries and veins. Of what are these walls composed? Would you find the same kind of muscle in the heart as in the vessels? What is the pacemaker of the heart? Consult your text.

Be familiar with the direction of blood flow throughout the body. Be able to describe how fetal circulation differs from postnatal circulation. Be able to trace blood from any part of the heart to any part of the body and back to the heart.

WRITTEN REPORT

On separate paper list the two chief differences between mammalian and the amphibian circulatory systems as illustrated by your study of the pig and the frog. How are these differences adaptive to an all-terrestrial or half-aquatic life?

PROJECTS AND DEMONSTRATIONS

1. Demonstrations of prepared slides of mammalian blood.

2. Prepared slide of artery and vein. Study prepared cross sections of an artery and a vein and observe the structural differences (see p. 72-73).

3. Capillary circulation in the tail fin of a goldfish or small minnow. Anesthetize a goldfish by placing it in a dish of 0.2% Chloretone for about 1 to 2 minutes; then wrap the fish in wet cloth, leaving the tail free. Place the fish on a glass slide, spread out the tail fin, and examine with low power of the microscope. Note the shape and size of the red corpuscles, and then compare with those of the frog and the human. How do you distinguish an arteriole from a venule?

Exercise 22B describes a method of observing capillary circulation in the frog's tongue.

4. Action of heart valves. Obtain a fresh sheep's heart with the roots of the great vessels uncut. Tie glass tubes in the aorta, the pulmonary artery, and the veins of the atria. Suspend the heart by clamping the glass tube tied in the aorta to a stand. Water poured into the atria will rise through the pulmonary artery and aorta. By means of a long pipette inserted through the veins of the atria, remove some of the water from the ventricles. The water will still remain at the same height in the aorta and pulmonary arteries because of the action of the semilunar valves. Empty the heart of water. With a long tube or pipette shoved down the aorta, place some water in the left ventricle. When the left ventricle is filled, the mitral valve prevents the water from entering the left atrium, and it will now rise up through the aorta.

5. *Other suggestions*. Other suitable demonstrations are given in Chapter 19, p. 328.

EXERCISE 20F
The nervous system

MATERIALS
Fetal pigs
Preserved brains

In your dissection of the nervous system of the fetal pig you will look first for the spinal nerves and plexuses and then for the sympathetic nerve trunks and some of their ganglia. These you will find in the neck region and in the body cavities. Then you will dissect out the brain and spinal cord, or, if your instructor prefers, you may study a preserved mammalian brain.

Consult your text* for the physiology and functions of the nervous system and be able to relate this material to your dissection.

SPINAL NERVES
There are 33 pairs of spinal nerves in the pig—8 cervical, 14 thoracic, 7 lumbar, and 4 sacral (Fig. 20-30).

Brachial Plexus
In your search for the arteries and veins in the foreleg you saw a group of tough white nerves extending into the arm. These are the sixth, seventh, and eighth cervical and the first thoracic nerves, which, with their interconnecting branches, form the brachial plexus.

☞ Clear away connective tissue and sever blood vessels if necessary to uncover as many of these nerves as possible and trace them back to their emergence from the vertebrae, being careful not to destroy any of the branches. Each nerve sends branches to join neighboring nerves.

Starting at the level of the first rib and moving anteriorly are three nerves into the foreleg that come from the seventh and eighth cervical and the first thoracic nerves—the ulnar, median, and radial nerves.

Ulnar. Most posterior of the brachial nerves. Follows the bend of the elbow to the underside of the foreleg and foot.

---

*See Hickman, C.P., Jr., L.S. Roberts, and F.M. Hickman. 1984. Integrated principles of zoology, ed. 7, St. Louis, The C.V. Mosby Co., pp. 761-777.

Median. Follows the brachial artery and vein along the inside of the leg.

Radial. Larger nerve. Lies underneath the median and passes into the deep muscles of the foreleg. By skinning the lateral side of the arm and shoulder and dissecting under the muscles (triceps) you may find the radial nerve where it crosses over to the lateral side of the foreleg and passes down the dorsal (radial) side of the foreleg and foot.

Subscapular Nerve. From parts of the sixth and seventh cervical nerves. Branches into the muscles under the scapula.

Suprascapular Nerve. Large nerve from the fifth and sixth cervical nerves to the muscles above the scapula.

You can probably also locate the fourth, third, and second cervical nerves, but the first is difficult to find.

Thoracic and Lumbar Nerves
The thoracic nerves lie in the intercostal spaces, running parallel with the intercostal arteries and veins. Trace one or two of these back to the vertebral column.

Posterior to the ribs in the dorsal body wall the first four lumbar nerves lie just under the peritoneum. They angle out posterolaterally from the vertebral column. Trace some of these.

Lumbosacral Plexus
The lumbosacral plexus is formed by the branches of the fifth, sixth, and seventh lumbar and the first sacral nerves; it sends large nerves into the hindlegs. To identify the nerves making up the plexus, you may have to clip the iliac artery and vein from the postcava and remove some of the muscle along the backbone. The **femoral nerve** and the **saphenous nerve** (which branches from the femoral) follow the arteries of the same names. The large **sciatic nerve** follows the internal iliac artery and vein into the deep sacral region. The nerve emerges with the vessels close to the backbone and travels around the end of the ischium to follow the femur (Fig. 20-31).

☞ If you wish to trace the sciatic into the leg, skin the dorsal side of the pelvis and thigh and lay

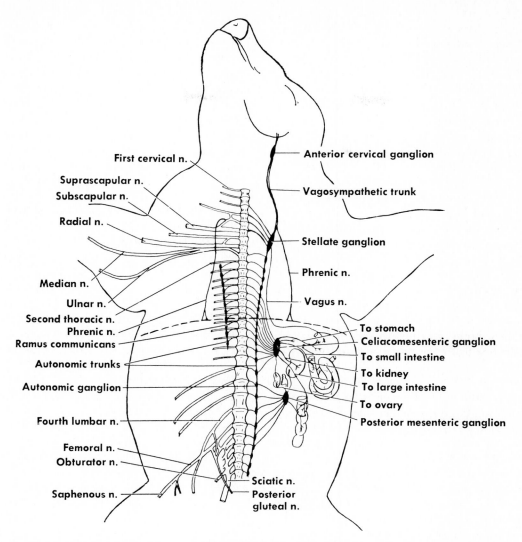

Fig. 20-30. Peripheral nerves. *Left,* Spinal nerves with short portion of sympathetic trunk and its rami communicantes; *right,* sympathetic trunk with its chief ganglia.

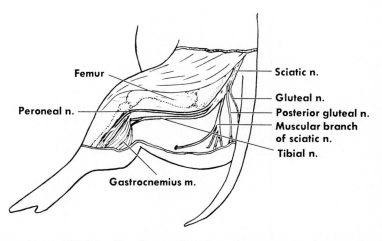

Fig. 20-31. Lateral view of left hindleg dissected to show sciatic nerve and some of its branches.

aside the layer of muscles along the backbone for 3 to 4 cm above the tail.

AUTONOMIC NERVOUS SYSTEM

The autonomic nervous system is concerned with the involuntary body functions that are not controlled by the cerebrum and do not ordinarily affect consciousness. It governs (1) heart muscle, (2) smooth involuntary muscle, such as that of the digestive tract, blood vessels, and the like, and (3) the secretions of various glands. The autonomic system is comprised of neurons whose cell bodies are aggregated into ganglia, some large and some very small. A **sympathetic trunk,** composed of ganglia connected by sympathetic nerve fibers, extends on each side of the vertebral column from the base of the cranium to the tail. Each ganglion is connected to one or more cerebrospinal nerves by branches, called **rami communicantes,** and sends peripheral branches to the viscera. The parasympathetic division of the autonomic system will not be dissected. Consult your text for its location and function.

Vagosympathetic Trunk

Each vagosympathetic trunk extends from the brain down the neck on the dorsomedial side of the carotid artery. It is made up of the **vagus (tenth cranial) nerve** and the **sympathetic trunk,** which are united in a common sheath and enter the thoracic cavity at the level of the first rib. The vagus nerve is a mixed nerve carrying sensory and motor fibers. It follows the esophagus to the abdominal cavity, sending branches to the heart, bronchi, stomach, and intestines. The sympathetic trunk continues along the vertebral column under cover of the peritoneum and is connected by rami communicantes to the spinal nerves.

☞ Clean away some of the peritoneum to uncover one of these trunks in the body wall.

Ganglia. There are several large ganglia in the sympathetic system that you may be able to find with some care. These are usually paired and appear as irregular whitish swellings along the trunks.

Anterior cervical ganglion. Small ganglion at the base of the skull near the division of the external and internal carotids.

Stellate ganglion. Made up of the posterior cervical and the first thoracic ganglia and located under cover of the first rib on the side of the trachea near the origin of the subclavian artery. It is at this ganglion that the vagus and sympathetic trunks separate. From this region, try to separate the two nerves and trace them both back to the skull. Now follow the vagus to the stomach and the sympathetic trunk into the dorsal thoracic wall.

Thoracic and lumbar ganglia. Segmentally arranged swellings on the main sympathetic trunks. Lift the trunk slightly to see the rami communicantes dip-ping into the muscle to connect with spinal nerves where they emerge from the vertebrae.

Celiacomesenteric ganglia. Large, elongated, irregular ganglia, one located on each side of the aorta at the origin of the celiac and anterior mesenteric arteries. They send branches to the stomach, intestine, kidney, and genital organs.

Posterior mesenteric ganglion (unpaired). Small ganglion on the posterior mesenteric artery. It sends branches to the intestine and to the ovaries or testes.

Do you see any pattern of similarity between the nervous systems of the pig and the frog? By observing charts or diagrams, compare the nervous system of the pig with that of the human being. What conclusions can you draw about the nervous system of vertebrates in general? About the nervous systems of the vertebrates versus that of the invertebrates?

BRAIN

Directions will be given for the dissection and study of the brain and spinal cord of the fetal pig. However, since the brain is sometimes too soft for much handling, your instructor may prefer to have you study the brain of another mammal, such as the sheep or dog. Similarities between these various brains are pronounced, and the description will be applicable to any.

The brain of a medium-sized adult pig weighs approximately 120 g; that of an adult white male weighs about 1360 g (3 pounds).

☞ Skin the head, neck, and back and remove the thick muscles from the back of the neck and along the vertebrae. Scrape the skull clean and identify the bones of the top of the cranium (Fig. 20-3).

To remove the top of the cranium, use the point of the forceps or the tip of the scalpel and begin with the suture between the nasals and frontals. Carefully lift and chip away bits of bone, being very careful not to puncture the membrane covering the brain underneath. Work carefully and remove all the bony covering. Then remove the tops from several of the vertebrae by clipping with scissors between the spinous process and transverse process on each side of each vertebra. This uncovers a portion of the spinal cord.

The brain and spinal cord are covered with protective membranes that carry blood vessels to nourish the brain. These are called the **meninges** and include the following:

Dura mater—tough outer membrane; tends to cling to the skull in places.

Arachnoid—clings to the pia mater; too thin to identify.

Pia mater—fine vascular membrane; adheres closely to the brain and spinal cord and contains many blood vessels that nourish the brain.

Cerebrospinal fluid flows in the space between the membranes and also in the cavities or ventricles of the brain and the central canal of the spinal cord.

☞ Carefully remove the **dura mater** from the brain. Now the brain will be very soft and quite easily damaged, but it can be removed from the brain case. Lift it gently and look for the cranial nerves on the ventral side where they pass into the floor of the braincase. Clip the nerves, leaving the stubs on the brain as long as possible. Cut the spinal cord 2.5 to 5 cm below the brain, leaving one or two stubs of spinal nerves in place. Now lift out the entire brain carefully and place it in a finger bowl of water.

Study the spinal cord, lifting it up to see the dorsal and ventral roots by which the spinal nerves emerge from the cord.

Dorsal Surface of Brain

Cerebrum (Cerebral Hemispheres). Two large oval lobes separated by the **longitudinal fissure** (Fig. 20-32, *A*). The cortex (outer layer) is made up of raised convolutions called **gyri,** separated by fissures called **sulci.** The cortex contains most of the gray matter of the brain. A groove, the **transverse fissure,** marks off the cerebrum posteriorly from the cerebellum.

Cerebellum. Lies behind the cerebrum and is somewhat triangular. Its folds and fissures are much finer than those of the cerebrum.

Medulla Oblongata. Small, lies behind and ventral to the cerebellum. It narrows posteriorly to continue as the nerve cord.

Corpus Callosum. Spread the lobes of the cerebrum apart to see at the bottom of the fissure a broad white band of nerve fibers passing between the hemispheres.

Corpora Quadrigemina (Midbrain). Spread the cerebrum and cerebellum apart gently to see the four small rounded knobs.

Ventral Surface of Brain

Olfactory Lobes. Extend anteriorly from the cerebral hemispheres (Fig. 20-32, *B*). Fit into depressions in the floor of the skull and send nerves to the nostrils.

Optic Chiasma. Crossing of the **optic nerves** posterior to the olfactory lobes.

Pituitary Gland (Hypophysis). Posterior to the optic chiasma a swelling called the **tuber cinereum** bears a stalk, the **infundibulum,** on which the **pituitary** is attached. The pituitary gland fits into a depression in the floor of the cranium. It often breaks off in dissection and may be found in the depression.

Pons. Broad raised area of transverse fibers posterior to the pituitary and joining the medulla posteriorly. Its fibers connect the two halves of the cerebellum and also connect the cerebellum with the cerebral cortex.

Note the **basilar artery,** a continuation of the vertebral arteries, in the midventral line of the brain and encircling some of the ventral structures.

Cranial Nerves

The cranial nerves will be seen only as stumps (Fig. 20-32, *B*). A hand lens may help in identifying them. See also the human brains on display in the laboratory. Some cranial nerves carry only sensory fibers to the brain from the eyes, nose, or ears. Some carry only motor fibers, which innervate the muscles or glands

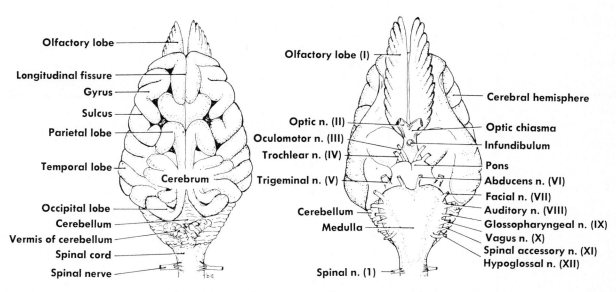

Fig. 20-32. Pig brain. **A,** Dorsal view. **B,** Ventral view.

of certain areas; others carry both sensory and motor fibers. All are paired.

Olfactory (I)—sensory; from the nose to the olfactory lobes.

Optic (II)—sensory; from the eyes to the optic chiasma.

Oculomotor (III)—motor; small; anterior to the pons; to the muscles of the eyeball.

Trochlear (IV)—motor; smallest; lateral to the oculomotor; to the muscles of the eyeball.

Trigeminal (V)—mixed; large; from the lateral aspect of the pons; sends three branches to the nose, eyelids, face, tongue, and muscles of the jaw.

Abducens (VI)—motor; small; posterior to the pons; to the muscles of the eyeball.

Facial (VII)—mixed; on the medulla; lateral to the abducens; to the facial muscles.

Auditory (acoustic) (VIII)—sensory; posterior to the facial nerve on the anterolateral aspect of the medulla; from the inner ear.

Glossopharyngeal (IX)—mixed; from the side of medulla; to the tongue and pharymx.

Vagus (pneumogastric) (X)—mixed; from the side of the medulla; passes into the thoracic and the abdominal cavities to the pharynx, larynx, heart, lungs, and stomach.

Spinal accessory (XI)—motor; from the lateral surface of the medulla and the spinal cord; to the muscles of the neck.

Hypoglossal (XII)—motor; from the ventral surface of the medulla; to the muscles of the tongue and neck.

Medical students for generations have used the following couplet to help them remember the order of the cranial nerves: "On Old Olympus' Towering Tops A Finn And German Viewed Some Hops." The first letter of each word stands for the initial letter of a nerve.

Sagittal Section of the Brain

On half sections of preserved brain (Fig. 20-33), find (1) gray and white matter; (2) corpus callosum—cut section, curved; (3) corpora quadrigemina—between the cerebrum and the cerebellum; (4) arbor vitae—white fibers, three-branched in shape, in the cerebellum; (5) third ventricle—cavity above the optic chiasma; and (6) fourth ventricle—small cavity with a very thin roof, dorsal in the medulla, under the cerebellum.

> Be able to identify and give the functions of the various parts of the central nervous system.

PROJECTS AND DEMONSTRATIONS

1. Models of mammalian eye and ear.

2. Fresh or preserved eyes for dissection. Eyes may be obtained fresh from slaughter houses or preserved from biological supply houses.

3. Nerve-muscle preparation. Pith a frog and remove the gastrocnemius muscle with the sciatic nerve connected to it. (Cut off the nerve near its connection to the sciatic plexus.) The muscle should be so dissected that part of the femur is left attached to it (instructions are found on p. 463).

Place the femur in a muscle clamp and the nerve on a glass plate. Keep moist with Ringer's solution. Stimulate the nerve by pinching it with forceps (mechanical), by touching it with a hot needle (heat), by putting salt crystals on it (chemical), and by submitting it to electric shock from an electronic stimulator. Note the effect of each kind of stimulus on the muscle.

4. Spinal reflex in the frog. See Exercise 24A, p. 413.

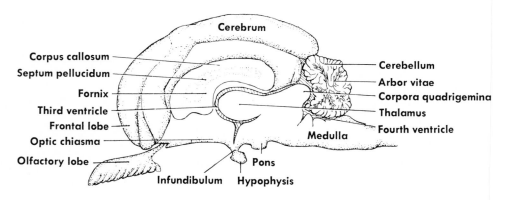

Fig. 20-33. Pig brain, sagittal section.

EXERCISE 20G
The respiratory system

 Draw down the tongue as you did when studying the mouth cavity. Remove the mandibles if you wish. Locate the opening into the **nasopharynx,** the space above the soft palate. Slit the soft palate and probe anteriorly to locate the nasal passages that lead to the **nostrils.**

Locate the small opening on each side of the roof of the nasopharynx. From these openings the **eustachian tubes** lead to the tympanic cavity of the middle ear.

Recall how the passageways for air and food cross in the pharyngeal cavity (Fig. 20-16). Locate again the **epiglottis** and note how it functions to close the **glottis** (open end of the larynx) when food is being swallowed.

The **larynx** is a cartilage-reinforced cylinder that contains the **vocal cords.** Displace the epiglottis and look into the glottis or laryngeal opening to see the **vocal folds (cords).** These are folds of flesh directed downward, each with a slitlike opening. Open the larynx ventrally to see them better.

The **trachea** is stiffened by a series of cartilage rings, which are incomplete dorsally where the trachea lies against the esophagus.

The **lungs** are unequally paired, the larger right lung having four lobes and the left lung two or three lobes. The right lung has an **apical lobe,** a **cardiac lobe,** a **diaphragmatic lobe,** and an **intermediate lobe,** which is median and notched to surround the postcava. The apical and cardiac lobes are usually fused on the left lung. The lungs are spongy and highly elastic.

The **bronchi** are branches of the trachea that extend into the lobes of the lungs. These branch repeatedly in the lungs to form, finally, microscopic **respiratory bronchioles,** which give off **alveolar ducts,** each of which terminates in a cluster of **alveoli** or air cells.

 Remove the heart and the large vessels, leaving stubs of the pulmonary vessels attached to the lungs. Remove the larynx, trachea, and lungs. Slit the larynx and trachea and follow the bronchi as far into the lungs as possible.

Trace a blood vessel into the lung tissue.

Cut a thin section from a lung and examine with a hand lens or dissecting microscope.

PROJECTS AND DEMONSTRATIONS

1. A "butcher's pull." Examine the fresh lungs, trachea, and heart of a small sheep or pig. Use a blowpipe to expand a lung. Trace one of the bronchi into the lung and follow some of its branches as far as possible into the lung tissue. Identify the pulmonary vessels. Trace a blood vessel as far as possible into the lung tissue. Why is the tissue so highly vascular? Cut a cross section from a large artery and a large vein and contrast their walls. Contrast the trachea with the esophagus. Where do you find the cartilage? Of what advantage is the cartilage?

PART THREE

ACTIVITY AND CONTINUITY OF LIFE

CHAPTER 21

MUSCLE PHYSIOLOGY

Contraction and movement are basic properties of all animal cells. It is evident in ameboid movement, cytoplasmic streaming, movement of cilia and flagella, spindle fiber contraction during mitosis, and muscle cells (where in muscle contraction it reaches its highest expression). There is a wide variety of muscle types in animals, and they perform a wide variety of functions, such as body movement, maintenance of posture, gastrointestinal tract movements, and circulatory movements. All, however, operate through the same basic sliding filament mechanism, utilizing proteins such as actin and myosin. These interact with calcium ions and ATP as the energy source to bring about shortening of the protein machinery.

Vertebrate skeletal muscle is composed of specialized muscle cells, called **fibers,** organized into sturdy, compact bundles. It is called **striated muscle** because it appears under the microscope as transversely striped, with alternating light and dark bands (Fig. 21-1). This is the type of muscle making up a large fraction of our body mass. It is attached to skeletal elements and is responsible for the movements of appendages, trunk, respiratory organs, eyes, mouthparts, and so forth. It is also called **voluntary muscle** because it is innervated by motor fibers from the spinal cord, which are under conscious cerebral control.

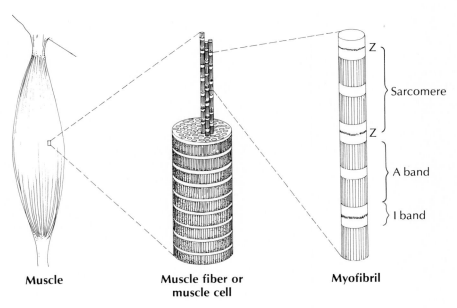

Fig. 21-1. Organization of skeletal muscle.

379

EXERCISE 21A
Contraction of glycerinated skeletal muscle

MATERIALS
Glycerinated muscle preparation*
Dissecting needles or glass needles
Sharp scissors and fine-pointed forceps
Petri dish
Microscope, compound with low and high power
Millimeter rule
Microscope slides and cover glasses
Medicine droppers

The muscle fibers that make up muscle tissue are themselves composed of numerous subunits called **myofibrils**—perhaps 1000 to 2000 myofibrils in each fiber, or muscle cell. The myofibrils are elongate strands of contractile protein that extend from one end of the fiber to the other. The myofibrils are subdivided into units called **sarcomeres**, which give the muscle its striated appearance. Within each sarcomere is an array of thick and thin **filaments**, consisting of bundles of myosin molecules and actin molecules, respectively.

Chemical and electron microscope studies have shown that the myosin molecules have globular heads that serve as cross-bridges to link thick and thin filaments. When the muscle is stimulated, and in the presence of magnesium, calcium, and ATP, the cross-bridges swing rapidly back and forth, pulling the filaments toward each other and shortening the sarcomere. This is the **sliding filament model** of muscle contraction, which is strongly supported by experimental evidence.

Glycerinated muscle preparations can be used to demonstrate and study muscle contraction. Small pieces of muscle have been extracted from a rabbit with 50% glycerol at low temperature, which removes soluble constituents of the muscle, leaving the fibrous contractile proteins intact. Studies with glycerinated muscle are important to scientists who need to control the composition of the intracellular environment without destroying the contractile machinery.

*Carolina Biological Supply Co. product no. 20-3525.

In this simple experiment you will be able to witness and measure the contraction of muscle fibers in the presence of essential ions and with ATP as a source of energy. Strips of skeletal muscle have been extracted with 50% glycerol, tied to a stick, and stored in a vial of glycerol. You are also provided with (1) 0.25% ATP in triply distilled water, (2) 0.25% ATP plus 0.05 M KCl plus 0.001 M $MgCl_2$ in distilled water, and (3) 0.05 M KCl plus 0.001 M $MgCl_2$ in distilled water. Glassware and dissecting instruments should be cleaned thoroughly and rinsed in distilled water. The KCl solution is used to restore the normal ionic environment of the muscle cell, since both potassium and chloride are major ions in intracellular fluid. Magnesium is an ion that is necessary to activate the enzyme (ATPase) that catalyzes the action of ATP.

☞ Pour the 50% glycerol from the vial into a Petri dish and cut the muscle into lengths of approximately 2 cm. Separate the fibers with dissecting needles into strands which are not more than 0.2 mm in diameter. Mount a fiber on a glass slide under a coverslip and examine with the compound microscope. Sketch and label in the proper space on p. 383.

Place several thin filaments on a glass slide in a minimum amount of glycerol, straighten them out parallel to one another, place on the stage of a dissecting microscope, and measure their lengths with a millimeter rule. Record in your report.

Flood the fibers with a few drops of the solution of ATP plus KCl and $MgCl_2$; observe carefully for about 30 seconds; repeat the measurements and calculate the degree of contraction. Examine the contracted fibers under a cover glass with the compound microscope. Make a sketch in the proper space on p. 383.

Repeat with clean slides and fresh fibers using only ATP. Repeat with KCl plus $MgCl_2$ but with no ATP. Explain the results in your report.

Physiology of the myoneural junction

MATERIALS

Two frogs per group
Scissors and needle
Strychnine sulfate, 1 mg/10 ml amphibian saline (strychnine sulfate is available from druggists)
Curare, 1 mg/10 ml amphibian saline* (CAUTION: dangerous drug!)
Electronic stimulator and electrodes

The contraction of muscle fibers is triggered by impulses arriving at a synapse where a motor nerve makes contact with the muscle fiber. This functional connection is called a **myoneural junction** (= motor endplate). At this narrow gap between nerve ending and muscle membrane the arriving nerve action potential stimulates the release of acetylcholine from synaptic vesicles in the synaptic bulb (Fig. 21-2). In less than a millisecond the acetylcholine diffuses across the 20 nm space, separating nerve and muscle membranes to combine with special receptor sites on the postsynaptic (muscle) membrane, leading to an abrupt drop in muscle membrane permeability. This creates a large muscle action potential. The electrical depolarization sweeps through the muscle fiber, generating a muscle contraction. It all happens within a few hundreths of a second.

There are a number of inhibitors and poisons that interfere with this elegant chain reaction. One such inhibitor is curare, long used by South American Indians as an arrowhead poison. An animal or human struck with such an arrow is soon paralyzed, though remaining fully conscious. Curare (the purified form

---

*d-tubocurarine chloride, product no. T2379, is available from Sigma Chemical Co.

is d-tubocurarine) competes with acetylcholine, the neurotransmitter substance, for receptor sites on the postsynaptic membrane, thus blocking muscle contraction. In this experiment you will demonstrate competitive inhibition of postsynaptic receptor sites and blockage of motor impulses across myoneural junctions.

Effect of Strychnine and Curare

☞ Single-pith a frog as described in Appendix B, VII, p. 461. Inject 1 ml of strychnine sulfate solution under the skin on the dorsal surface of the frog. When strychnine convulsions occur, inject 1 ml of tubocurarine (curare) solution. Observe the effects during a period of 15 to 20 minutes, making notes on what you observe.

Strychnine is used in this experiment to block inhibitory synapses in the central nervous system. When this happens, the slightest stimulus sets off an enormous chain reaction of neuronal activity. In the normal animal this catastrophic spread of excitation is prevented by inhibitory activity, which modulates all central nervous system activity.

Effect of Curare after Ligaturing One Leg

☞ Single-pith another frog and expose a sciatic nerve by cutting away the skin from around the thigh. Place a ligature *under* the nerve and around the thigh of the frog. Tie tightly. Inject 1 ml of strychnine as before and observe the results. When convulsions occur, inject 1 ml of tubocurarine solution as before.

Wait several minutes for the effect of the curare. The tied leg should now be the only part of the frog

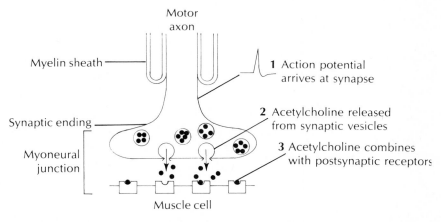

Fig. 21-2. A myoneural junction, showing how acetylcholine (the transmitter substance) released from synaptic vesicles diffuses across the synaptic cleft to bind with receptors on the postsynaptic membrane of the muscle cell.

to show convulsive behavior. If an electronic stimulator has been provided, stimulate the exposed nerve of the *untied* leg with single shocks. (The instructor will explain the correct voltage and duration settings on the stimulator.) Repeat with the nerve of the tied leg. Now directly stimulate the muscle of the untied leg. Repeat with the muscle of the tied leg. Explain all of your observations in your report on p. 384.

Make a sketch below of the muscle fibers as they appear under high power. Label.

Record your measurement of muscle fiber length before and after the addition of ATP plus KCl and MgCl$_2$.

_____

Make a sketch below of the muscle fibers as they appear after the addition of ATP plus KCl and MgCl$_2$.

What happens when only ATP is applied? Explain. _____

_____

_____

What happens when only KCl and MgCl$_2$ are applied? Explain. _____

_____

_____

PHYSIOLOGY OF THE MYONEURAL JUNCTION

EFFECT OF STRYCHNINE AND CURARE

Briefly describe the effect of the strychnine injection. _____

_____

_____

Explain why curare quieted the convulsive frog. _____

_____

_____

Explain why this experiment would not have worked had the spinal cord been destroyed. _____

_____

_____

EFFECT OF CURARE AFTER LIGATURING ONE LEG

Briefly describe the results of this experiment. _____

_____

_____

Why does the tied leg still show convulsive behavior after the curare injection? _____

_____

_____

Does the muscle of the *untied* leg respond when electrically stimulated indirectly through the nerve? _____

Of the *tied* leg? Explain. _____

_____

_____

Does the muscle of the *untied* leg respond to direct electrical stimulation? _____ Of the *tied* leg? _____

Explain. _____

_____

_____

CHAPTER 22

CIRCULATION AND

RESPIRATION

EXERCISE 22A

Study of human blood

NOTE: It is suggested that students work in pairs for all three parts of this exercise. One student of the pair will be the subject for Part I and the other member of the pair the subject for Parts II and III.

MATERIALS

For Parts I, II, and III:
  ABO and Rh blood typing kit (offered by several biological supply houses, for example Carolina Biological Supply Company's kit no. 70-4029 or Ward's Natural Science Establishment kit 36 W 5509)
  Lancets and cotton for sterilizing fingers
  70% alcohol
  Tallquist booklets
  Petri dishes
  Medicine droppers
  Clean, blank microscope slides
  Filter paper
  Wright's stain and/or Camco Quik stain*
  Distilled water
  Microscope with oil immersion lens
  Immersion oil
Additional materials for Parts IV and V:
  Coagulation capillary tubes (1 to 2 mm in diameter, 10 to 15 cm in length)
  Toothpicks
  Microhematocrit centrifuge
  Heparinized capillary tubes
  Wax to seal tubes
  Sphygmomanometer
  Stethoscope

Blood is a complex liquid suspension composed of plasma and suspended formed elements that include red blood cells (erythrocytes), white blood cells (leukocytes), and platelets (thrombocytes). It is a medium that serves many transportation functions, including the movement of nutrients, the transport of oxygen and carbon dioxide between the cells and the respiratory organs, the movement of wastes to the organs

---

*The latter is available from Scientific Products, 1210 Waukegan Rd., McGaw Park, Ill. 60085, product no. B4130.

of excretion, and the shuttling of hormones from endocrine glands to target organs. The blood protects the body from the invasion of foreign bodies through phagocytosis and the production of antibodies. It also distributes body heat and provides for its own (and the body's) protection through hemostasis (mechanisms that prevent blood loss).

In this exercise you will demonstrate some of the unique physiological properties of blood by examining your own blood and comparing the values you obtain with the norm for the human population.

I. ABO AND RH BLOOD GROUP SYSTEMS

The human red blood cell contains more than two dozen antigens on its surface. If blood containing one or more of these antigens is transfused into a recipient whose blood contains antibodies to the donor's blood antigens, a serious agllutination reaction may occur. Agglutination is the clumping of red blood cells together. Fortunately only two of these antigen-antibody systems are potentially troublesome in transfusions; these are the ABO system and the Rh system.

The ABO blood group system, discovered by Karl Landsteiner in 1900, is the best known of the blood antigens. It is an inherited immune system in which the antigens A and B are inherited as dominant genes. A person with genes *A/A* and *A/O* develops A antigen and is said to have type A blood. If the genotype is *B/B* or *B/O*, the person develops B antigen on the red blood cells and has type B blood.

The peculiar feature of the ABO system (Table 22-1) is that the person's plasma will *always* contain the noncomplementary antibodies to the blood cell antigens. This is unlike most antigen-antibody responses; normally antibodies develop only after an antigen is introduced into the system. Thus people with type A blood always carry anti-B antibodies in their plasma, and those with type B blood always carry anti-A antibodies. Type AB persons have both A and B antigens but have *neither* anti-A nor anti-B antibodies. Type

Table 22-1 The ABO blood groups

Blood group phenotype	Blood group genotype	Antigens	Antibodies	Occurrence in:	
				Whites	Blacks
A	A/A or A/O	A	anti-B	41%	22%
B	B/B or B/O	B	anti-A	10%	29%
AB	A/B	AB	none	4%	4%
O	O/O	none	anti-A and anti-B	45%	45%

O persons have no antigens on their red blood cells but have *both* anti-A and anti-B antibodies in their plasma. Thus the blood group name (O, A, B, or AB) identifies the *antigen* content of the blood. Type O persons are called "universal donors" because, having no antigens, their blood can be transfused into a person with any blood type. Type AB people are called "universal recipients" because they have no antibodies to the A or B antigens. But in practice blood types are always carefully matched for transfusions. (There are, in fact, other weak antigens present that make the matching of donor and recipient an important requirement for modern blood banking.)

The Rh blood group system is named for the rhesus monkey in which it was first discovered (again by Karl Landsteiner, 40 years after his discovery of the ABO system!). The Rh system is complex. More than 40 different Rh antigens are known, but only one of these, the D antigen, is so strongly antigenic as to always require close donor-recipient matching. People having the D antigen are called Rh +, and those lacking it are Rh −.

Unlike those in the ABO system, Rh − people do not have the corresponding antibody in their plasma *unless* they have been exposed to the D antigen by a prior transfusion. Then a subsequent transfusion of Rh + blood (containing the D antigen) into the sensitized Rh − person may result in a serious immune reaction. Therefore blood banks always insist on typing of donors for Rh factor as well as for ABO. Another aspect of the Rh factor, which is of great significance to expectant mothers, is the potential incompatibility between mother and fetus. A discussion of this subject will be found in your textbook.

☞ Obtain a blood test card, bottles of antiserums, an alcohol-soaked swab, a disposable, sterile lancet, and an instruction sheet accompanying the antiserum kit. Follow these directions closely.

In brief, you will place drops of the anti-A, anti-B, and Rh serums on the blood test card, then add a drop of fingertip blood to the circles adjacent to the antiserums. The blood and serum in adjacent circles are then mixed together and observed for occurrence of agglutination. Record the results of your own blood

type in the blank provided on p. 391. When class results have been tabulated on the board, enter the percentages of persons having each blood type in the table on p. 391.

PREPARATION FOR PARTS II AND III

Parts II and III should be performed in sequence, using blood collected from a single finger puncture. Collect your materials before making the finger puncture: a clean lancet and fresh alcohol-soaked swab (or cotton), Tallquist paper and scale for the hemoglobin test, and clean slides for preparing the blood smear. Read over the outline carefully before proceeding. One student of the pair will be responsible for the hemoglobin determination, and the other will have responsibility for preparing the blood smear.

☞ Clean the fingertip with an alcohol-soaked swab and puncture the fingertip with a fresh lancet (never use a lancet twice). Immediately place a drop of blood on the Tallquist paper and place a second drop of blood on a clean slide about 1 to 2 cm from one end. You may want to collect a third drop of blood on another slide to have separate blood smears for you and your partner to study. While one student performs the hemoglobin test on the Tallquist paper, the other prepares the blood smear.

II. AN ESTIMATION OF HEMOGLOBIN CONTENT

☞ As soon as the blood is absorbed into the Tallquist filter paper and has lost its glossy appearance (do not wait until it is dried), match its color with that of the Tallquist scale by moving the test paper up and down behind the scale, and read the percentage of the matching color.

One of the functions of red blood corpuscles is the transport of oxygen, which is dependent on hemoglobin, a red pigment. Hemoglobin is a complex protein combined with an iron-containing compound called heme. Males tend to have more hemoglobin than females, averaging 16 g per 100 ml of whole blood, whereas females average 14 g. According to the Tallquist standard 15.6 g of hemoglobin per 100 ml of blood equals 100%. What is your percent? To ascertain the grams of hemoglobin per 100 ml of your blood, multiply the percent shown on the chart by 15.6.

Anemia, which is a deficiency in the amount of circulating hemoglobin, may be caused by too few erythrocytes or by too little hemoglobin in the individual erythrocytes.

NOTE: Although the Tallquist method is a quick and simple means of estimating the blood hemoglobin, it is not considered accurate enough to be used by clinical laboratories, having a possible error of approximately 20%.

Record the percentage reading for your blood on p. 391, and also the number of grams of hemoglobin per 100 ml. Record the latter also on the blackboard. If time permits, determine the average for the males and for the females in your laboratory section.

III. STUDY OF HUMAN BLOOD CELLS

As soon as the drop of blood has been placed on the slide (or slides), lay the slide down on the table surface with the drop toward you. Bring one end of another slide in contact with the first, slanting toward you at a 45-degree angle. Draw it toward the drop of blood till it just touches the drop. The blood should spread laterally along the edge of the slide. Now push the slide *forward* so that it draws the blood along evenly and thinly over the length of the horizontal slide to form a thin, uniform smear. Wave gently in the air for a moment to speed drying, then set aside to complete air drying. You are now ready to stain the slide.

☞ **Method 1.** Using the slide or slides on which you have prepared blood smears, and which are thoroughly dry, use a pipette to cover the smear with Wright's stain, counting the number of drops used. Wait 1 to 2 minutes, then add an equal number of drops of distilled water. Let stand 2 or 3 minutes (bloods vary; some may take 5 minutes or even more). Wash the slide twice by dipping it into successive dishes of clean distilled water. Stand the slide on end to dry. A coverslip is not necessary. Study with an oil immersion lens.

Method 2 (quick method.* After the blood smear has air dried, add 10 to 12 drops of Camco Quick stain. Leave 30 seconds, then without draining off the stain, lay the slide into a dish of distilled water and rock the dish to rinse.

---

*G.L. Humason, 1972. Animal tissue techniques. San Francisco, W.H. Freeman and Co.

Rinse again in clean water, then carefully blot (*do not rub*) with two sheets of filter paper. Stand on end and allow to air dry without a coverslip. This method stains the cytoplasm an orange pink and the nuclei purple. Study the cells with an oil immersion lens.

If you prefer to cover the slides, use only a synthetic or neutral balsam mounting medium.

For a description of the types of blood cells, see p. 68.

DRAWING

Sketch and label on p. 391 examples of as many types of blood cells as you can identify. Use colored pencils if you wish.

IV. SOME OPTIONAL BLOOD TESTS
Coagulation Time

Blood clotting, or **coagulation,** is a protective device of the body for preventing hemorrhage from a wound. It is a complex physiological process by which fibrinogen, a large plasma protein, is converted into an insoluble threadlike protein called fibrin, which entangles the blood cells, forming a gel-like clot. Read about the entire process in your text.*

The process of coagulation (clotting) should not be confused with agglutination (clumping), as seen in the exercise on blood typing.

☞ **Slide method.** Perform a finger puncture (p. 386) and place a drop of blood on a clean slide, noting and *recording the exact time the blood was drawn*. At 30-second intervals draw the tip of a clean toothpick through the drop as though you were trying to pick up the blood on the end of the toothpick. Repeat until you can pick up a fine thread of clotted blood that stretches from the slide to the slightly raised toothpick. This is the coagulation point. *Record the time.* Sometimes the entire mass forms a gel, which can also be considered the coagulation point. The normal time between the first drop and the first thread is from 2 to 8 minutes.

☞ **Capillary tube method.** Perform a finger puncture (p. 386) and wipe off the first drop of blood. When the next drop appears, *record the time*, and place a capillary tube in the drop, holding the tube lower than the drop, so that it becomes nearly filled with blood. At each 30-second interval break off a small section of the tube. Continue until a small fibrin thread forms between the two ends of the tube. This is the coagulation point. *Record the time.*

---

*See Hickman, C.P., Jr., L.S. Roberts, and F.M. Hickman. 1984. Integrated principles of zoology, ed. 7. St. Louis, The C.V. Mosby Co., pp. 701-702.

Hematocrit (Micromethod)

The **hematocrit** is the volume of red blood cells in whole blood. It is obtained by centrifuging whole blood, then measuring the relative proportion of cells to fluid (plasma). For example, a hematocrit of 46 means that in every 100 ml of whole blood there are 46 ml of blood cells and 54 ml of plasma. The normal hematocrit for males is 47.0 ($\pm$7) and for females 42.0 ($\pm$5).

 Perform a finger puncture (p. 386). Wipe away the first two drops, then allow a *large* drop to accumulate. Insert one end of a heparinized capillary tube into the blood drop. Holding the tube horizontally or slightly tilted downward, allow it to fill by capillary action to within 1 cm from the end of the tube. Hold a finger over the open end of the tube, withdraw the tube from the blood drop, and seal the blood-filled end with wax.

Place the sealed tube into one of the grooves of the microhematocrit centrifuge, *with the sealed end of the tube at the outer circumference of the centrifuge*. Spin the centrifuge by hand a few times to make sure the tubes are properly placed, then close the lid and centrifuge for 5 minutes.

Remove the tube and determine the hematocrit ratio by using the hematocrit scale. If no scale is available, measure the height of the red cell column and the total height of the blood column. To find the percentage of cells, divide the height of the cells by the total height and multiply by 100.

Record your hematocrit on p. 394. If time permits, compute the class average for both males and females.

V. BLOOD PRESSURE DETERMINATION

Blood pressure is the force the flowing blood exerts against the walls of the arteries. When the wall of the left ventricle contracts and forces blood into the aorta and its branches, the pressure in the arteries increases. This maximum pressure is called the **systolic pressure.** When the heart rests—when the left ventricle is relaxing and the aorta is decreasing in diameter—the pressure in the arteries is reduced and is called the **diastolic pressure.** Many factors, such as exercise, state of mind, age, or even the time of day, can influence the blood pressure.

Both systolic and diastolic pressure are measured by a device known as a **sphygmomanometer** (sfig'mo-man-om'et-er), which consists of an inflatable arm cuff, an inflating bulb for inflating the cuff, a screw valve for controlling the rate of inflation or deflation, and a manometer or a calibrated gauge for reading the pressure. A stethoscope is used to detect the sounds of the heartbeats. Briefly, the cuff, placed around the arm just above the place where the brachial artery branches into the radial and ulnar arteries, is inflated until the pressure stops the circulation in the area. Air is slowly released until a thumping sound detected by the stethoscope indicates that blood is again flowing in the artery. The pressure reading at this point is the systolic pressure. As more air is released, the sound dies away, and that point is the diastolic pressure.

Have your partner seated comfortably near a table on which the bared left arm can rest on a level with the heart. Fold the cuff around the arm halfway between the shoulder and elbow and fasten it, not too tight nor too loose.

Locate the brachial artery, just above and to the right of the bend in the elbow, by feeling the pulse with the fingers. Place the stethoscope earpieces in your ears and the stethoscope head over the brachial artery. Close the pressure valve and squeeze the bulb to inflate the cuff until the manometer registers about 150 mm Hg. If you can still hear a beat, inflate the cuff until the sound disappears.

Slowly open the valve to release air gradually until you can hear a tapping sound for at least two consecutive beats. *Record the pressure (systolic) at this point.* Continue releasing air slowly. The sounds will become more distinct, then finally disappear altogether. *Record the pressure (diastolic) at the instant the sound ceases.*

If you miss the change in sounds at any point, deflate the cuff entirely and wait 2 or 3 minutes before trying again. It is well to take two or three readings and record the average.

Record your blood pressure and that of your partner on p. 394.

NORMAL RANGES IN BLOOD PRESSURE

Age	Female		Male	
	Systolic	Diastolic	Systolic	Diastolic
20	100-130	60-85	105-140	60-80
30	102-135	60-88	110-145	68-92
40	103-150	65-92	110-150	70-94
50	110-165	70-100	115-160	70-98
60	115-175	70-100	115-170	70-100

A STUDY OF BLOOD

Name _____

Date _____ Section _____

I. BLOOD TYPES

What is your blood type? _____

Determine the percentage of persons of each type in your laboratory section and, if there is time, in all the sections combined:

Blood group	Your section (%)	All sections (%)
O		
A		
B		
AB		

Why is it important to know your blood type? _____

_____

To what type might you be able to give a transfusion? _____

From what type might you be able to receive a transfusion? _____

II. ESTIMATION OF HEMOGLOBIN

What is your Tallquist reading? _____ % _____ g Hb/100 ml

What is the class average? _____ % _____ g Hb/100 ml

In your laboratory section what is the average g Hb/100 ml for males? _____

For females? _____

III. STUDY OF BLOOD CELLS

Sketch and label as many types of blood cells as you can identify.

IV. BLOOD TESTS
Coagulation time

What was the coagulation time for your blood?_____

What was the average coagulation time for your section? _____

What was the maximal time in your section? _____

What was the minimal time in your section? _____

What is the difference between coagulation and agglutination? _____

_____

_____

_____

What is the importance of coagulation? _____

_____

_____

What is the source of thromboplastin? _____

_____

What is its function? _____

_____

What is thrombin, and what is its function? _____

_____

_____

What is hemophilia? _____

_____

Hematocrit

What is your hematocrit? _____

In your laboratory section what is the average hematocrit for males? _____

For females? _____

How do these values compare with normal hematocrit values? _____

_____

How does your hematocrit compare with your hemoglobin estimate? _____

_____

_____

If a hematocrit is low, would you expect to find the hemoglobin content high or low? _____

Why? _____

_____

If the hemoglobin content of blood is below normal in an individual, would the hematocrit necessarily be

below normal also? _____ Why or why not? _____

_____

_____

V. BLOOD PRESSURE

What is your blood pressure? _____ Your partner's? _____

Why is high (systolic) blood pressure often associated with arteriosclerosis (hardening of the arteries)?

_____

_____

_____

If you undergo physical exercise strenuous enough to increase your pulse, how might this affect your blood

pressure? _____

_____

_____

EXERCISE 22B
Capillary circulation in the frog

MATERIALS
Frogs

Microscopes, 100× magnification

Frog boards with cork surface (see Appendix B, p. 462)

5% urethane (ethyl carbamate)—about 100 ml poured in bottom of widemouthed gallon jar (one jar enough for 10 students)

Pins

Polyethylene wash bottle containing water

Adrenaline chloride, 1:1000 in dropping bottle

Methyl salicylate (oil of wintergreen) in dropping bottle

Capillaries, the extensive network of small vessels that permeate the body's tissues, are the *raison d'etre* of the circulatory system. Through these tiny tubes, seldom more than 1 mm in length and only 10 to 25 μm in diameter, flow the oxygen and nutrients body cells require. So vast is the network that any single body cell is not more than 2 or 3 cells away from a capillary. High arterial pressures are required to force blood through the millions of capillaries because resistance to fluid flow increases dramatically as vessel diameters narrow. Despite their delicate dimensions and appearance, capillaries are remarkably tough structures that cannot be easily ruptured with surges of pressure.

Blood flow to tissues depends on need, and control is achieved by the alternate constriction and dilation of vessels. Most important are the arterioles because even small changes in their diameters can effect large changes in resistance to blood flow. Most arterioles receive sympathetic nerves, which release noradrenaline at their endings in smooth muscle fibers surrounding the arterioles. Increased sympathetic discharge to these arterioles will constrict them, squeezing off blood flow to the capillaries beyond. Nervous control ensures that only a limited number of capillaries are open at one moment, for if all opened at once, the blood pressure would drop dramatically, impairing the crucial flow of blood to the brain. Indeed, only 3% to 5% of all capillaries are open at any time, and not more than 7% of the total blood volume is contained within capillaries.

Where arterioles enter the capillary bed, precapillary sphincters are located that are under local rather than nervous control. These sphincters, together with the smallest arterioles and the capillaries, are immersed in the metabolic environment of the tissues they serve. Active tissues use up oxygen and produce metabolites (for example, CO_2 and H^+) that cause the sphincters to dilate, increasing blood flow and flushing out metabolites while bringing in needed oxygen.

In this exercise you will examine the capillary circulation of the frog's tongue and observe blood cells circulating through arterioles, capillaries, and venules.

I. STUDY OF CAPILLARY CIRCULATION
☞ Anesthetize a frog by placing it for several minutes in a jar containing 5% urethane at a depth of 2 or 3 cm. When the frog is unresponsive, rinse with tap water and place it ventral side down on the cork board. Extend the tongue and spread it over the opening in the board. Pin the edges to the board so that the tongue is flat and spread evenly. Keep the tongue wet with water and wrap the frog's body with a wet paper towel. Place the board on the microscope stage and position it so that the substage light passes through the hole and tongue.

Study the capillary circulation with low power. You should see a beautiful vascular network and be able to distinguish erythrocytes (red blood cells) coursing rapidly through the vessels. Identify arterioles, capillaries, and venules by their diameters and direction of blood flow. Note the relative spread of cell movement in the vessels. Can you distinguish the effect of the heartbeat in any of the vessels by rhythmic changes in flow velocity? Does flow rate become smooth in the smaller vessels? Why do cells travel faster in some vessels than in others? Is flow rate in large vessels faster in the center or along the edges of the vessel? Look closely at the individual erythrocytes. What is their shape? Do they contain nuclei? Compare with human erythrocytes. Can the erythrocyte bend? Look for white blood cells, which sometimes can be seen as small, colorless spheres moving slowly along the periphery of vessels. Sometimes they stick to vessel walls. What is the function of white blood cells?

Now examine the capillary bed more closely. Can you locate the precapillary sphincters? Are all of the capillaries open to flow? Does blood always flow in the same direction through capillaries? Note whether some capillaries open or close over a period of several minutes. Why is shifting flow of this kind necessary?

Make estimates of the diameters of arterioles, capillaries, and venules by relating to the dimensions of the red blood cells. A frog's erythrocytes are about 20 μm long, 15 μm wide, and 5 μm thick. Estimate the relative thickness of vessel walls.

☞ Make a sketch of a selected field, labeling all vessels with names and diameters. When that is completed, proceed to the experiments that follow.

II. ACTION OF ADRENALINE

☞ While watching through the microscope, have your partner apply 1 or 2 drops of 1:1000 adrenaline to the surface of the tongue.

Do vessels constrict as predicted? Do the precapillary sphincters change diameter? How is blood flow through the capillary bed altered? Record again the diameters of vessels that you measured before adrenaline was applied. When your observations are complete, thoroughly rinse the tongue with water.

III. ACTION OF METHYL SALICYLATE

☞ When the circulation has returned to normal, apply 1 drop of methyl salicylate.

What is the response of the vascular bed? Does local circulation increase or decrease? Which vessels change diameter? Why is methyl salicylate (oil of wintergreen) sometimes used in liniments?

Return the frog to its holding tank, in which it will eventually recover. Wash and dry the cork board. Clean the microscope stage.

CAPILLARY CIRCULATION IN THE FROG

Name _____

Date _____ Section _____

1. **Study of capillary circulation**
   Answer all of the questions in this section in brief essay form. Use the other side of this sheet if necessary. Prepare a labeled sketch of the circulation you see, using a separate sheet of paper.

2. **Action of adrenaline**

   What is the effect of adrenaline on vessel diameter? _____

   _____

   Do all vessels respond alike? Explain. _____

   _____

   Describe the effect of adrenaline on blood flow in the capillary bed. _____

   _____

   Record here the diameters of vessels measured before and after application of adrenaline.

Vessel	Diameter before adrenaline	Diameter after adrenaline

   What is the effect of adrenaline on local capillary delivery of oxygen and nutrients? _____

   _____

   Adrenaline does not constrict all vessels in the body; vessels of the heart (coronary circulation) and of skeletal

   muscle are dilated. Of what advantage is this? _____

   _____

3. **Action of methyl salicylate**

   What is the effect of this drug on the circulation? _____

   _____

   Why is this drug often used in liniments? _____

   _____

EXERCISE 22C
Small mammal respiration

MATERIALS
Gerbil or half-grown rat, 35 to 75 g

Small mammal respirometer ("Small animal metabolism studies kit" from Carolina Biological Supply Co., product no. 68-2000; Ward's Natural Science Establishment product no. 14 W 4235)

Soda lime

Bubble solution (soap bubble solution from a toy store)

Spacer: $4\frac{1}{2}$ inch length of doweling, $1\frac{1}{2}$ inch diameter

Black cloth, 1 foot square (to minimize disturbance to animal)

The animal is less apt to struggle during measurements if the metabolism chamber is placed in the cage with the gerbil or rat for a day or two before the experiment, allowing the animal to become familiar with it.

Be careful not to asphyxiate the animal by leaving it too long in the respirometer. Remember that the animal is consuming only oxygen from air containing 21% oxygen at the start. In time, with no air replenishment, there will be little but nitrogen left in the chamber.

Most animals with which we are familiar consume oxygen to support cellular energy metabolism and release carbon dioxide as a waste product of metabolism: they are **aerobic organisms.** (Strictly speaking, anaerobic organisms—those that exist in oxygen-free environments—are for the most part primitive forms, such as certain bacteria. However, aerobic organisms, ourselves included, are able to metabolize anaerobically for brief periods of heavy muscular activity.)

The energy an animal consumes per unit of time is referred to as its **metabolic rate.** There are three principal ways to measure metabolic rate. We could measure the energy value of all the food an animal eats, subtract from this the energy value of all excreta (mostly feces and urine), and accept the difference as

energy metabolism. The method is cumbersome and requires a long measurement period. Or we could place the animal in a special apparatus that would measure total heat production (calorimetry). This method too is cumbersome and technically difficult. Finally, we can do what most physiologists do: measure the animal's oxygen consumption. The rationale for relating oxygen consumption to energy metabolism is that the amount of heat produced for each liter of oxygen consumed remains almost constant, no matter what the animal is eating. If the animal eats pure carbohydrates, it produces 5.0 kilocalories (kcal) of heat per liter of oxygen consumed. If it eats pure protein, it might produce as little as 4.5 kcal per liter of oxygen. Since most animals in fact eat mixtures of carbohydrate, fat, and protein, we can accept an average value of 4.8 kcal. The largest possible error from using this figure is 6%.

The exercise described here will enable you to measure the oxygen consumption of a gerbil or small rat using a commercially available metabolism chamber (Fig. 22-1). The principle of operation is simple: oxygen is consumed and carbon dioxide released by the animal. Because the carbon dioxide is absorbed by soda lime as it is released, the removal of oxygen from the air reduces the volume of gas space within the respiration chamber. This volume change is measured by timing the movement of a soap bubble within a volumetric pipette.

☞ Cover the bottom of the plastic metabolism chamber with a thin layer of soda lime.

Soda lime will absorb all expired carbon dioxide and differentially allow for the measurement of oxygen consumption only. Why is this necessary?

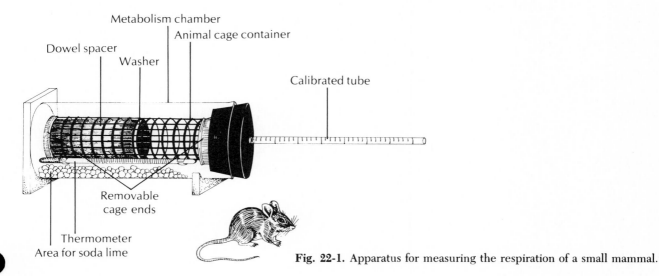

Fig. 22-1. Apparatus for measuring the respiration of a small mammal.

Metabolism chamber
Animal cage container
Dowel spacer
Washer
Calibrated tube
Removable cage ends
Thermometer
Area for soda lime

Weigh the gerbil or rat and record the weight. Place the animal in the metabolism cage together with the wooden spacer (for minimizing physical activity). Place a centigrade thermometer in the clamps provided in the cage, place the cage in the chamber, and close by inserting the rubber stopper. Cover the chamber with the piece of black cloth provided and permit temperature to equilibrate for about 10 minutes. Following equilibration, wet the inside of the calibrated 5 ml pipette with soap bubble solution; this will reduce the possibility of the soap bubble seal drying out and breaking during migration. Insert the tube into the hole of the rubber stopper of the metabolism chamber and seal by applying a drop of bubble solution to the end of the pipette. This may best be accomplished by touching the end of the pipette with a finger moistened in the bubble solution. Carefully record on p. 401 the time necessary for the meniscus of the soap bubble to transverse a distance along the pipette equivalent to exactly 5 ml. Repeat this technique until the measured time intervals appear consistent. Probably no more than 8 runs will be necessary.

If your readings are not consistent look for the following problems: (1) leaks in the system, (2) insufficient CO_2 absorbent, (3) dirty or blocked pipette, (4) animals variably active in respirometer, or (5) changing temperature within chamber.

Calculate the volume of oxygen consumed per minute by dividing the total number of ml consumed during 4 *consistent* runs (20 ml) by the total length of time in seconds required to make the 4 measurements. Then:

ml oxygen consumed per minute =

$$\frac{20 \text{ ml}}{\text{total time in seconds}} \times 60 \text{ seconds}$$

In your calculations you must correct the measured volume of oxygen consumed in milliliters per minute to standard temperature and pressure (STP). It is convention in all respiration studies to express oxygen consumed at 0° C and 760 mm Hg pressure. Use the following formula:

O_2 consumed/minute$_{STP}$ =

$$O_2 \text{ consumed/minute}_{OBSERVED} \times \frac{BP}{760} \times \frac{273°}{T° C + 273°}$$

in which:

BP = barometric pressure in mm Hg
T° C = temperature inside metabolism chamber

Calculate heat production (energy metabolism) by your animal by assuming that a normal animal releases 4.8 kcal for every liter of oxygen consumed. To do this, simply multiply your corrected oxygen consumption (ml/minute) by 0.0048. This yields the number of kilocalories of heat produced by the animal per minute. Since basal metabolic rate determinations are based on a one-hour interval, the number of kilocalories produced must be multiplied by 60. Thus:

kcal/hour = ml O_2 consumed/minute$_{STP}$ × 0.0048 × 60

Finally, calculate the amount of food the gerbil or rat must eat to maintain a steady body weight. We know that 1 L of oxygen consumed is equivalent to about 4.8 kcal of energy production. Using the figure you obtained in your experiment for kilocalories produced each hour by your animal, and assuming the metabolizable energy value of the animal's food to be 3 kcal/g, calculate the amount of food it must eat each day.

RESPIRATION

Name _____

Date _____ Section _____

Explain why it was necessary to use soda lime in the experiment. _____

_____

_____

What would be the effect on your results if chamber temperature were not in equilibrium before starting the

experiment? _____

_____

Record below the time in seconds for the animal to consume 5 ml of oxygen for each run.

	seconds	chamber T		seconds	chamber T
Run 1	_____	_____	Run 5	_____	_____
Run 2	_____	_____	Run 6	_____	_____
Run 3	_____	_____	Run 7	_____	_____
Run 4	_____	_____	Run 8	_____	_____

Total of 4 most consistent measurements = _____ seconds

Calculate the milliliters of oxygen consumed per minute (show calculation)

ml O_2 consumed per minute =

Record barometric pressure. _____ Record average chamber temperature. _____

Calculate oxygen consumption corrected to standard temperature and pressure (write out complete formula and answer).

O_2 consumed per minute$_{STP}$ =

Calculate heat production by your animal (show calculation).

kcal/hr =

Calculate below the amount of food the animal must eat to maintain unchanging body weight (show calculation).

Food (g) consumed per day =

Comment on sources of error in this experiment. _____

_____

_____

If you were to repeat this experiment, what would you do to obtain more accurate results? _____

_____

_____

CHAPTER 23

DIGESTION AND EXCRETION

EXERCISE 23A
Distribution of digestive enzymes*

Among the vertebrates and many invertebrates, digestion is largely or completely an extracellular process through which complex foodstuffs are reduced to simple organic units suitable for absorption. The digestive tract of a cockroach (Fig. 23-1) is like that of most vertebrates: a tube modified into subdivisions that serve specialized digestive functions. These functions are food reception, conduction and storage, internal trituration (grinding), digestion, absorption, and conduction and formation of feces. Digestive enzymes are rather sharply localized along the digestive tract. In this exercise the activity of specific enzymes in different regions of the cockroach digestive tract will be determined through the use of suitable substrates.

---

*Modified from Hoar, W.S., and C.P. Hickman, Jr. 1983. A laboratory companion for general and comparative physiology, ed. 3. Prentice-Hall, Inc. Englewood Cliffs, N.J.

MATERIALS

Cockroaches; giant American cockroaches of the genus *Blaberus* are recommended and are available from Carolina Biological Supply Co., product no. L725. (American cockroaches *Periplaneta* are quite satisfactory, but a larger number will be required.)

Dissecting instruments
Tissue grinder
Pasteur pipettes and rubber bulbs
Small test tubes and test tube rack
Boiling water bath
Glass marking pencil
1 ml graduated pipettes
30° C water bath
Photographic film (exposed, developed, washed, and dried)
Insect saline (7.5 g NaCl per liter of solution)
Spot plates
pH test paper (such as Hydrion)
Olive oil
Phenolphthalein
0.001 N HCl or NAOH
0.005 N NAOH
Other solutions, the directions for which are given in Appendix B, pp. 464-465

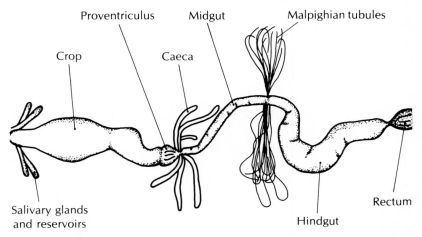

Fig. 23-1. Cockroach digestive tract. (From William S. Hoar, Cleveland P. Hickman, Jr., A LABORATORY COMPANION FOR GENERAL AND COMPARATIVE PHYSIOLOGY, 3rd Ed., © 1983, p. 181. Reprinted by permission of Prentice-Hall, Inc., Englewood Cliffs, N.J.)

The following parts of the cockroach digestive tract are to be tested for the presence of hydrolytic enzymes: salivary glands and reservoirs, crop, midgut with attached caeca, and hindgut. All four regions of the gut should be tested for the presence of sucrase, amylase, protease and lipase. Tabulate your results on the board so that all students have class results available for the report.

☞ Remove the digestive tracts of 3 cockroaches that have been anesthetized with carbon dioxide. If *Periplaneta* is used instead of *Blaberus*, use 6 to 8 animals. Cut the anus free from the body wall with scissors. Remove the complete digestive tract of the cockroach by holding the animal in insect saline and pulling off the head and attached gut with forceps. Should the gut break between the proventriculus and midgut, open the abdomen ventrally and dissect out the posterior gut portion.

Refer to the diagram and identify the salivary glands, crop, proventriculus, midgut with attached caeca, hindgut with attached Malpighian tubules, and rectum.

☞ Slit open the gut (except the salivary glands) and, holding it with a pair of forceps, vigorously rinse it in a dish of insect saline to remove any contained food; vigorous rinsing is important. Isolate the four regions of the gut to be studied and collect the separate parts into separate depressions of a spot plate. Label the depressions with a wax pencil to prevent confusion. Now grind each region in a tissue grinder. CAUTION: handle the grinder carefully; it is expensive glassware. Transfer salivary glands and reservoirs to the grinder tube with forceps and add 1 ml of insect saline. Insert the grinder pestle and grind the tissue thoroughly to break down cells and release enzymes into solution. Remove the pestle and allow the tissue to settle. Transfer the saline extract by Pasteur pipette into a small test tube and label. Bring the volume of the extract to 1.0 ml with insect saline. Wash out the tissue grinder and repeat this procedure with the other 3 gut regions.

Before proceeding with the experimental sections, you need to make a suitable denatured enzyme control for the enzyme tests.

☞ From each of the 4 extracts you have prepared, pipette out 0.1 ml and transfer to a fifth small test tube. This tube will contain 0.1 ml of each extract combined (0.4 ml total). Heat the combined extract in a beaker of boiling water for 10 to 15 minutes to denature the enzymes present. Cool and bring the volume to about 1 ml with insect saline. Label.

This is your enzyme control. Why is it needed?

Do not confuse this denatured control with the four tissue extracts. You are now ready to test the tissue extracts for enzyme activity.

1. Sucrase

☞ Place in each of 6 test tubes 1 ml of 5% sucrose (the substrate). Transfer about 0.2 ml of each of the extracts to the first four tubes. Label. To the fifth tube add about 0.2 ml of the denatured enzyme as a control. In the sixth tube add about 0.2 ml of the 0.5% invertase. Incubate all tubes in a 30° C water bath for 1 hour.

After incubation test each tube for the presence or absence of reducing sugars by adding 1 drop of extract-substrate mixture to a small test tube containing 4 drops of water and 2 drops of Benedict's solution. Heat for 10 minutes in a beaker of boiling water. A reddish precipitate indicates the presence of a reducing sugar or sugars, such as glucose and fructose. What is meant by the term reducing sugar? Estimate visually the relative amounts of precipitate to compare sucrase activities of the different tissues.

2. Amylase

☞ Place 1 ml of freshly prepared (boiled) 1% starch solution in each of six test tubes. Add 0.2 ml of each extract to the first four tubes, the denatured control to the fifth, and amylase control (0.5% amylase) to the sixth. Label the tubes. Incubate in the water bath at 30° C. Immediately, and at 10-minute intervals thereafter, place a drop of extract-substrate mixture on a porcelain spot plate and add one drop of iodine solution. Blue color indicates starch. A red-violet color appears when the starch is hydrolysed to dextrins. Once you get a positive test for dextrins, make a test on the extract-substrate mixture for reducing sugar (in this case maltose) by adding 1 drop to a small test tube containing 4 drops of water. Test for reducing sugar with Benedict's solution as described for sucrase.

3. Proteases

☞ Place drops of the enzyme preparations and the denatured enzyme control on the gelatin (emulsion) surface of uniformly blackened, exposed, developed, washed, and dried photographic film. Also add a drop of 0.5% trypsin to the surface as a positive control. Label each spot. Put the film in a moist chamber (e.g., a covered Petri dish containing a piece of moistened paper toweling) and leave overnight. Wash the film. If the gelatin has been digested by protease, the silver particles in the film will be easily removed, leaving a clear spot. It is best to spot more than one strip of film so that they may be washed and examined at intervals (several hours) for a better estimate of differences in protease activity.

4. Lipases

☞ Dilute 0.2 ml of the extract to 5 ml with water and determine pH with pH test paper (such as Hydrion) or with a pH meter. Neutralize if necessary with dilute (0.001 N) HCl or NaOH. Add 1 ml of olive oil and 10 drops of bile salt solution. Shake vigorously and quickly transfer 1 ml of the mixture to another small test tube, add a drop of phenolphthalein and titrate with 0.005 N NaOH. Record number of drops (or volume of titer) needed to turn the indicator pink. Incubate the remaining extract at 25 to 30° C for 1 hour. Remove a 1 ml sample and titrate as before. If time permits, repeat the titration of another 1 ml sample of the incubating mixture 1 hour later. As digestion proceeds, fats are hydrolyzed to fatty acids, and more NaOH is required to neutralize them. Run the lipase control and the denatured extract in parallel with the other extracts, using this procedure.

WRITTEN REPORT

Tabulate your results and those of the class on pp. 407-408 and answer all of the questions.

DISTRIBUTION OF DIGESTIVE ENZYMES

Name _____

Date _____ Section _____

Tabulate your results and average class results below. Use a scale of 0 to + + + + to indicate relative enzyme concentration (0 for absence, + for weak reaction, and + + + + for strongest reaction).

	Your results				Average class results			
	sucrase	amylase	proteases	lipases	sucrase	amylase	proteases	lipases
Salivary glands and reservoirs								
Crop								
Midgut and caeca								
Hindgut								

Write a brief interpretation of the results. _____

_____

_____

_____

_____

_____

What is the function of the cockroach's salivary glands? _____

_____

What is the function of the crop? _____

_____

Why is it important to rinse out all gut contents before making enzyme tests? _____

_____

What area of the cockroach gut is the major digestive area? _____

To what region of the vertebrate gut would this compare? _____

_____

Write a brief statement on sources of error in your experimental procedure. What would you do differently were you to repeat the experiment? _____

_____

_____

_____

_____

EXERCISE 23B
Glomerular filtration in the mudpuppy*

MATERIALS

Mudpuppy (*Necturus maculosus*), one per student pair, available from several dealers in living material

Anesthetic: urethrane, 5% and 0.12% (ethyl carbamate) or MS222, 0.2% and 0.007% (tricaine methanesulfonate, available from Carolina Biological Supply Co. or as "Finquel" from Ayerst Laboratories)

Enamel tray with thin foam rubber pad in bottom

Dissecting instruments: scissors and forceps

String or suture thread

Cotton

Amphibian Ringer's solution

Stereoscopic dissecting scope, 20× to 60× magnification

Illuminator

Ultraviolet lamp, high-intensity longwave (optional) (e.g. Fisher product no. 11-992-2 or A.H. Thomas product no. 6283-K10)

2% Evans blue dye

2% fluorescein sodium

Syringe, ½ or 1 cc with 23-gauge needle

Stopwatch (optional)

The urine of vertebrates, and many invertebrates as well, is formed by a process of **filtration** in which water and small solute molecules are filtered from the blood. Then, as the filtrate passes through long tubules, physiologically important substances (such as glucose, water, and some ions) are selectively **reabsorbed** while some other substances (potassium, hydrogen ions, and foreign substances such as penicillin) are **secreted** into the urine. The final urine that enters the bladder is quite different in composition from the glomerular filtrate. These three processes, filtration, reabsorption, and secretion, are performed by different parts of a delicate organ called a **nephron.** Many of these units operate together in the kidney; each human kidney contains more than a million nephrons, whereas the kidney of lower vertebrates may contain only several hundred or several thousand.

Urine formation begins in the **glomerulus**, a tuft of capillaries enclosed by a double-walled cup called the renal (or Malpighian) corpuscle. This is the filter. It receives its blood supply from the **afferent arteriole,** which is in turn supplied by a renal artery. Amphibian glomeruli are large and relatively few in number as compared with those in the human kidney. In the mudpuppy they are so large that they can be seen with the naked eye. Furthermore, the mudpuppy kidney's position in the abdomen is readily accessible to viewing. In this exercise you will observe the mudpuppy's glomeruli microscopically and follow the

*Modified from Hoar, W.S. and C.P. Hickman, Jr. 1983. A laboratory companion for general and comparative physiology, ed. 3. Prentice-Hall, Inc. Englewood Cliffs, N.J.

course of an injected dye as it enters the glomeruli, flows down the long tubule, and then appears in the ureter and subsequently in the urinary bladder.

☞ Submerge a mudpuppy in 5% urethrane or 0.2% MS222. When the animal becomes unresponsive to pinching (after about 5 minutes), remove and place it on its back in an enamel tray (a dissecting pan may be substituted) containing 0.12% urethane or 0.007% MS222 to maintain anesthesia. Slip a slide box or other support beneath one end of the tray to elevate it so that the anesthetic covers the mudpuppy's head but not the abdomen. It is helpful to place a thin (0.5 cm thick) pad of foam rubber under the animal to prevent it from slipping about.

☞ Make two ventral and longitudinal incisions extending from just anterior to the hindlimbs and 5 to 6 cm long on each side of the abdomen (Fig. 23-2).

The ventral abdominal vein lies in the midline of the ventral body wall. Your incisions should be about 1.5 cm on each side of the midline, but do not cut through the midline.

☞ Now pass ligatures (pieces of thread about 10 inches long) under the central strip as shown in Fig. 23-2 and tie tightly. Remove the central strip.

Observe the underlying viscera. The elongated gonads cover the kidneys. Pull the gonad to one side and hold it in place with a wad of cotton soaked in Ringer's solution. Observe the long, flat, brown kidneys. Look for a row of red dots near the medial edge of the kidney; these are the gomeruli. Shine a strong light on the kidney and study with your dissecting microscope. Sketch the kidney as you see it (p. 411) and label as directed. Compare several glomeruli. Does the blood flow equally well in all of them? Can you see material entering the tubules from the glomeruli? Describe.

☞ Now expose the heart by cutting a neat window in the body wall as shown in Fig. 23-2. Carefully examine the skin first to see the pulsating heart beneath before making your incision. Cut close to the body wall and do not injure the heart or blood vessels.

You are now ready to make the dye injection. The circulation will be displayed especially well if the preparation is illuminated with an ultraviolet (UV) light. The room lights should be turned off, but if you are using a high-intensity UV lamp, it is not necessary to completely darken the room. Be cautious not to look directly at the UV lamp. If no UV lamp is avail-

409

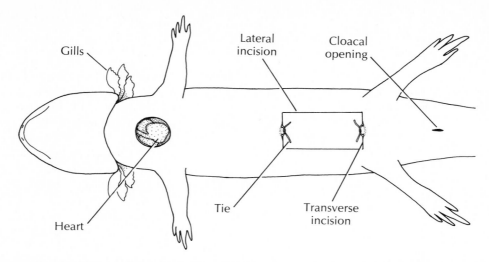

Fig. 23-2. Ventral view of anesthetized mud puppy, showing incisions. (From William S. Hoar, Cleveland P. Hickman, Jr., A LABORATORY COMPANION FOR GENERAL AND COMPARATIVE PHYSIOLOGY, 3rd Ed., © 1983, p. 181. Reprinted by permission of Prentice-Hall, Inc., Englewood Cliffs, N.J.)

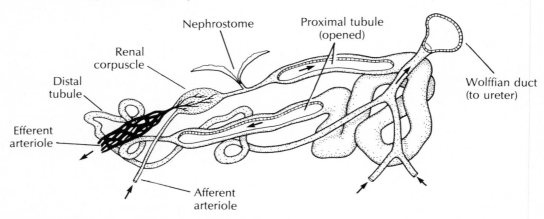

Fig. 23-3. Nephron of *Necturus*, diagrammatic. (From William S. Hoar, Cleveland P. Hickman, Jr., A LABORATORY COMPANION FOR GENERAL AND COMPARATIVE PHYSIOLOGY, 3rd Ed., © 1983, p. 181. Reprinted by permission of Prentice-Hall, Inc., Englewood Cliffs, N.J.)

able, you can still see the dye movement with ordinary light.

☞ While one student watches the kidney with the microscope, the other will inject 0.15 ml of 2% fluorescein dye into the heart. *Start timing.* Record the time when the dye enters the gills, passes the aorta, enters the renal arteries, and first appears in the glomeruli. How quickly do the capsules fill? Record the time the dye enters the tubules.

Watch the movement of the dye through the tubules. If you observe one tubule, you will notice that the dye disappears from view, then reappears again near the glomerulus. Consult the diagram in Fig. 23-3 and explain what has happened. The tubules are harder to see than the glomeruli, but they can be distinguished by their faint yellowish color. Does the dye enter some tubules more quickly than others?

Locate the ureter (an opaque tube lying along the lateral edge of the kidneys). Note the time when the dye enters the ureter. Note also the time when the dye enters the bladder and cloaca.

Look for fluorescein dye in the capillaries along various parts of the viscera. Does the dye escape from these vessels? Compare with the glomeruli. What does this tell you about the permeability of the glomeruli as compared to ordinary tissue capillaries?

☞ After 15 to 20 minutes repeat the cardiac injection using 0.15 ml of 2% Evans blue dye and continue the observations on the glomeruli as before.

Evans blue combines with the serum protein albumin. Does this dye pass the glomerular filter? Explain your observation.

When you are finished, dispose of your animal as directed by the instructor. It will not awaken.

In the space below sketch the mud puppy kidney as it appears to you when viewed with the dissecting microscope. Label kidney, glomeruli, renal arteries, gonad (ovary or testis), and urinary bladder.

Describe blood flow through the glomeruli and comment on whether or not all glomeruli receive equal amounts

of blood. _____

_____

_____

_____

_____

Do blood cells pass the glomerular filter? Explain why or why not. _____

_____

_____

_____

Record below the seconds after injection that the fluroescein dye first appears in these structures:

Seconds after injection

Gills   _____

Renal arteries   _____

Glomeruli   _____

Tubules (first appearance)   _____

Distal tubule (reappearance)   _____

Ureter   _____

Urinary bladder   _____

Do the glomeruli function intermittently? _____

What do you conclude about the permeability of the gomerular capillaries to small molecules such as fluorescein

dye? _____

_____

_____

Are the capillaries of the glomerulus more or less permeable to small molecules than ordinary tissue capillaries?

_____

_____

Describe what happens when you inject the Evans blue and compare its behavior with that of fluorescein. __

_____

_____

_____

CHAPTER 24

N E R V O U S S Y S T E M

EXERCISE 24
Spinal reflexes

MATERIALS

Frog
Aquarium or large bucket of water
Scissors and dissecting needle
25% acetic acid
1% sulfuric acid
Pipettes for diluting sulfuric acid
Syracuse watch glasses
Ring stand and clamps
400 ml or 600 ml beakers for rinsing frog
Electronic stimulator and bare copper wires (this portion of exercise may be omitted if electronic stimulators are not available)

The physician who sharply taps the tendon below your kneecap to evoke a knee jerk or taps the wrist to rotate your hand, is testing the status of a most fundamental level of movement control: the spinal reflex arc. Indeed, the knee jerk reflex is among the simplest of reflexes because it involves only two neurons. The hammer tap suddenly stretches a tendon, which in turn stretches an extensor muscle in the thigh. A spindle organ in the muscle is activated and fires off an action potential in a sensory fiber that extends to the spinal cord. Here it synapses with a motor neuron passing back to the extensor muscle. Stimulated, the muscle contracts, and the lower leg snaps forward.

What are the components of this reflex? The hammer tap is the **stimulus.** The muscle spindle organ is a stretch **receptor,** and the sensory nerve is the **afferent** portion of the reflex. The motor nerve is the **efferent** link, and the muscle is the **effector** (Fig. 24-1). The cell body of the sensory neuron is located in the dorsal root ganglion just outside the spinal cord, but its axon continues into the spinal cord, where it synapses with dendrites of the motor neuron. From its cell body in the ventral part of the gray matter of the spinal cord, the motor neuron sends an axon to the muscle. The entire pathway is the **reflex arc.**

The stretch reflex is important in many routine movements, but most reflex movements involve more

components. Limb movement requires the cooperative action of various muscles. Consider the illustration showing reflex connections of two antagonistic muscles. When the extensor muscle is stretched, its sensory neurons reflexly signal the motor neurons that cause contraction. At the same time branches of the sensory neuron activate inhibitory neurons in the spinal cord that synapse with motor neurons passing to the flexor muscle of the same limb (Fig. 24-1). Thus when the extensor contracts, the antagonistic flexor relaxes, increasing the effectiveness of the extensor muscle. Without this **reciprocal inhibition**, the two muscles would work against each other.

Let us carry our analysis one step further. Imagine a dog that steps on a thorn. It quickly withdraws its foot with a yelp, throwing its body weight suddenly on the opposite leg. But our dog does not collapse on its face, because a strong extensor reflex is set up in the supporting leg. This happens because at the same

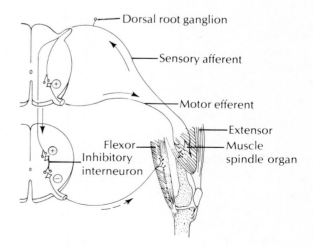

Fig. 24-1. Components of an extensor reflex, showing how contraction of the extensor muscle is accompanied by reciprocal inhibition of the antagonistic flexor muscle.

moment the insulted leg is flexed, signals pass across the cord to motor neurons on the opposite side causing activation of extensor muscles and inhibition of flexor muscles in the supporting leg.

The extraordinary thing is that such complex reflexes completely bypass the central nervous system. Certainly our thorn-pierced dog is aware of what has happened because other sensory neurons convey messages to the brain for its consideration. But the response was automatic and would have been accomplished just as successfully by a dog lacking a cerebrum.

Behavior of the Normal Frog

☞ Place a frog on the floor and observe its normal posture and movements. Flip it onto its back and note how quickly it rights itself.

The frog is aware of its abnormal position because of pressure on its back, abnormal position of its limbs, stimuli from its eyes, and balance organs in its ears.

☞ Put the frog in a water-filled aquarium or bucket and note normal swimming activity.

Will it right itself if turned over? What receptors are especially important in the righting response in this situation?

☞ Place the frog on a board and tilt it in various directions.

What is its response? Would stretch receptors in muscles be important in corrective movements?

Behavior of the Decerebrate Frog

☞ Decerebrate the frog by inserting scissors between the jaws and *just* behind the eyes (Fig. 24-2). Cut off the head with one quick stroke.

Your frog now lacks a cerebrum (the optic lobes, cerebellum, and medulla are still intact); you have removed all conscious activity and any possibility that it can feel pain.

Now observe the frog's activity as it recovers from decerebration. Is it as spontaneously active as the normal frog? Does it still breathe? Will it hop if prodded from behind? Place the frog in the aquarium. Will it swim spontaneously? If prodded? Will it right itself? Are eyes essential for righting?

☞ Place a small square of paper towel saturated with 25% acetic acid on the right side of the frog's back.

Note the resulting reflex response. What is the purpose of this reflex? Wash off the frog.

☞ Hold the frog's right leg while repeating the acetic acid stimulus. Note the response and consider the innervation that makes this response possible. Rinse the acid off the frog.

Behavior of the Spinal Frog

☞ Destroy the remaining brain tissue by pushing a needle through the dorsal midline of the skull and twisting the needle about. *Be careful not to destroy the spinal cord in the neck region.*

Brain destruction will be followed by a period of flaccidity (spinal shock) that may last for several minutes. In mammals spinal shock may last for days or even weeks. Observe the gradual recovery of spinal reflexes for 10 to 15 minutes, noting especially limb position, return of muscle tone, and response to stimulation. Does the frog still breath?

Compare reactions of the spinal frog with those of a normal frog. Note spontaneous movements (or absence thereof), response to being tilted on a board, and position when placed in water. Does the frog swim if prodded? Note its reactions to being pinched on the foot or skin of the leg and to being placed on its back. Can the spinal frog right itself?

☞ Suspend the spinal frog with a bent pin through the lower jaw or by clamping the lower jaw in a femur clamp attached to a ring stand. Prepare a series of four watch glasses containing, respectively, 0.05%, 0.15%, 0.3%, and 0.4% sulfuric acid. Measure the frog's response to each dilution, beginning with the weakest concentration. To do this, hold the watch glass so that the longest toe is immersed in the acid. Time

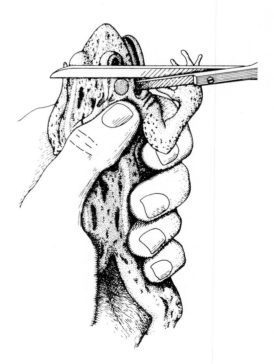

Fig. 24-2. How to decerebrate a frog.

the withdrawal response. If no response occurs within 2 minutes, consider the strength ineffective. Wash the foot in a large beaker of water. Wait 1 minute before trying the next strength. Record the responses on p. 418.

What is the effect of increasing strength of stimulus? Can you determine a "threshold level" for the stimulus? The period between application of stimulus and beginning of an overt response is termed the **latent period.** Does this change? **Irradiation** in the nervous system happens when stronger and stronger stimuli produce ever stronger responses. Did you see this happen?

☞ Apply a small square of paper towel moistened with 25% acetic acid to the abdomen of the suspended frog.

Note the response and promptly wash the frog by dipping in a beaker of fresh water. Apply pieces of acid paper to other points on the body. Are these reflexes coordinated in a purposive manner?

☞ Wrap two thin copper wires from an electronic stimulator around the *left* leg of a frog near the foot. The wires should be about 1 cm apart. Immerse the toes of the *right* foot in 0.5% sulfuric acid and note the latent period for the response. Wash off the acid and after 3 minutes stimulate the left leg with weak tetanizing current as the right foot is again immersed in the acid. If the right foot is not withdrawn in 30 seconds, discontinue stimulation and wash the foot.

What you have observed is inhibition of a reflex. Based on what was said about reflexes in the introduction, can you suggest why a pain stimulus applied to one leg causes inhibition of a flexor response to pain in the opposite leg?

☞ Destroy the spinal cord of the frog by pushing a dissecting needle down the vertebral canal.

What is the frog's posture and behavior now? Immerse the foot of the "double-pithed" frog in 0.4% sulfuric acid. Does the frog respond?

When you are finished, drop the frog in a waste can. Clean the dissecting instruments and wash up the glassware.

SPINAL REFLEXES

Name _____

Date _____ Section _____

Behavior of the normal frog

Describe the response of the frog to being placed on its back on the floor, inverted while swimming, and tilted

on a board. _____

_____

_____

_____

Behavior of the decerebrate frog

Compare spontaneity of activity in the decerebrate and normal frog. _____

_____

_____

Does the decerebrate frog breathe? _____ Why? _____

_____

Compare swimming behavior of the decerebrate frog with that of a normal frog. _____

_____

_____

Describe what happened when you applied acetic acid to the frog's back. _____

_____

What happened when you held the frog's right leg while applying acid to the frog's right abdomen? _____

_____

What can you say about the level of integrative behavior in the decerebrate frog? _____

_____

Behavior of the spinal frog

How long would you estimate spinal shock to last in a frog following spinal transection? _____

Does the frog breathe? _____ Explain. _____

_____

417

Contrast the behavior of the spinal frog with the decerebrate frog. Consider spontaneity, ability to swim, response to tilting, and response to painful stimuli. _____

_____

_____

_____

Effect of stimulus strength
Record the response time for each strength of sulphuric acid.

0.05% _____ ; 0.15% _____ ; 0.3% _____ ; 0.4% _____ .

Summarize the results of this experiment, commenting on threshold level, latent period, and irradiation in the nervous system. _____

_____

_____

_____

Describe the response of the spinal frog to acid applied to the abdomen. _____

_____

Were there areas of the body where you could evoke no response? _____ Why? _____

_____

Describe what happened when you immersed the frog's right foot in 0.5% sulfuric acid. _____

_____

What was the response when you repeated this test on the right while stimulating the left leg with an electrical stimulus? _____

_____

Offer an explanation for these results. _____

_____

_____

Describe the responsiveness of the double-pithed frog. _____

_____

CHAPTER 25

ANIMAL BEHAVIOR*

Ideally any study of the behavior of animals should be conducted in the field, where the animals can be observed for long periods of time under natural conditions. However, many means have been devised by animal behaviorists for studying behavior under controlled conditions in the laboratories.

One afternoon is a short period for making such observations and coming to any decisive conclusions. However, either of the following experiments should give the student some insight into the study of the behavior of the particular animal studied. Neither exercise requires any equipment not ordinarily found in the biology laboratory.

*See Hickman, C.P., Jr., L.S. Roberts and F.M. Hickman. 1984. Integrated principles of zoology, ed. 7. St. Louis, The C.V. Mosby Co., pp. 812-835.

EXERCISE 25A
Habituation in earthworms

MATERIALS
Earthworms
Plastic or glass Petri dishes with lids
Pencils, pens, tongue depressors, or toothpicks*

Preparation

Lumbricus terrestris is usually a convenient species of earthworm to use, but smaller species, such as the manure worm *Eisenia* or the garden worm *Allolobophora*, are quite satisfactory.

Habituation is a simple form of learning. In this case the animal learns not to respond to a specific stimulus when that stimulus is repeatedly presented. Habituation is not the same thing as muscular fatigue or sensory adaptation, which can also cause a reduction or cessation of response.

☞ Place an earthworm into each of two Petri dishes and cover the dishes. One worm is for experimentation; the other is to be observed as a control animal. *Keep the dishes in a dim light.* Leave the worms undisturbed for a period of 10 minutes or more to become acclimated to the dishes. When they have become relatively inactive, begin the experiment.

It is well to remember that earthworms breathe through their moist skin and should not be allowed to become dry. Be sure the skin is moist before you put them into the Petri dishes.

Now test the response each worm makes to the vibration of its dish by dropping an object onto the center of the dish from a height just sufficient to cause vibration of the dish and to elicit a reponse from the animal. (The object could be a pencil, pen, or wooden applicator dropped end-first onto the center of the lid. The weight of the object and the height from which it is dropped depend on the type of dish used.) Note the type of response each worm makes to the vibration. After 15 seconds repeat the stimulation and note the response. These are trial stimulations to familiarize the experimenter with the type of response to expect.

Experiment

Now begin the experiment. Use exactly the same type of stimulation, that is, drop the same object from the same distance at regularly measured intervals of 15 seconds. The timing may be done with a stopwatch or by measured counting. Subject *only the experimental animal* to the stimulus of vibration, but note

---
*A little experimentation may be necessary to find the proper weight object to drop, depending on whether glass or plastic containers are used.

each time the reaction of both worms. After each trial record whether or not the worm responds and note also any changes of intensity in the responses. Also note any other occurrence during the experiment, such as a slamming door or the like, that might affect the result. Repeat the stimulus for a total of 50 trials, or until the animal no longer gives an observable response.

On p. 423 record the number of observable responses to each successive group of 10 trials, and calculate the percentage of responses. (Seven responses out of 10 trials would be 70%.) Note any changes in the number of responses in each successive group of 10 trials. Results of all the teams in the class may be combined to produce average scores.

Fatigue and Adaptation

At the completion of the experiment on habituation, you may test the animals for **fatigue** and for **sensory adaptation.** If some stimulus other than the one you used, such as pinching the worm or blowing air on it, elicits the same response as did the vibrations, you can assume that the animal is not fatigued. If a greater intensity or duration of vibration than the one you used elicits the same response that the vibration did, the sensory system has not become adapted. For testing **memory** in these worms, see project 1, which follows.

REFERENCES

Edwards, C.A., and J.R. Lofty. 1972. Biology of earthworms. London, Chapman & Hall, Ltd., Chap. 11.
Ratner, S.C., and L.E. Gardner. 1975. Habituation in earthworms. In Price, E.O., and A.W. Stokes. Animal behavior in the laboratory, ed. 2. San Francisco, W.H. Freeman & Co.

PROJECTS AND DEMONSTRATIONS

1. Memory in earthworms. Worms that have been habituated to vibrations in Exercise 25A may be tested for memory by repeating the 50-trial experiment several hours later or a day or two later. If in these subsequent tests fewer stimuli are required to habituate the animals to the vibration, this would indicate that memory was involved.

2. A maze for studying learning ability in earthworms. A maze can be constructed of glass or plastic tubing in the form of a T or Y. The inside diameter of the tubing should be 0.5 to 2 cm, depending on the size of the earthworms used. Construct first a short **T** or **Y** piece from larger tubing into the openings of which the straight arms and stem of the maze can be inserted. Such a plan allows the worm ample room in which to make the turns into the arms. The inside of the tubing should be kept damp and should be cleansed thoroughly after each use to prevent the worms from following previous tracks.

Into one arm of the maze introduce the electrodes (negative goals), resting on a piece of rubber sheet, and connected to a 1 to 1.5 v battery. The other arm of the maze leads into a box or beaker of damp moss and soil (the positive goal), which has been covered to exclude the light.

Preliminary trials should be made without the electrodes or soil container to see whether the worm has any natural preference for turning right or left. If it does, insert the electrodes into the preferred arm when making the learning tests. If the worm refuses to move along through the tube, it can be gently stimulated from the rear with a camel's hair brush. Many trials may be necessary before the worm learns the conditioned response of avoiding the electric shock. The trials can be repeated later to see if memory is involved. They could also be repeated with the goals reversed.

EXERCISE 25B
Aggressive behavior and the dominance hierarchy in crickets

MATERIALS
For each team:
   3 or 4 marked male crickets, previously kept isolated
   An aquarium or other container
   Sand
   Ruled notebook paper

Crickets belong to the insect order Orthoptera, family Gryllidae. Field and house crickets, genus *Acheta (Gryllus)*, are common and can be collected in season and reared in the laboratory.* House crickets can be obtained from biological supply houses.

Both males and females possess a pair of posterior **cerci** that can be used to detect earthborne vibrations, but the female is easily distinguished from the male by a long, needlelike **ovipositor** located between the cerci. In both sexes there is a **tympanum** located on the inside of the tibia of each foreleg, which detects airborne vibrations.

The males produce a characteristic chirping, known as **stridulation,** produced by rubbing a special "scraper" on one wing against a ridge, or "file," on the other.

The house cricket *Acheta domesticus* is less aggressive than the various species of field cricket, so it is somewhat less suitable for this exercise; but although vigorous fighting is rare among members of this species, they can be used successfully.

Social **hierarchies** among male crickets are established by certain aggressive actions by means of which one male gains **dominance** over another male. At some point during an aggressive encounter one opponent usually retreats (**submission**). An **aggressive encounter** between male crickets may begin with two crickets approaching each other and touching antennae, or with one cricket lashing an antenna against some part of the body of the other. If neither cricket retreats as a result of the antennae lashing, the encounter may be continued with kicking with the hindlegs, active chasing, stridulation, grappling with forelegs and mandibles, or butting with the head until one member is forcibly overturned. The retreat of one of the opponents may occur following any of these types of encounter. The loser of an encounter may retreat entirely or may return at once to reengage the winner.

By watching a small group of male crickets and recording the encounters and the number of wins and losses of each member of the group, a hierarchy of dominance can be established. This does not preclude the possibility that, if the group were watched over a longer period, the order of dominance might later change.

☞ Each team of students will be provided with three or four male crickets that have previously been marked for identification with dabs of different color enamel or nailpolish applied with a toothpick and that have been kept separated in isolation cages for the past several days to a week and provided with food and water. Identify the crickets as A, B, C, and D (or, if you prefer, merely identify them by color). Place them together in a container, such as a 5- or 10-gallon aquarium, the bottom of which is covered with 2 to 3 cm of sand. Begin observing and recording their behavior immediately.

Record as an encounter only those instances in which some part of one cricket touches some part of another. If B and D have an encounter, note the types

---

*See p. 422 for rearing instructions.

of aggressive action used, which animal is the winner and which the loser, or whether it was a "no-decision" encounter. Watch and record for 45 minutes, or for as long as your instructor designates. For rapid recording use a sheet of ruled notebook paper marked into five vertical columns. Jot down for each encounter (1) the pair involved, (2) the winner, (3) the loser, (4) no-decision (if appropriate), and (5) the type of aggressiveness that seemed to terminate the encounter. An example is shown on p. 425.

> At the end of the period tally up on p. 425 the total encounters, wins, and losses of each cricket—in other words, the number of times each was dominant or submissive. From the results can you determine any order of hierarchy within the group?

REFERENCES

Alexander, R.D. 1961. Aggressiveness, territoriality and sexual behavior in field crickets (Orthoptera: Gryllidae). Behavior **17**:130-223.

Dingle, H. 1975. Agonistic behavior and the social behavior of crickets. In Price, E.O., and A.W. Stokes. Animal behavior in laboratory and field. San Francisco, W.H. Freeman and Co.

Polt, J.M. 1971. Experiments in animal behavior. Am. Biol. Teach. **38**:475-477.

Thompson, D.E. 1977. The common house cricket. Carolina Tips **40**(15):59-61 (Dec. 1). (Publication of the Carolina Biological Supply House, Burlington, N.C.)

PROJECTS AND DEMONSTRATIONS

1. Rearing the house cricket. Crickets can be reared in any clean container containing about 3 cm of clean dry sand and having a glass or screen covering over the top. For the deposit of eggs, provide a dish of *moist* sand covered with a wire screen that has a mesh large enough to admit the ovipositors but small enough to prevent the adults from cannibalizing the eggs. The cages should be provided with food such as dry dog food, chicken mash, oatmeal, or occasional pieces of fresh apple, pear, or lettuce, and a source of water such as a watering bottle with a cotton wick. Crickets should be kept in a somewhat moist environment.

At copulation the female mounts the male and the male orients its cerci upward, produces a spermatophore, and deposits it in the female seminal receptacle. In nature the female lays eggs in moist sand or earth in late fall, and the nymphs emerge in the spring. A female may lay 2000 to 3000 eggs in her lifetime. In the laboratory at room temperature the eggs will usually hatch in 2 to 3 weeks. The yellowish or white eggs, always laid singly, are small and banana shaped. The newly hatched nymph, the same size and color as the egg, is difficult to see with the unaided eye.

A series of molts precedes maturity, which may be reached in 4 to 12 weeks, depending on temperature, food, moisture, population density, and so on. Adults kept at 27° C usually live about 3 months.

2. Territorial behavior. Territorial behavior in crickets may be observed by placing several marked male crickets together in a cage that has the bottom covered with sand and that is provided with food and water. Supply one or two artificial crevices or burrows in the form of empty matchboxes with open ends, or upturned cups from an egg carton, with one side raised. Observe daily (as in Exercise 25B) to note the dominance order, and allow a week or two for the establishment and maintenance of territories.

Another method is to set up several containers, each with one matchbox and one male. Allow several days for the resident to become established and then introduce a second male in each container.

What form does territorial behavior take? How long does it take for territories to become established? Do encounters occur? How does an established animal defend its home?

A female might be introduced after territories are established and the behavior of both males and female observed.

3. Sensory perception in the cricket. Polt (1971) describes several experiments on sensory perception. One is to place wooden blocks, colored white, black, and shades of gray, into a cage of crickets to determine their preference for brightness. Color preference can be observed by using blocks of various colors and odor preference by using different odors applied to blocks of a single color. In any case, take periodic counts of the number of individuals on each block.

HABITUATION IN EARTHWORMS

Name _____

Date _____ Section _____

Record the number of observable responses seen in each successive group of 10 stimuli:

	Experimental animal		Control animal	
	Number of responses	Percentage of responses	Number of responses	Percentage of responses
1				
2				
3				
4				
5				

Total class percentages for each successive group of trials:

Experimental animal 1 _____ 2 _____ 3 _____ 4 _____ 5 _____

Control animal 1 _____ 2 _____ 3 _____ 4 _____ 5 _____

1. Was there a change in the *frequency* of response to vibration with repeated stimulation? _____

2. Was there a change in the *intensity* of the responses with repeated stimulation? _____

3. Reviewing again the definition of habituation, do you think that either animal became habituated to the vibrations? _____

4. Give an example of habituation in your own life experience or in the life of a pet. _____

_____

_____

_____

_____

_____

_____

_____

DOMINANCE HIERARCHY IN CRICKETS

Name _____

Date _____ Section _____

On a separate sheet of ruled paper record the result of each encounter between a pair of males and the type of aggressive act that seemed to terminate the encounter. Here is an example:

Pair of males	Winner	Loser	No decision	Type of aggression
CD	C	D		Kicking with hindlegs
AD	–	–	X	Antennating
AB	B	A		Grappling with forelegs

Summary of the encounters of each individual:

Male	Total encounters	Won	Lost	No decision
A				
B				
C				
D				

From these data, can you distinguish which male stands at the top of the hierarchy? _____
Rearrange the list to show the order of hierarchy in the space below:

1. Which males experienced the most encounters, the ones at the top of the hierarchy or the ones at the

bottom? _____

2. What relationship do you see between an animal's position in the hierarchy and its total number of en-

counters? _____

CHAPTER 26

GENETICS*

Heredity is the transmission from generation to generation of physiological, physical, and psychological factors that cause offspring to resemble parents. The basis of the mechanism of heredity lies in the behavior of the chromosomes and the genes. The genes are believed to be arranged in a linear fashion on the chromosomes. Since there are two sets of chromo-

*See Hickman, C.P., Jr., L.S. Roberts, and F.M. Hickman. 1984. Integrated principles of zoology, ed. 7. St. Louis, The C.V. Mosby Co., pp. 890-934.

somes, there are also two sets of genes. If the two genes for a given character are alike, the organism is pure, or **homozygous,** for that character; if the two genes are unlike, the organism is hybrid, or **heterozygous,** for that character. Whenever the two genes for a character are unlike, one of the genes usually gives expression to the visible character and is called **dominant;** the unexpressed gene is called **recessive.** Organisms of any size usually have both homozygous and heterozygous characters in their makeup.

Inheritance in the fruit fly *Drosophila*

MATERIALS
Drosophila
    Culture bottles with wild-type flies
    Culture bottles with F_1 of vestigial-winged × wild-type flies
    Culture bottles with F_1 of long-winged ebony-bodied × vestigial-winged gray-bodied flies
    Culture bottles with F_1 of white-eyed male and red-eyed female
    Stock cultures of vestigial-winged and white-eyed for identification
Culture bottles with prepared medium*
Cotton stoppers
Fine camel's hair brushes
Etherizing bottles*
Reetherizers*
Ether
Hand lenses or dissecting microscopes

Since 2 or 3 weeks is required to complete a genetic experiment, it is usually convenient to run the experiments simultaneously with other laboratory exercises. The first matings will have been made previously by the instructor and the resulting F_1 generation will be provided for class use.

If time does not permit the running of these experiments by the students, the instructor may prefer to make the second crosses also and provide the F_2 generations for examination and counting.

THE WILD-TYPE DROSOPHILA

Examine a culture of the wild type *Drosophila*. These will show the collection of trait expressions generally found in wild flies. Try to locate in the bottle the larvae, pupae, eggs, and flying adults.

Etherizing the Flies. Shake a few of the flies into an etherizing bottle and insert the stopper quickly. To kill the flies for leisurely examination, leave them for several minutes in the bottle. To anesthetize them, leave them for only a minute or less, and as soon as they are immobilized, empty them onto a piece of white paper.

☞ Kill some of the wild-type flies now for examination. Use a soft brush for handling them.

Examining the Wild-type Flies. Using a hand lens or dissecting microscope, note the size, shape, and color of the flies. Study the wings, noting their size, shape, and color. Notice the color of the compound eye. Examine also the legs, antennae, and balancers. You may consider the expression of traits found in the wild type of flies to be the base line, and any different

*See Appendix B, p. 458.

expressions of these traits to be mutants. A gene for wing development is located on chromosome II in *Drosophila*. The occurrence of greatly reduced, or "vestigial," wings is a recessive mutant condition. A gene for eye color is located on chromosome I. White eye color (from lack of pigmentation) is a recessive mutant condition. A great many such mutations are known to occur fairly regularly in *Drosophila*. You may also examine some flies that show these mutant characteristics.

Distinguishing the Sexes. The abdomen of the female is larger, more pointed, and less pigmented than that of the male. It also bears several pigmented stripes evenly spaced on the dorsal side (Fig. 26-1). The male abdomen is more blunt, and the posterior portion of the dorsal side is dark. On the ventral side the males have an area of dark bristles around the genital opening near the posterior end, and the females do not. With a soft brush, separate the males into one pile and the females into another.

Procedure for Experiments

Each team will receive a bottle that contains the offspring from a mating between a male and female that were different in one easily recognizable characteristic. The parents were a homozygous long-winged fly (wild type) and a fly with the mutant nonfunctional (vestigial) wings. We call these offspring the F_1 (first filial) generation. The parents (P_1) of these offspring were removed from the bottle before the emergence of the young to prevent their mating with any of the offspring.

Various symbols are used to designate common mutant characteristics, for example, *e* denotes ebony, a shiny black body; *se* stands for sepia, an eye color; *w* for white eyes; *v* for vermilion eyes; and *vg* for vestigial wings. Using *vg* for the vestigial-winged parent and *Vg* for the long-winged one, one can state this first mating as the following:

Cross I	P_1:	*Vg/Vg*	×	*vg/vg*
	Gametes:	*Vg* *Vg*		*vg* *vg*

What genotype(s) would you expect in these offspring? What phenotype(s)?

Mating the F_1 Flies

☞ Anesthetize the flies and divide them according to sex and type of wing. Are there any with vestigial wings? Why?

Record the results of your count under Breeding Experiment 1A on p. 437.

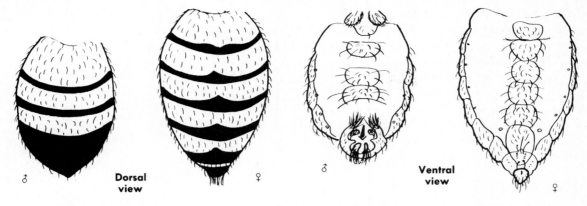

Fig. 26-1. Sexual differences in abdomen of *Drosophila* as shown in dorsal and ventral views. In each view: *left*, male; *right*, female.

☞ Select a healthy-looking male and virgin female* and place them in a fresh culture bottle. Do not let them fall onto the moist culture medium. Insert a cotton plug in the bottle and lay the bottle on its side. Label the bottle with your name, the date, and the appropriate data.

The mating you have just made may be represented symbolically as:

$$P_2: Vg/vg \times Vg/vg$$

What kinds of gametes can result from this union? Among the offspring what genotypes do you expect? What phenotypes? What proportion of each would you expect?

Eggs should appear within a few days after mating; the larvae appear in about 5 days, and the pupae in 7 or 8 days. The flies should emerge in about 11 or 12 days.

☞ Examine the bottle daily to be sure the parents are alive, replacing them if necessary. On the eighth day remove the parents from the bottle. Two weeks after mating etherize (kill) this generation and tabulate your count under Breeding Experiment 1B on p. 437. Place team counts in the box, and class totals at the right.

Other Suggested Breeding Experiments

In addition to the preceding mating, the following are suggested, but other crosses may be substituted, depending on the material available. Space for tabulation of results is found on pp. 438-440.

Making a Testcross. Because the long-winged condition is dominant over other wing types, long-winged flies may be either homozygous or heterozygous. A testcross, using a homozygous recessive, can be used

to determine the genotype of an individual. Half the teams will receive long-winged flies known by the instructor to be homozygous, and the other half will receive flies known by the instructor to be heterozygous. Each group will mate long-winged with vestigial-winged flies and in 2 weeks record the results on p. 438 and attempt to determine, by the proportions of the offspring, which group had homozygous parents.

Dihybrid Cross to Show Random Assortment. From the F₁ flies of a cross between a homozygous long-winged ebony-bodied fly *(Vg/Vg e/e)* and a homozygous vestigial-winged gray-bodied fly *(vg/vg E/E)*, mate a male and female and note the results. Compare the results obtained by the different teams. You should get the following phenotypes: long-gray, long-ebony, vestigial-gray, and vestigial-ebony. Does the ratio approach 9:3:3:1? Record on p. 439.

Sex-linked Inheritance. In the crosses listed above, it makes no difference in either the F₁ or F₂ generation whether a given characteristic is introduced by the male or by the female. However, some recessive characters are apparently carried on the X chromosome and there is no corresponding allele on the Y chromosome. White eyes in *Drosophila* is such a character.

Half the teams will receive the F₁ generation of a mating between a white-eyed female *(w/w)* and a red-eyed male *(W/—)*. Count and record the results on p. 440. Then mate a white-eyed male *(w/—)* and a red-eyed female *(W/w)* from this generation and 2 weeks later record the results.

The rest of the class will receive the F₁ generation of a mating between a red-eyed female *(W/W)* and a white-eyed male *(w/—)*. Count and record the results on p. 440. From this generation mate a red-eyed male *(W/—)* and a red-eyed female *(W/w)* and 2 weeks later record the results.

---

*Females do not mate for the first 12 to 18 hours after emergence from the pupa. To obtain virgins, remove all flies from a culture bottle and within 12 hours remove all flies again. Females of this second group are assumed to be virgin.

PROJECTS AND DEMONSTRATIONS

1. Demonstration of genetics slides:

a. SEX/LINKED INHERITANCE IN DROSOPHILA. This slide will show individuals to the F$_2$ generation.

b. SLIDE SHOWING COMMON MUTANTS IN DROSO-PHILA.

2. Demonstration of giant chromosomes of Drosophila. Slides to demonstrate the giant chromosome in *Drosophila* are made from salivary gland smears that are stained in a certain way to bring out the details of the chromosomes.

When the head of the fruit fly is pulled off, the salivary glands remain attached to the head and may be removed and treated in the same manner as onion root tip cells (p. 30).

EXERCISE 26B
Human inheritance

MATERIALS
PTC taste test papers
Blood-typing materials*
   Clean slides
   Cotton
   Alcohol, 70%
   Wax pencils
   Disposable lancets
   Anti-A and Anti-B serum
Members of the class

In this exercise you will be able to identify in yourself some of the common heritable traits, to compare your results with those of the rest of the class (a sample of the population), and to learn some biometrical methods for using and interpreting the statistics you compile. Biometry is the science of statistics as applied to biological observations.

SOME MONOHYBRID TRAITS
The study of human inheritance is complicated by the fact that most inherited characteristics are influenced by the interaction of more than one pair of genes, sometimes many pairs. However, studies have shown many physical characteristics in which variations in a single pair of genes will result in two distinct phases of expression, or **phenotypes.** Following are several of these mendelian traits that you will be able to recognize in yourself or in your friends. It is always well to remember, however, that modifying or cumulative factors may be involved in the development of many traits.

You can easily determine your phenotype in each of the following examples. Your phenotype will be the expression of either a **dominant** or a **recessive gene.** If the gene is recessive, you know that you are **homozygous** for that trait—you must be carrying two recessive genes. However, if your phenotype is that of a dominant gene, you may be either homozygous or

---

*These will be needed only if blood typing was not done in Exercise 22A. Instructions are given on pp. 385-386.

heterozygous for that trait; that is, you may carry two dominant genes or one dominant and one recessive gene. For example, a straight chin is a dominant characteristic; a receding chin is the expression of a recessive character. If you have a receding chin, you can describe your genotype as cc. But if your phenotype is the straight chin, you do not know whether your genotype is C/C or C/c; so you may indicate this by writing $C/—$, using the dash (—) to represent the unknown gene. The letters used to represent the characters below were chosen arbitrarily. Other authors might use other symbols.

☞ In the table on p. 441 you may indicate for each character your phenotype, your possible genotype, the number of persons in your laboratory section that have each phenotype, and the percentage they make up of the entire class. At your next meeting when you are given the figures for the other sections, record these totals and percentages also.

Widow's Peak. When the hairline dips down to a point in the center of the forehead, it is called a widow's peak. This condition is caused by a dominant gene *(W)*, whereas the continuous hairline is recessive *(w)*. (The hairline may have been altered if the gene for baldness has affected the front part of the head.) (See Fig. 26-2, *A*.)

Convex (Roman) Nose. A high convex bridge, often called a Roman nose, seems to be dominant over a straight or concave nose. The letter *N* may be used to express the dominant gene.

Dimpled Chin. A distinct depression or dimple in the chin apparently results from a dominant gene *(D)*. The depth of the dimple is probably controlled by multiple factors.

Rolling the Tongue. Some persons can roll the tongue into a U-shape when it is extended. Others cannot do this. The ability to roll the tongue results from a dominant gene *(R)* (Fig. 26-2, *F*).

Taster of PTC. The ability to taste phenylthiocarbamide seems to be inherited as a simple men-

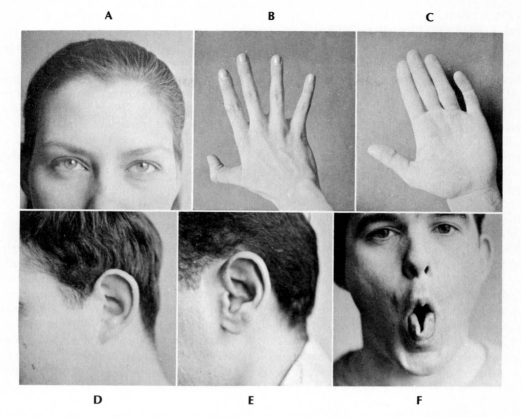

Fig. 26-2. Some inherited characteristics. **A,** Widow's peak. **B,** Hypermobility of thumb joint. **C,** Little finger bent toward other fingers. **D,** Attached small earlobe. **E,** Free earlobe. **F,** Ability to roll tongue.

delian character. Taste a piece of treated paper. If you taste nothing, chew the paper. If you are a taster, you will know it—the taste will be very distinct. If you merely detect the taste of the paper itself, or if you are doubtful about the taste, you are a nontaster and are homozygous for a recessive gene *(t)*.

Hypermobility of the Thumb Joints. Loose-jointedness, or the ability to put the thumb out of joint, is an inherited characteristic caused by a dominant gene *(H)* (Fig. 26-2, *B*).

Bent Little Finger. Lay your hands flat on the table, muscles relaxed. In some persons a dominant gene *(B)* causes the little fingers to bend toward the ring fingers (Fig. 26-2, *C*). Straight little fingers result from homozygous recessive genes *(b/b)*.

Long Palmar Muscle. Clench your fist and flex your hand. Now feel the tendons in your wrist. If there are three tendons there, you have the long palmar muscle and are homozygous for a certain recessive gene *(l)*. If there are only two, you lack this muscle and have the dominant gene *(L)*.

Color of the Eyes—Pigmented Iris. The inner lining of the iris usually appears a deep blue or purple. If there is no pigment in the outer layer of the iris, the eyes appear blue or sometimes gray as a result of the reflection of this purple lining. This lack of pigment is caused by a homozygous recessive gene *(p)*. If pigment is deposited in the outer layer, it tends to mask the blue and result in brown or hazel or green eyes, depending on the kind and amount of the pigment deposited. Other genes are responsible for the nature and density of the pigment, but the presence or absence of the pigment depends on the presence or absence of the dominant allele of the gene *(P)*. Consequently, brown, hazel, or green eyes (pigmented iris) are dominant to blue or gray eyes (nonpigmented iris).

Free Ear Lobes. Free ear lobes *(E)* are dominant over attached small ear lobes *(e)* (Fig. 26-2, *D* and *E*).

Blood Groups. Blood groups are inherited. The determination of blood groups involve a gene *(I^A)* that produces the A antigen, a gene *(I^B)* that produces the B antigen, and a gene *(i)* that produces neither A nor B antigens. The genes I^A and I^B are both dominant to the gene *i* but intermediate when both are present. A person carrying either I^A/I^A or I^A/i ($I^A/\text{—}$) has type A blood. One carrying I^B/I^B or I^B/i ($I^B/\text{—}$) has type B blood. The combination I^A/I^B produces type AB blood, and i/i produces type O blood. Record your phenotype and genotype.

APPLICATION OF STATISTICAL METHODS TO GENETICS (BIOMETRY)
Calculating Gene Frequencies

Knowing the percentage of homozygous recessives in any population, one can compute both the gene frequency in that population and the distribution of the genes into the three possible genotypes by using the Hardy-Weinberg law. This law states that in a large stable population not undergoing mutation, immigration, emigration, or chance (genetic drift) and where there is random mating, the population tends to remain in genetic equilibrium in which the gene frequencies are not appreciably altered from generation to generation. So if we know the gene frequencies of one generation, we can compute the frequencies of the next. The Hardy-Weinberg formula, $p^2 + 2pq + q^2$, is based on the 1:2:1 ratio of the mendelian monohybrid.

Let us suppose, for example, that 16% of the students in your laboratory section (which we will consider as a sample population) are unable to roll the tongue. Since this is a recessive trait, this 16% must be homozygous recessives (indicated by the genotype r/r). We want to find out (1) the proportion (frequency) of the alleles R and r in this population and (2) the proportion (frequency) of the genotypes R/R and R/r.

Letting p = the frequency of the R genes and q = the frequency of the r genes:

$$p + q = 1$$

or, raised to the second power:

$$(p + q)^2 = 1$$

or

$$p^2 + 2pq + q^2 = 1$$

It follows, then, that q^2 = the proportion of the recessive trait observed (r/r) (in this case 16%) and that:

$$q = \sqrt{0.16} = 0.4$$

or 40% recessive genes (r). If $p+q = 1$, and $q = 0.4$, then $p = 0.6$, or 60% dominant genes (R). Now we know that the gene frequency of this hypothetical population is 40% recessive alleles and 60% dominant alleles.

Knowing that $q = 0.4$ and $p = 0.6$, we can now substitute these values into the formula:

$$p^2 + 2pq + q^2 = (R/R + 2R/r + r/r)$$
$$p^2 = (0.6)(0.6) = 0.36, \text{ or } 36\% \ R/R$$
$$q^2 = (0.4)(0.4) = 0.16, \text{ or } 16\% \ r/r$$
$$2pq = 2(0.6)(0.4) = 0.48, \text{ or } 48\% \ R/r$$

Thus in this sample population we would have 36% homozygous dominants, 48% heterozygotes, and 16% homozygous recessives, with a gene frequency of 60% dominant and 40% recessive alleles for the genes controlling the ability to roll the tongue.

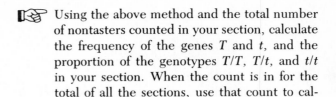 Using the above method and the total number of nontasters counted in your section, calculate the frequency of the genes T and t, and the proportion of the genotypes T/T, T/t, and t/t in your section. When the count is in for the total of all the sections, use that count to calculate again. Record your results on p. 442.

In a sample made by L. H. Snyder of nearly 4000 persons the ratio of tasters to nontasters was found to be 7:3; in other words, 70% were tasters and 30% were not. How closely do your results resemble Snyder's? Which set of your figures is closer to his? Is there a logical explanation for this?

Chi-square Test of Significance

How closely do the results of a test such as the one above agree with those predicted in a total population? Obviously in any sample of the population there is almost sure to be some deviation from the expected ratio, because of the operation of the laws of chance. Certainly the size of the sample is likely to have some effect on the accuracy of figures. The larger the sample, the greater the chance of obtaining the expected ratio. For example it would not be at all unusual for a couple to have three sons and one daughter, but if out of 400 children born 300 were reported to be boys, you might expect either an error in the figures or some most unusual cause. How great a deviation, then, is considered "normal," that is, would have a high probability of occurring entirely because of chance? This is a problem frequently encountered by scientists: how to determine whether the deviations they observe in their experimental results are significant or not.

Let us take an example. Suppose two groups of students were given the PTC test. In one group of 85 students, 53 were tasters and 32 were nontasters. In the other group of 30 students 27 were tasters and 3 nontasters. In the sample made by L. H. Snyder 70% were tasters and 30% were nontasters. Using this ratio as the expected value, how do our tests compare?

The deviation (d) is the difference between the observed value (o) and the expected value (e) and can be written

$$d = o - e$$

In the group of 85 students:

	Tasters		Nontasters	
Observed (o)		53		32
Expected (e)	$.7 \times 85$ =	59.5	$.3 \times 85$ =	25.5
Deviation (d)	$53 - 59.5$ =	-6.5	$32 - 25.5$ =	6.5

In the group of 30 students:

	Tasters		Nontasters	
Observed (o)		27		3
Expected (e)	0.7×30 =	21	0.3×30 =	9
Deviation (d)	$27 - 21$ =	6	$9 - 3$ =	6

The deviation in one group is 6.5 and in the other 6. Is this a significant figure in either case? How does one evaluate the results of such an investigation?

Statisticians have devised many mathematical tests for evaluating observational data. One simple formula, the **chi-square test,** is very convenient for testing the significance of deviations, and is applicable to many genetic experiments, whether dealing with two classes (in this case the two phenotypes, tasters and nontasters) or more, such as peas that have three phenotypes—red, pink and white blossoms, or blood types having four classes—A, B, AB, and O.

Again, using o for the observed value, e for the expected value, d for the deviation, and the Greek letter sigma Σ, meaning "the sum of," we find that the chi-square (χ^2) value is the following:

$$\chi^2 = \Sigma \frac{d^2}{e}$$

So let us return to our figures:

	Tasters		Nontasters	
o	53		32	
e	59.5		25.5	
d	6.5		6.5	
d^2	42.25		42.25	
$\frac{d^2}{e} =$	$\frac{42.25}{59.5} =$	0.71	$\frac{42.25}{25.5} =$	1.65

$\Sigma \dfrac{d^2}{e} = 0.71 + 1.65 = 2.36$ (the chi-square value)

Determining the chi-square value for the group of 30:

	Tasters		Nontasters	
o	27		3	
e	21		9	
d	6		6	
d^2	36		36	
$\frac{d^2}{e} =$	$\frac{36}{21} =$ 1.71		$\frac{36}{9} =$ 4	

$\Sigma \dfrac{d^2}{e} = (1.71 + 4) = 5.71$

Now we know the χ^2 values of each group and that each group dealt with two classes (two phenotypes). It is customary to refer to the number of independent classes in a chi-square table as "degrees of freedom." When dealing with only two classes in a sample of, say, 100 and knowing that 60 of them belong to one phenotype, we automatically know that 40 of them belong to the other; so there is one independent class, thus one degree of freedom. If we were dealing with three phenotypes and knew the number of individuals exhibiting two of them, we would automatically know the number in the third class; so there would be two independent classes, or degrees of freedom. The degrees of freedom are always one less than the number of classes involved.

Tables of chi-square values have been worked out that give the probability that certain deviations would occur by chance alone. Table 26-1 shows a portion of such a table. Statisticians have decided arbitrarily that if a deviation has a chance probability as great as or greater than 0.05 (1 in 20), it will not be considered statistically significant. In Table 26-1 you will notice that, for one degree of freedom, the chi-square value given that has 1 chance in 20 is 3.84. In our larger group we found the chi-square value to be 2.36, which has a probability greater than 1 in 20, in fact, even greater than 1 chance in 10 of occurring entirely by chance. So the deviation here is not significant and can be ignored. Our results agree well enough with predicted results.

In the smaller group our chi-square value is 5.71, which would have a smaller chance (between 1-in-20 and 1-in-100) of actually happening normally. So we need to reexamine these results. If it is found that there has been no error in conducting the experiment or in recording the results, then perhaps the smaller size of the group prevented getting the predicted results. Note that although the actual deviations in these two experiments were similar (6.5 and 6), they had quite a different significance in the smaller group than in the larger one.

Table 26-1 Probabilities for certain values of chi-square

Degrees of freedom	0.50 (1 in 2)	$p = 0.20$ (1 in 5)	$p = 0.10$ (1 in 10)	$p = 0.05$ (1 in 20)	$p = 0.01$ (1 in 100)	$p = 0.001$ (1 in 1000)
1	.46	1.64	2.71	3.84	6.64	10.83
2	1.39	3.22	4.60	5.99	9.21	13.82
3	2.37	4.64	6.25	7.82	11.34	16.27
4	3.36	5.99	7.78	9.49	13.28	18.46
5	4.35	7.29	9.24	11.07	15.09	20.52
6	5.35	8.56	10.64	12.59	16.81	22.46
7	6.35	9.80	12.02	14.07	18.48	24.32
8	7.34	11.03	13.36	15.51	20.09	26.12
9	8.34	12.24	14.68	16.92	21.67	27.88
10	9.34	13.44	15.99	18.31	23.21	29.59

Based on a larger table in Fisher, R.A. 1946. Statistical methods for research workers, ed. 10. Edinburgh, Oliver & Boyd, Ltd.

☞ Using (1) the number of nontasters observed in your laboratory section and (2) the number observed in the combined sections, find the chi-square values and determine whether the observed results agree essentially with the expected ratio of 7:3. Is there a significant deviation? Record the results on p. 442.

☞ If there is time use the tabulated results of the blood-typing test and compare this sample with the frequencies reported in the United States, given below. How many degrees of freedom are involved here?

	O	A	B	AB
Blacks	45%	22%	29%	4%
Whites	45%	41%	10%	4%

Genetic Variability in Humans

No two persons are exactly alike in their genetic constitution, with the exception of identical twins, who are the nearest alike of any two individuals. Even in identical twins certain features may be mirrored in the two members rather than duplicated.

☞ Try the following class exercise to demonstrate just how individual each person is, using only the physical characteristics that you recorded earlier on p. 441. Record the results in the table on p. 443.

One person will be selected to stand and read his list of characteristics. As the first phenotype on the list is read, all members of the class who share this characteristic will also stand. When the second characteristic is read, only those who also share this phenotype will remain standing; the others will sit down. Continue the reading until only the reader remains standing—an individual genetically different from all others in the class. Have one member of the class count the number of persons left standing for each trait. How many traits must be compared before all members are seated except the reader? Repeat this by having three or four other persons read their lists.

EXERCISE 26C
Problems in genetics

To understand the fundamentals of genetics, there is no better way than working out genetic problems. In answering these questions, make free use of your textbook. Be sure to show how you arrived at your answers. Use capital letters to represent the dominant gene; its recessive allele should be indicated by the same lower case letter. Start with the genotypes of the parents and then form the gametes of the parents of the cross. Next, form all possible crosses by the Punnett square. Finally, show the phenotypic ratio. Place your answers in the blanks provided and your explanations on other paper.

1. Usually brown eyes in humans are dominant to blue eyes.
   a. The child of brown-eyed parents has blue eyes. What is the genotype of each parent? _____
   b. If a blue-eyed man marries a brown-eyed woman whose mother had blue eyes, what proportion of their children might be expected to have blue eyes? _____

2. In dogs wire hair is dominant to smooth hair. If a homozygous wire-haired dog is bred to a smooth-haired dog:
   a. What will be the hair condition in F_1? _____
   b. If litter mates of this cross are bred, what will be the hair condition? _____

   c. If a dog of the F_1 of a is crossed to a smooth-haired dog, what are the possible results? _____

3. a. If each parent has a genotype of $A/a\ B/b$, what will be their gametes? _____
   b. What proportion of their offspring could be expected to have the genotype $a/a\ b/b$? _____

4. The ability to taste the drug phenylthiocarbamide (PTC) is attributable to a dominant gene. A nontaster man marries a taster woman whose father was a nontaster:
   a. What will be the expected genotypes of their four children? _____
   b. What could have been the genotype of this woman's mother? _____

5. The gene for yellow coat color in mice is lethal in a homozygous condition. Yellow coat is dominant to gray coat.
   a. What will be the ratio of viable phenotypes in a cross between two yellow-coated mice? _____
   b. What are the phenotypes and their ratio in a cross between yellow coat and gray coat? _____

6. In humans, suppose brown eyes are dominant over blue eyes and right-handedness is dominant over left-handedness. A brown-eyed right-handed man

marries a blue-eyed right-handed woman. If their first child is blue eyed and left handed, what are the genotypes of the parents? _____

7. Two sets of factors, called complementary, are responsible for flower color in sweet peas. These genes may be represented by *P* for purple pigment and *E* for the activator of *P*. These two dominant genes must be present to get purple color. If either one or both of these genes are absent, the flower color is white.

a. If two white varieties of genotypes *P/P e/e* and *p/p E/E* are crossed, what is the color of the F_1? ___
_____

b. If members of F_1 are crossed, what will be the genotypes and phenotypes? _____

8. In humans, hemophilia (failure to clot) is sex-linked and recessive. If a normal woman whose father had hemophilia marries a normal man whose father also had hemophilia, what can be expected in their children, two boys and two girls? _____
_____

9. Suppose a certain disease in humans is caused by a dominant gene with 10% penetrance. If a heterozygous man with this disease married a normal woman, what percentage of their children would be expected to have the disease? _____

10. In Ayrshire cattle, a genotype of *A/A* produces a mahogany color, whereas a genotype of *a/a* is red; however, the genotype *A/a* is mahogany in males and red in females. If a calf born to a mahogany cow is red, what is its sex? _____

Name _____

Date _____ Section _____

BREEDING EXPERIMENT 1—MONOHYBRID CROSS

A. Mating homozygous long-winged fly *(Vg/Vg)* with vestigial-winged fly *(vg/vg)*

Resulting F_1 generation: long-winged males _____ vestigial-winged males _____

long-winged females _____ vestigial-winged females _____
B. Mating long-winged male *(Vg/vg)* and long-winged female *(Vg/vg)* of F_1 generation

Resulting F2 generation:

Team	Male		Female		Class totals
	Long-winged	Vestigial-winged	Long-winged	Vestigial-winged	
1					Long-winged _____
2					Vestigial-winged _____
3					Ratio _____
4					Ratio expected _____
5					Male expected _____
6					Male _____

Female _____

Ratio _____

Ratio expected _____

BREEDING EXPERIMENT 2—TESTCROSS

Mating long-winged (*Vg/Vg* or *Vg/vg*) with vestigial-winged (*vg/vg*) to determine genotype of the long-winged parent

Predicted proportions of offspring if:

Long-winged parent is homozygous   Long-winged _____
                                                       Vestigial-winged _____

Long-winged parent is heterozygous   Long-winged _____
                                                       Vestigial-winged _____

Team	Male		Female	
	Long-winged	Vestigial-winged	Long-winged	Vestigial-winged
A1				
A2				
A3				
A4				
A5				

Actual count:

  A. Total long-winged _____

       Total vestigial-winged _____

  Conclusion: The long-wingd parent was

_____

Team	Male		Female	
	Long-winged	Vestigial-winged	Long-winged	Vestigial-winged
B1				
B2				
B3				
B4				
B5				

  B. Total long-winged _____

       Total vestigial-winged _____

  Conclusion: The long-winged parent was

_____

BREEDING EXPERIMENT 3—DIHYBRID CROSS

Mating long-winged, gray-bodied male *(Vg/vg E/e)* and long-winged, gray-bodied female *(Vg/vg E/e)* (from F₁ generation from union of long-winged, ebony-bodied male *[Vg/Vg e/e]* and vestigial-winged, gray-bodied female *[vg/vg E/E]*)

Team	Male				Female			
	Long-gray	Long-ebony	Vestigial-gray	Vestigial-ebony	Long-gray	Long-ebony	Vestigial-gray	Vestigial-ebony
1								
2								
3								
4								
5								
6								
Total								

Total from all teams participating

Long-gray _____ Long-ebony _____ Vestigial-gray _____ Vestigial-ebony _____

Ratio? _____ Ratio expected? _____

BREEDING EXPERIMENT 4—SEX-LINKED INHERITANCE

A. First cross
   P₁: Red-eyed male $(W/—)$ × white-eyed female (w/w)

Team	F₁ males		F₁ females	
	Red-eyed	White-eyed	Red-eyed	White-eyed
A1				
A2				
A3				
A4				
Total				

Expected ratio _____

Actual ratio _____

B. First cross
   P₁: White-eyed male $(w/—)$ × red-eyed female (W/W)

Team	F₁ males		F₁ females	
	Red-eyed	White-eyed	Red-eyed	White-eyed
B1				
B2				
B3				
B4				
Total				

Expected ratio _____

Actual ratio _____

A. Second cross
   P₂: White-eyed male $(w/—)$ × red-eyed female (W/w)

Team	F₂ males		F₂ females	
	Red-eyed	White-eyed	Red-eyed	White-eyed
A1				
A2				
A3				
A4				
Total				

Expected ratio _____

Actual ratio _____

B. Second cross
   P₂: Red-eyed male $(W/—)$ × red-eyed female (W/w)

Team	F₂ males		F₂ females	
	Red-eyed	White-eyed	Red-eyed	White-eyed
B1				
B2				
B3				
B4				
Total				

Expected ratio _____

Actual ratio _____

440

Name _____

Date _____ Section _____

Phenotypes and their frequency

	Characteristics	Your phenotype	Possible genotype	Your section		All sections	
				No.	%	No.	%
1.	Widow's peak						
	Continuous hairline						
2.	Convex nose						
	Straight or concave nose						
3.	Dimpled chin						
	Chin not dimpled						
4.	Can roll tongue						
	Cannot roll tongue						
5.	Taster of PTC						
	Nontaster						
6.	Hypermobility of thumb						
	Normal mobility						
7.	Bent little finger						
	Straight little finger						
8.	Long palmar muscle						
	Muscle absent						
9.	Pigmented iris						
	Nonpigmented iris						
10.	Free ear lobes						
	Attached ear lobes						
11.	Blood type A						
	Blood type B						
	Blood type AB						
	Blood type O						

Gene and genotype frequencies

Population	Phenotype		Gene frequency		Genotypes		
	% tasters	% nontasters	% T	% t	% T/T	% T/t	% t/t
Your section							
All sections							
United States	70	30					

THE CHI-SQUARE TEST

My section

	Tasters	Nontasters
o		
e		
d		
d^2		
$\Sigma \dfrac{d^2}{e} = \chi^2 =$		

All sections

	Tasters	Nontasters
o		
e		
d		
d^2		
$\Sigma \dfrac{d^2}{e} = \chi^2 =$		

What do the χ^2 values you have computed indicate about the results of these two experiments?

GENETIC VARIABILITY IN HUMANS

Record in each column the number of persons in the class whose phenotype for each trait agrees with that of the reader.

Trait number	1st reader	2nd reader	3rd reader	4th reader	5th reader
1					
2					
3					
4					
5					
6					
7					
8					
9					
10					
11					

Find the average number of characteristics that it was necessary to consider before a genetic difference was found.

_____

If a similar survey was made within any group of relatives of the same number as your class, would you expect

similar or different results? _____ Explain.

CHAPTER 27

FIELD TRIPS TO STUDY HABITATS AND LIVING ANIMALS

ECOLOGICAL RELATIONSHIPS OF ANIMALS*

MATERIALS

Collecting jars or plastic bags
Covered pails
Dip nets
Butterfly nets
Taxonomic keys
Rubber hip boots
Trowel or spade
Hand lens
Insect killing jars
Minnow bucket
Thermometer
Squares of cloth
White tray for sorting
Checklist of regional fauna (if available)

The study of the complex interrelationship between organisms and their environment is called **ecology,** an important part of the study of animals and plants. The **physical environment** includes such forces as temperature, light, moisture, soil, and gravity; the **biotic environment** refers to the influence organisms exert on each other and the interrelationships that exist between the various living groups of the environment, such as the density of population, the competition for food and shelter, the predatory-prey relationship and the parasite-host relationship, and the like.

Ecology has given rise to many concepts. One of the most important concepts is that of the so-called community. An **ecological community** is the association that exists between plants and animals in a specified region and forms a more or less unified biological system. A **habitat** refers to the place where an organism lives, or in another sense it may refer to the place

occupied by the community. Thus within the brook rapids community, where a considerable number of different animals live together, certain species of caddis fly larvae have their cases located in crevices of submerged stones (a habitat), the planarian flatworm lives on the undersurface of stones (a habitat), the darters live in the open water (a habitat), and the water striders skim the surface of the water (still another habitat).

You should have the opportunity of a field trip early in the fall as a sort of preliminary survey of living animals; in the spring, if time is available, you should have one or two trips in which ecological factors are stressed along with the collecting.

INSTRUCTIONS FOR FIELD WORK

The class will be divided into teams, each team under the supervision of an instructor or assistant. Each team will consist of several collectors and a recorder. The recorder should record in some detail the location and physical dimensions of each habitat visited, weather, season, date, time of day, duration of the collecting period, and any other geological, biotic, and physical aspects of each habitat that distinguish it. As animals are collected, the recorder must note the species (if unknown, it will be placed in a labeled container to be identified later), relative scarcity or abundance of each species, and where it is found. The recorder should also draw a sketch of the collection area, noting prominent features and landmarks.

Other team members will locate specimens, pointing out each discovery to the recorder. Collectors should be provided with equipment appropriate to the habitats to be visited, such as dip net, trowels, magnifying lens, and so on. If a specimen cannot be identified in the field, it should be collected and

---

*See Hickman, C.P., Jr., L.S. Roberts, and F.M. Hickman. 1984. Integrated principles of zoology, ed. 7, St. Louis, The C.V. Mosby Co., pp. 996-1011, 1021-1041.

placed in a numbered container for identification in the laboratory later. Aquatic specimens are placed in bottles or buckets containing a small amount of water; terrestrial animals are put in dry bottles. It may be prudent to assign one member of the team the responsibility of seeing that collected specimens are properly numbered and cared for. Avoid overenthusiastic collecting; take no more than a few specimens of each unidentified species to the laboratory for identification.

Team members should stay close together and work the chosen habitat systematically and thoroughly before proceeding to another area. After completing the collecting, return to the laboratory to identify the specimens, using guidebooks and keys made available by your instructor. Sufficient time should be left for this important work, either at the end of the period or on the following day (most specimens can be kept alive for at least a day or two in a refrigerator). Information gathered by the recorder of each team should be transferred to class data sheets and made available to everyone.

IMPORTANT NOTE

Respect the Animals' Homes! When you turn over a rock or log to look under it, replace it exactly as you found it. Many of the creatures under it depend on a moist habitat for survival. Do not destroy their security.

Do Not Deplete a Population! Large classes visiting the same areas year after year and bringing large numbers of salamanders or other animals back to the laboratory soon play havoc with natural populations and with the ecological balance of an area. Take a sample to identify, if necessary. Return other animals unharmed to their natural habitat. Keep notes on what and how many you saw in the field.

SELECTION OF HABITATS

The purpose of this field trip is to determine the variety, relative abundance, and adaptations of animals living in a specific habitat. The habitats or communities selected will depend on local climatic and geophysical conditions. If you are located near a seacoast, you may wish to visit a mud flat, sand flat, tidal pool, or rocky coastline. Each of these communities is characterized by certain vegetative and physical features that determine the kinds of animal communities that can flourish there and the adaptations that they bear. If you live in or near an arid or desert region, your field trip obviously would stress such an area. Your instructor may wish to do a thorough study of a single habitat, stressing the diversity of life forms found there and the anatomical and physiological adaptations

of the animals collected. Alternatively, the instructor may wish to visit as many habitats as possible within the alloted time, stressing how distinct habitat conditions affect animal inhabitants. Following are some suggested areas for study.

Plot Sampling.

For this commonly used sampling procedure, an area of manageable size is selected, usually square or rectangular, although circles or other shapes can be used. Once the plot is marked out and characterized as to its dominant vegetation and geophysical features, search it thoroughly for all forms of animal life, as well as for animal burrows, animal signs, ant nests, and so forth. Note where different kinds of animals are found and their abundance. Plot sampling can be applied to several of the habitat studies that follow.

Abandoned Field. Old abandoned pastures, where stock grazing is very restricted or absent, pass through an interesting succession of plant communities, such as annual herbs to weeds and broom sedge to shrubs to trees. Animal life, at first not abundant on freshly abandoned pastureland, increases in numbers and variety during plant succession toward the natural climax vegetation.

Invertebrates living on plant foliage may be collected by sweeping: swing an insect net repeatedly through the vegetation. To remove collected insects, first close the bag of the net by gripping it with the hand. Then insert a bottle into the net and transfer the insects to the bottle. An insect killing jar may be used for this purpose.

Many insects can be collected from shrubs and bushes by spreading a white cloth under the plant and beating the plant with a stick. Insects will drop onto the cloth where they can be collected. Even small trees can be sampled when shaken suddenly and vigorously by a few robust students. When doing this, hold a sheet above the heads of the shakers.

Ground macroinvertebrates living in leaf litter and debris are best collected by shaking a handful of litter onto a white sheet or cloth. Animals can be detected by their movement and picked up with forceps. A more thorough alternative is to dig up a square of sod with vegetation and soil intact and place it into a tray where the animals can be separated and collected. Captured animals can be killed in 70% ethanol.

Birds can be censused visually with the help of binoculars, but small mammals are much too elusive and secretive to be counted by so direct an approach. They must be snap-trapped or live-trapped using procedures described in manuals on field ecology. Such procedures are time consuming and generally impractical for general zoology classes.

Forest Floor. A temperate forest having a mat of leaves and leaf mold, fallen limbs, and decaying logs and stumps supports more niches and microhabitats than any other type of terrestrial environment. Careful study of such a region may reveal an astonishing assemblage of animal life, including salamanders, frogs, reptiles, small mammals, snails, slugs, millipedes, centipedes, numerous insects, spiders, annelids, and crustaceans, as well as a rich microfauna (protozoa, flatworms, nematodes, tardigrades, and so forth).

Rocky Habitat. A region with many drifting rocks, such as may be found on a hillside, is often rich in many animal forms. Snakes, lizards, salamanders, beetles, larval forms, and worms are among the animals one can usually collect.

Ponds and Marshes. Ponds differ from lakes in having vegetation over most of the bottom. Floating and emergent vegetation is common around a pond's margin, forming extensive areas of marsh. The most important collecting tool is a water net or dip net. First examine the pond along its edges, looking for water snakes, frogs, tadpoles, amphibian eggs (if it is springtime), water mites, and insect larvae and nymphs. Then skim the net through the water, scraping against the vegetation, but do not disturb the bottom at this stage. Transfer the contents to a white tray and examine for small amphibians, leeches, snails, aquatic insects, bryozoans, and small crustaceans.

Using hip boots, wade into the pond and sweep the submerged vegetation. Also turn over lily pads and examine emergent vegetation for snails, sponges, ectoprocts, and insect eggs. Finally, scoop up small samples of bottom mud and litter. Let the water drain from the net and then examine the inhabitants; pick them out by hand or with forceps as they clamber out of the mud and drop them into a tray for examination. Here you can expect to find snails, nematodes, annelid worms, and insect larvae.

Fish, newts, aquatic salamanders and numerous other aquatic forms can be collected effectively with a minnow net or small shore seine. Seines should be cleaned and hung out to dry after use.

Streams and Brook Rapids. Freshwater streams contain several distinct habitats. The character of the bottom and its inhabitants depends on the velocity of the water current. Where the gradient is steep, water flow is rapid (>50 cm/second) and the bottom comprises cobble and large boulders, together with trapped pebbles and some litter. This is a rapids or riffle habitat where one should find caddis fly larvae, mayfly and stonefly naiads, hellgrammites, snails, and beetle larvae and adults. One may also find ectoprocts, sponges, flatworms, clams, and perhaps crayfish in more protected areas. Most animals hide beneath stones. Working upstream, pull stones from the water and turn them over quickly before the animals drop off. Examine carefully, since many animals are flattened against the stones and are of cryptic coloration. A useful collecting technique is to hold a net downstream to trap the forms that are dislodged and swept away as stones are moved.

In less swift streams, gravel and sand pools form. Where the current is negligible, suspended materials settle to form mud-bottom pools. The latter is an especially fertile area, harboring burrowing mayfly naiads, aquatic annelids, damselfly and dragonfly naiads, red midge fly larvae, water boatmen, and several other forms. Collecting in these pools is accomplished much the same as from a pond bottom.

General Summary of Field Trip Collections.
After returning to the laboratory, complete the identification of the collected animals. If you cannot carry the identification to species, record the lowest taxon to which a species can be assigned (for example, class, order, or family). The class should draw up a checklist of all forms collected at each habitat, noting the relative abundance of each.

An instructive adjunct to this exercise that your instructor may ask of you is to comment on the anatomical and physiological adaptations of the inhabitants of each habitat. These include adaptations for food gathering, breathing, swimming, support, and clinging in currents and various adaptations for defense.

WRITTEN REPORT

Write a report based on the field notes and on the checklist assembled by the class. First prepare a habitat or community profile, indicating its location, the weather and season, temperature, and geophysical and biotic characteristics. Then list the animals found, their classification, and, if your instructor requests it, some principal adaptations of the more abundant forms collected.

REFERENCES
The following is a selection of guidebooks and keys that may be useful for the identification of animals collected on field trips. Not listed here are many excellent state and regional keys. Most of these guidebooks, as well as a listing of primary reference sources for the identification of any animal from any part of the world, may be found in the three volumes edited by Sims and Hollis:

Sims, R.W., ed. 1980. Animal identification: a reference guide. Vol. 1, Marine and brackish water animals. Vol 2, Land and freshwater animals (not insects). London, British Museum (Natural History).

Hollis, D., ed. 1980. Animal identification: a reference guide. Vol. 3, Insects. London, British Museum (Natural History).

GENERAL FIELD GUIDES AND TAXONOMIC KEYS

Blair, W.F., A.P. Blair, P. Brodkorb, F.R. Cagle, and G.A. Moore. 1968. Vertebrates of the United States, ed. 2. New York, McGraw-Hill Book Co.

Collins, H.H., Jr. 1959. Complete field guide to American wildlife: east, central and north. New York, Harper & Row, Publishers.

Eddy, S., and A.C. Hudson. 1961. Taxonomic keys to the common animals of the north central states. Minneapolis, Burgess Publishing Co.

Edmondson, W.T., ed. 1959. Ward and Whipple's freshwater biology, ed. 2. New York, John Wiley & Sons, Inc.

Needham, J.G., and P.R. Needham. 1962. A guide to the study of freshwater biology, ed. 5. San Francisco, Holden-Day, Inc.

Pennak, R.W. 1978. Freshwater invertebrates of the United States, ed. 2. New York, John Wiley & Sons.

MARINE ENVIRONMENT

Gosner, K.L. 1971. Guide to identification of marine and estuarine invertebrates: Cape Hatteras to the Bay of Fundy. New York, John Wiley & Sons, Inc.

Gosner, K.L. 1978. A field guide to the Atlantic seashore: invertebrates and seaweeds of the Atlantic coast from the Bay of Fundy to Cape Hatteras. Boston, Houghton Mifflin Co.

Kaplan, E.H. 1982. A field guide to the coral reefs of the Caribbean and Florida. Boston, Houghton Mifflin Co.

Light, S.F., R.I. Smith, F.A. Pitelka, D.P. Abbott, and F.M. Weesner. 1967. Intertidal invertebrates of the central California coast. Berkeley, University of California Press.

Morris, R.H., D.P. Abbott, and E.C. Haderlie. 1980. Intertidal invertebrates of California. Stanford, Calif., Stanford University Press.

TAXONOMIC KEYS FOR SPECIFIC ANIMAL GROUPS
Molluscs

Abbott, R.T. 1974. American seashells, ed. 2. New York, Van Nostrand Reinhold Co.

Burch, J.B. 1962. How to know the eastern land snails. Dubuque, Ia., Wm. C. Brown Co., Publishers.

Keen, A.M. 1971. Marine molluscan genera of western North America, ed. 2. Stanford, Calif., Stanford University Press.

Keen, A.M., and J.H. McLean. 1971. Sea shells of tropical west America. Stanford, Calif., Stanford University Press.

McLean, J.H. 1969. Marine shells of southern California. Los Angeles, Los Angeles County Museum of Natural History.

Morris, P.A. 1973. A field guide to shells of the Atlantic and Gulf coasts and the West Indies, ed. 3. In W.J. Clench, ed. Boston, Houghton Mifflin Co.

Rice, T. 1971. Marine shells of the Pacific Northwest. Edmonds, Wash., Ellison Industries.

Segmented Worms

Brinkhurst, R.O., and B.G. Jamieson. 1972. Aquatic oligochaetes of the world. Toronto, Toronto University Press.

Hartman, O. 1968. Atlas of the errantiate polychaetous annelids from California. Los Angeles, University of Southern California, Allan Hancock Foundation.

Sawyer, R.T. 1972. North American freshwater leeches, exclusive of the Piscicolodae, with a key to all species. Ill. Biol. Monogr. No. 46.

Spiders and their Kin (Chelicerate Arthropods)

Comstock, J.H. 1948. The spider book. Ithaca, N.Y., Comstock Publishing Co.

Gertsch, W.J. 1979. American spiders, ed. 2. New York, Van Nostrand Reinhold Co.

Headstrom, R. 1973. Spiders of the United States. New York, A.S. Barnes and Co., Inc.

Kaston, B.J. 1972. How to know the spiders, ed. 2. Dubuque, Ia., Wm. C. Brown Co., Publishers.

Levi, H.W., L.R. Levi, and H.S. Zim. 1968. A guide to spiders and their kin. Golden Nature Guide, New York, Golden Press.

Insects

Borror, D.J., and R.E. White. 1970. A field guide to the insects of America north of Mexico. Boston, Houghton Mifflin Co.

Chu, H.F. 1949. How to know the immature insects. Dubuque, Ia., Wm. C. Brown Co., Publishers.

Dillon, E.S., and L.S. Dillon. 1972. A manual of common beetles of eastern North America. New York, Dover Publications, Inc. 2 vols.

Ehrlich, P.R., and A.H. Ehrlich. 1961. How to know the butterflies. Dubuque, Ia., Wm. C. Brown Co., Publishers.

Helfer, J.R. 1963. How to know the grasshoppers, cockroaches, and their allies. Dubuque, Ia., Wm. C. Brown Co., Publishers.

Holland, W.J., 1968. The moth book: a popular guide to a knowledge of the moths of North America. New York, Dover Publications, Inc.

Jaques, H.E. 1947. How to know the insects. ed. 2. Dubuque, Ia., Wm. C. Brown Co., Publishers.

Jaques, H.E. 1951. How to know the beetles. Dubuque, Ia., Wm. C. Brown Co., Publishers.

Klots, A.B. 1951. A field guide to the butterflies of North America, east of the great plains. Boston, Houghton Mifflin Co.

Merritt, R.W., and K.W. Cummins. 1978. An introduction to the aquatic insects of North America. Dubuque, Ia., Kendall/Hunt Publishing Co.

Milne, L., and M. Milne. 1980. The Audubon Society field guide to North American insects and spiders. New York, Alfred A. Knopf, Inc.

Needham, J.G., and H.B. Heywood. 1929. A handbook of the dragonflies of North America. Springfield, Ill., Charles C Thomas, Publisher.

Otte, D. 1981. The North American grasshoppers. Vol. 1: Acrididae (Gomphocerinae and Arcidinae). Cambridge, Mass., Harvard University Press.

Pyle, R.M. 1981. The Audubon Society field guide to North American butterflies. New York, Alfred A. Knopf, Inc.

Swan, L.A., and C.S. Papp. 1972. The common insects of North America. New York, Harper & Row, Publishers.

Wiggins, G.B. 1977. Larvae of the North American caddisfly genera (Trichoptera). Toronto, University of Toronto Press.

Fishes

Breder, C.M., Jr. 1948. Field book of marine fishes of the Atlantic coast from Labrador to Texas. New York, G.P. Putman's Sons.

Eddy, S. 1969. How to know the freshwater fishes. Dubuque, Ia., Wm. C. Brown Co., Publishers.

McClane, A.J. 1978. Field guide to freshwater fishes of North America. New York, Holt, Rinehart & Winston.

Miller, D.J., and R.N. Lea. 1972. Guide to the coastal marine fishes of California. California Department of Fish and Game, Fish Bull. No. 157.

Stokes, F.J. 1980. Handguide to the coral reef fishes of the Caribbean and adjacent tropical waters including Florida, Bermuda, and the Bahamas. New York, Lippincott and Crowell, Publishers.

Ursin, M.J. 1977. A guide to fishes of the temperate Atlantic coast. New York, E.P. Dutton.

Amphibians and Reptiles

Bishop, S.C. 1943. Handbook of salamanders. Ithaca, N.Y., Comstock Publishing Co.

Cochran, D.M. and C.J. Goin. 1970. The new field book of reptiles and amphibians. New York, G.P. Putnam's Sons.

Conant, R. 1975. A field guide to reptiles and amphibians of eastern and central North America, ed. 2. Boston, Houghton Mifflin Co.

Ernst, C.H., and R.W. Barbour. 1972. Turtles of the United States. Lexington, University of Kentucky Press.

Smith, H.M. 1946. Handbook of lizards of the United States and of Canada. Ithaca, N.Y., Comstock Publishing Co.

Stebbins, R.C. 1966. A field guide to western reptiles and amphibians. Boston, Houghton Mifflin Co.

Wright, A.H., and A.A. Wright. 1949. Handbook of frogs of the United States and Canada, ed. 3. Ithaca, N.Y., Comstock Publishing Co.

Wright, A.H., and A.A. Wright. 1957. Handbook of snakes of the United States and Canada, Ithaca, N.Y., Comstock Publishing Co. 2 vols.

Birds

Bull, J., and J. Farrand, Jr. 1977. The Audubon Society field guide to North American birds, eastern region. New York, Alfred A. Knopf, Inc.

Harrison, C. 1978. A field guide to the nests, eggs, and nestlings of North American birds. New York, Collins.

Headstrom, R. 1949. Birds' nests. New York, Ives Washburn, Inc.

Peterson, R.T. 1980. Field guide to the birds, east of the Rockies. ed. 2. Boston, Houghton Mifflin Co.

Robbins, C.S., B. Brunn, H. Zim, and A. Singer. 1966. Birds of North America: a guide to field identification. Racine, Wis., Western Publishing Co., Inc.

Udvardy, M.D.F. 1977. The Audubon Society field guide to North American birds, western region. New York, Alfred A. Knopf, Inc.

Mammals

Booth, E.S. 1971. How to know the mammals. Dubuque, Ia., Wm. C. Brown Co., Publishers.

Burt, W.H., and R.P. Grossenheider. 1976. A field guide to the mammals, ed. 3. Boston, Houghton Mifflin Co.

Hall, E.R., and K.R. Kelson. 1959. The mammals of North America, New York, The Ronald Press Co. 2 vols.

Hamilton, W.J., Jr., and J.O. Whitaker, Jr. 1979. Mammals of the eastern United States, ed. 2. Ithaca, N.Y., Cornell University Press.

Murie, O.J. 1954. A field guide to animal tracks. Boston, Houghton Mifflin Co.

Whitaker, J.O., Jr. 1980. The Audubon Society field guide to North American mammals. New York, Alfred A. Knopf, Inc.

A KEY TO THE MAJOR

ANIMAL TAXA

The numbers in parentheses refer back to the couplets from which these couplets were reached, making it possible to work backward if a wrong choice is made.

1 Each animal composed of a single cell, some in colonies; chiefly microscopic . . . . **Subkingdom Protozoa** (one-celled animals)   2
Many-celled animals; mostly macroscopic   6

2(1) Cilia or ciliary organelles in some stage; usually two types of nuclei; contractile vacuole(s) and cytostome usually present . . . . **Phylum Ciliophora** (ciliates)
Not ciliated; usually one type of nucleus   3

3(2) Locomotion by pseudopodia or flagella or both; not spore forming . . . . **Phylum Sarcomastigophora**   4
Locomotion mainly by gliding or flexion; endoparasitic; usually produce spores . . . . **Phyla Apicomplexa, Microspora, Myxozoa,** and **Acetospora**

4(3) Locomotion by pseudopodia; body shape variable; body naked or with internal or external skeleton . . . . **Subphylum Sarcodina** (amebas)
Locomotion by flagella; body shape not variable   5

5(4) One or more flagella in adult; single nucleus **Subphylum Mastigophora** (flagellates)
Cilium-like organelles in longitudinal rows covering body; two to many nuclei of one type; all parasitic . . . . **Subphylum Opalinata**

6(1) Body with numerous pores; body radially symmetrical or irregular, with one or more large openings (oscula); no mouth or digestive tract; skeleton of spicules, spongin fibers, or both . . . . **Phylum Porifera** (sponges)   7
Body without numerous pores or oscula; usually symmetrical in outline; mouth and digestive tract usually present   10

7(6) Skeleton of calcareous spicules (with one to four rays); body and osculum may be bristly; asconoid, syconoid, or leuconoid canal system; marine **Class Calcispongiae**
Skeleton of siliceous spicules, spongin fibers, or both   8

8(7) Internal skeleton of siliceous spicules and spongin fibers with outer encasement of calcium carbonate; leuconoid; small group associated with coral reefs . . . . **Class Sclerospongiae**
Skeleton of siliceous spicules, spongin, or both, but no outer encasement of calcium carbonate   9

9(8) Three-dimensional and six-rayed spicules, often united to form network resembling spun glass; spongin absent; syconoid or leuconoid; often cylindrical or funnel shaped; marine . . . . **Class Hyalospongiae** (glass sponges)
Spicules not six rayed; spongin present or absent; leuconoid; freshwater or marine . . . . **Class Demospongiae**

10(6) Body with radial or biradial symmetry   11
Body with bilateral symmetry (or asymmetry)   21

11(10)   Body mostly soft and gelatinous; cylindrical, umbrella shaped, or somewhat spherical; body parts usually in divisions of four, six or eight 12
Body usually hard and spiny or with leathery skin; body parts in divisions of five; tentacles branched when present . . . . **Phylum Echinodermata** 17

12(11)   Body cylindrical or umbrella shaped; mouth or rim of umbrella usually encircled with unbranched tentacles; nematocysts (stinging cells) present; mostly marine . . . . **Phylum Cnidaria** (coelenterates) 13
Symmetry biradial; one pair of tentacles (not encircling mouth) or none; eight radial rows of ciliated comb plates . . . . **Phylum Ctenophora** (comb jellies, sea walnuts) 16

13(12)   Body a gelatinous medusa in bell or umbrella shape; free swimming 14
Body a cylindrical polyp, usually sessile or attached; single or colonial; tentacles surrounding the mouth 15

14(13)   Medusa usually possessing a velum; usually four to eight radial canals . . . . **Class Hydrozoa** (hydromedusae)
Medusa usually larger (2 to 20 cm or more) and lacking a velum; fringed oral lobes; highly branched radial canals; scalloped margins . . . . **Class Scyphozoa** ("true" jellyfish)

15(13)   Polyp typically small, often in branching colonies with more than one type of polyp; gastrovascular cavity not divided by septa . . . . **Class Hydrozoa** (hydroids)
Polyps usually larger; gastrovascular cavity divided by septa; no medusa stage in life history . . . . **Class Anthozoa** (sea anemones and corals)

16(12)   With tentacles retractile into sheaths . . . . **Class Tentaculata**
Without tentacles . . . . **Class Nuda**

17(11)   Body without arms; body with hard endoskeleton of calcareous plates, or with soft leathery skin containing calcareous spicules 18
Body with branched or unbranched arms; body with hard endoskeleton of calcareous plates 19

18(17)   Body with rigid endoskelton, globular or flattened, with movable spines . . . . **Class Echinoidea** (sea anemones, sea biscuits, sand dollars)
Saclike body with leathery skin; elongated in mouth-anus axis; mouth surrounded by branching, contractile tentacles; minute calcareous spicules embedded in skin . . . . **Class Holothuroidea** (sea cucumbers)

19(17)   Body with five movable, branched, and feathery arms; stalked or free-swimming; mouth and anus on oral surface, which is directed upward . . . . **Class Crinoidea** (sea lilies, feather stars)
Body with unbranched arms except in some brittle stars; oral surface directed downward 20

20(19)   Arms not sharply set off from central disc; ambulacral grooves with tube feet on ventral side of each arm . . . . **Class Asteroidea** (sea stars)
Arms sharply set off from central disc; no ambulacral grooves . . . . **Class Ophiuroidea** (brittle stars)

21(10)   Body soft and vermiform; not segmented; leaf shaped, ribbon shaped, or threadlike; digestive tract surrounded by a type of connective tissue (i.e., acoelomate, or without a body cavity) 22
Body cavity present 27

22(21)   Body long, soft, and contractile; long eversible proboscis; digestive tract with anus . . . . **Phylum Rhynchocoela** (ribbon worms) 23
Body flattened dorsoventrally; gastrovascular cavity with no anus (or without a digestive tract) . . . . **Phylum Platyhelminthes** (flatworms) 24

23(22)   Proboscis with stylets . . . . **Class Enopla**
Proboscis without stylets . . . . **Class Anopla**

24(22) Body of scolex and usually numerous proglottids (pseudosegments); increasing in size posteriorly (some tapewormos have only one proglottid); usually long and ribbonlike; parasitic; digestive tract lacking; may have suckers or hooks on scolex . . . . **Class Cestoda** (tapeworms)

Body without scolex and proglottids, with mouth and incomplete digestive tract . . 25

25(24) Mostly free-living; ciliated epidermis; no attachment organ(s) . . . .
. . . . **Class Tubellaria** (planarians)

Parasitic; nonciliated syncytial tegument; attachment organ(s) present 26

26(25) Usually oral and ventral suckers present; no hooks; first host a mollusc, final host a vertebrate . . . . **Class Trematoda** (digenetic flukes)

Posterior attachment organ with hooks, suckers, or clamps; single host, usually skin or gills of fishes . . . . **Class Monogenea** (monogenetic flukes)

27(21) Body cavity not lined with peritoneum (i.e., a pseudocoel); body mostly cylindrical and vermiform 28

Body cavity lined with peritoneum (a true coelom) 30

28(27) Digestive tract absent; with anterior retractile proboscis armed with hooks; parasitic . . . . **Phylum Acanthocephala** (spiny-headed worms)

Digestive tract with anus; cuticle pronounced; locomatory cilia mostly absent 29

29(28) Elongate and cylindrical; lateral lines present; cilia absent; circular muscles in body wall absent . . . . **Phylum Nematoda** (roundworms)

Microscopic; corona with ciliated discs; internal jaws . . . .
. . . . **Phylum Rotifera** (rotifers)

30(27) Body with some evidence of segmentation 31

Body usually not segmented 39

31(30) Body vermiform and segmented throughout; setae, parapodia, or both often present; no jointed appendages . . . . **Phylum Annelida** (segmented worms) 32

Not as above 34

32(31) Setae present on each somite 33

No setae (except *Acanthobdella*); no parapodia; segments with many annuli; clitellum present . . . . **Class Hirudinea** (leeches)

33(32) Many setae on each somite; parapodia or fleshy lateral appendages present (may be reduced); clitellum absent . . . . **Class Polychaeta** (sandworms, tubeworms)

Few setae on each somite; no parapodia; clitellum present . . . .
. . . . **Class Oligochaeta** (earthworms)

34(31) Body segments often combined into functional tagmata; jointed appendages; chitinous exoskeleton . . . . **Phylum Arthropoda** (joint-footed animals) 35

Body segmentation mostly evident in developmental stages; internal skeleton of cartilage or bone 51

35(34) Jointed appendages on most somites 36

Jointed appendages only on part of somites 38

36(35) Head, thorax, and abdomen present, but head and at least part of thorax fused; two pairs of antennae; appendages mostly biramous and specialized for different functions; mostly aquatic and gill breathing . . . . **Class Crustacea** (crustaceans)

Head present, but somites similar on rest of elongate body 37

37(36) Each somite with one pair of legs; dorsoventrally flattened body . . . .
. . . . **Class Chilopoda** (centipedes)

Each somite usually with two pairs of legs; subcylindrical body . . . .
. . . . **Class Diplopoda** (millipedes)

38(35) Head, thorax, and abdomen distinct; three pairs of legs on thorax; one pair of antennae; wings (one or two pairs) often present . . . . **Class Insecta** (insects)

Head completely fused with thorax; four pairs of legs; no wings; no antennae . . . .
. . . . **Class Arachnida** (spiders)

39(30)   Animal with shell of some form or body covered with scales or spicules; with or without tentacles on head; or shell absent or internal but with one or two pairs of tentacles on head, or 8 or 10 prehensile arms around head 40

Animal without shell, tentacles, or prehensile arms 47

40(30)   Shell of two valves arranged in dorsal and ventral position to each other; stalk or peduncle for attachment . . . . **Phylum Brachiopoda** (lamp shells)

Shell of one valve, or of two lateral valves, or of dorsal plates; or shell absent or reduced and internal; ventral muscular foot or fleshy arms; unsegmented . . . .

. . . . **Phylum Mollusca** (molluscs) 41

41(40)   Shell circular and limpet-shaped; mantle cavity with five or six pairs of gills; radula present . . . . **Class Monoplacophora**

Shell not circular or limpet shaped 42

42(41)   Shell of eight dorsal plates; radula present . . . . **Class Polyplacophora**

Shell of one or two valves, or shell absent or internal 43

43(42)   Shell of two lateral valves with ligamentous hinge; muscular foot present; head reduced; no tentacles or radula . . . . **Class Bivalvia** (bivalves)

Shell of one piece or absent or internal 44

44(43)   Shell tubular and open at both ends; head absent; mouth with tentacles and radula . . . . **Class Scaphopoda** (tooth shells)

Shell not as above 45

45(44)   Shell usually coiled or spiraled (uncoiled or absent in some); head with radula, one or two pairs of tentacles and one pair of eyes . . . .

. . . . **Class Gastropoda** (snails, slugs, nudibranchs)

Shell not as above 46

46(45)   Shell coiled (nautiluses), absent (octopuses), or reduced and internal (squids and cuttlefishes); head well developed, with radula and eyes; foot modified into large, prehensile arms or tentacles; one or two pairs of gills . . . .

. . . . **Class Cephalopoda** (nautiluses, squids, cuttlefishes, and octopuses)

Shell and head absent; wormlike body; mantle usually covered with scales or spicules; radula usually absent; foot represented by pedal groove . . . .

. . . . **Class Solenogastres** (solenogastres)

47(39)   Body wormlike 48

Body not wormlike (but torpedo shaped, sac shaped, or lance shaped) 49

48(47)   Long (10 to 85 cm or more), very slender; lives in stiff chitinous tube; body in three parts; short forepart with long ciliated tentacles, long trunk, and short segmented opisthosoma; no gill slits . . . . **Phylum Pogonophora** (beard worms)

Long, slender body divided into proboscis, collar, and trunk; paired pharyngeal gill slits on trunk; burrows in mud or sand . . . .

. . . . **Phylum Hemichordata** (acorn worms, tongue worms)

49(47)   Adults saclike and sedentary; single or colonial; body covered with tunic containing cellulose and with two siphons at one end; pharynx perforated with gill slits and adapted for filter feeding; free-swimming larva with notochord (stomochord) in tail . . . . **Subphylum Urochordata** (tunicates)

Body not saclike or covered with tunic; no siphons 50

50(49)   Body slender and torpedo shaped, with lateral and caudal fins; mouth with bristles or spines; planktonic . . . . **Phylum Chaetognatha** (arrow worms)

Body lance shaped; lateral musculature in conspicuous V-shaped segments; notochord and dorsal nerve cord extend length of body . . . .

. . . . **Subphylum Cephalochordata** (amphioxus)

51(34)   Body fishlike 52

Body not fishlike 55

52(51)   Without true jaws, scales, or paired fins   . . .   **Superclass Agnatha (Cyclostomata)**   53

With jaws and (usually) paired appendages; notochord persistent or replaced by vertebral centra . . . . **Superclass Gnathostomata**   54

53(52)   Suctorial mouth with horny teeth; seven pairs of gill pouches . . . .

. . . . **Class Petromyzontes** (lampreys)

Terminal mouth with four pairs of tentacles; 5 to 15 pairs of gill pouches . . . .

. . . . **Class Myxini** (hagfishes)

54(52)   Skeleton cartilaginous throughout life; ventral mouth; placoid scales or no scales

. . . . **Class Chondrichthyes** (sharks, rays, skates, and chimeras)

Skeleton mostly bony; body primarily fusiform but variously modified; ganoid, cycloid, or ctenoid scales; tail usually homocercal; terminal mouth with many teeth

. . . . **Class Osteichthyes** (bony fishes)

55(51)   Body covered usually with horny epidermal scales and sometimes body plates; paired limbs, usually with five toes, or limbs absent; no gills . . . .

. . . . **Class Reptilia** (snakes, turtles, crocodiles)

Body not covered with horny scales   56

56(55)   Epidermis with feathers on body and scales on legs; forelimbs (wings) usually adapted for flying; toothless horny beak . . . . **Class Aves** (birds)

No feathers, wings, or beak   57

57(56)   Skin naked, often moist, slimy, and sometimes warty; limbs often with webbed feet but no claws or nails . . **Class Amphibia** (frogs, toads, salamanders, newts, caecilians)

Body covered with hair, but reduced in some; integument with sweat, sebaceous, and mammary glands; young nourished by mammary glands; limbs variously modified for specialized functions . . . . **Class Mammalia** (mammals)

APPENDIX B

LABORATORY AIDS

CONTENTS

   I. Care of living animals 455
  II. Aids to microscopic and macroscopic
      study 455
 III. Narcotizing methods 456
 IV. Some culture methods 457
  V. Aquariums and terrariums 458
 VI. Some preservation methods 460
VII. Solutions and reagants 461
VIII. Some miscellaneous methods 462
 IX. Some sources of living material 465

I. CARE OF LIVING ANIMALS
Care of Purchased Cultures

1. As soon as a live culture arrives, remove the lid and aerate the contents by forcing air back and forth through the liquid with a new or very clean pipette.

2. The contents may be poured into a shallow bowl, allowed to settle, and then examined on the stage of a binocular dissecting microscope. Planaria may be put into wide dishes or enamel pans. Use live materials as soon as possible.

3. Cover the dish with a glass plate or put a lid on the jar loosely to prevent evaporation but allow aeration. Store away from direct sunlight or extremes of heat or cold. Optimum temperature for most invertebrates is about 21° C (70° F).

4. Use only new or chemically clean pipettes for removing samples. Warn students often against any contamination of the culture.

5. Avoid agitation of the culture that would scatter the animals. Look for amebas and hydras on the bottom and sides of the container; *Volvox* and *Euglena* will concentrate on the lighted side; paramecia are usually found near the surface.

6. If it is necessary to add or change the water, use only spring, pond, or well water or well-aged tap water. Never use distilled or fresh tap water.

7. More detailed instructions for the care of cultures are usually enclosed with the cultures by the biological supply companies that sell them.

Care of Living Frogs, Turtles, and Crayfish

Live frogs may be kept in an aquarium jar or tank, with wet paper toweling in the bottom, or in a terrarium. They may be fed flies or other small insects. Add a dish of water, place a glass cover over the tank, and keep in a cool place.

Frogs can be kept for several weeks without feeding if placed in jars with moist toweling in the bottom and kept in the refrigerator. Check daily and do not allow the toweling to dry out.

Turtles, crayfish, or frogs can be kept in a tank or aquarium having a sloping floor with a pool in the lower part. There should be a means of maintaining the water in the pool at a constant level, with a spray of water over the sloping floor. For crayfish some shelter should be provided, such as rocks, old coffee mugs, or pieces of tile in the pool. The water should be dechlorinated.

II. AIDS TO MICROSCOPIC AND MACROSCOPIC STUDY
Maintaining Live Microscopic Forms on Microscope Slides

Several methods can be used to prevent evaporation and so to keep protozoans or other microscopic forms alive on a slide for several hours or even days. This permits demonstration slides to be kept over for use over several laboratory periods. Such slides should be kept in a relatively cool place.

1. **Vaseline Ring.** Using a hypodermic syringe filled with melted or softened petroleum jelly, make a ring on a slide, add the desired culture material, and cover with a coverslip. The depth of the ring may vary to accommodate different-sized forms or amounts of fluid.

2. **Sealed Coverslip.** Place a drop of the culture on a microscope slide in the usual way, apply a coverslip, and then use fingernail polish, melted petroleum jelly, or ordinary 3-in-1 machine oil to seal the edges of the coverslip. If algae are present in the culture, they will provide oxygen for the animal life.

3. **Hanging Drop.** Place a drop of the cluture on

a coverslip; then invert the slip over a deep-well depression slide, forming a hanging drop. By sealing the coverslip with nail polish, melted petroleum jelly, or oil, you can retard evaporation. Air in the depression cavity supplies oxygen for the animals in the hanging drop.

4. Double Drop. A method of keeping protozoans and the like for some time and at the same time quieting them enough for study under an oil immersion objective is as follows. Place a small drop of culture on a slide, avoiding the presence of sand grains or large pieces of detritus that would raise the coverslip. Then on a coverslip, place an equal-sized drop of 1% solution of agar-agar (liquefied in a water bath at about 40° C). Invert the drop, exactly centered, over the culture drop. As the two drops merge, the jelly sets and forms many tiny water spaces in which the animals become confined. Organisms can live one-half to several hours on such a slide. For marine forms, use seawater to make up the agar solution; for parasitic forms, use a 0.7% sodium chloride solution.

Holding Small Invertebrates for Observation

Use a small piece of plastic wrap such as Saran Wrap as a coverslip over a drop of liquid on a slide. This holds the organism quiet, retards evaporation, and allows the organisms to be returned unharmed. Larger forms may be mounted between two pieces of the wrap and then placed on a slide. The preparation can then be turned over and viewed on both sides.

Positioning Small Preserved Specimens for Stereomicroscopic Study

1. Specimens can be held in almost any desired position by pressing gently onto a bit of Plasticene or modeling clay.

2. Partially fill a small culture dish or Stender dish with clean washed sand (preferably rounded rather than sharp grains). Add just enough alcohol to cover the specimen. Push the specimen into the sand just enough to hold it in the position desired for study.

3. Small specimens can also be secured in a dissecting dish by heating the wax and pressing the appendages gently into it so that the animal is held firmly as the wax cools.

Aids for Handling Microscopic and Macroscopic Organisms

To make a **very fine pipette** for handling individual microscopic organisms, draw out to a fine point a piece of glass tubing 2 to 3 mm in diameter. Fit the other end of the tube with a 2 to 3 cm length of rubber tubing and close the rubber tube with a short glass rod. These can be made into various sizes and drawn out to various widths to fit different types of organisms.

Microdissection needles can be made of no. 00 insect pins mounted in dowel handles or held in a purchased needle holder. Or the blunt end of a minute insect pin can be inserted into the small end of a Pasteur pipette and flamed to constrict the glass opening and so attach the pin.

A microloop for transfer of paramecia or other protozoans can be made of 10 cm of standard-gauge platinum wire. Insert one end in a handle and bend the free end around a 2 mm diameter metal rod. Fix the free end of the loop by hammering it gently. Pipette some concentrated culture onto a depression slide, immerse the loop just under the surface of the fluid, locate the organisms, and lift gently. The loop can pick up about 0.025 ml of fluid and several paramecia. The size of the loop can be varied for other types of organisms.

Vital and Supravital Stains

These stains are applied to organisms without previous fixation. Vital staining does not kill the organisms, but supravital staining eventually kills them.

Brilliant Cresyl Blue will stain many cellular structures. Make a 1% stock solution in absolute alcohol, and for use, dilute it 1:10 with water. Place a drop on a slide and let the alcohol evaporate; add a drop of culture medium and seal on a coverslip. Let stand for 5 to 30 minutes.

A mixture of 1 ml of a 10% aqueous solution of China Blue and 4 to 6 drops of 6.5% aqueous solution of phloxine-rhodamine will stain some structures while the animals are alive and others after they die by drying. Place on a slide a very small drop of ciliate culture and an equal drop of the stain and spread lightly.

China Blue, Nigrosin, or Opal Blue can also be used as 10% solutions. Allow the drop of culture and the drop of stain to air-dry on the slide, or make a hanging drop and seal. Nigrosin, 10%, is a good relief stain against a background of which ameboid movement and contractile vacuole activity show up well. When allowed to dry, ciliary rows, trichocysts, and the pellicle of ciliates and some flagellates can be demonstrated.

Janus Green B in dilutions of 1:10,000 or more stains mitochondria.

III. NARCOTIZING METHODS
For Small Invertebrates, Larvae, and the Like

1. Place the organisms in a 0.1% solution of Chloretone (chlorobutanol 1,1,1-trichloro-2-methyl-2-propanol, Sigma Chemical Co. product no. T5138) and allow them to remain until quiet. If kept moist, the animals will revive when returned to water.

2. Warm water, 37° to 38° C, if properly regulated,

will relax worms, clams, and so on, without stopping the dorsal vessel. This is good to use when pegging clams. 48° C is lethal.

3. Freshwater worms and the like may be narcotized by placing them in a watch glass or Petri dish with water to barely cover them and blowing ether or chloroform vapor over them. (Instructions for ether-vapor tube are given later.)

4. Annelids, cnidarians, molluscs, echinoderms, and the like may be slowly narcotized by being placed in a very small amount of the water in which they were collected. Allow them to settle and expand and then gradually add any one of the following: magnesium sulfate crystals (Epsom salt); menthol crystals; saturated solution of magnesium sulfate or menthol; or 70% to 95% alcohol. Add the liquid drop by drop. It may take several hours to render some forms insensitive.

5. Earthworms may be placed in 0.2% chlorobutanol (Chloretone) until relaxed or immersed in 0.08% tricaine methanesulfonate (MS222) for 1½ to 2 hours.

6. Echinoderms can be relaxed in 7% $MgCl_2$ made up in tapwater. It will take about 40 minutes. Anesthetized animals can be restored to activity by bathing in running seawater.

7. Urochordates can be placed in 1% chloroform in seawater and allowed to sit until relaxed. Or, with the animals in a small amount of seawater, sprinkle menthol crystals over the water and leave for 1 to 4 hours.

8. For rotifers phenylephrine (Neo-Synephrine) is suggested. It is obtained from the Sigma Chemical Co. (product no. P6126) or at drug stores as the hydrochloride in a 1% solution and is used as a 0.1% to 0.5% solution. It works best in slightly acid waters.

9. For protozoans a 1% nickel sulfate solution paralyzes cilia; 1% potassium (or sodium) iodide solution is said to prevent myoneme constriction in *Stentor* (and presumably other ciliates); and tobacco smoke in a test tube inverted over a drop of culture (or hanging drop inverted over the test tube) is useful. Methylcellulose constricts movement but does not narcotize.

10. Other suggested narcotizing agents:
   a. Carbon dioxide added as ordinary charged water to fluid containing animals (particularly cnidarians and echinoderms) or as CO_2 gas bubbled directly into the container
   b. Chloral hydrate, 10% solution, added drop by drop until organism is fully extended
   c. Clove oil, a few drops scattered on surface of water
   d. Magnesium chloride, 2.5% solution of the hexahydrate in tap water
   e. Magnesium sulfate, saturated solution, added very gradually until the organism is relaxed (for marine organisms)
   f. Tricaine methanesulfonate (MS222) (0.33 g/L in the ambient medium) for cold-blooded animals
   g. Eucaine; prepare a solution of β-eucaine hydrochloride (1 g), 90% alcohol (10 ml), and distilled water (10 ml), and add drop by drop to culture of microorganisms

For Vertebrates

Frogs. Two methods are suggested for anesthetizing frogs.

1. Immerse the frogs in 2% MS222.
2. Inject 1 ml of 5% urethane per 30 g of animal weight into the dorsal lymph sac of each frog.

Small Mammals. For rats and mice inject 0.1 to 0.2 ml/100 g body weight of 25% to 50% solution of urethane peritoneally. For rabbits inject 1 to 2 g of urethane (1g/ml solution) per kg body weight by marginal ear vein. Inject slowly, about 1 g every 15 minutes, until surgical anesthesia is reached.

Ether-vapor Tube for Narcotizing Small Animals

Place a little ether in a small bottle or test tube. Close with a two-hole stopper fitted with two bent glass tubes that reach to just above the ether in the bottle. Warm the bottle in your hands and then blow gently through one tube while directing the ether vapor with the other. Chloroform may be substituted when needed.

IV. SOME CULTURE METHODS
Amebas

To 100 ml of distilled water in a finger bowl add 3 kernels of boiled wheat. Inoculate with amebas. Every 2 weeks replace part of the water and, if bacterial growth is not too heavy, add another grain of boiled wheat. Keep in diffused light and moderate temperature (around 23.8° C, or 75° F).

Paramecia, Mixed Protozoans, *Stentor*

For paramecia and mixed protozoans fill a finger bowl two-thirds full of distilled water and add 4 kernels of boiled wheat or rice and 15 to 20 pieces of boiled timothy hay 1 to 2 cm long. Inoculate immediately. If bacterial film forms over water surface, break it up. Allow plenty of diffused light. For *Stentor* prepare as above, inoculate with mixed protozoans, and when the culture is rich, add the *Stentor*.

Brine Shrimp (*Artemia*)

Brine shrimp eggs can be obtained from any biological supply house and hatched in a day or two. They are widely used as food for aquarium animals. Use a shallow rectangular pan or tray of glass, plastic, or enamel (not metal). Cut a divider of glass, Plexiglas, or wood that will extend across the width of the pan, fitting snugly against the sides of the pan but lacking ½ to 1 inch of reaching the bottom. Fill the pan three-quarters full of salt water (natural or artificial seawater or simply a tablespoon of sodium chloride into a quart of tap water). The divider should extend above the surface of the water.

Add about ¼ to ½ teaspoon of dry eggs to the water in one side of the pan; they will float on top of the water and be confined to one side of the pan. Place an opaque cover over the pan except for 1 inch or so at the end farthest from the eggs. The hatched nauplii will be attracted to the light, swim under the divider, and cluster at the open end, where they are easily siphoned off into a brine shrimp net or finger bowl. At 23.8° to 26° C (75° to 80° F) they will hatch in 24 to 48 hours. Two or three trays started daily on a rotating basis will keep a constant supply available for daily feeding.

If the larvae are to be fed to freshwater forms, they should be rinsed well first.

If the larvae are to be kept, they should be removed to another container of saltwater where they are not crowded and fed small amounts of yeast suspension or green one-celled algae daily. The water should be changed often to prevent fouling. Though aeration is not necessary, it is helpful in keeping successful cultures.

Preparation for *Drosophila* Metamorphosis (Exercise 14C)
Banana-agar Medium

Water	575 ml
Agar-agar	20 g
Brewer's yeast	35 g
White corn syrup	125 ml
Very ripe banana, crushed	225 ml
Mold inhibitor	0.5 g

Add the agar-agar to the water and bring to a boil. Stir in the other ingredients. Boil gently for 10 minutes, stirring constantly. Cool slightly.

Place a thick layer of banana-agar medium on a clean glass microscope slide. Allow to solidify and then add a drop of yeast suspension made by dissolving a bit of yeast in a little water. Place the prepared slide in a small glass or plastic tube or vial. Add a pair of fruit flies and stopper with a cotton plug (Fig. 14-18). Keep the slide in a horizontal position, agar side up. Slides so prepared can be taken out at intervals, studied, and returned to the vials. If the medium tends to become dry, place a piece of moistened filter paper in the vial under the slide. When adding anesthetized flies, watch to see that they do not become entangled in the medium before they recover fully.

To have all stages available at a single laboratory period, start new vials daily, beginning about 2 weeks before the laboratory.

Preparations for Fruit Fly Experiments (Exercise 26A)

There are several good media for the culture of fruit flies. Two satisfactory ones are the banana-agar medium just given and the one described here:

Water	1600 ml
Agar-agar (granulated)	30 g
White corn syrup	50 ml
Molasses	50 ml
Cornmeal	100 ml
Raisins (ground up)	80 g
Dry yeast	30 g
Mold inhibitor	1 g

Add the agar-agar to the water and bring to a boil. Mix all the other ingredients and stir into the boiling water. Boil 10 minutes, stirring constantly. Using a funnel, pour ½ to 1 inch of the medium into the bottom of sterilized half-pint or quarter-pint milk bottles or glass vials. Add a double fold of paper toweling to each bottle. Cool. Just before using the bottles, add a drop of yeast suspension made by dissolving a bit of yeast in a little water. Stopper the bottles with a wad of cotton sheathed in cheesecloth.

Etherizing Bottles. Fit a bottle (same size as the culture bottles) with a cork. Attach a wad of cotton to the small end of the cork by means of a wire thrust through the cork or by folding cheesecloth around the cotton and the cork. Pour a few drops of ether onto the cotton.

Reetherizers. Reetherizers can be made by taping a piece of blotter or a bit of cotton into the bottom of a Petri dish and adding a few drops of ether. Invert over the flies if they start to revive while being counted.

When adding etherized flies to the culture bottles, place them on a piece of paper in which a crease has been made down the center, and let them slide down the fold. Lay the bottles on their sides, being careful that the flies do not drop into the soft medium. Check from time to time to see that the flies are all right.

V. AQUARIUMS AND TERRARIUMS
Starting a Freshwater Aquarium

A 5- or 10-gallon glass tank is recommended, but use whatever is available. Clean thoroughly with soap and ammonia water and rinse several times. Place where a strong diffused light is available (north or east exposures are good). Add enough aquarium sand or gravel to cover the bottom an inch or two deep, but be sure that it has first been washed in running water until all debris is removed. Then fill the tank three-quarters full of water (if chlorinated, let it stand a day or two before adding animals).

Add healthy plants to help oxygenate the water and to provide protection for small fishes. The roots of *Vallisneria*, *Sagittaria*, and the like should be spread out, covered to the crown, and pressed down to anchor. Stalks of *Elodea*, *Colomba*, *Myriophyllum*, and so on can be weighted down with small stones. Now add water to within an inch of the top. Unless dead or decaying matter is present, the water will be clear in a day or two. Provide a glass cover for the tank.

Animals can be added after several days. Avoid animals that require running water and those that uproot plants and stir up the bottom (turtles, for example). Do not crowd the tank. A good rule for fish is 1 inch of fish to each gallon of water. Snails are good scavengers, and up to a dozen pond snails can be kept in a 6-gallon aquarium. Small native fish and newts are interesting, and a clam will illustrate filter feeding, but be sure its siphons are working, since a dead animal soon fouls the water.

Examine the tank daily, trim off dead leaves, and remove any dead animals immediately. Avoid overfeeding because decaying food may foul the water and kill the animals.

Microaquariums can be prepared in almost any small container. Instructions are on p. 109.

Starting a Marine Aquarium

A marine aquarium should be in operation for at least 2 weeks before it is stocked with animals, so begin your preparations early. For a permanent aquarium the larger the size the better; a minimum of 10 gallons is recommended, although smaller containers are suitable for temporary holding (see below). The aquarium can be all glass, or Plexiglas, or even a plastic-lined wooden container, but absolutely *no metal* should be exposed to the seawater.

An under-gravel filter that recirculates water by air-lift should cover the bottom of the tank and about 3 inches of well-washed calcareous pebbles or limestone (Dolomite is ideal) should be spread over the filter (avoid the siliceous gravel usually sold at aquarium shops). The gravel particles must be large enough to avoid clogging the slits in the filter. For burrowing animals the gravel can be covered with well-washed, coarse beach sand or extremely well-washed crushed oyster shell (available in some feed stores). This combination of the filter and the gravel aerates the water through an air-lift tube and keeps the water circulating through the gravel, which filters out organic matter. Here various bacteria decompose the organic particles, converting toxic ammonia and urea to nitrites and nitrites to nitrates, which are nontoxic.

If desired, an auxiliary outside filter equipped with filter floss and charcoal can be used along with the under-gravel filter. This is a necessity if the tank is to be heavily loaded with animals.

Filtered natural seawater can be used if available, but quite satisfactory artificial seawater can be made up from good quality synthetic salts such as Instant Ocean or Rila Marine Mix. Mixing tanks can be any nonmetal container; large plastic trash containers, if sturdy, are satisfactory.

When filling the aquarium pour the water onto a dish or piece of glass placed over the gravel to avoid stirring up the bottom. When the setup is complete and the filters are running, wait 2 weeks to establish a bacterial population before adding animals. About every 4 to 6 weeks one-fourth of the water should be siphoned off and replaced with fresh salt solution (oftener if the tank is heavily loaded). A glass tank cover will slow down evaporation. Replace evaporated water with fresh tap water.

Optimum temperatures for tropical animals lie between 21° and 29° C; for North Atlantic or Pacific species it lies between 7° and 18° C. Do not place the aquarium in a window; it needs standard, diffuse light.

Many invertebrates such as sea anemones, corals, sea cucumbers, sea stars, sea squirts, crabs, and the like can live for months, even years, in such a properly cared for aquarium. It is important not to overload the tank. About 18 to 20 medium-sized invertebrates can be handled in a 20-gallon tank, but less if fish are added. Filter feeders can be fed freshly hatched brine shrimp; carnivores can be fed small bits of fish, clam, or shrimp.

More information is available from such sources as Aquarium Systems, Inc., Eastlake, Ohio, or the Carolina Biological Supply Company, Burlington, North Carolina, or Powell Laboratories Division, Gladstone, Oregon. For additional reading we suggest the following:

Diehl, F.A., J.B. Feeley, and D.G. Gibson. 1971. Experiments using marine animals. Eastlake, Ohio, Aquarium Systems, Inc.

Rudloe, J. 1971. The erotic ocean: a handbook for beachcombers. New York, World Publishing Co.

Spotte, S.H. 1970. Fish and invertebrate culture: water management in closed systems. New York, John Wiley & Sons, Inc.

Straughan, R.P.L. 1970. The salt-water aquarium in the home, ed. 2. New York, A.S. Barnes & Co.

Small Marine Aquariums

Small culture or demonstration aquariums can be made from wide-mouthed gallon jars filled with natural or artificial seawater and filtered with small inexpensive plastic outside filters of the two-compartment type that use filter floss and charcoal in one of the compartments. A series of such jars can be set up using three-way valves and can be connected to a compressed air outlet or an aquarium pump.

Starting a Terrarium

A terrarium can be any watertight container of suitable size from a dishpan or battery jar to an aquarium or terrarium fitted with a glass cover. Larger containers are usually easier to plant and maintain. A north window location and 18° to 22° C temperature is ideal for woodland terrariums; a sunny spot and a 27° to 32° C temperature is ideal for desert forms. Avoid heavy soil as plants need root ventilation. Provide a false bottom of hardware cloth and cover it with gravel, or else cover the bottom of the terrarium with gravel.

For a woodland terrarium use one part sand and three parts humus mixed thoroughly and spread over the gravel. Moisten till it clings together loosely but does not cake. Planted with mosses, lichens, liverworts, wood ferns, and the like, it is suitable for newts, toads, salamanders, frogs, chameleons, small snakes, snails, and insects. Provide a glass cover to conserve moisture. For a semiaquatic terrarium a container may be added for a small pool and the woodland soil built up around it, higher at one end. Some aquatic and semiaquatic animals can then be added.

The desert terrarium needs only an inch or two of course sand in the bottom, slightly dampened and covered with a half inch or so of desert sand. It can be covered with a screen. Cactus roots should be dampened before planting and the sand sprinkled lightly. Scorpions, lizards, snakes, spiders, and other such species may be added.

Most terrarium problems are caused by excessive moisture or temperature. If it is too moist, remove the cover for a part of each day. When plants become crowded, trim them or thin them out.

VI. SOME PRESERVATION METHODS
Preserving Invertebrates

Sponges. Fix in 90% or absolute alcohol for 24 hours; then store in 70% to 90% alcohol. (Formalin or water solutions cause maceration.)

Flatworms. Drop Bouin's fluid over well-expanded specimens, wash in 70% alcohol, and store in 70% alcohol.

Nematodes. Fix in very hot 70% alcohol or 5% formalin. When they have cooled, store in the same kind of fluid with 5% glycerin added.

Crustaceans. Kill by adding formalin to their water up to 5% strength. Then store in 70% alcohol or, better, in 20% solution of glycerin in 70% alcohol.

Insects and Spiders. Drop into 70% alcohol. Store in 20% solution of glycerin in 70% alcohol.

Earthworms. Kill with chloroform water or narcotize in 0.2% Chloretone or 0.08% MS222. Store in 5% formalin or 70% alcohol.

Most Molluscs, Echinoderms, Annelids, Brachiopods, Cnidarians, and so on. Narcotize slowly until they are well relaxed (p. 457). Store in 5% formalin or 70% alcohol. Sea urchins should have a small hole made in the shell or in the peristomial membrane to let in preservative. Sea cucumbers should be grasped firmly with blunt forceps just behind the tentacles to prevent their retraction and then plunged into 5% formalin for a few minutes; then inject 70% alcohol into the cloacal opening and through the body wall; preserve in 70% alcohol.

Hemichordates and tunicates should be narcotized slowly and preserved as above. Pelagic tunicates need not be narcotized; they can be placed in seawater made up to 5% formalin.

Amphioxus can be plunged directly into Bouin's fixative without prior narcotization and left for 24 hours before being transferred to 70% alcohol.

Methods for narcotizing are given on pp. 456-457.

Preserving Color in Aquatic Animals*

Have *ready* at the collection site the following: wide-mouthed 1- to 3-gallon (4- to 12-liter) plastic jars covered with foil or painted black; 20 to 50 ml plastic syringe with no. 15 or 20 needle; strong scissors or scalpel; large forceps with tips wrapped in plastic tape; shallow tray; 1-gallon (4-liter) semitransparent plastic jug marked in pints for rapid mixing; plastic funnel; concentrated formalin; an antioxidant soluble in formalin, such as butyl hydroxytoluene (BHT, Ionol-40; Shell Chemical Co.); roll of plastic wrap; ice chest and ice to hold fixing jars; string tags for labeling specimens.

At the site prepare fixing solution of 10% formalin (in fresh or marine water) and add BHT 1:200. Shake for 30 seconds and put into cooler to chill. Inject fresh specimens with the formalin solution in the area of the central nervous system and then in the body cavity (or slit abdomen). Label and place in fixing jar. When jar is filled (but not crowded), add fixing solution to within ¼ inch of the top, cover with plastic wrap, and screw on the lid. Keep in the cooler for at least 36 hours, shaking gently every 24 hours.

Exposure to air and heat fades colors, so transfer to small specimen jars rapidly and fill the jars with buffered 10% formalin or 40% isopropyl alcohol with BHT added. Store out of direct sunlight and heat.

Preparation of Preserved Animals for Dissection or Demonstration

Preserved frogs, fish, and the like may be removed from the preservative and placed in water the day

---

*White, D.A., and E.J. Peters. 1969. A method of preserving color in aquatic vertebrates and invertebrates, Turtox News 47(9)(Dec.).

before use. They may then be kept moist in individual plastic bags and reused for several laboratory periods. Preserved earthworms will keep overnight in trays if a saturated solution of borax is poured over them each time they are left.

Neutralizing Formalin Fumes

Formalin-preserved materials may be rinsed in water and then in a dilute solution of ammonium hydroxide (strong enough to have a decided ammoniacal odor) just before using. Large specimens may have to be rinsed again during a laboratory period.

Or soak the specimens for 15 to 30 minutes in an aqueous solution containing 5% urea and 1% ammonium phosphate. Rinse in water before soaking and again before returning the specimens to formaldehyde.

Care of Embalmed Pigs and Other Specimens

Embalmed specimens that are showing signs of drying out may be moistened in a solution of:

Carbolic acid crystals	30 g
Glycerin	250 ml
Water	1000 ml

VII. SOLUTIONS AND REAGENTS

PERCENTAGE RULE FOR PREPARATION OF SOLUTIONS

Using a ratio and proportion relationship, you will find that the quantity of drug or chemical to be used in solution divided by the total amount of solution to be made up equals the percentage wanted divided by 100.

For example, to make up 3 liters of 0.4% hydrochloric acid solution:

$$\frac{X}{3000} = \frac{0.4}{100}$$

$$100X = 1200$$

$$X = 12$$

Thus, to make up 3 L of 0.4% HCl, use 12 ml of HCl and enough water (2988 ml) to make the 3 L of solution.

Acetic Acid Solution (10%)

To 10 ml glacial acetic acid add 90 ml distilled water.

Acetocarmine Stain (Belling's Solution)

Heat to boiling a 45% solution of glacial acetic acid; add an excess of powdered carmine—more than will go into solution. Cool and filter through a coarse filter paper. Prepare under a hood to prevent inhaling fumes.

Acidified Methyl Green

Methyl green	1 g
Distilled water	100 ml
Glacial acetic acid	1 ml

Albumin Solution

While stirring constantly, add the slightly beaten white of an egg to about 300 ml of boiling water. Boil for a moment and then cool. Commercially prepared dried albumin may be substituted for egg whites.

Alcohol Dilution

For diluting 95% stock ethyl alcohol to other desired percentages, use a number of milliliters of alcohol equal to the percentage desired and then add the number of milliliters of distilled water equal to the percentage desired subtracted from 95%. For example:

To dilute 95% alcohol to 80% alcohol, use 80 ml of 95% alcohol and 15 ml (95 minus 80) of distilled water.

To dilute 95% to 25% alcohol, use 25 ml alcohol and 70 ml water.

To dilute 70% to 30% alcohol, use 30 ml of 70% alcohol and 40 ml of water.

Bouin's Fixative

Saturated aqueous picric acid (about 1 g will dissolve)	75 ml
Formalin (40% formaldehyde)	25 ml
Glacial acetic acid	5 ml

Add the acetic acid just before use. After fixation of tissues wash in ethyl alcohol, 45%, or stronger until the yellow color is removed.

Chloretone Solution (0.2%)

Dissolve 0.2 g Chloretone in 100 ml distilled water.

Congo Red–Yeast Mixture

In a large test tube combine about 2 g of active dry yeast and a pinch of Congo red with 10 to 15 ml of distilled water and boil gently for a few minutes. Cool before using. If necessary, the mixture can be used without boiling—just let it stand until thoroughly dissolved and slightly thickened—but such a preparation must be made up fresh each day.

Congo Red–Milk Mixture

Combine equal quantities of rich paramecium culture (centrifuged if necessary) and milk stained with a little Congo red. In an hour or so all stages of digestion should be demonstrated by the indicator dye.

Formaldehyde Solutions

The commercial stock formalin, usually purchased by the gallon or drum, is 40% pure strength formaldehyde solution. To make a 5% formalin solution, use 5 ml stock formalin and 95 ml distilled water. For 10% formalin, use 10 ml stock formalin to 90 ml water.

Formalin–Alcohol–Acetic Acid Solution (FAA)

A good general fixative and preservative for plant and animal preparations is formalin–alcohol–acetic acid solution (FAA). It penetrates rapidly, fixes large volumes of tissue compared to the volume of liquid, and lasts indefinitely; it is good for museum specimens.

Alcohol (ethyl or grain), 70%	85 ml
Formalin (commercial strength)	10 ml
Glacial acetic acid	5 ml

Glassware Cleaning Solution

Potassium bichromate $K_2Cr_2O_7$, technical grade	11 g
Water	50 ml
Sulfuric acid, H_2SO_4	50 ml

Heat water to dissolve the bichromate, cool, and then add the acid cautiously.

India Ink (for Paramecia and Other Ciliates)

Do not use the commercial fluid inks, since they contain chemicals toxic to protozoans. Instead, rub a solid ink stick (used for photographic retouching) in a small volume of water to produce a black suspension.

KAAD Mixture (for Killing Insect Larvae)

Kerosene	10 ml
Glacial acetic acid	20 ml
95% ethyl alcohol	70-100 ml
Dioxane	10 ml

Larvae should be ready to transfer to alcohol for storage in ½ to 4 hours. For soft-bodied larvae, such as maggots, the amount of kerosene should be reduced.

Methylcellulose (10%) (for Slowing Protozoans)

Add 10 g methylcellulose to 50 ml water. Bring to a boil, cool, and let stand for 30 minutes. Add cold distilled water to make 100 ml.

Sodium carboxymethylcellulose, 2%, can also be used. Bring 100 ml distilled water to a boil and then add slowly 2 g sodium carboxymethylcellulose.

Protoslo is a suitable methylcellulose product available from Carolina Biological Supply Co. With any of the above, make a small ring on a slide and add a drop of culture to the center of the ring and cover.

Methylene Blue Solution

Dissolve 1 g methylene blue crystals in 2 liters 0.6% NaCl solution.

Noland's Stain for Flagella and Cilia

Moisten 20 mg of gentian violet with 1 ml distilled water; add 80 ml of a saturated solution of phenol in water; then add 20 ml of formalin (40% formaldehyde); and finally, add 4 ml glycerin. Mix these constituents together and add a drop of the solution to the drops of culture to be examined.

Potassium Hydroxide Solution (2%)

Dissolve 2 g potassium hydroxide in 100 ml distilled water.

Physiologically Balanced Ringer's Solutions

For Amphibians and Fish

Sodium chloride	6.5 g
Potassium chloride	0.14 g
Calcium chloride	0.12 g
Sodium bicarbonate	0.2 g
Distilled water	1 L

For Mammals, Birds, and Reptiles

Sodium chloride	9.0 g
Potassium chloride	0.42 g
Calcium chloride	0.24 g
Sodium bicarbonate	0.2 g
Distilled water	1 L

For Marine Invertebrates

Sodium chloride	23.4 g
Potassium chloride	0.73 g
Calcium chloride	1.12 g
Magnesium chloride	5 g
Sodium bicarbonate	0.21 g
Distilled water	1 L

(Ordinary seawater is suitable for marine dissection work, if available.)

Saline Solutions

For Amphibians and Other Cold-blooded Forms. Mix 6.5 g NaCl with distilled water to make 1 L.

For Mammals, Birds, and Reptiles. Mix 9 g NaCl with distilled water to make 1 L.

For Insects. Add 7.5 g NaCl to distilled water to make 1 L.

For Crayfish. Add 18 g NaCl to distilled water to make 1 L.

For Marine Invertebrates. Use artificial seawater or Ringer's solution for marine invertebrates (just given).

VIII. SOME MISCELLANEOUS METHODS
How to Pith a Frog

Pithing means destroying the brain and the spinal cord. An anesthetic such as ether or chloroform may affect some tissue adversely. Pithing does not.

Single Pithing. With the left hand, hold the frog firmly in the region where the hindlegs join the trunk. Bend the frog's head downward with the index finger of the left hand until the head is almost at right angles to the trunk (Fig. 1). Find the region where the skull joins the backbone. The foramen magnum is here. The region is just behind the posterior level of the tympanic membranes. In this region press a shallow groove in the skin with the side of the dissecting needle. Then into the midline of this groove, thrust a sharp-pointed dissecting needle a short distance, turn the point forward, and push it into the brain cavity. Twist the point around, thus destroying the brain. Be careful not to force the needle; if it is in the cranium, it will find little resistance.

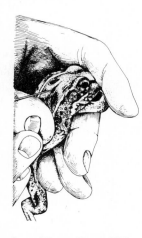

Fig. 1. Pithing a frog. (From Hoar, W.S., and C.P. Hickman, Jr. 1983. A laboratory companion for general and comparative physiology, ed. 3. Reprinted by permission of Prentice-Hall, Inc., Englewood Cliffs, N.J.)

Double Pithing. If the spinal cord is also to be destroyed, partly withdraw the needle and, through the same hole in the skin, push the point caudally into the neural canal of the vertebral column. When this is done correctly (be sure the needle is pushed the full length, as it tends to become caught against the vertebrae), the hindlegs will at first stiffen and then become completely limp and unresponsive to stimulation.

Frog Nerve-Muscle Preparation (p. 374)

Skin the lower half of a double-pithed frog (see above for pithing and p. 313 for skinning directions). Place the animal in a tray and keep moist with Ringer's solution. Lift the tip of the urostyle and cut around it (Fig. 2, *A*), avoiding injury to the underlying nerves. Locate the sciatic nerve (Fig. 2, *B*) and free it carefully from surrounding fascia and arteries. Lift the nerve with a glass hook, sever it near the vertebral column, and lay it back over the gastrocnemius muscle. Now cut off the leg at the proximal end of the femur; then loosen and cut the Achilles tendon (Fig. 2, *C*). Finally, cut away the tibia-fibula, thus removing the foot and completing the preparation (Fig. 2, *D*). Keep in cool aerated Ringer's solution until needed.

Making a Frog Board

A frog board for demonstrating capillary action in the tongue or web of the foot can be made from a piece of balsa wood or cork board 25 × 10 × 0.5 cm. Make a hole in the board about 1 cm in diameter and far enough from the end to coincide with the stage aperture of the microscope. The wrapped frog is held to the board by cloth ties or masking tape, the foot is

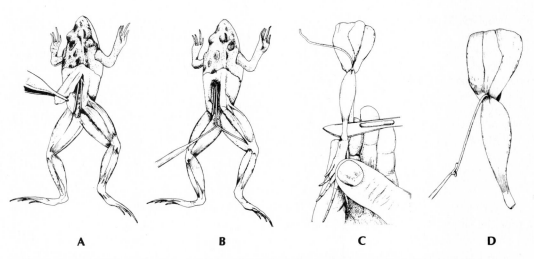

A   B   C   D

Fig. 2. How to make a frog nerve-muscle preparation. (From Hoar, W.S., and C.P. Hickman, Jr. 1983. A laboratory companion for general and comparative physiology, ed. 3. Reprinted by permission of Prentice-Hall, Inc., Englewood Cliffs, N.J.)

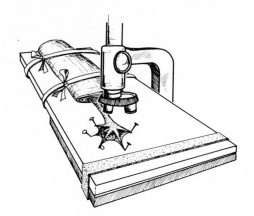

Fig. 3. Frog board.

pinned over the hole, and the board is then positioned on the microscope stage and taped in place (Fig. 3).

Killing Molluscs for Fresh Dissections
(Chapter 12)

Freshly killed molluscs are best for dissection. Kill freshwater or terrestrial pulmonates and freshwater mussels by keeping them for 24 hours or so in closed-stoppered bottles entirely full of water. When the oxygen is used up, the asphyxiated animals will be fully extended and insensitive.

Place marine molluscs in a small amount of water and add magnesium sulfate (crystals or saturated solution) or alcohol very gradually. Avoid disturbing the water or the animals will contract. It may take several hours for them to become insensitive. Cephalopods die very quickly when removed from the water and do not need narcotizing.

Osmometer Assembly (Exercise 3A, Part 2)

Glass tube, length 4 feet
Dialysis tubing, flat width 1 inch, length 5 inches
Rubber stopper, single hole, no. 1 size
Dialysis tubing closure (Carolina Biological Supply Co., no. 68-4240)
Quart size mason jar
35% white Karo syrup solution
Ring stand with clamp
Wax pencil
Meter stick
Thread

Soak the dialysis tubing in water for a few minutes to make it pliable. Roll the end of the tubing between the fingers. Carefully open the end and stretch gently so as not to tear it. Insert the rubber stopper, wrap thread securely around the tubing, and tie. Fold the opposite end back about ½ inch and clamp with closure (Fig. 3-1). Add syrup solution through the hole in the stopper using a Pasteur pipette. Fill to capacity.

Insert the glass tubing carefully. Lower it into the mason jar filled with water and wait until the liquid comes above the stopper before beginning the experiment.

Solutions for Exercise 3B—Action of Enzymes

1. 0.25% Starch Solution. Make a paste by stirring 2.5 g of soluble potato starch in 50 ml of distilled water. Add this paste slowly to 500 ml of boiling salt solution containing 0.35 g NaCl. Allow to boil for 5 minutes with constant stirring. Cool and dilute to 1 liter.

2. 0.25% α-Amylase from Bacillus Subtilis (0.125 g/50 ml). Prepare just before needed. Fifty ml should suffice a laboratory of 12 student pairs. α-amylase is available from Sigma Chemical Co.

3. McIlvaine's Buffer. Buffer solutions of pH 3.4, 7.0, and 8.0 are prepared by mixing specified proportions of two stock solutions.

Stock solution A: 0.1 M citric acid, anhydrous, 19.212 g made to 1000 ml with distilled water.

Stock solution B: 0.2 M sodium phosphate, dibasic, anhydrous 28.396 g (if 7 hydrate, use 53.628 g) made up to 1000 ml with distilled water.

Prepare working buffers as follows (check pH with a pH meter):

  pH 3.4: 14.3 parts of Sol. A and 5.7 parts of Sol. B.
  pH 7.0: 3.53 parts of Sol. A and 16.47 parts of Sol. B.
  pH 8.0: 0.55 parts of Sol. A and 19.45 parts of Sol. B.

4. I-KI Solution. Dissolve 15 g potassium iodide in 500 ml water, add 5 g iodine slowly, stirring to dissolve. Make up to 1 liter with water. Store in a brown bottle.

Solutions for Exercise 23A, Distribution of Digestive Enzymes

Insect Saline. 7.5 g NaCl/L of solution.

5% Sucrose Solution. 5 g sucrose/100 ml insect or amphibian saline.

0.5% Invertase (Sucrase) in Saline. 0.5 g invertase/100 ml saline.

0.5% Amylase in Saline. 0.5 g amylase/100 ml saline.

0.5% Lipase in Saline. 0.5 g lipase/100 ml saline.

0.5% Trypsin in Saline. 0.5% trypsin/100 ml saline.

Bendict's Solution

Dissolve 173 g sodium citrate and 100 g anhydrous sodium carbonate in 600 ml of hot distilled water.

Dissolve 17.3 g cupric sulfate in 150 ml distilled water.

Slowly, with constant stirring, add the cupric sulfate solution to the sodium citrate–sodium carbonate solution. Dilute with distilled water to make 1 liter.

1% Starch Solution

Dissolve 1 g soluble starch in 10 ml water. Add 90 ml saturated NaCl solution and bring to a boil while stirring. Cool and store refrigerated.

Iodine Solution. 2% potassium iodide in distilled water. Add enough elemental iodine to color deep yellow.

Bile Salt Solution. 1 g sodium choleate (ox bile extract)/100 ml water, available from Sigma Chemical Co., product no. S9875.

0.1 N Sodium Hydroxide. 4 g/L. Prepare 0.005 and 0.001 N solutions by appropriate dilution.

IX. SOME SOURCES OF LIVING MATERIAL

United States

Ann Arbor Biological Center, Inc., P.O. Drawer M1128, Ann Arbor, MI 48106

Carolina Biological Supply Co., Burlington, NC 27215 *or* Powell Laboratories Division, Gladstone, OR 97027

Central Scientific Co., 2600 South Kostner Ave., Chicago, IL 60623

Coe-Palm Biological Supply House, 1130 North Milwaukee Ave., Chicago, IL 60622

College Biological Supply Co., 21707 Bothell Way, Bothell, WA 98011 *or* 8857 Mt. Israel Road, Escondido, CA 92025

Connecticut Valley Biological Supply Co., Inc., Valley Road, Southampton, MA 01073

Fisher Scientific Co., Educational Materials Division, 711 Forbes Ave., Pittsburgh, PA 15219

Frey Scientific Co., 905 Hickory Lane, Mansfield, OH 44905

Graska Biological Supplies, Inc., P.O. Box 2322, Oshkosh, WI 54901

Gulf of Maine Biological, P.O. Box 538, Brunswick, ME 04011

Gulf Specimen Co., Inc., P.O. Box 237, Panacea, FL 32346

LaPine Scientific Co., 6001 South Knox Ave., Chicago IL 60629 *or* 375 Chestnut St., Norwood, NJ 07648 *or* 920 Parker St., Berkeley, CA 94710

Wm. A. Lemberber Assoc., P.O. Box 222, Germantown, WI 53022

Macmillan Science Co., Inc. (Turtox), P.O. Box 20765, Chicago, IL 60620

Mogul-Ed, P.O. Box 2482, Oshkosh, WI 54901

NASCO, 901 Jamesville Ave., Fort Atkinson, WI 53538

National Biological Laboratories, P.O. Box 579, Fairfax, VA 22030

Nova Scientific Corporation, P.O. Box 500, Burlington, NC 27215

Pacific Bio-Marine Laboratories, Inc., P.O. Box 536, Venice, CA 90291

Parco Scientific Co., P.O. Box 595, Vienna, OH 44473

Powell Laboratories Division, Gladstone, OR 97027

Rockland, P.O. Box 316, Gilbertsville, PA 19525

Sargent-Welsh, 7300 North Linder Ave., Skokie, IL 60076

Sea Life Supply, 740 Tioga Ave., Sand City, CA 93955

Southern Biological Supply Co., P.O. Box 68, McKenzie, TN 38201

Ward's Natural Science Establishment, Inc., P.O. Box 1712, Rochester, NY 14603

Canada

Arbor Scientific Co. Ltd., 1840 Mattawa, Missisauga, Ont. LX4 1L1

Boreal Laboratories, Ltd., 1820 Mattawa, Missisauga, Ont. L4X 1K6

NASCO Educational, 58 Dawson Road, Quelph, Ont. N1H 6P9

Northwest Laboratories, Ltd., P.O. Box 6100, Station C, Victoria, B.C., V8P 5L4

Ward's of Canada, P.O. Box 113, Port Credit, Ont. L5G L46

INDEX

A

Abdomen
  of crayfish and lobsters, 214
  of fetal pig, muscles of, 346-347
  of grasshopper, 226, 228-229
  of honeybee, 229
  of horseshoe crab, 208
  of spider, 211-212
Abdominal artery of crayfish and lobster, 220
Abdominal cavity
  of fetal pig, 349-350
  of frog, 320
  of yellow perch, 301-304
Abdominal muscle of fetal pig, transverse, 346
Abduction in mammals, 341
ABO blood groups, 385-386
Aboral surface
  of sea star, 251-252
  of sea urchin, *264*
Acanthobdella, 191
Acantocephala, *170*, 171
Acetabulum(a)
  of bladder flukes, 157
  of *Clonorchis*, 156
Acetylcholine, action of, *381*
Acheta domesticus, 421
Aciculum of parapodium, 193
Acoelomate animals, 143-162
Acolosoma, 202
Acontia of sea anemone, 137
Acris gryllus, classification of, 79
Acropora, 138
Actinoped amebas, 90
Actinophrys, 85, 90
Actinopoda, 85
Actinosphaerium, 85, 90
Adaptation
  of earthworms, 420
  of frogs, 308
Adduction in mammals, 341
Adductor muscle
  of fetal pig, 348
  of frog, 314
  of mussel, 177, 178
Adipose tissue
  in blood vessels, *73*
  connective, 64, *65*
Adrenal arteries of fetal pig, 367
Adrenal veins of fetal pig, 366
Adrenaline, 396

African sleeping sickness, 97
Aggression in crickets, 421-422
Agnatha, 272, 283
Alcohol
  to clean microscope lens, 11
  to fix
    insects, 236
    spiders, 212
Alcyonaria, 123
Alcyonium, 123
Alimentary canal
  of earthworm, 201-202
  of freshwater oligochaetes, 202
Alleles of genes, 46
Allolobophora, 420
Amaroucium larvae, 275
Ambulacral ossicles of sea star, 253
Ambulacral plates of sea urchin, 263
Ambulacral spines of sea star, 252-253
Ambystoma, classification of, 79
Ameba(s), 87, *88*
  actinoped, 90
  parasitic, 89-90
  shelled, 90
Amebocytes of sponges, *117*
American termites and *T. campanula*, symbiotic mutualism of, 96
American toad, 79
Ammocoetes larva, 283, *284*, *285*, 286-287
Amoeba, 85, 86-89
Amoeba proteus, 86, 87
Amphiarthrosis, 340
Amphibia, 272, 307-333
Amphibian(s)
  small intestine of, cross section of, 72
  taxonomic key to, 78-80
Amphiblastula larvae of sponge, 115, 117
Amphioxus, *276*, 277-279
Amphipoda, 223
Amphitrite, 194, 195
Amphiuma, 22, 72
  classification of, 79
Ampulla of sea star, 251, 253
Amylase, 404
α-Amylase, action of, 43-44
Anal pore of ciliates, 106
Analis, spinal cord of, *71*
Ancylostoma duodenale, 167
Anemone, sea, external structure of, 136-*137*
Animal(s)
  acoelomate, 143-162
  classification of, 77-81

Italicized numbers refer to illustrations.

Animal(s)—cont'd
  dissection of, making, 1
  ectothermic, 328
  living
    care of, 455
    study of, 445-449
  pseudocoelomate, 163-174
  radiate, 123-141
  social, honeybee as, 229
Animal behavior, 419-425
Animal hemisphere of frog eggs, 57
Animal kingdom, key to chief phyla and classes of, 78
Animal pole of frog egg, 55
Animal taxa, major, key to, 450-454
Annelid(s), 191-206
Annelida, 191-206
Annuli of leech, 204
Anodonta, 175, 176
Anopheles mosquito, 102, 103
Anostraca, 221
Ant, carpenter, 245
Antedon, 250, 269
Antenna(e)
  of arthropods, 213
  of centipedes, 224
  of crayfish and lobsters, 213, 214, 216
  of grasshopper, 226
  of honeybee, 226
  of millipedes, 225
Antenna cleaner of honeybee, 228
Antenna comb of honeybee, 228
Antennal glands of crayfish and lobster, 221, 223-224
Antennary arteries of crayfish and lobster, 220
Antennules of crayfish and lobsters, 214, 216
Anthozoa, 123
Antipathes, 123
Anura, classification of, 79
Anus
  of Ascaris, 164, 165
  of centipede, 224
  of clamworm, 192
  of crayfish and lobster, 214, 220
  of earthworm, 196, 200
  of frog, development of, 57, 58
  of grasshopper, 229
  of horseshoe crab, 208
  of leech, 204
  of mussel, 180
  of sea cucumber, 267
  of sea urchin, 263
  of spider, 212
  of squid, 189
Aorta
  of ammocoetes, 284-285
  of amphioxus, 277, 279
  of fetal pig, 362-363
  of mussel, 179
  of shark, 295
  of yellow perch, 304
Aortic arches
  of earthworm, 198, 200
  of frog, 327
Aperture(s)
  of mussel, 178
  of snail
    genital, 187
    of shell, 186

Apex of shell, snail, 186
Aphid, corn-root, 247
Apicomplexa, 85-112
Apis, 226-227, 228-229
Aplysia, 175
Aponeurosis in mammal, 341
Apopyles of sponge, 115
Appendages
  of arthropod, jointed, 207
  of crayfish and lobster
    biramous, 214, 215
    dissection of, 215-216
  of crustacean, biramous, 213, 215
Appendicular skeleton
  of frog, 313
  of mammal, 336-339
Aquariums, 458-460
Aquatic mandibulate, 213-224
Aquiferous pore of squid, 188
Arachnids, collecting and preserving, 212
Arachnida, 207
Arachnoid in fetal pig, 372
Arbacia, 250, 264
Arcella, 85, 90
Archenteron, 53, 54, 57
Architeuthis, 188
Arctobolus, 225
Arenicola, 195
Areolar connective tissue, 64, 65
Argiope, 207, 210
Argiope aurantia, 210
Argiope trifasciata, 210
Aristotle's lantern in sea urchin, 263
Arm(s)
  of Aurelia, oral, 134
  of sea star, cross section of, 255
  of squid, 188
Arolium of grasshopper, 228
Artemia, 124, 223
Arterial system
  of frog, 326-327
  of shark, 295
  of yellow perch, 304-305
Arterioles in capillary circulation, 395
Artery(ies)
  of ammocoete, 285
  of amphioxus, 277
  of crayfish and lobster, 220
  of fetal pig
    of head, shoulders, and forelimbs, 362-363, 364-365
    of trunk and hindlegs, 367-368
  section through, 72-73
  of shark, 295
Arthropod, 207-247
Arthropoda, 207-247
Articulation of bones in mammal, 340
Artiodactyla, 335
Ascaris, 27, 164-166, 167, 202
  female, transverse section of, 166
  oogenesis and fertilization in, 48
Ascaris lumbricoides, 164
Ascaris megalocephala, 164
Ascaris megalocephala bivalens, 30
Ascaris suum, 164
Ascaris egg, fertilized
  maturation of, 48
  mitosis in, 30

Aschelminthes, 163
Ascidians, 272-275
Asconoid canal system of sponge, 118
Asexual reproduction
    of hydra, 127
    of sponge, 117
    of *Volvox*, 95
Astacus, 213
Asterias, 251-258
Asteroidea, 250-252, 253-*254*, 255-258
Asterozoa, 250
Astral rays, 28
Astrangia, 123, 137, 138-139
Astrosclera, 113
Asymmetron, 276
Atrium of heart
    of fetal pig, 368
    of mammal, 361
    of sea squirt, 275
    of yellow perch, 304
Aulophorus, 202
Aurelia, 123, *133*-135
    structure of, *134*
Auricle of mussel heart, 179
Auricularia, 249
Autonomic nervous system
    of fetal pig, 372
    of frog, 332
Aves, 272
Axial skeleton
    of frog, 311-312
    of mammal, 336
Axons, 70

B

Back of fetal pig, muscles of, 346-347
Bacteria on stained slide, 22
Baermann apparatus, *169*
Balanus, 222
Barnacle, 222
Basal disc
    of hydra, 124
    of sea anemone, 136
Basal plate of coral, 138
Bdelloura, 147, 148
Bedbug, *245*
Beetle, ladybird, 244
Behavior
    animal, 419-425
    of *Aurelia*, 133
    of bivalve, 176
    of brittle star, 260
    of clamworm, 192
    of comb jelly, 139
    of coral, 138
    of crayfish and lobster, 213
    of earthworm, 196
    of frog, 308, 414-415
    of garden spider, 210
    of grasshopper, 226
    of horseshoe crab, 208-210
    of hydra, 124
    of hydroid colony, 128
    of hydromedusae, 129-130
    of leech, 204
    of *Paramecium*, 104
    of sea cucumber, 267-268

Behavior—cont'd
    of sea star, 251
    of sea urchin, 262
    of snail, 185
    of squid, 188
Benedict's test, 44
Bent little finger, inheritance of, 431
Berlese funnel, *212*, 235
Bicarotid artery of fetal pig, 363-364
Biceps brachii of fetal pig, 345
Biceps femoris muscle
    of fetal pig, 347
    of frog, 315
Bilateral animals, body cavity of, 163
Bilateral symmetry of acoelomate, 143
Binary fission
    of ameba, 89
    of paramecium, 107
Binomial nomenclature, Linnaean system of, 77
Bioluminescence
    in Chaetopterus, 195
    in *Cyridina*, 224
Bird, crop and stomach of, 355
Bivalent homologs, 46
Bivalve, live, dissection of, 181
Bivalvia, 175, 176-184
Blaberus, 404
Bladder
    of crayfish and lobster, 221
    of fetal pig, urinary, 355
    of frog, urinary, 321
    of yellow perch
        swim, 301-303
        urinary, 304
Bladder fluke, 157
Blastocoel, 53, 57
Blastoderm, 53
Blastomere in cleavage stage, 53
Blastopore, 53, 57
Blastostyle of *Obelia*, 129
Blastula, whitefish, mitosis in, 30
Blastula stage, 53
    of frog eggs, 57
Blatto, 101
Blepharisma, 86, 105, 107
Blood, 67
    of fish, 305
    of horseshoe crab, 208
    of human, 385-389
    of molluscs, 179
Blood cell(s)
    of arthropod, 223
    red
        damaged, *17*
        Shüffner's dots in, *103*
Blood flow in earthworm, 203
Blood groups, inheritance of, 431
Blood platelets, 67, 68
Blood pressure, 388-389
Blood sinuses of grasshopper, 229
Blood smears, invertebrate, 203
Blood trypanosome, 96, 97
Blood vessels
    in earthworm, 202, 203
    dorsal, 198, 200
Blowfly, metamorphosis of, 234

Body(ies)
   cell, nerve, 70
   fat, in grasshopper, 229
   paramylum, in euglena, 94
   segmented, of arthropod, 207
Body cavity of bilateral animals, 163
Body covering of euglena, 94
Body wall
   of *Ascaris*, 165
   of *Clonorchis*, 156
   of fetal pig, 349
   of sea anemone, 136
Bone, 64, *66-67, 339-340*
   compact, 66
   decalcified, examination of slide of, 73
   long, longitudinal section of, examination of slide of, 73
   spongy, 66
Bone marrow cavity, 66
Bony fish, 299-305
Book gills of horseshoe crab, 208
Brachial arteries of yellow perch, 304
Brachial plexus nerve of fetal pig, 370
Brachialis muscle of fetal pig, 345
Brachiocephalic muscle of fetal pig, 344
Brachiocephalic trunk of fetal pig, 363-364
Brain
   of ammocoetes, 284, 286
   of crayfish and lobster, 221
   of fetal pig, 372-373, *374*
   of frog, 332-333
Branchial arteries of shark, 295
Branchial heart of squid, 189
Branchiopoda, 221
Branchiostoma, 271, 276, 283
Breathing by frog, 308
Brine shrimp, 223
Brittle star, *259-260*
Brood chamber of mussel, 178
Brood pouches of mussel, 179
Brownian movement, 34-35
Buccal cavity of amphioxus, 276-277
Buccal membrane of squid, 188
Buccal podia of sea urchin, 263
Budding
   of hydra, 124, 127
   of medusa, 129, 134
   of tapeworm, 159
Bufo americanus, classification of, 79
Bufo woodhousei fowleri, classification of, 79
Bufonidae, classification of, 79
Bulbourethral gland of fetal pig, 357
Bullfrog, classification of, 80
Burrowing polychaetes, 195
Bursa of brittle star, 259
Busycon, 175, 187
Butterfly, 227
   swallowtail, *239*

C

Caddis fly, *240*
Calcifibrospongia, 113
Calcispongiae, 113, 114-119
Cambarus, 207, 213, 216
Canal systems of sponge, 113, 118
Canaliculi, 66-67
Capillary circulation in frog, 395-396

Carapace
   of crayfish and lobster, 214
   of horseshoe crab, 208
Cardiac chamber of crayfish and lobster, 220, 221
Cardiac muscle, *69*
Cardiac region of crayfish and lobster, 214
Cardinal venous system of yellow perch, 305
Carotid arch of frog, 327
Carotid trunk of fetal pig, 363-364
Carpenter ant, *245*
Carpi radialis muscles of fetal pig, 345
Cartilage, 64, *66*
   elastic, 64
   of squid, 188, *189-190*
Carybdea, 123
Cassiopeia, 123, 135
Castes of honeybee society, 229
Cat, skeleton of, *338*
Caudata, classification of, 79
Caudal artery of fetal pig, 368
Caudal vein of fetal pig, 366
Caudofoveata, 175
Cecropia, 234
Cecum(a)
   of crayfish and lobster, 220
   of fetal pig, 353
   of grasshopper, gastric, 229
   of sea star, 253
Celiac artery
   of fetal pig, 367
   of yellow perch, 305
Celiacomesenteric ganglia in fetal pig, 372
Cell(s), 21-26
   of arthropod, blood, 223
   of *Ascaris*, longitudinal muscle, 165
   of *Aurelia*, sex, 134
   division of, 26-30
   of earthworm, chloragogue, 200
   frog egg, follicular layers of, 55
   function of, 33-44
   germ
      maturation division of, 45-62
      primordial, 45
   goblet, *63*
      in trachea cross section, 73, *74*
   human
      blood, 387
         red, damaged, 17
         Shüffner's dots in, 103
   of hydra, 127
   liver, 15, 22-23, 24-25
      fine structure of, 23-26
   muscle, smooth, 25
      in mitosis, 28
   of planarian, flame, 149
   of pseudocoelomate, flame, 165
   of sea star
      coelomic, phagocytosis of, 256
      egg, 22
   squamous epithelial, stained, 22-26
   of squid, pigment, 188
   structure of and division of, 21-32
   of supporting tissues, 64
   in *Volvox*
      colonies of, 94
         sperm, 95
         differentiation of, 94

Cell body, nerve, 70
Cell membranes of frog eggs, 55
Cellular level of organization, 113
Cellular structure of sponge, 115-117
Centipede, 224, 225
Central nervous system of frog, 332
Centriole, 28
Centromere, 28
Cephalaspidomorphi, 272, 283-290
Cephalic vein of fetal pig, 362
Cephalization in acoelomate, 143
Cephalobus, 167
Cephalochordata, 271, 276-279
Cephalopoda, 175, 188-190
Cephalothorax
    of crayfish and lobster, 214
    of horseshoe crab, 208
    of spider, 211
Ceratium, 85, 97
Cercariae of fluke, 157
Cercus
    of grasshopper, 228
    of insect, 239
Cerebellum of fetal pig, 373
Cerebral ganglia of earthworm, 200-201
Cerebropleural ganglia of mussel, 180
Cerebrospinal fluid of fetal pig, 372
Cerebrum of fetal pig, 373
Cerianthus, 123
Ceriantipatharia, 123
Cerithidea californica, 157
Cervical artery of fetal pig, 364-365
Cervical ganglion in fetal pig, 372
Cestoda, 143, 159-162
Chaetoderma, 175
Chaetogaster, 202
Chaetonotus, 170-171
Chaetopleura, 175
Chaetopterus, 194
    bioluminescence in, 195
Chela(e)
    of crayfish and lobster, 213, 215
    of horseshoe crab, 208
Chelicera(e)
    of horseshoe crab, 208
    of spider, 211
Chelicerata, 207
Chelicerates, 207, 208-212
Chelipeds of crayfish and lobster, 213, 215
Chemotaxis, 146
    in paramecia, 106
Chest, muscles of, of fetal pig, *343-345*
Chiasmata, 46
Chilaria of horseshoe crab, 208
Chilomonas, 85
Chilopoda, 207, 224, 225
Chin, dimpled, inheritance of, 430
Chinch bug, *243*
Chironex, 123
Chi-square test of significance, 432-434
Chitin
    of arthropod, 207
    of crayfish and lobster, in exoskeleton, 214
Chloragogue cells of earthworm, 200
Chlorohydra viridissima, 124
Chloroplasts
    in euglena, 93, 94
    in *Volvox* cells, 94

Choanocytes of sponge, 113, 115, 116, 117
Chondrichthyes, 272, *291-297*
Chordata, 271-281
Chordates, 271
Chorioallontoic sac of pregnant pig, 359
Chorus frog, classification of, 79
Chromatids, 28
Chromatin granules
    in frog eggs, 55
    in sea star egg cell, 22
Chromatophores, 72, 188
    in fish scales, 305
Chromosomes, 26
    in anaphase, 29
    in first meiotic division, 49
    haploid number of, 45
    in meiosis, 46
    in metaphase, 29
    of *Paramecium*, 105
    in prophase, 28
Cicada, *242*
Cilia
    of ammocoetes, pharynx of, 287
    of *Chaetonotus*, 170
    of paramecium, on stained slide, 107
    of *Vorticella*, 108
Ciliary action
    of clam gill, effect of temperature on, 181
    of *Paramecium*, 105
    of snail intestine, 187
Ciliary feeder, 194
Ciliary movement
    in frog mouth cavity, 322
    of nephridium, 203
Ciliates, *107*, 108-109
    color in, 107
Ciliophora, 85-112
Ciona, 272
Circular muscles
    of earthworm, 201, 202
    of planarian, 148
Circulation, 385-402
    in ammocoete, *284-285*
    in cockroach, 234
    in fetal pig
        postcaval venous, 366
        pulmonary, 365
    in freshwater clam, 179
    in polychaete, 195
Circulatory system
    of amphioxus, 277
    of crayfish and lobster, 220
    of earthworm, 200
    of fetal pig, 360-370
    of frog, 325-329
    of grasshopper, 229
    of mollusc, open, 175
    of mussel, 179-180
    of sea squirt, 274
    of shark, 294-295
    of yellow perch, *304-305*
Circumpharyngeal connectives of earthworm, 201
Cirripedia, 222
Cirrus(i)
    of clamworm, 192
    of parapodium, 192
Cisternae, 26

Cladorcera, 221
Clam; *see also* Mussel
  freshwater, *176, 177, 178, 179*
  marine, siphon of, 176
Clamworm, 192, *193-195*
Classes, 77
  chief, of animal kingdom, key to, 78
Classification
  of animals, 77-81
  of insects, 236-247
Claws of grasshopper, 228
Cleavage, total, 53
Cleavage stage, 53
  of frog eggs, 55-58
Cliona, 113
Clitellum
  of earthworm, 196
  of leech, 204
Cloaca
  of *Ascaris*, 165
  of frog, 319-320, 321
  of sea cucumber, 268
  of shark, 294
Clonorchis, 143, *156*-158
Clymenella, 195
Clypeus of grasshopper, 226
Cnidaria, 123-139
Cnidarians, polymorphism among, 130
Cnidocils of hydra, 125
Cnidocytes
  of hydra, 124, 125-126
  of *Obelia*, 129
Coagulation time, 387
Cocacobrachialis muscle of fetal pig, 345
Coccidia, 101
Coccidians, 86
Coccidiosis, 103
Cockroach
  circulation in, 234
  digestive tract of, 403
  as host of *Gregarina*, 101
Cocoon
  of earthworm, 203
  observation of, 234
Coelom
  of earthworm, 198, 201
  of mussel, 179-180
  true, 163, 175
Coelomic cavity of mussel, 179
Coelomic cells, phagocytosis of, 256
Coenobita clypeatus, 224
Coenosarc of *Obelia*, 129
Coleoptera, 233, *244*
Collagen in *Ascaris*, 164
Collagenous fibers in section of skin of frog, 71
Collar of squid, 188
Collection
  of arachnids, 212
  of insects, 236-247
Collembola, 230, *246*
Collencytes of sponge, 117
Colloidal system, 34
Colloids, sol-gel transformation of, 37
Colony(ies)
  of coral, 138
  hydroid, 127-129
    behavior of, 128

Colony(ies)—cont'd
  of *Obelia*, preserved, 128
  phytomonids, 97
  of sea squirt, 274
  of *Volvox*, 94-95
    daughter, 94, 95
Colpoda, 86
Columella
  of coral, 138
  of snail, 186-187
Columnar epithelium
  in amphibian intestine, 72
  pseudostratified, 63
  simple, 62-63
Comb of honeybee, 228
Comb jelly, 139
Comb rows, 139
Compact bone, 66
Compound eyes
  of crayfish and lobster, 213, 214
  of grasshopper, 226
  of honeybee, 226
Conjugation
  of ciliates, 108
  of paramecium on stained slide, 107-108
Connective tissue
  adipose, 64, 65
  in amphibian intestine, 72
  areolar, 64, 65
  dense, 64
  fibrous
    in blood vessel, *73*
    in trachea cross section, 73
  loose, 64
  skin of frog, 71
  in trachea cross section, 73, *74*
Connectives, circumpharyngeal, of earthworm, 201
Contractile vacuole
  of ameba, 89, 90
  of *Paramecium*, 104-*105*, 107
  of *Vorticella*, 108
Contraction of glycerinated skeletal muscle, 380
Convex nose, inheritance of, 430
Copepoda, 222
Copulation, earthworm, 199, *200*
Copulatory organs in crayfish and lobster, 215
Copulatory spicules of vinegar eel, 167
Coral
  soft, 138, 139
  stony, 137, *138*, 139
  zoantharian, 139
Coral skeleton, hard, 138
Corallite, 138
Corallum of coral, 138
Corn-root aphid, 247
Corpora quadrigemina in fetal pig, 373
Corpus callosum in fetal pig, 373
Corpuscle(s)
  of frog, renal, 321
  red and white, 67, *68*
Corynactis, 139
Costal grooves, 78
Costocervical trunk in fetal pig, 362, 364-365
Cowper's glands of fetal pig, 357
Coxa of grasshopper, 228
Crab, 223
  hermit, 224

Crab—cont'd
  horseshoe, 208, *209-210*
Crane fly, *239*
Cranial crests, 78
Cranial nerves in fetal pig, 373-374
Craniata, 271
Craspedacusta, 129
Craspedacusta sowerbye, 130
Crayfish, 213, *214, 215*, 216-219, 220-221
Crest(s)
  cranial, 78
  olfactory, of squid, 188
Cricket frog, classification of, 79
Crickets, behavior of, 421-422
Crinoidea, 250, 269-270
Crinozoa, 250
Cristae, 25
Crop
  of bird, 355
  of earthworm, 200
  of grasshopper, 229
Crossed reflexes of ascidians, 274
Crossing over in meiosis, 46
Crusis muscle of frog, extensor, 316
Crustacea, 207, 213-224
Crustacean(s), 222
  biramous appendages of, 213
Crystalline style of mussel, 180
Crystalloid solution, 34
Ctenoid scales of yellow perch, 299
Ctenophora, 139-141
Cuboid epithelium
  simple, 62
  in skin of frog, 71
  in trachea cross section, 73
Cubozoa, 123
Cucumaria, 250, 267
Culture methods, 457-458
Curare, 381
Currents, water, and sponges, 119
Cutaneous maximus of fetal pig, 343
Cuticle
  of *Ascaris*, 164, 166
  of clamworm, 193
  of earthworm, 197, 201
  of *Philodina*, 170
Cyclops, 222
Cypridina, bioluminescence of, 224
Cysts
  in grasshopper testis, lobe of, 47
  of trichina worm, 169
Cytopharynx
  of *Paramecium*, 104, 107
  of *Vorticella*, 108
Cytoplasm
  of *Ascaris* egg, 30
  of epithelial cells, 21
  of sea star egg cell, 22
Cytoproct of ciliates, 106
Cytostome
  of *Paramecium*, 104, 107
  of *Vorticella*, 108

D

Daphnia, 124, 221, 223
Daughter colonies of *Volvox*, 94, 95
Decapoda, 223

Decerebrate frog, behavior of, 414
Dehybrid cross in fruit fly, 429
Deltoid muscle of fetal pig, 345
Demospongiae, 113, *118*
Dendraster, 264
Dendrites, 70
Dendrocoelopsis vaginatus, 144
Dendrocoelum, 144
Dentalium, 175
Dense connective tissue, 64
Denticles of dogfish shark, 292
Deposit feeder, 194
Depressant, effect of, on frog heart rate, 328
Dermal ostia of sponge, 115
Dermaptera, *244*
Dermis of sea star, 253
Dero, 202
Desmognathus fuscus, classification of, 79
Deuterostome group, 249-270
Dextral snail, 186
Diarthrosis, 340
Diastolic pressure, 388
Diblastic hydra, 127
Diemictylus viridescens, classification of, 79
Differentially permeable membrane, 35
Differentiation, cell, in *Volvox*, 94
Difflugia, 85, 90
Diffusion, 34-37
Digastric muscle of fetal pig, 344
Digenetic flukes, 156-158
Digestion, 403-412
  by hydra, 124-125
  by mussel, 180
Digestive glands
  of crayfish and lobster, 220
  of fetal pig, 353-355
  of mussel, 180
Digestive system
  of *Ascaris*, 165
  of *Clonorchis*, 156
  of crayfish and lobster, 220-221
  of earthworm, 199
  of fetal pig, 348-355
  of frog, 319-320
  of grasshopper, 229
  of mussel, 180
  of planarians, 147
  of sea cucumber, 268
  of sea squirt, 274
  of sea star, 253
  of shark, 292
  of yellow perch, 303
Digestive tract
  of cockroach, *403*
  of earthworm, 198
  of fetal pig, 352-355
Digits of frog, 308
Digital muscles of fetal pig, 346, 348
Dilator muscles of earthworm, 200
Dimpled chin, inheritance of, 430
Dinoflagellates, 97
Dioecious hydra, 127
Diopatra, 194
Diphyllobothrium, 143
Diphyllobothrium latum, 159
Diplogaster, 167
Diplopoda, 207, *225*

Diptera, 233, 234, *239*
Dipylidium caninum, 159
Direct reflexes in ascidians, 274
Dissection(s)
    animal, making, 1
    of bivalve, live, 181
    of crayfish and lobster, appendages of, 215-216
    of fetal pig, 348-349, *360*
        muscles of, 341-343
    of frog, 319
        muscles of, 313
        for pituitary gland collection, 58
    of sea star, 253, *254-255*
    of shark, 292-295
Diverticula in planarians, 147
Dividing stages of paramecium, 109
Division, cell, 21-32
    germ, maturation, 45-52
Dobson fly, *241*
Dogfish shark, *291-293*, 294-297
Dominance hierarchy in crickets, 421-422
Dorylaimus, 167
Dracunculus medinersis, 169
Dragonfly, *242*
Drone, honeybee, 229
Drosophila, 235, 428-430
    metamorphosis of, 229-234
Ductus arteriosus of fetal pig, 365
Dugesia, 144-149
Duodenum
    of shark, 292
    of yellow perch, 303
Dura mater in fetal pig, 373
Dusky salamander, classification of, 79
Dyad, chromosome, 46, 49

E
Ear lobes, free, inheritance of, 431
Earthworm, 196-*197*, *198, 199, 200, 201-203*
    habituation of, 420-421
    parasites of, 169
Earthworm copulation, 199, *200*
Earwig, *244*
Ecdysis, 208
Echinoderm, 241-270
Echinodermata, 53, 249-270
Echinoidea, 250, *261-262, 263-264*
Echinozoa, 250
Ecology, 445-449
Ectoderm, 53, 54, 57
Ectodermal epidermis of earthworm, 201
Ectoneural system of sea star, 254
Ectoplasm
    of ameba, 87
    of *Euglena*, 94
    of *Paramecium*, 104
    of *Vorticella*, 108
Eel, vinegar, 167
Eggs; *see also Ova*
    of *Ascaris*, 165
        fertilized
            maturation of, 48
            mitosis in, 30
    of earthworm, 197, 199
    of fluke, 157
    of frog, 321

Eggs—cont'd
    of frog—cont'd
        in cleavage stages, 55-57
        in ovary, 55-59
        poles of, 55
    of horseshoe crab, 208
    of insect, 234-235
    isolecithal-type, 53
    of *M. hirudinaceus*, 171
    mature, 49
    of nematomorphs, 171
    of sea star, unfertilized, *22*
    of snail, 187
    of tapeworm, 160
    telolecithal-type, 55
    of tunicates, 274
Eimeria, 101
Eimeria stiedae, 103
Eimeria tenella, 103
Eisenia, 420
Eisenia foetida, 203
Ejaculatory duct in *Ascaris*, 165
Ejaculatory tube of grasshopper, 229
Elastic cartilage, 64
Elastic fibers
    in blood vessels, 73
    in section of skin of frog, 71
Elliptio, 176
Elytron of insects, 239
Embryo, whitefish, mitosis in, *29*
Embryology, 45-59
    of sea star, *54*
        early, 53-55
    of sponge, 119
Emulsion, 34
Endocrine system of frog, 333
Endoderm, 53, 54, 57
Endoplasm
    of ameba, 87
    of *Paramecium*, 104
Endoplasmic reticulum, 25
    of liver cell, 23
Endopods in crayfish and lobster, 215
Endoskeleton
    of echinoderm, 249
    of sea cucumber, 268
    of sea star, 253
    of sea urchin, 263
Endostyle
    of ammocoete, 284
    of amphioxus, 277
    of sea squirt, 274
Entamoeba, 89, 90
Enterobius vermicularis, *168*, 169
Entosphenus, 283
Enzyme(s)
    action of, 43-44
    digestive, 403-408
    of spider, 210
Enzyme activity, effect of pH on, 44
Epaxial muscles of yellow perch, 300
Ephemeroptera, 230, *240*
Ephyrae, 134
Epiboly, 57
Epicranium of grasshopper, 226

Epidermis
   of *Ascaris*, 166
   of comb jelly, 141
   of crayfish and lobster, 216
   of earthworm, 198, 201
   of hydra on stained slide, 127
   of *Obelia*, 129
   of planarian, 148
   of sea urchin, 263
Epidinium, 86
Epiglottis of fetal pig, 352
Epigynum of spider, 211
Epimorphosis, 230
Epithelial cells, squamous, 21-22
   stained, 22-26
Epithelial tissue, 61-*62*, 63-64
Epitheliomuscular cells of hydra, 127
Epithelium, 61-*63*, *64*
   columnar, in amphibian intestine, *72*
   cuboid
     in section of skin of frog, 71
     in trachea cross section, 73
   pseudostratified, in trachea cross section, 73, *74*
   simple squamous, in blood vessel, *73*
   squamous, in amphibian intestine, *72*
   stratified squamous, in section of skin of frog, 71, *72*
ER; *see* Endoplasmic reticulum
Erythrocytes, 67, *68*
   frog, 395
Esophageal bulb of vinegar eel, 167
Esophageal ganglion of crayfish and lobster, 221
Esophagus
   of ammocoete, 284
   of *Clonorchis*, 156
   of crayfish and lobster, 220
   of earthworm, 200
   of fetal pig, 353
   of grasshopper, 229
   of mussel, 180
   of shark, 292
   of vinegar eel, 167
   of yellow perch, 303
Eubranchipus, 221, 223
Eucoelomate, protostome, group, 175-247
Eudorina, 97
Euglena, 85, 93
   stained slide of, 94
Euglenoid, 93
Euglenoid movement, 93
Euplates, 105
Euplectella, 113, *118*
Eurycea bislineata, classification of, 79
Eurycea longicauda longicauda, 77
   classification of, 79
Eurypterida, 207
Eustachian tube of frog, 319
Evisceration in sea cucumber, 269
Evolution of arthropod, 215
Excretion, 403-412
   by planarian, 148
   by sea squirt, 274
Excretory canal(s)
   of *Ascaris*, 166
   of planarian, 148
   of tapeworm, 159, 160

Excretory pore
   of *Ascaris*, 165
   of *Clonorchis*, 156
Excretory system
   of acoelomate, 143
   of *Ascaris*, 165
   of crayfish and lobster, 221
   of earthworm, 200-201
   of frog, 321
   of mussel, 180
   of yellow perch, 304
Excretory vesicle of *Clonorchis*, 156
Excurrent aperture in mussel, 178
Exopods in crayfish and lobster, 215
Exoskeleton
   of arthropod, 207
   of bivalve, 176
   of crayfish and lobster, 214, 216
   of grasshopper, 226
   of horseshoe crab, 208
   of mollusc, 175
   of spider, 210
Extension in mammal, 341
Extensor muscles
   of crayfish and lobster, 219
   of fetal pig
     carpi obliquus, 346
     carpi radialis, 345
     digital, of hindlimb, 348
   of frog, crusis, 316
Extracellular digestion
   by hydra, 125
   by planaria, 147
Eyes
   of ammocoete, 284
   of clamworm, 192
   color of, inheritance of, 431
   of crayfish and lobster, 213, 214
   of frog, development of, 57
   of garden spider, 210, 211
   of grasshopper, compound, 226
   of honeybee, compound, 226
   of horseshoe crab, 208
   of planarian, 145
   of snail, 185, 187
   of squid, 188
   of tadpole, development of, 58
Eyespot of sea star, 252, 255

F

Face of fetal pig, muscles of, *343-345*
Facets of eye in crayfish and lobster, 214
Fairy shrimp, 221, 223
Fallopian tubes of fetal pig, 358
Families, 77
Fang
   of centipede, poison, 224
   of spider, terminal, 211
Fascia in mammal, 340-341
Fasciola, 143
Fasciolaria, 187
Fat body in grasshopper, 229
Fat tissue, 64, *65*
Feather star, *269-270*
Feeder
   ciliary and deposit, 194

Feeder—cont'd
  filter, 176, 194
    cephalochordate as, 277
Feeding
  by ameba, 88-89
  by brittle star, 260
  by ciliate, 106
  by clam, 180
  by euglena, 94
  by hydra, 124-125
Feeding mechanism of freshwater clam, *178*
Feeding reaction of frog, 308
Feet, tube
  of brittle star, 259
  of sea urchin, 262-263
Felis, 77
Femoral arteries of fetal pig, 367
Femoral veins of fetal pig, 366
Femur of grasshopper, 228
Fertilization in *Ascaris*, 48
Fertilized *Ascaris* egg, mitosis in, 30
Fertilized undivided ovum, 53
Fetal pig, 335-375
Fetal sac in pregnant pig, 359
Fibrocartilage, 64
Fibrous connective tissue
  in blood vessels, *73*
  in trachea cross section, *73*
Field work, 445-449
Filter feeder, 176, 194
  cephalochordate as, 277
  clam as, 180
Filtration, glomerular, in mudpuppy, 409-412
Fin
  of amphioxus, 276
  of dogfish shark, 291
  of squid, lateral, 188
  of yellow perch, 299
Fin rays of yellow perch, 299
Finger, little, bent, inheritance of, 431
Firefly luminescence, 235
Fish
  bony, 299-305
  telecost, skeletal musculature of, *300*
Fission
  binary, of paramecium, 107
  longitudinal, of euglena, 94
Fixation of insects, 236
Flagella in *Volvox* cells, 94
Flagellate, 93
Flagellate protozoan, *96*
Flagellum
  of *Euglena*, 93
  of trypanosome, 97
Flame bulbs, ciliated, of planarian, 148
Flame cells of planarian, 149
Flame cells in pseudocoelomate, 165
Flatworm, 143
Flea, *245*
  water, 221
Flexion in mammal, 341
Flexor muscles
  of crayfish and lobster, 220
  of fetal pig
    carpi radialis, 345
    carpi ulnaris, 346
    deep digital, 348

Florometra, 250, 269
Floscularia, 170
Fluke, 143
  bladder, 157
  digenetic, 156-158
  liver, of human, 156-158
  living, observation of, 157-158
  lung, 157
Fly, *227, 239, 240, 241*
  fruit
    inheritance of, 428-430
    metamorphosis, vial for, 233
  tsetse, trypanosome transmitted by, 97
Follicular cell layers of frog eggs, 55
Food vacuoles
  of ameba, 88
  of ciliate, 106
  of *Vorticella*, 108
Foot
  of clam, 176
  of mollusc, 175
  of mussel, 178
  of *Philodina*, 170
  of snail, 185, 187
Foramen ovale of fetal pig, 365
Foraminiferan, 85, 90
Forebrain
  of ammocoete, 284
  of frog, 332
Foreleg
  of honeybee, 228
  muscles of, 345-346
Forelimb
  of fetal pig
    arteries of, 362-365
    veins of, 362
  of frog, 313
Fowler's toad, classification of, 79
Free ear lobes, inheritance of, 431
Free-living nematodes, 167
Freeze-fracture preparation, 15, 16, *24*
  of liver cells, 23-24
Freshwater clam, *176, 177, 178, 179*
Freshwater jellyfish, 130
Freshwater mussel, 176-184
Freshwater oligochaete, *202*
Freshwater sponge, 118
Frog, 303-333
  behavior of, 414-415
  capillary circulation in, 395-396
  classification of, 79
  development of, 55-59
  egg of
    in cleavage stages, 55-57
    in ovary, 55-59
    poles of, 55
  pithed leopard, 157
  pithing of, 462-463
  pituitary gland of, obtaining, 58
  skin of, section through, 71-72
  trematode as parasite of, *158*
Frog board, 463-464
Frog nerve-muscle preparation, 463
Frog ova, 55
Frons of grasshopper, 226
Fruit fly
  inheritance in, 428-430

Fruit fly—cont'd
    metamorphosis of, vial for, *233*
Fumigant, using safely, 236
Funnel
    Berlese, *212, 235*
    of squid, 188
        retractor muscles of, 189

G

Gallbladder of yellow perch, 303
Gametocyst of *Gregarina*, 101
Gametogenesis, 45-59
Gammarus, 223
Ganglion(a)
    of crayfish and lobster, 221
    of earthworm, 200-201
    of fetal pig, 372
    of mussel, 180-181
Garden spider, *210, 211-212*
Gastric artery of fetal pig, 367
Gastric ceca of grasshopper, 229
Gastric filament of *Aurelia*, 134
Gastric mill of stomach of crayfish and lobster, 221
Gastric muscle of crayfish and lobster, 219
Gastric pouches of *Aurelia*, 134
Gastrocnemius muscle
    of fetal pig, 348
    of frog, 315
Gastrodermis
    of comb jelly, 141
    of hydra on stained slide, 127
    of *Obelia*, 129
Gastrohepatic artery of fetal pig, 367
Gastroliths in crayfish and lobster, 220
Gastropoda, 175, 188-187
Gastrotricha, *170*-171
Gastrovascular cavity
    of comb jelly, 141
    of hydra, 125
    of *Obelia*, 129
    of planarian, 147
    of radiate, 123
    of sea anemone, 136
Gastrovascular system of *Aurelia*, 134
Gastrozooids of Portuguese man-of-war, 130
Gastrula stage, 53, 54
    of frog eggs, 57
Gelation, 37
Gemmules of sponge, 117
    freshwater, 119
Gena of grasshopper, 226
Gene frequencies, calculation, 432
Genetics, 427-443
Geniohyoid muscle of fetal pig, 344
Genital aperture of snail, 187
Genital opening of crayfish and lobster, 214
Genital operculum of horseshoe crab, 208
Genital plates of sea urchin, 263
Genital pore(s)
    of crayfish and lobster, 215
    of fluke, 157
    of horseshoe crab, 208
    of leech, 204
    of tapeworm, 159
Genital tube of vinegar eel, 167
Genus, 77
Geotaxis, 106
    effect of, on snail movement, 186

Germ cells
    maturation division of, 45-52
    primordial, 45
Germ layers
    in frog egg, 57
    in gastrula, 53-54
    in radiate, 123
Gill
    of clam, effect of temperature on ciliary action of, 181
    of crayfish and lobster, 216
    of frog, development of, 57
    of horseshoe crab, book, 208
    of mollusc, 175
    of mussel, 178, *179*
    of sea star, skin, 251, 252
    of sea urchin, 263
    of squid, 189
    of tadpole, 58
    of yellow perch, 301
Gill chamber of crayfish and lobster, 214
Gill slits
    of dogfish shark, 291
    pharyngeal, 271
Gill unit of shark, 295
Gizzard of earthworm, 200
Gland
    of ammocoete, subpharyngeal, 284, 287
    of crayfish
        antennal, 223-224
        digestive, 220
    digestive, 180
    of fetal pig
        digestive, 353-355
        pituitary, 373
        prostate and Cowper's, 357
        salivary, 351
        thymus, 350
        thyroid, 351
    of fluke
        Mehlis', 157, 158
        vitelline, 157
    of frog
        pituitary, obtaining, 58
        poison, 71
    parotoid, 78
    of snail, mucous, 187
    of tapeworm
        Mehlis', 159
        vitelline, 159
Gland cells of hydra, 127
Glass sponge, 117-*118*
Glochidia larvae, 181
Glochidium of mussel, 180
Glomerular filtration in mudpuppy, 409-412
Glomerulus, 409
Glottis of frog, 320
Glutathione, 125
Gluteus muscle
    of fetal pig, medius, 347
    of frog, 315
Glycera, 195
Glycogen granules of liver cells, 23
Gnathobase of horseshoe crab, 208
Gnathostomata, 272
Goblet cells, *63*
    in trachea cross section, 73, *74*
Golgi complex, 25, 26

Gonads
    of *Aurelia*, 134
    of crayfish and lobster, 220
    of grasshopper, 229
    of hydra, 124
    of jellyfish, 130
    of mussel, 180
    of sea anemone, 137
    of sea cucumber, 268
    of sea star, 253-254
Gonangia of *Obelia*, 127, 128, 129
Gonidia of *Volvox* colony, 95
Gonionemus, 123, 124, 129-130, 134
Gonium, 97
Gonophores of Portuguese man-of-war, 130
Gonopore
    of centipede, 224
    of millipede, 225
    of *Obelia*, 129
Gonotheca of *Obelia*, 129
Gontonemus, 133
Gonyaulox, 97
Gordius, 171
Gorgodera, 157
Gorgoderina, 157
Gorgonia, 123, 138
Gorgonocephalus, 250, *259*
Gracilis muscle
    of fetal pig, 347
    of frog, 314
Gradual metamorphosis, 230
Granules
    in frog eggs, chromatin, 55
    of *Noctiluca*, luminescent, 97
Grasshopper
    as host of *Gregarina*, 101
    lubber, *226-229*
    structure of, *230*
    testis of, spermatogenesis in, *47-48*
Gravid proglottids of tapeworm, 159, 160
Gravity, effect of, on snail movement, 186
Gray treefrog, classification of, 80
Green frog, classification of, 80
Green glands of crayfish and lobster, 221
Gregarina, *101*-102
Gregarines, 86
Ground substance in epithelial cells, 21
Growth period of oogenesis, 48
Gryllidae, 421
Gryllus, 421
Gullet of *Paramecium*, 104
Gymnodinium, 97
Gyrinophilus parphyriticus, classification of, 79
Gyrodactucas, 143

H

Habituation in earthworm, 420-421
Haematoloechus, 157
Hairs
    of crayfish and lobster, 221
    of spider, sensory, 210
Halocynthia, 273
Haltere of insect, 239
Haploid number of chromosomes, 45
Hard coral skeleton, 138
Hardy-Weinberg formula, 432

Haversian systems, 66, 67
Head
    of clamworm, 192
    of crayfish and lobster, 214
    of fetal pig
        arteries of, 362-365
        muscles, long, lateral and medial of, 345
        veins of, 362
    of grasshopper, 226
    of honeybee, 226-227
    of snail, 185, 187
    of squid, 188
        retractor muscles of, 189
Heart(s)
    of ammocoete, 284
    of frog, *327*
    of grasshopper, 229
    of mammal, 361-362
    of mollusc, 175
    of mussel, 179
    of pig, *368-369*
    of sea squirt, 274
    of squid, 189
    of yellow perch, 304
Heart rate of frog, effect of stimulants and depressants on, 328
Heart urchin, 262
Heartbeat
    of *Daphnia*, 223
    of frog, observing, 328
Hectocotyly, 188
Helicodiscus, 186
Helix, 175, 185
Hematocrit, 388
Hemimetabola, 230
Hemiptera, 230, *243*, 245
Hemispheres of frog eggs, 57
Hemocoel of grasshopper, 229
Hemocyanin in horseshoe crab, 208
Hemoglobin content, 386-387
Hemolymph in crayfish and lobster, 220
Hepatic portal system
    of fetal pig, 366, *367*
    of frog, 226
Hepatic portal vein
    of ammocoete, 285
    of fetal pig, 366
    of shark, 294-295
Hepatopancreas in crayfish and lobster, 219, 220
Heredity, 427
Hermit crab, 224
Heterodora, 167
Heteromeyenia, 118
Hindbrain
    of ammocoete, 284, 286
    of frog, 332
Hindlegs
    of fetal pig
        arteries of, 367-368
        muscles of, 347-349
    of honeybee, 228
Hindlimbs of frog, 313
Hinge ligament of mussel, 177
Hip of fetal pig, muscles of, 346-347
Hirudinea, 191, 204
Hirudo, 191
Hirudo medicinalis, 204

Histology, 61
    of earthworm cross section, 201-220
    of frog, urogenital organs of, 321-322
Hogs, trichina worms in, 168
Holoblastic cleavage, 53
Holometabola, 230-233
Holophytic nutrition by euglena, 94
Holothuroidea, 250, 267-269
Homalozoa, 250
Homarus, 207
Homarus americanus, 213
Homo sapiens, 77
Homologs, 46
Homology, serial, in crayfish appendages, 215
Homoptera, *242, 247*
Honeybee, 225-227, *228-229*
Hooks of tapeworm, 159
Hookworm, 167-168
Horse, parasites of, intestinal, 164
Horsehair worms, 171
Horseshoe crab, 208, *209-210*
Human
    blood of, 385-389
    genetic variability in, 434
    liver fluke of, 156-158
    parasites of, intestinal, 164
    skeleton of, *339*
    trichina worm in, 168
Human inheritance, 430-434
Hump, visceral, of snail, 187
Hyaline cap of ameba, 87
Hyaline cartilage, 64, 65
Hyalonema, 113
Hyalospongiae, 113
Hydra, 123, 124, *125, 126,* 127
Hydranths of *Obelia*, 127, 128
Hydrocaulus of *Obelia*, 128
Hydroid
    colony, 127-129
        behavior of, 128
    solitary, 124-127
Hydromedusa, 129-130
Hydrorhiza of *Obelia*, 128
Hydrotheca of *Obelia*, 129
Hydrozoa, 123, 124-132
Hydrozoan medusae, 133
Hyla, classification of, 80
Hylidae, classification of, 79
Hymenoptera, *241, 245*
Hypaxial muscles of yellow perch, 300
Hypermobility of thumb joint, inheritance of, 431
Hypnotoxin of hydra, 126
Hypophysis of fetal pig, 373
Hypostome
    of hydra, 124, 125
    of *Obelia*, 129

I

Ichthyomyzon, 283
Iliac arteries of fetal pig, 367
Iliac veins of fetal pig, 366
Iliacus muscle of fetal pig, 347
Iliofibularis muscle of frog, 315
Immature proglottids, 159
Inclusions in epithelial cells, 21
Incurrent aperture in mussel, 178

Incurrent canals of sponge, 115
Infection by parasitic nematode, 167
Infraspinatus muscle of fetal pig, 345
Infundibular cartilages of squid, 188
Inheritance
    in fruit fly, 428-430
    human, 430-434
Ink sac of squid, 189
Insect(s), 225-247
Insecta, 207
Instar, 230
Interphase of mitosis, 27-28
Interstitial cells of hydra, 127
Intestinal roundworm, *164*-166
Intestine
    of amphioxus, 277
    of *Ascaris*, 165, 166
    of crayfish and lobster, 220
    of earthworm, 200
    of fetal pig, 349-350, 353
    of grasshopper, 229
    of mussel, 180
    of sea star, 253
    small, amphibian, cross section of, 72
    of vinegar eel, 167
    of yellow perch, 303
Intracellular digestion
    by hydra, 125
    by planarian, 147
Invertebrate blood smears, 203
Involuntary muscle; *see* Smooth muscle
Isolecithal-type egg, 53
Isopoda, 223
Isoptera, 230, *242*

J

Jaws of clamworm, 192
Jelly, comb, 139
Jelly polyps, 130
Jellyfish, 127-128
    freshwater, 130
    scyphozoan, 135
    true, 133-135
Jointed appendages of arthropods, 207
Jugular veins of fetal pig, 362
Julus, 225

K

KAAD to fix insects, 236
Kerona, 124
Kidney, 409
    of ammocoete, 284
    examination of slide of, 73
    of fetal pig, 350, 355, *356*
    of frog, 321
    of mussel, 180
    of shark, 294
    of yellow perch, 304
Kingdom, animal, key to chief phyla and classes of, 78

L

Labial palps
    of honeybee, 226
    of mussel, 178, 180
Labium
    of grasshopper, 226

Labium—cont'd
  of honeybee, 226
  of millipede, 225
Labrum
  of centipede, 224
  of grasshopper, 226
  of millipede, 225
Lacuna, 64, 66
Ladybird beetle, *244*
Lamella(e), 66
  of horseshoe crab, 208
  of mussel gills, 178, 179
  of shark, 295
Lamprey, 283-290
Lampsilis, 176
Land snail, *185-187*
Larva(e)
  Amaroucium, 275
  ammocoete, 283, *284, 285*, 286-287
  ascidian, 274-275
  of *Aurelia*, planula, 134
  of echinoderm, *249*
  of fruit fly, 233
  glochidia, 181
  of horseshoe crab, 210
  of *Obelia*, planula, 127
  of sponges, 117
    amphiblastula, 115
  of trichina worm, 169
Larval stages of brine shrimp, 223
Larval threads, 181
Lateroneural vessel of earthworm, 200
Latissimus dorsi muscle of fetal pig, 345
Laurer's canal in flukes, 157
Learning ability in earthworm, 420-421
Leech, 204
Legs
  of crayfish and lobster, 213, 214, 215
  of grasshopper, 227, 228
  of honeybee, 227, 228
  of spider, 211
Leishmania, 85
Leodia, 264
Leopard frog
  classification of, 80
  pithed, 157
Lepidoptera, 233, *239*
Lepus, 222
Leuconoid canal system of sponge, 118
Leucosolenia, 113, 118
Leukocytes, 67, *68*
Life cycle
  of cecropia and polyphemus moth, 234
  of fluke, 156, 157
  of *Gregarina*, 101-102
  of horseshoe crab, 208
  of *Obelia*, 127-128
Ligament, hinge, of mussel, 177
Limax, 187
Limifosser, 175
Limulus, 148, 207, 208, *209-210*
Linnaean system of binomial nomenclature, 77
Lipasis, 405
Lips of Ascaris, 164
Lithobius, 207, 224, *225*

Liver
  cells of, *15*, 22, 23-25
    fine structure of, 23-26
  of crayfish and lobster, 220
  of fetal pig, 353
  of rat, 16
  of shark, 292
  of yellow perch, 303
Lizard, spinal cord of, *71*
Lobopodia of ameba, 87
Lobster, 213-221
Locomotion
  of ameba, 87-88
  of annelid, 191
  of brittle star, 260
  of clamworm, 192
  of earthworm, 196
  of *Euglena*, 93
  of frog, 308
  of leech, 204
  of paramecium, 104
  of planarian, 146
  of sea urchin, 263
  in *Volvox* colonies, 94
Loligo, 175, 188-190
Long bone, *339*
  longitudinal section of, examination of slide of, 73
Long head muscle of fetal pig, 345
Long monaxons of sponge, 117
Longissimus dorsi muscle of fetal pig, 346-347
Longitudinal fission by euglena, 94
Longitudinal muscle cells of *Ascaris*, 165
Longitudinal muscle layer of earthworm, 201, 202
Longitudinal muscles of planarian, 148
Long-tailed salamander, classification of, 79
Loose connective tissue, 64
Lorica of rotifer, 170
Lubber grasshopper, *226-229*
Lumbar arteries of fetal pig, 367
Lumbar ganglion in fetal pig, 372
Lumbar nerve of fetal pig, 370
Lumbar vein in fetal pig, 366
Lumbosacral plexus nerve of fetal pig, 370-372
Lumbricus, 191, *196-197, 198, 199, 200*, 201-203
Lumbricus terrestris, 420
Lumen of intestine, frog, 72
Luminescence, firefly, 235
Luminescent granules of *Noctiluca*, 97
Lung flukes, 157
Lung of snail, 187
Lymph, 67
Lymphatic system of frog, 327
Lytechinus, 250, 261

M

Macracanthorhynchus hirudinaceus, 171
Macrogametes of *Volvox* colonies, 95
Macromeres, 57
Macronucleus
  of *Paramecium*, 105
    on stained slide, 107
  of *Stentor*, 108
  of *Vorticella*, 108
Macrophage, 17
Madreporite plate
  of brittle star, 259

Madreporite plate—cont'd
  of sea star, 255
  of sea urchin, 263
Maggot, 223
Malacostraca, 223
Malaria, 102
Malaria parasites, 86
  ring stage of, 102, *103*
Malpighian tubules of grasshopper, 229
Mammal respiration, 399-400
Mammalia, 272, 335-375
Mandible(s)
  of arthropod, 213
  of centipede, 224
  of crayfish and lobster, 216
  of grasshopper, 226
  of honeybee, 226
  of millipede, 225
Mandibular gland of fetal pig, 351
Mandibular muscles of crayfish and lobster, 219
Mandibulata, 207
Mandibulate, 207
  aquatic, 213-224
  terrestrial, 224-235
Mantle
  of bivalve, 176
  of mollusc, 175
  of mussel, 177-178
  of sea squirt, 272
  of snail, 187
  of squid, 188
Mantle cavity
  of mollusc, 175
  of mussel, 178
  of squid, 189-*190*
Manubrium of jellyfish, 130
Marbled salamander, classification of, 79
Marine clams, siphons of, 176
Marrow of bone, 339
  cavity, 66
Masseter muscle of fetal pig, 344
Mastax of Philodina, 170
Mastigophora, 85, 93-100
Mastigophoran, 93
Mating reaction of ciliate, 108
Maturation division of germ cells, 45-52
Maturation of fertilized *Ascaris* egg, *48*
Maturation period of oogenesis, 48
Maturation process, 45
Mature ovum, *49*
Mature proglottids, 159-160
Maxilla(e)
  of arthropod, 213
  of centipede, 224
  of crayfish and lobster, 216
  of grasshopper, 226
  of honeybee, 226
  of millipede, 225
Maxillipeds
  of centipede, 224
  of crayfish and lobster, 214, 215-216
Mayfly, *240*
Mealworm as host of *Gregarina*, 101
Meandrina, 138
Mecoptera, *241*
Mediastinum of fetal pig, 350

Medulla oblongata in fetal pig, 373
Medusa, 123, 127-128, 129, 133
  budding, 134
Mehlis' gland
  of fluke, 157
  of tapeworm, 159
Meiosis, 26, 45-52
Mellita, 250, 264
Membrane; *see* specific membrane
Mercenaria, 176
Merlia, 113
Merostomata, 207
Merozoites of *Plasmodium*, 102, 103
Mesenteric artery, anterior, of fetal pig, 367
Mesenteric ganglion, posterior, in fetal pig, 372
Mesentery
  examination of slide of, 73
  of sea anemone
    incomplete, 137
    primary, 136
Mesoderm, 54, 57
Mesodermal of earthworm, 201
Mesodon, 186
Mesoglea
  of hydra on stained slide, 127
  of *Obelia*, 129
Mesohyl of sponge, 115
Mesonephroi of yellow perch, 304
Mesothorax, 227
Metabolic rate, 399
Metamerism, 191
Metamorphosis, 58
  of *Drosophila*, 229-*233*, 234
Metanauplius, 223
Metaphase of mitosis, 28-29
Metathorax, 227
Metazoan, cells of, 21
Methyl salicylate, action of, 396
Metridium, 123, 137, 138
Microaquarium, 109
Microgametes of *Volvox* colonies, 95
Micromere, 57
Micronucleus
  of *Paramecium*, 105
    on stained slide, 107
  of *Vorticella*, 108
Microscope, 7-20
Microspora, 86
Microstomum, 149
Microtriches of *Clonorchis*, 156
Midbrain
  of fetal pig, 373
  of frog, 332
Millipede, 225
Miracidia of fluke, 157
Mitochondria, *25*
  of liver cell, 23
Mitosis, 26, 27-30
Mnemiopsis, 139
Molecules, motion of, 34
Molgula, 271, 272-*273*, 274-275
Mollusc, 175-190
Mollusca, 175-190
Monaxons of sponge, 117
Monhystera, 167
Monochus, 167

Monocystis, 103
Monoecious hydra, 127
Monogenea, 143
Monohybrid traits, 430-431
Monoplacophora, 175
Monostyla, 170
Mopalia, 175
Morula stage, 53
  of frog eggs, 57
Mosquito, 227
  Anopheles, 102, 103
Moth, polyphemus, 234
Motor end plates, examination of slide of, 73
Motor nerve endings on skeletal muscle, examination of slide of,
    73
Mouth
  of *Ascaris*, 164
  of clamworm, 192
  of *Clonorchis*, 156
  of earthworm, 196, 199
  of hydra, 124, 125
  of leech, 204
  of mussel, 178, 180
  of *Obelia*, 129
  of *Paramecium*, 104
  of planarian, 147
  of sea anemone, 136
  of sea star, 253
  of sea urchin, 263
  of snail, 187
  of squid, 188
  of tadpole, 58
  of vinegar eel, 167
  of yellow perch, 299, 301
Mouth cavity
  of fetal pig, 351-352
  of frog, ciliary movement in, 322
Mouthparts of frog, *313*
Mucous gland of snail, 187
Mud snail as host of circariae, 157
Mudpuppy
  classification of, 79
  glomerular filtration in, 409-412
  incisions in, *410*
Muscle(s)
  of arthropod, striated, 207
  of crayfish and lobster, 219, 220
  of earthworm, 198, 201, 202
    dilator, 200
  of fetal pig, superficial, *342*
  of mussel, 177, 178
  palmar, long, inheritance of, 431
  of planarian, 148
  of sea cucumber, 268
  skeletal, *68*, 69
    of frog, 313-*314*, *315*-316
    motor nerve endings on, examination of slide of, 73
  smooth, *68*, 69
    in amphibian intestine, *72*
    in blood vessel, *73*
    in section of skin of frog, 71
    in trachea cross section, 73
  of squid, 189
Muscle cells
  of *Ascaris*, longitudinal, 165
  smooth, *25*
    in mitosis, *28*

Muscle layers of *Clonorchis*, 156
Muscle physiology, 379-384
Muscle tissue, *68*, 69
Muscle-nerve preparation, frog, 463
Muscular system
  of crayfish and lobster, 219-220
  of fetal pig, 340-348
  of yellow perch, *300, 301*
Muscular-nutritive cells of hydra, 127
Mussel, freshwater, 176-184
Myelin sheath, 70
Mylohyoid muscle of fetal pig, 344
Myofibrils, 380
Myoneural junction, physiology of, 381-382
Myosin filaments in *A. proteus*, 87
Myotomes of yellow perch, 300
Myriapods, 224-225
Mysis, 223
Mytilus, 176
Myxini, 272
Myxosoma, 86
Myxozoa, 86

N
Nacre of snail shell, 186
Nacreous layer of mussel, 177
Naiad, 230
Narcotizing methods, 456-457
Nares, external, of tadpole, 58
Nasolabial groove, 78
Nauplius, 223
Neck
  of fetal pig, muscles of, *343*-345
  of tapeworm, 159
Neck region of fetal pig, 350-351
Nectophores of Portuguese man-of-war, 130
Nector americanus, 167-*168*
Necturus, 72, 410
Necturus maculosus, classification of, 79
Nematocysts, 139
  of *Aurelia*, 134
  of coral, 138
  of hydra, 124, 125, 127
  of *Obelia*, 129
Nematoda, 157, 164-169
Nematode
  free-living, 167
  parasitic, 167-169
  soil, 167
Nematomorpha, 171
Nemertina, 143
Neomenia, 175
Neopilina, 175
Nephridia of earthworm, 198, 200
Nephridial canals of *Paramecium*, 104-105
Nephridiopores
  of earthworm, 197, 200
  of leech, 204
Nephridium, 203
Nephron, 409, *410*
Nephrostome of earthworm, 200, 203
Nereis, 191, 192-*193*, 194-195, 197, 225
Nerve cell body, 70
Nerve cord(s)
  of *Ascaris*, 166
  of earthworm, ventral, 201, 202
  of tapeworm, 160

Nerve cord(s)—cont'd
  tubular, dorsal, 271
Nerve endings, motor, on skeletal muscle, examination of slide of, 73
Nerve fibers, 70
  in blood vessel, 73
Nerve net of hydra, 127
Nerve processes, 70
Nerve ring of sea star, 255
Nerve-muscle preparation, frog, 463
Nerves
  of *Bdelloura*
    lateral, 148
    transverse, 148
  of earthworm, 200
  of sea star, radial, 255
  section through, 72-73
Nervous system, 413-418
  of ascidian larvae, 275
  of crayfish and lobster, 221
  of earthworm, 200-201
  of fetal pig, 370-*371*, 372-374
  of frog, *331*-333
  of grasshopper, 229
  of mussel, 180-181
  of planarian, 148
  of sea star, 254-255
Nervous tissue, 69-71
Neural tube stage of frog egg, 57-58
Neurals, of earthworm, lateral, 202
Neurolemma, 70
Neuromuscular system of sea squirt, 272-274
Neuron, *70*
Neuropodium of parapodium, 192
Newt, classification of, 79
Nickel sulfate for quieting paramecia, 108
Noctiluca, 97
Nomenclature, binomial and trinomial, 77
Nonstriated muscle; *see* Smooth muscle
Nose, convex, inheritance of, 430
Nosema, 86
Notochord, *271*
  of amphioxus, 277
  of frog egg, 57
Notopodium of parapodium, 192
Nuclear envelope, *24*
  of egg cell of sea star, 22
Nucleolus, 25
  of liver cell, 23
  of sea star egg cell, 22
Nucleoplasm of liver cell, 23-25
Nucleus(i)
  of ameba, 89
  of cells
    epithelial, 21
    liver, 23-25
    of sea star egg, 22
  of egg, *Ascaris*, 30
  of *Paramecium*, 105
  of spermatozoa, 48
  of *Vorticella*, 108
  of zygote, 49
Nudibranch snail, 187
Nutrition, holophytic, by euglena, 94
Nutritive-muscular cells of hydra, 127
Nymph, 230
  of grasshopper, 229

O
Obelia, 123, 124, 127-129
Oblique muscles
  of fetal pig, 346
  of frog, external, 313
Obliquus muscle of fetal pig, extensor carpi, 346
Ocelli
  of centipede, 224
  of grasshopper, 226
  of honeybee, 226
  of millipede, 225
  of scallop, 181
Octopus, 175
Odonata, 230, *242*
Olfactory crest of squid, 188
Olfactory lobes in fetal pig, 373
Oligochaeta, 191, 196-203
Oligochaetes, freshwater, *202*
Onchosphere of tapeworm, 160
Oocyte
  primary, 48
  secondary, *49*
Oogenesis in *Ascaris*, 48
Opalina, 85
Opalinata, 85
Operculum
  of horseshoe crab, genital, 298
  of tadpole, development of, 58
  of yellow perch, 299
Ophioderma, 259, *260*
Ophiopholis, 259
Ophioplocus, 259
Ophiopluteus, *249*
Ophiothrix, 259
Ophiura, 250
Ophiura albida, 259
Ophiuroidea, 250, *259-260*
Ophthalmic artery of crayfish and lobster, 220
Opisthaptor of bladder fluke, 157
Opisthoma of horseshoe crab, 208
Optic chiasma in fetal pig, 373
Oral arms of *Aurelia*, 134
Oral disc
  of sea anemone, 136
  of *Vorticella*, 108
Oral groove
  of ciliate, 106
  of *Paramecium*, 104
    on stained slide, 107
Oral hood of amphioxus, 277
Oral plate in frog, development of, 57
Oral shields of brittle star, 259
Oral sucker of *Clonorchis*, 156
Orb web of garden spider, 210
Orconectes, 213
Orders, 77
Organelles of epithelial cells, 21
Orthoptera, 230, 422
Osculum of sponge, 113, 114
Osmometer, *36*, 464
Osmoregulation
  by ameba, 89, 90
  by euglena, 94
  by *Paramecium*, 104-105
  by planarian, 148
Osmosis, 35-37

Ossicles
    of echinoderm, 256
    of sea cucumber, 268-269
    of sea star, ambulacral, 253
Osteichthyes, 272, 299-305
Osteoclasts, 67
Ostium(a)
    of crayfish and lobster, 220
    of mussel heart, 179
    of sea anemone, 137
    of sponge, dermal, 115
Ostium tubae of shark, 294
Ostracoda, 222
Ovarian system of fetal pig, 367
Ovarian vein in fetal pig, 366
Ovary(ies)
    of *Ascaris*, 165, 166
    of crayfish and lobster, 220
    of fetal pig, 358
    of fluke, 157
    of frog, 321
        eggs in, 55-59
    of grasshopper, 229
    of hydra, 127
    of mussel, 180
    of shark, 294
    of tapeworm, 159
    of vinegar eel, 167
    of yellow perch, 303
Oviducts
    of *Ascaris,* 165, 166
    of crayfish and lobster, 220
    of earthworm, 199
    of fluke, 157
    of grasshopper, 228, 229
    of shark, 294
    of tapeworm, 159
Ovipositor of grasshopper, 228
Ovoviviparous vinegar eel, 167
Ovum(a); *see also* Eggs
    fertilized undivided, 53
    of frog, 55
    mature, *49*
    of shark, 294
    of *Volvox* colonies, 95
    unfertilized, 53

P

Pallial cartilage, of squid, 188
Pallial cavity of mussel, 178
Pallial line of mussel, 177, 178
Pallial muscle of mussel, 177, 178
Pallium of mollusc, 175
Palmar muscle, long, inheritance of, 431
Palp(s)
    of clamworm, 192
    of honeybee, labial, 226
    of mussel, labial, 178, 186
Pancreas of fetal pig, 355
Panulirus, 213
Paper wasp, *241*
Papilla(e)
    of leech, sensory, 204
    of shark, renal, 294
Paragonimas, 158
Paragordius, 171

Paramecium, 86, 104, *105*-108
    dividing stages of, 109
Paramylum bodies in euglena, 94
Parapodium of clamworm, 192, *193*
Parasites
    of frog, *158*
    malaria, 86
    nematode as, 167-169
    ring stage of, 102, *103*
    roundworm as, 164
    trematode as, 156
Parasitic amebas, 89-90
Parastichopus, 250, 267
Parenchyma, 156
    of acoelomate, 143
    of flatworm, 148
Parenchymula larvae of sponges, 117
Parietal arteries in fetal pig, 367
Parietal vein in fetal pig, 366
Parotid gland in fetal pig, 351
Parotoid glands, 78
Pauropoda, 224, 297
Pauropus, 207
Pecten, 181
Pectineus muscle of fetal pig, 348
Pectoral girdle
    of frog, 313
    of mammal, 336-339
Pectoralis muscle of frog, 313
Pedal ganglia of mussel, 180
Pedicel of spider, 211
Pedicellaria(e)
    of echinoderm, 256
    of sea star, 251-252
    of sea urchin, 263
Pedipalp(s)
    of horseshoe crab, 208
    of spider, 211
Pelecypoda, 176-184
Pellicle
    of *Euglena*, 94
    of *Paramecium*, 104
        on stained slide, 107
    of *Vorticella*, 108
Pelomyxa, 86
Pelvis girdle
    of frog, 313
    of mammal, 339
Pen of squid, 189-190
Penis
    of fetal pig, 356
    of grasshopper, 229
Peranema, 85, 93
Perca, 299-305
Perch, yellow, 299-305
Pereiopods in crayfish and lobsters, 215
Pericardial membrane of mussel, 178
Pericardial sac of mussel, 179
Pericardial sinus in crayfish and lobster, 216, 220
Pericardium
    of crayfish and lobster, 220
    of mussel, 178
Periosteum, 66
Periostracum
    of mussel, 177
    of snail shell, 186

Periplaneta, 101, 404
Peristalsis
  in frog, 322
  in mammal, 355
Peristome
  of sea anemone, 136
  of sea urchin, 263
  of *Vorticella*, 108
Peristomial membrane of squid, 188
Peristomium
  of clamworm, 192
  of earthworm, 196
Peritoneum
  of earthworm, 198, 201
  of fetal pig, 349
  of sea star, 253
  of shark, 292
Perivitelline space
  of *Ascaris* egg, 30
  of oocyte, 49
  of zygote, 53
Perkinsea, 86
Perkinsus, 86
Permeable membrane, differentially, 35
Peroneus muscle
  of fetal pig, 348
  of frog, 316
Perophora, 272
Petromyzon, 283
pH, effect of, on enzyme activity, 43, 44
Phagocytes, 88
Phagocytosis, 88-89
  of sea star coelomic cells, 256
Pharyngeal chamber of flatworm, 48
Pharyngeal gill slits, 271
Pharyngeal sheath of planarian, 147
Pharynx
  of ammocoete, sections through, 287
  of amphioxus, 277
  of clamworm, 192
  of *Clonorchis*, 156
  of earthworm, 200
  of fetal pig, 351-352
  of flatworm, 148
  of planarian, 145, 147
  of sea anemone, 136
  in sea squirt, 274
  of yellow perch, 301
Phenotypes, 430
Philodina, 170
Phototaxis, 106, 146
Phrenic veins of fetal pig, 366
Phrenicoabdominal arteries of fetal pig, 367
Phrenicoabdominal vein in fetal pig, 366
Phylum(a), 77
  chief, of animal kingdom, key to, 78
Physa, 157, 175, 186, 187
Physalia, 123
Physalia pelagica, 130
Physiology, muscle, 379-384
Phytomastigophorea, 85
Phytomonids, colonial, 97
Pia mater in fetal pig, 372
Pickerel frog, classification of, 80
Pig
  fetal, 335-375

Pig—cont'd
  parasites of, intestinal, 164
  skeleton of, *338*
  uterus of, pregnant, *359-360*
Pigment in ciliate, 107
Pigment cells of squid, 188
Pinnules of crinoids, 269-270
Pinocytosis by ameba, 90
Pinworm, *168*, 169
Piriformis muscle of frog, 315
Pithed leopard frog, 157
Pithing of frog, 462-463
Pituitary gland
  of fetal pig, 373
  of frog, obtaining, 58
Placenta in pregnant pig, 359
Placobdella, 191
Placoid scales on dogfish shark, 292
Planarian, 143, 144, *145-155*
Plankton, 134
Plantar tubercle, 78
Planula larvae
  of *Aurelia*, 134
  of *Obelia*, 127
Plasma, 67
Plasma membrane, *21*
  of *Ascaris* egg, 30
  of cells, epithelial, 21-22
  of *Paramecium*, 104
  of sea star egg cells, 22
Plasmalemma, 86-87
Plasmodium, 101, *102-103*
Platasterias latiradiata, 250
Platelets, blood, 67, *68*
Platyhelminthes, 143
Platyias, 170
Plecoptera, 230, *243*
Plectus, 167
Pleodorina, 94, 97
Pleopods of crayfish and lobster, 215
Plethodon, classification of, 79
Pleurobrachia, 139, *140*
Pleuron(a)
  of crayfish and lobster, 214
  of fetal pig, 350
Plot sampling, 446-447
Pneumatophore of Portuguese man-of-war, 130
Pneumostome of snail, 187
Podia, buccal, of sea urchin, 263
Poison canal of honeybee, 229
Poison fang of centipede, 224
Poison glands in section of skin of frog, 71
Polar body, 49
Pollen, gathering of, by honeybee, 228
Poles of frog egg, 55
Polian vesicles in sea cucumber, 268
Polyaxons of sponge, 117
Polycelis, 144
Polychaeta, 191, 192-195
Polychaete, 194-195
Polygyra, 185, 186
Polymorphism among cnidarian, 130
Polyp, 123
  of coral, 138
  jelly, 130
Polyphemus moth, 234

Polyplacophora, 175
Polystoma, 143, 157
Pond snail, 187
   as host of cercariae, 157
Pons in fetal pig, 373
Porifera, 113, 122
Portal system
   of fetal pig, hepatic, 366, *367*
   of frog, 326
Portal vein, hepatic
   of ammocoete, 285
   of shark, 292-295
Portuguese man-of-war, 130
Postanal tail, 271
Postcava of frog, 326
Postcaval venous circulation in fetal pig, 366
Precaval vein of fetal pig, 362
Pregnant pig uterus, *359-360*
Preservation of arachnid, 212
Preservation methods, 460-461
Primordial germ cells, 45
Prismatic layer
   of mussel, 177
   of snail shell, 186
Proboscis
   of *M. hirudinaceus*, 171
   of Rhynchocoela, 143
Procambarus, 213
Processes, nerve, 70
Proglottid
   gravid, 159, 160
   immature, 159
   mature, 159-160
   of tapeworm, 159
Pronephric kidney of ammocoete, 284
Pronucleus(i)
   of *Ascaris* egg, 30
   female, 49
   male, 49
Prophase of mitosis, 28
Propylene phenoxytol to store spiders, 212
Prosobranch snails, 187
Prosopyles of sponge, 115
Prostate gland of fetal pig, 357
Prostomium
   of clamworm, 192
   of earthworm, 196
Proteases, 404-405
Proteins
   as enzymes, 43
   in epithelial cells, 22
Prothorax, 227
Protochordata, 271
Protochordate, 271-281
Protonephridia of planarian, 148
Protonephridial system of *Clonorchis*, 156
Protoplasmic level of organization, 85
Protoplasmic strands in *Volvox* colonies, 94
Protopod in crayfish and lobster, 215
Protostome eucoelomate group, 175-206, 207-247
Protozoa, 85-112
Protozoan
   cells in, 21
   flagellate, *96*
   functions of, *88*
   sarcodine, 89

Protozoea, 223
Protractor muscle of mussel, 177, 178
Pseudocardinal teeth of mussel, 177
Pseudocoel, 163
   of *Ascaris*, 165, 166
Pseudocoelomate animals, 163-174
Pseudocris nigrita, classification of, 79
Pseudopodia, 87
Pseudostratified columnar epithelium, 63
Pseudostratified epithelium in trachea cross section, 73, *74*
Pseudotriton ruber, classification of, 79
Psoas major muscle of fetal pig, 348
Psocoptera, *243*
Ptychodiseus, 97
Pulmocutaneous arch of frog, 327
Pulmonary circulation of fetal pig, 365
Pulmonate, *185, 186*
   shell-less, 187
Pupa, fruit fly, 233
Pycnogonida, 207
Pyloric chamber of crayfish and lobster, 220

Q

Quadriceps femoris muscle of fetal pig, 347
Quadrula, 176
Queen honeybee, 229

R

Radial canals
   of *Aurelia*, 134
   of sea cucumber, 268
   of sea star, 255
   of sponge, 114, 115
Radial chambers of sea anemone, 136
Radial nerve
   of fetal pig, 370
   of sea star, 255
Radial symmetry
   of radiate, 123
   tetramerous, 133
Radialis muscles of fetal pig, 345
Radiate animal, 123-141
Radiolarian, 90
Radula
   of mollusc, 175
   of snail, 185, 187
Ragworm; *see* Clamworm
Rana, 77
   classification of, 80
Rana catesbeiana, 307
   classification of, 80
Rana pipiens, 58, 157, 307
Ranidae, classification of, 79
Rat
   liver, *16*
   trichina worms in, 168
Rays
   of sea star, 251
   of yellow perch, fin, 299
Reagents, 461-462
Rectum of grasshopper, 229
Rectus abdominis muscle
   of fetal pig, 346
   of frog, 313
Rectus femoris muscle of fetal pig, 347
Rectus internus muscles of frog, 314

Red blood cells
  damaged, *17*
  Shüffner's dots in, 102, *103*
Red corpuscles, 67, *68*
Red salamander, classification of, 79
"Red tides," 97
Reef-building coral, 137
Reflexes
  in ascidian, 272-273
  spinal, 413-418
Regeneration in planarian, 146-*147*
Renal arteries of fetal pig, 367
Renal corpuscles of frog, 321
Renal papilla of shark, 294
Renal portal system of frog, 326
Renal portal veins
  of shark, 295
  of fetal pig, 366
Renilla, 123, 138
Reproduction
  of *Amoeba*, 89
  of *Aurelia*, 134
  of euglena, 94
  of fluke, 156-157
  of hydra, 127
  of planarian, 147-148
  of sea squirt, 274
  of sponge, 117
  of *Volvox*
    asexual, 95
    sexual, *95*-96
Reproductive cells in *Volvox* colonies, 94
Reproductive system
  of *Ascaris*, 165
  of crayfish and lobster, 220
  of earthworm, 198-199
  of fetal pig, 356-359
  of frog, 321
  of grasshopper, 229
  of mussel, 180
  of sea cucumber, 268
  of sea star, 253
  of yellow perch, 303
Reptilia, 272
Respiration, 385-402
Respiratory system
  of fetal pig, 375
  of frog, 320
  of mussel, 178-179
  of sea cucumber, 268
  of sea squirt, 274
  of shark, 295
  of yellow perch, *301*
Retractor muscles
  of mussel, 177, 178
  of squid, 189
Rh blood groups, 385-386
Rhabdites of planarian, 148
Rhabditis, 167, 169
Rhabdodermella, 114
Rhizopoda, 85
Rhizostoma, 123
Rhomboideus muscle of fetal pig, 345
Rhopalium of *Aurelia*, 134
Rhynchocoela, 143
Ribbon worm, 143

Ribosomes, 25
Ribs of mammal, 336
Righting reaction of frog, 308
Ring canal
  of *Aurelia*, 134
  of sea cucumber, 268
  of sea star, 255
Ring stage of malarial parasite, 102, *103*
Rolling tongue, inheritance of, 430
Romalea, 207, *226*-229
  spermatogenesis in, 47
Rostrum of crayfish and lobster, 214
Rotation in mammal, 341
Rotifera, *170*
Roundworm, 164
  intestinal, *164*-166
Rows, comb, 139
Ruminant, stomach of, 355

S

Saccoglossus, 275
Salamanders, classification of, 79
Saleus muscle of fetal pig, 348
Salientia, classification of, 79
Salivary glands of fetal pig, 351
Sampling, plot, 446-447
Sand dollar, *262*
  live, 264
Sandworm; *see* Clamworm
Saprophagous nematodes, 167
Sarcodina, 85, 86-92
Sarcodine, 89-90
Sarcodine protozoan, *89*
Sarcomastigophora, 85-112
Sarcomeres, 380
Sartorius muscle
  of fetal pig, 347
  of frog, 313
Scalene muscles of fetal pig, 345
Scales
  of dogfish shark, placoid, 292
  fish, chromatophores in, 305
  of yellow perch, 299
Scallop, 181
Scaphopoda, 175
Scarabeidae, 171
Schistostoma, 143, 158
Schizogony, 101, 102, 103
Sciatic nerve of fetal pig, 370-372
Sclerosepta of coral, 138
Sclerospongiae, 113, 117
Scolex of tapeworm, 159
Scolopendra, 224
Scorpion fly, *241*
Scutigerella, 207
Scypha, 113, 114-*116*, 117-119
Scyphistomae of *Aurelia*, 134
Scyphozoa, 123, 133-135
Scyphozoan jellyfish, 135
Scyphozoan medusa, 133
Sea biscuit, *262*
Sea cucumber, *267*-269
Sea lily, 269-270
Sea squirt, 272, *273*-275
Sea star, 251, *252*-*254*, 255-258
  development of, 53-55

Sea star—cont'd
  egg cell of, 22
  embryology of, *54*
    early, 53-55
Sea urchin, *261-262, 263-264*
Segmentation in annelid, 191
Segmentation cavity, 53, 57
Segmented bodies of arthropods, 207
Segments of spider, 211
Selectively permeable membrane, 36
Semimembranosus muscle
  of fetal pig, 348
  of frog, 315
Seminal grooves of earthworm, 197
Seminal receptacle
  of crayfish and lobster, 214
  of earthworm, 197, 198
  of fluke, 157
  of grasshopper, 229
  of vinegar eel, 167
Seminal vesicles
  of *Ascaris*, 165
  of earthworm, 198
  of fetal pig, 357
  of fluke, 157
Seminiferous tubules of frog, 321
Semipermeable membrane, 36
Semitendinous muscle
  of fetal pig, 348
  of frog, 314
Sense organs
  of *Bdelloura*, 148
  of comb jelly, 139
  of crayfish and lobster, 221
  of frog, 333
  of sea star, 255
Sensory cells of hydra, 127
Sensory hairs of spider, 210
Sensory papilla(e) of leech, 204
Sensory perception in cricket, 422
Sepia, 175
Septa of earthworm, 198
Septal filaments of sea anemone, 137
Serial homology in crayfish appendages, 215
Serratus muscle of fetal pig, ventral, 345
Seta(e)
  annelid, isolation of, 203
  of earthworm, 196, 201
  of freshwater oligochaetes, 202
  of parapodium, 192
Sex(es)
  in fruit fly, distinguishing, 428, *429*
  in *Volvox*, development of, 94
Sex cells of *Aurelia*, 134
Sex-linked inheritance in fruit fly, 429
Sexual reproduction; *see also* Reproduction
  by hydra, 127
  by sponge, 117
  in *Volvox* colony, 95, 96
Shank of frog, muscles of, 315-316
Shark, dogfish, *291-293, 294-297*
Shell
  of *Ascaris* egg, 30
  of bivalves, 176
  of mollusc, 175
  of mussel, 177
    opening, 177-179

Shell—cont'd
  of oocyte, 49
  of snail, *186*
Shelled amebas, 90
Shell-less pulmonate, 187
Shipworm, 181
Short monaxons of sponge, 117
Shoulders of fetal pig
  arteries of, 362-365
  muscles of, *343-345*
  veins of, 362
Shrimp
  brine, 223
  fairy, 221, 223
Shüffner's dots in blood cell, *103*
Silverfish, 246
Simple epithelium, 62-64
Simple squamous epithelium in blood vessel, *73*
Sinistral snail, 186
Sinus(es)
  of crayfish and lobster, 216, 220
  of grasshopper, blood, 229
Siphon
  of marine clam, 176
  of sea squirt, 272
Siphonaptera, *245*
Siphonoglyph(s)
  of coral, 138
  of sea anemone, 136
Skeletal elements, *118*
Skeletal muscle, *68*, 69, *379*
  of fetal pig, 340-348
  of frog, 313-*314*, *315-316*
  glycerinated, contraction of, 380
  motor nerve endings on, examination of slide of, 73
Skeletal musculature of telecost fish, *300*
Skeletal system of yellow perch, 300
Skeleton, 66
  of frog, 311, *312*, 313
  of hard coral, 138
  of mammal, 336-340
  of radiate, hydrostatic, 123
  of sponge, 117-118
Skin
  of frog, section through, 71-72
  of planarian, 145
  of shark, *292*
Skin gills of sea star, 251, 252
Skull
  of frog, 311
  of mammal, 336, *337*
Sleeping sickness, African, 97
Slimy salamander, classification of, 79
Slits, gill
  of dogfish shark, 291
  pharyngeal, 271
Slug, 187
Small intestine, amphibian, cross section of, 72
Smears, blood, invertebrate, 203
Smooth muscle, *68*, 69
  in amphibian intestine, 72
  cells of, 25
    in mitosis, *28*
  layer of, in blood vessel, 73
  in section of skin of frog, 71
  in trachea cross section, 73

Snail
  land, *185-187*
  mud, as host of cercariae, 157
  pond, as host of cercariae, 157
Snail eggs, 187
Social animal, honeybee as, 229
Soft coral, 138, 139
Soil nematode, 167
Solation, 37
Solengastres, 175
Sol-gel transformation of colloids, 37
Solitary hydroid, 124-127
Solutions, 461-462
  crystalloid or true, 34
Somasteroidea, 250
Somatic cells in *Volvox* colonies, 94
Somites of freshwater oligochaete, 202
Species, 77
Sperm cells of *Volvox* colonies, 95
Sperm ducts
  of crayfish and lobster, 220
  of tapeworm, 159
  of yellow perch, 303
Sperm entrance of oocyte, 48
Spermatheca of grasshopper, 229
Spermatic arteries of fetal pig, 367
Spermatic vein of fetal pig, 366
Spermatid, 46-47
  in grasshopper testis, 48
Spermatocytes
  primary, 45
    in grasshopper testis, 47
  secondary, 46
    in grasshopper testis, 48
Spermatogenesis, 45-48
  in grasshopper testis, *47-48*
Spermatogonia, 45
  in grasshopper testis, 47
Spermatophores of squid, 188
Spermatozoa, 48
  of earthworm, 197
  of fluke, 157
  of frog, 321
  in grasshopper testis, 48
  of hydra, 127
  of mussel, 180
Sphaeridia of sea urchin, 263
Sphygmomanometer, 388
Spicule strew, 119
Spicules
  of *Ascaris*, 164, 165
  of sponge, 113, 114
  of vinegar eel, copulatory, 167
Spider, garden, *210, 211-212*
Spinal cord, *71*
  of frog, 332-333
Spinal nerves
  of fetal pig, 370-372
  of frog, 331-332
Spinal reflexes, 413-418
Spines
  of brittle star, 251
  of sea star, 251, 252-253
  of sea urchin, 262
Spinnerets of spider, 211
Spinodeltoid muscle of fetal pig, 345
Spiny-headed worm, 171

Spiracles
  of centipede, 224
  of dogfish shark, 291
  of grasshopper and honeybee, 227, 229
  of millipede, 225
  of spider, tracheal, 211
  of tadpole, development of, 58
Spiral valve in shark, 292
Spirobolus, 207, 225
Spirostomum, 105, 108
Spisula, 176
Spleen of fetal pig, 350
Splenic artery of fetal pig, 367
Splenius muscle of fetal pig, 345
Sponge, 113-122
Spongia, 113
Spongilla, 118
Spongin of sponge, 113
Spongocoel of sponge, 113, 114, 115
Spongy bone, 66
Sporogony, 101
Sporozoea, 85, 101-104
Sporozoites
  of *Gregarina*, 101
  of *Plasmodium*, 102
Spotted salamander, classification of, 79
Spring peeper, classification of, 80
Spring salamander, 79
Springtail, *246*
Squalus, 291-293, 294-297
Squamous epithelial cells, 21-22
  stained, 22-26
Squamous epithelium
  in amphibian intestine, 72
  simple, 62
    in blood vessel, *73*
  stratified, *64*
    in section of skin of frog, 71, 72
"Squash technique" to demonstrate mitosis, 30
Squid, 188, *189-190*
Stains for electron microscope viewing, 14
Stars
  brittle, *259-260*
  feather, *269-270*
  sea, 251, *252-254*, 255-258
Statistics and genetics, 432-434
Statocyst
  of comb jelly, 141
  of crayfish and lobster, 221
Stellate ganglion of fetal pig, 372
Stelleroidea, 250
Stenostomum, 149
Stentor, 86, 105, 108
Stentor coerulus, 107, 108
Sternal artery in crayfish and lobster, 220
Sternocephalic muscle of pig, 344
Sternohyoid muscles of fetal pig, 343-345
Sternomastoid muscle of fetal pig, 344
Sternothyroid muscle of fetal pig, 344
Sternum
  of crayfish and lobster, 214
  of frog, 312
  of mammal, 336
Stigma
  of *Euglena*, 94
  in *Volvox* cells, 94
Stimulants, effect of, on frog heart rate, 328

Stimulus(i)
  response of paramecia to, 106
  response of planarian to, 146
Sting of honeybee, 229
Stomach(s)
  of *Aurelia*, 134
  of bird, 355
  of crayfish and lobster, 220
  of fetal pig, 353
  of grasshopper, 229
  of mussel, 180
  of ruminant, 355
  of sea star, 253
  of yellow perch, 303
Stone canal of sea star, 255
Stonefly, *243*
Stony coral, 137, *138*, 139
Stratified epithelium, 63-64
  squamous, *64*
    in section of skin of frog, 71, 72
Strew, spicule, 119
Striated muscle, 379; *see also* Skeletal muscle
  of arthropod, 207
Strongylocentrotus, 250, 261
Strongylocentrotus purpuratus, 263
Strychnini, 381
Stylaria, 202
Style, crystalline, of mussel, 180
Subclavian artery of fetal pig, 364-365
Subclavian vein of fetal pig, 362
Subgenital pit of *Aurelia*, 134
Sublingual glands of fetal pig, 351, *352*
Submaxillary gland of fetal pig, 351
Subneural vessel of earthworm, 200, 202
Subpharyngeal ganglia of earthworm, 201
Subpharyngeal gland in ammocoete, 284, 287
Subscapular nerve of fetal pig, 370
Subspecies, 77
Sucker(s)
  of bladder fluke, 157
  of *Clonorchis*, 156
  of frog, development of ventral, 57
  of leech, 204
  of tapeworm, 159
Sucrase, 404
Supporting tissue, 64-67
Suprabranchial chamber of mussel, 178
Supraesophageal ganglia of crayfish and lobster, 221
Suprascapular nerve of fetal pig, 370
Supraspinatus muscle of fetal pig, 345
Suspensions, 34
Swallowtail butterfly, *239*
Swim bladder of yellow perch, 301-303
Swimmerets of crayfish and lobster, 213, 214, 215
Syconoid sponge, 114, *115*-119
Symbionts of hydra, 124-125
Symbiotic mutualism of *T. campanula* and American termite, 96
Symmetry
  of acoelomate, bilateral, 143
  of echinoderm, 249
  radial, tetramerous, 133
  of radiate, radial, 123
Symphyla, 207, 224
Synapsis, 46
Synarthrosis, 340
Systemic arch of frog, 327
Systemic heart of squid, 189

Systolic pressure, 388
Syzygy, 101

T

Tactile hairs of crayfish and lobster, 221
Tadpole; *see* Frog, development of
Tadpole stage, 58
  early development of frog to, *56*
Taelia, 123
Taenia, 143, 159-160
Taeniarhynchus saginata, 159
Tagelus, 175
Tagmata of arthropod, 207
Tail
  of gastrotrich, 171
  postanal, 271
Tail fan of crayfish and lobster, 214
Tailpiece of horseshoe crab, 208
Tapeworm, 143, 159-162
Tarsus
  of grasshopper, 228
  of insect, 239
  of spider, 211
Taste, inheritance of, 430-431
Taxa, animal, major, key to, 450-454
Taxonomic key to certain amphibians, 78-80
Taxonomy, 77
Teeth
  of mussel, 177
  of sea urchin, 263
  of yellow perch, 301
Tegument of *Clonorchis*, 156
Telecost fish, skeletal musculature of, *300*
Telolecithal-type egg, 55
Telophase of mitosis, 29
Telson
  of crayfish and lobster, 214
  of horseshoe crab, 208
Temperature
  effect of, on ciliary action of clam gill, 181
  and enzymes, 43
Tendon in mammal, 341
Tenebrio, 101
Tensor fasciae latae muscle of fetal pig, 347
Tentacle(s)
  of *Aurelia*, 134
  of clamworm, 192
  of hydra, 124, 125
  of *Obelia*, 129
  of sea anemone, 136
  of sea cucumber, 267
  of snail, 185, 187
  of squid, 188
Tentacle sheath of comb jelly, 139
Teredo, 175, 181
Teres major muscle of fetal pig, 345
Tergum of crayfish and lobster, 214
Terminal discs, 78
Terminal fang of spider, 211
Termite, *242*
  American, and *T. campanula*, symbiotic mutualism of, 96
Terrarium, 458-460
Terrestrial mandibulate, 224-235
Test
  of sea squirt, 272
  of sea urchin, 263
Testcross in fruit fly, 429

Testis(es)
    of *Ascaris*, 165
    of crayfish and lobster, 220
    of earthworm, 199
    of fetal pig, 356
    of fluke, 156
    of frog, 321
    of grasshopper, 229
        spermatogenesis in, *47-48*
    of hydra, 127
    of mussel, 180
    of shark, 294
    of tapeworm, 159
    of vinegar eel, 167
    of yellow perch, 303
Tetrad of chromosomes, 46, 49
Tetrahymena, 86
Tetramerous radial symmetry, 133
Thalassicolla, 85
Thecae of coral, 138
Thigh of frog, muscles of, 314-315
    ventral, 313-314
Thigmotaxis, 106, 146
Thoracic artery
    of crayfish and lobster, ventral, 220
    of fetal pig, internal, 364
Thoracic cavity of fetal pig, 350-351
Thoracic ganglion of fetal pig, 372
Thoracic nerve of fetal pig, 370
Thoracic vein of fetal pig, internal, 362
Thorax
    of grasshopper, 226, 227
    of honeybee, 227
Threadworm, 171
Thrips, *240*
Thrombocytes, 67
Thumb joint, hypermobility of, inheritance of, 431
Thymus gland of fetal pig, 350
Thyone, 350, 367
Thyrocervical artery of fetal pig, 364
Thyroid gland of fetal pig, 351
Thysanoptera, *240*
Thysanura, 230, *246*
Tibia of grasshopper, 228
Tibialis muscle
    of fetal pig, anterior, 348
    of frog, 316
Tiger salamander, classification of, 79
Tissue, 61
    adipose, in blood vessels, 73
    combined into organs, 71-75
    connective
        adipose, 64, 65
        in amphibian intestine, 72
        areolar, 64, 65
        dense, 64
        loose, 64
        in section of skin of frog, 71
        in trachea cross section, 73, *74*
    epithelial, 61-64
    fibrous connective
        in blood vessel, 73
        in trachea cross section, 73
    function of, 61-76
    muscle, *68*, 69
    nervous, 69-71
    structure of, 61-76

Tissue—cont'd
    supporting, 64-67
    vascular, 67-68
Tissue fluids, 67
Tissue level of organization, 123
Toads, classification of, 79
Toes of *Philodina*, 170
Tongue
    of fetal pig, 351-352
    of frog, 319
    of honeybee, 226
    rolling, inheritance of, 430
Total cleavage, 53
Trachea, cross section through, 73
Tracheal spiracle of spider, 211
Tracheal system, *233*
    of arthropod, 207
    of grasshopper, 229
Tracheal tubes of spider, 211
Transverse canal of tapeworm, 160
Transverse muscle of fetal pig, abdominal, 346
Transverse nerves of *Bdelloura*, 148
Transversus muscle of frog, 313
Trapezius muscle of fetal pig, 345
Treehopper, *242*
Trematoda, 143, 156-158
Trematode as parasite, 156
    of frog, *158*
Triceps brachii of fetal pig, 345-346
Triceps muscle of frog, femoris, 314-315
Trichina worm, 168-169
Trichinella spiralis, 168-169
Trichocysts of *Paramecium*, 105-106
    on stained slide, 107
Trichodina, 124
Trichonympha, 85, 96
Trichoptera, *240*
Triclad, 144
Trilobita, 207
Trinomial nomenclature, 77
Triradiates of sponge, 117
Trochanter of grasshopper, 228
Trophozoite, 101
True frog classification, 79
Trunk(s)
    of fetal pig
        arteries of, 367-368
        pulmonary, 365
    of frog, muscles of, ventral, 313
    of planarian, anterior and posterior, 147
Trypanosoma, 85
Trypanosoma brucei, 96-97
Trypanosoma lewisi, 97
Trypanosome, blood, 96, 97
Tsetse fly, trypanosomes transmitted by, 97
Tube feet
    of brittle star, 259
    of sea star, 251, 252, 253, 255
    of sea urchin, 262-263
Tubercle, plantar, 78
Tubeworm, *194*
Tubifex, 191, 202
Tubipora, 138
Tubularia, 123
Tunic of sea squirt, 272
Tunicate, 272
Turbatrix aceti, 167

Turbellaria, 143, 144-155
Turbellarian, *144*
Two-lined salamander, classification of, 79
Tylenchus, 167
Tympanum of grasshopper, 228
Typhlosole
    of earthworm, 198, 200, 202
    of mussel, 180

U

Ulnar nerve of fetal pig, 370
Ulnaris lateralis muscles of fetal pig, 346
Umbilical vein of fetal pig, 349, 366
Umbo of mussel, 177
Urethra of fetal pig, 358
Urinary bladder
    of fetal pig, 355
    of frog, 321
    of yellow perch, 304
Urinary system of fetal pig, 355-356
Urine of crayfish and lobster, 221
Urochordata, 271, 272-275
Urodela, classification of, 79
Urogenital organs of frog, histology of, 321-322
Urogenital system
    of fetal pig, 355, *357-360*
    of frog, 320, 321
    of shark, 292, 294
    of yellow perch, *303*
Uropods of crayfish and lobster, 215
Uterus(i)
    of *Ascaris*, 165, 166
    of fetal pig, 357
    of fluke, 157
    of pregnant pig, *359-360*
    of tapeworm, 159
    of vinegar eel, 167

V

Vacuole
    of ameba
        contractile, 89, 90
        food, 88
    of ciliate, food, 106
    of *Paramecium*
        contractile, 104-*105*
            on stained slide, 107
    of *Vorticella*
        contractile, 108
        food, 108
Vagina
    of *Ascaris*, 165
    of fetal pig, 358
    of grasshopper, 229
    of tapeworm, 159
Vagosympathetic trunk of fetal pig, 372
Valve(s)
    of bivalves, 176
    of shark, spiral, 292
Variability, genetic, in humans, 434
Vas deferens, 356
    of fluke, 156
    of grasshopper, 229
Vas efferens of fluke, 156
Vascular tissue, 67-68
Vascular-water system
    of sea cucumber, 268

Vascular-water system—cont'd
    of sea star, 251, 252-253, 255
Vastus muscle of fetal pig, 347
Vegetal hemisphere of frog egg, 57
Vegetal pole of frog egg, 55
Vein(s)
    of ammocoete, 285
    of amphioxus, 278
    of fetal pig, *361*, 362
        pulmonary, 365
        umbilical, 349
    section through, 72-73
    of shark, 294-295
    of squid, 189
Velum
    of honeybee, 228
    of jellyfish, 130
Vena cava
    of fetal pig, anterior, 362
    of frog, posterior, 326
Venous circulation, in fetal pig, postcaval, 366
Venous system
    of frog, *325-326*
    of shark, 294, 295
    of yellow perch, 305
Ventilation by mussel, 179
Ventricle of heart
    of fetal pig, 368-369
    of mammal, 362
    of mussel, 179
    of yellow perch, 304
Venus, 175
Vertebral artery of fetal pig, 364
Vertebral column
    of frog, 312
    of mammal, 336
Vertebrata, 272
Vesicle
    of ameba, water-expulsion, 89, 90
    of *Ascaris*, seminal, 165
    of *Clonorchis*, 156
    of earthworm, seminal, 198
    of euglena, water-expulsion, 94
    of fluke, seminal, 157
    of *Paramecium*, water-expulsion, 104-105
Vinegar eel, 167
Visceral ganglia of mussel, 180-181
Visceral hump of snail, 187
Visceral mass of mussel, 178
Visceral muscle; *see* Smooth muscle
Vitelline gland
    of fluke, 157
    of tapeworm, 159
Vitelline membrane of zygote, 53
Vocal cords in fetal pig, 375
Voluntary muscle, 379; *see also* Skeletal muscle
Volvox, 85, 94-97
Vorticella, 86, 105, 108
Vulva
    of *Ascaris*, 164, 165
    of fetal pig, 358
    of vinegar eel, 167

W

Wasp, paper, *241*
Water currents and sponge, 119
Water flea, 221

Water tubes of mussel gills, 178
Water-expulsion vesicle
   of ameba, 89, 90
   of euglena, 94
   of *Paramecium*, 104-105
Water-vascular system
   of sea cucumber, 268
   of sea stars, 251, 252-253, 255
Web, orb, of garden spider, 210
White corpuscles, 67, 68
Whitefish blastula, mitosis in, 30
Whitefish embryo, mitosis in, 29
Whorl of shell, snail, 186
Widow's peak, inheritance of, 430
Wild-type *Drosophila*, 428-430
Wings
   of grasshopper, 227
   of honeybee, 227-228
Wolffian duct of shark, 294
Wood frog, classification of, 80
Worker honeybee, 229
Worm
   horsehair, 171
   operating on, 146-147
   ribbon, 143
   spiny-headed, 171
   trichina, 168-169
Wuchereria bancrofti, 169

X

Xylol to clean microscope lens, 11

Y

Yellow perch, 299-305
Yolk duct of fluke, 157
Yolk gland of fluke, 157
Yolk plug in gastrula stage, 57

Z

Zoantharia, 123
Zoantharian coral, 139
Zoea, 223
Zooids
   sea squirt, 274
   in *Volvox* colonies, 94
Zoomastigophorea, 85
Zygote, 53
   of fluke, 157
   of hydra, 127
   of mussel, 180
   nucleus of, 49
   of *Volvox*, 95